Agrochemicals

Biographical Details

Professor R. J. Cremlyn, BSc, PhD(Wales), PhD(Cantab.), DSc(Wales), CChem, FRSC

Richard Cremlyn was educated at Queen Elizabeth Grammar School, Carmarthen, South Wales; Swansea University College; and Trinity Hall, Cambridge. He graduated in Chemistry from Swansea in 1950 and stayed on there to carry out research in steroid chemistry under Professor C. W. Shoppee, FRS, obtaining his PhD degree in 1953. He was awarded a University of Wales Fellowship for postdoctoral research at Cambridge in nucleotide chemistry under Professor Lord Todd, FRS, and obtained the Cambridge doctorate in 1956. Dr Cremlyn then joined ICI (Plant Protection Division) at Jealott's Hill Research Station, where he was involved in the synthesis of a wide range of organic compounds as potential pesticides and in structure–activity studies. In 1959, he was appointed lecturer in Organic Chemistry at Brunel College of Technology (now Brunel University) and in 1961 became Senior Lecturer in Organic Chemistry at Hatfield Polytechnic, being promoted Principal Lecturer (1964), Reader (1980), Professor of Organic Chemistry (1987), and Head of Department (1990).

Professor Cremlyn's major research interests are the chemistry and biological activity of organosulphur and phosphorus compounds, especially as crop protection agents. He has published 130 research papers and 24 review articles on the mode of action of pesticides, together with four books (three on organic chemistry and one on pesticides).

Professor Cremlyn was awarded the DSc degree by the University of Wales in 1981, and has acted as External Examiner for the BSc degree in Agricultural Chemistry at Wye College (London University) and for numerous PhD degrees at Universities and Polytechnics.

Agrochemicals

Preparation and Mode of Action

R. J. Cremlyn
Hatfield Polytechnic, Hertfordshire, UK

JOHN WILEY & SONS

Chichester · New York · Brisbane · Toronto · Singapore

Other Wiley Editorial Offices

John Wiley & Sons, Inc., 605 Third Avenue,
New York, NY 10158-0012, USA

Jacaranda Wiley Ltd, G.P.O. Box 859, Brisbane,
Queensland 4001, Australia

John Wiley & Sons (Canada) Ltd, 22 Worcester Road,
Rexdale, Ontario M9W 1L1, Canada

John Wiley & Sons (SEA) Pte Ltd, 37 Jalan Pemimpin 05-04,
Block B, Union Industrial Building, Singapore 2057

Library of Congress Cataloging-in-Publication Data:

Cremlyn, R. J. W. (Richard James William)
    Agrochemicals : preparation and mode of action / R. J. Cremlyn.
        p.   cm.
    Includes bibliographical references and index.
    ISBN 0-471-92955-7 (cloth)—ISBN 0-471-92992-1 (paper)
        1. Pesticides. I. Title.
    SB951.C68   1990
    632′.95—dc20   90-22242
                                        CIP

A catalogue record for this book is available from the British Library

Typeset by APS, Salisbury, Wilts.
Printed and bound in Great Britain by Biddles Ltd., Guildford, Surrey

Contents

action. Uracils. Triazoles. Pyridazines. Thiadazines and miscellaneous heterocycles. Bipyridinium compounds and their mode of action. Miscellaneous compounds—diphenyl ethers and organophosphorus, sulphonylureas and imidazolines. Herbicide safeners. Resistance of weeds towards herbicides

14: Some Novel Methods of Insect Control   315

Natural and synthetic sex attractants, e.g. gyptol, medlure, methyl eugenol; other semiochemicals, e.g. aphid alarm and mosquito oviposition phermones; food lures. Chemical repellents, e.g. dimethyl phthalate. Antifeedants—triazenes, triphenyltins, trichlorophenoxyethanol. Naturally occurring compounds, e.g. polygodial, warburganal. Chemosterilants—alkylating agents such as aziridines, e.g. apholate, or other compounds such as bisulphan; and antimetabolites—5-fluorouracil. Arthropod hormones —ecdysones and juvenile hormones—javabione. Synthetic juvenile hormone mimics, e.g. methoprene; diflubenzuron and other benzoylphenylureas. Triazine insect growth inhibitors, e.g. cyromazine. Precocenes. Microbial insecticides from bacteria, e.g. *Bacillus thuringiensis*; viruses, and fungi

15: Pesticides in the Environment   341

Introduction—the food chain, problems of pollution by chemical pesticides; copper and mercury fungicides, arsenicals, dinitrophenols. 2,4,5-T, paraquat, dithiocarbamates. Organophosphates, carbamates, and organochlorine compounds. The widespread pollution by organochlorine compounds, including polychlorobiphenyls. Special features of organochlorine compounds—resistance to chemical and biological degradation and accumulation along food chains. Effects on fish and birds, and possible toxicity to man. Anticoagulant rodenticides, e.g. difenacoum, bioaccumulation problems. Synthetic pyrethroid insecticides. Future measures to reduce environmental pollution by pesticides

16: Future Developments   361

The introduction of more selective pesticides, including behaviour-controlling chemicals. New, more active systemic fungicides and herbicides. The need for effective antibacterial and antiviral agents. The design of new insecticides based on natural products. New more selective herbicides, PGRs, and chemical sterilants to inhibit pollination. Fungicidal compounds not acting directly on the pathogen. The impact of biotechnology on crop protection. Importance of new formulations and the development of integrated pest management programmes

Preface

The book retains the general format adopted in *Pesticides*, published in 1978. It has been amended and expanded to reflect the changes that have occurred since that date.

The text discusses the growth in the application and sophistication of chemical pesticides or agrochemicals which has been particularly rapid since the end of World War II.

The physiochemical factors and biochemical reactions important in pesticides are discussed as an introduction to the subsequent chapters dealing with the major chemical groups used to control different kinds of pests. The various methods of formulation and application of pesticides are also covered; there is now much research directed towards discovering new, improved formulations.

The greater awareness of the dangers of environmental pollution arising from the widespread use of chemical pesticides is reflected in the growing emphasis placed on safer and more selective chemicals. In most countries, candidate chemicals have to satisfy increasingly stringent legislative criteria before they can be marketed as pesticides; the effects of pesticides on the environment are discussed together with future developments. Areas currently of special interest include new specific systemic fungicides, selective herbicides against grass weeds, chemicals controlling the growth of insects and plants, insect antifeedants, and microbial insecticides.

The discovery of new biochemical target sites for pesticides is a major objective, which has been given added impetus by the introduction of sulfonyl-urea herbicides and triazole fungicides. The spectacular exploitation of the insecticidal properties of natural pyrethrum has led to renewed interest in natural products as a route to novel insecticides.

Important developments have occurred in the use of pheromones in insect monitoring and control; application of integrated pest management programmes is becoming of increasing significance in order to reduce the dangers of pest resistance and environmental pollution.

Biotechnology is already making an appreciable impact on crop protection and will be the major weapon of control in the twenty-first century.

It is hoped that the book will prove valuable to all interested in chemical pesticides, and especially to undergraduates in universities, polytechnics, and colleges studying for degrees in agriculture, agricultural chemistry, applied biology, and applied chemistry. It should also prove of use to students taking courses for National Diplomas and Certificates in Agriculture. The author wishes to thank Dr Kirkwood and Miss Jenny Cossham of John Wiley & Sons Ltd for their help during the passage of the book through the press. My thanks are also due to those firms who allowed me to reproduce photographs of some of their pesticides in action, and to Miss Jane Fordham, Mrs Pam Meiklejohn, and Mrs Pauline Gouldthorpe for their expert typing of the manuscript.

R. J. Cremlyn

Chapter 1

Introduction

Pesticides are chemicals designed to combat the attacks of various pests on agricultural and horticultural crops. They fall into three major classes: insecticides, fungicides, and herbicides (or weed killers). There are also rodenticides (for control of vertebrate pests), nematicides (to kill microscopic eelworms), molluscicides (to kill slugs and snails), and acaricides (to kill mites). The term agrochemicals is broader and includes chemicals which enhance the growth and yield of the crop, but excludes large-scale inorganic fertilizers.

Pesticides may also be divided into two main types, namely contact or non-systemic pesticides and systemic pesticides. Contact or surface pesticides do not appreciably penetrate plant tissues and are consequently not transported, or translocated, within the plant vascular system. The earlier insecticides, fungicides, and herbicides were of this type; their disadvantages are that they are susceptible to the effects of weathering (wind, rain and sunlight) over long periods and new plant growth will be left unprotected and hence open to attack by insect and fungal pests. The early agricultural fungicides were, therefore, protectant fungicides—in other words they are designed to prevent the development of the fungal spores, but once the fungus has become established and the infection starts to ramify through the plant tissues such non-systemic fungicides possess little eradicant action and usually cannot halt the infection.

It is not surprising that the earlier pesticides were non-systemic in character because these are easier to discover, since phytotoxicity does not present such a great problem as it does in the case of systemic pesticides where the chemical comes into intimate contact with the tissues of the host plant. The danger of the fungicide causing damage to the host plant is especially formidable in the case of systemic fungicides, since here both host and pest are plants and for success the chemical must show selective toxicity to the fungus. The severity of these problems is reflected in the long time that elapsed before commercial systemic fungicides appeared on the market.

The effectiveness of many relatively insoluble surface fungicides depended on the fungal spores dissolving sufficient toxicant from the mixture to be killed, but,

1

on the other hand, there must not be enough soluble toxicant present to appreciably damage the host plant.

In contrast, many of the more recent pesticides are systemic in character—these can effectively penetrate the plant cuticle and move through the plant vascular system. Examples are provided by the phenoxyacetic acid selective herbicides (1942), certain organophosphorus insecticides like schradan (1941), and the more recently discovered systemic fungicides like benomyl (1967) and hexaconazole (1986).

Systemic fungicides are also sometimes termed plant chemotherapeutants and can not only protect the plant from fungal attack but also cure or inhibit an established infection. They are little affected by weathering and will also confer immunity on all new plant growth.

Pests can be divided into various groups. In the plant kingdom, characterized by the ability of the organism to photosynthesize carbohydrates from air and water with the aid of the green pigment chlorophyll, higher plants growing where man does not want them are termed weeds and are important pests. Of the lower plants, algae are not generally of great importance as pests, although in some circumstances, e.g. in lakes and other slow-moving water, excessive algal growth or 'bloom' may cause considerable damage and require treatment with chemicals (algicides).

Fungi or non-photosynthetic plants cannot obtain their nutrients from air and water since they do not have chlorophyll; consequently they feed directly on decaying plant or animal matter (saprophytic fungi) or on living plants or animals (parasitic fungi). There are thousands of different species of fungi mainly found in soil—some, like yeasts, are unicellular while others are composed of a network of branched filaments (hyphae). A number of fungi are serious pests attacking both living crop plants and also crops in storage.

Several bacteria are causal agents of plant disease, although they are not nearly as important as the phytopathogenic fungi. Bacteria can be observed under the microscope and can be classified according to their shape; thus a spherical bacterium is termed a *coccus* while a rod-shaped one is a *bacillus*.

Viruses, like bacteria and fungi, attack plants and animals and some species cause significant plant diseases. Viruses form a distinct category of living organism because they are not true cells. Unlike bacteria they are too small (100–300 Å) in diameter to be observed with an ordinary microscope, but they can be revealed under the electron microscope—each virus consists of a single strand of DNA or RNA surrounded by a protective coat of protein.

Several higher animals (vertebrates) are important pests, e.g. mice, rats, and rabbits; another group of pests is represented by the true insects (arthropods) which are invertebrates. The latter possess three pairs of legs and the adult body has three parts; the arachnids (mites and ticks) differ from the true insects in having no distinct division of the body into three parts and they also usually have four pairs of legs. In the lower orders of animals, certain nematodes, parasitic worms often with unsegmented bodies, are important crop pests.

HISTORICAL ASPECTS

Ever since the dawn of civilization man has continually endeavoured to improve his living conditions; in his efforts to produce adequate supplies of food man has been opposed by the ravages wrought by insect pests and crop diseases. The blasting mentioned by the prophet Amos (760 BC) was the same cereal rust disease that is still responsible for enormous losses. The father of botany, Theophrastus (300 BC), described many plant diseases known today such as scorch, rot, scab, and rust. There are also several references in the Old Testament to the plagues of Egypt for which the locust was chiefly responsible, and even today locusts cause vast food losses in the Near East and Africa.[1,2]

The major pests inhibiting the growth of agricultural crops are insects, fungi, and weeds, and the idea of combating these pests by the use of chemicals is not new.

Sulphur was known to avert diseases, as well as insects, before 1000 BC and its use as a fumigant was mentioned by Homer. Pliny (AD 79) advocated the use of arsenic as an insecticide and by the sixteenth century, the Chinese were applying moderate amounts of arsenic compounds as insecticides. In the seventeenth century the first naturally occurring insecticide, nicotine—from extracts of tobacco leaves—was used to control the plum curculio and the lace bug. Hamberg proposed mercuric chloride as a wood preservative (1705) and a hundred years later Prévost described the inhibition of smut spores by copper sulphate.[3]

These early discoveries may probably be ascribed to a mixture of acute observation following trial and error application, together with strong undertones of superstition.

Fossil evidence preserved in rocks indicates that the agents of plant disease were operating long before the appearance of man on the earth. The tales of woe about the ravages of blights, mildews, and plagues are prominent in the earliest written records. There are many such references in the Bible, when plant diseases and plagues were seen to be visited upon man by God as a punishment for his sin. Although rather contradictory to the Christian concept of a loving God, this idea lingered on in subsequent centuries. It was certainly held in Hungary when during the eleventh and thirteenth centuries the cereal crops were ravaged by disease attacks. Those disasters not emanating from the fury of God were considered to be due to the actions of evil spirits—witches, goblins, and hobgoblins—which opposed the good forces in the world. Such demonic forces were held responsible, right up to the eighteenth century, for damage to agricultural crops caused by insect pests and diseases, and magical prescriptions to control pests abound in the early agricultural literature. It was not until the middle of the nineteenth century that systematic scientific methods began to be applied to the problem of controlling agricultural pests.

About 1850 two important natural insecticides were introduced: rotenone from the roots of the derris plant and pyrethrum from the flower heads of a species of chrysanthemum. These are still widely used insecticides (see Chapter 4). About this time, too, soap was used to kill aphids and sulphur was used as a fungicide on peach trees. A mixture of sulphur with lime to soften it, later called lime sulphur, was first suggested by Weighton (1814), and in 1902 it was observed that lime sulphur was effective against apple scab.[4] Also a treatise by Forsyth (1841) described a combined wash composed of tobacco, sulphur, and unslaked lime to control insects and fungi. During the nineteenth century new inorganic materials were introduced for combating insect pests; for instance, an investigation into the use of new arsenic compounds led in 1867 to the introduction of an impure copper arsenite (Paris Green) for control of Colorado beetle in the state of Mississippi, and in 1892 lead arsenate was used for control of gipsy moth. By 1900 Paris Green was used so extensively as an insecticide that it caused the introduction of the first State legislation governing the use of insecticides in the United States of America.[5] The Irish Potato Famine of 1845-9 provides an illustration of what can occur when a staple food crop is stricken by a disease against which there is no known defence. The potato crop was virtually destroyed by severe attacks of the fungal disease known as potato late blight, resulting in the deaths of more than a million people (some 12% of the population) from starvation and the emigration of a million and a half people, chiefly to the United States of America.[6,7]

A large number of fantastic theories were propounded to account for the epidemic and a distinguished chemist of the time, Dr Lyon Playfair, was consulted. He did not realise, however, that the causal agent of the disease was a fungus *Phytophthora infestans*, whose spores were capable of very rapid reproduction so that a whole field of potatoes could be destroyed virtually overnight. Some ten years after the potato famine, evidence was accumulated, largely from the researches of the Rev. M. J. Berkeley in England and Anton de Bary in Germany, that indicated that the potato disease was due to a parasitic fungus, though another decade had elapsed before this idea was widely accepted. One valuable chemical treatment for the control of pathogenic fungi, like potato blight and vine mildew, was discovered accidentally by Millardet in 1882.[7] A local custom of the farmers in the Bordeaux district of France was to daub the roadside vines with a mixture of copper sulphate and lime in order to discourage pilfering of the crop. At this time the crops of the French vineyards were being destroyed by the downy mildew disease and Millardet observed that although the vines away from the road were heavily infested with mildew, those alongside the road which had been treated with the mixture were relatively free from the disease. Millardet subsequently carried out further experiments which established the effectiveness of Bordeaux mixture (copper sulphate, lime, and water) against vine mildew. The mixture was widely applied, the disease was arrested, and Millardet became a national hero.

This success stimulated the search for other chemical pesticides and the succeeding years witnessed the successful introduction of new materials containing copper, mercury, or sulphur. In addition, during this period, the manufacture of equipment for effectively applying these materials to the crop was begun.

Just as the search for fungicides to control phytopathogenic fungi was stimulated by the threat of famine, so the development of industrial fungicides was stimulated by the growth of railways and the consequent need to protect wooden railroad ties (some 3200 ties per mile of track) from rotting. This posed a tremendous challenge and meant that a successful chemical treatment would be guaranteed a profitable return and numerous patents were issued covering proprietary products containing creosote and salts of copper, mercury, and zinc for this purpose.[7]

Many of the well-known poisons have been applied at one time or another for the control of insects and other pests. They were sometimes quite successful, although the hazards to the operators were great. Cyanide, generally as hydrogen cyanide gas, had been used as a fumigant in buildings to kill bedbugs and wood-boring beetles, and also in California from 1886 onwards against scale insects on citrus trees. Tents were placed over the trees and hydrogen cyanide generated inside. Initially this treatment proved a considerable success but after a time failures became apparent. These resulted from the development of resistant strains of the insect—this was the first reported example of resistance to an insecticide. In 1897 formaldehyde was introduced as a fumigant and in 1913 organomercurials were first used as fungicidal seed dressings against cereal smut and bunt diseases.[7]

In 1896 a French farmer applying Bordeaux mixture to his vines noticed that it caused the leaves of yellow charlock in the vicinity to turn black. This fortuitous observation is probably the origin of the idea of selective herbicides. A little later it was discovered that by spraying a solution of iron sulphate on to a mixture of cereals and dicotyledonous weeds, only the weeds were killed. During the next decade many other inorganic compounds, such as copper sulphate, ammonium sulphate, and sulphuric acid, were found to exhibit selective herbicidal action at suitable concentrations.[8]

In 1912 W. C. Piver developed calcium arsenate as a replacement for Paris Green and lead arsenate, which soon became important for controlling the boll weevil on cotton in the United States of America. By the early 1920s the extensive application of arsenical insecticides caused widespread public dismay because treated fruits and vegetables were sometimes shown to contain poisonous residues. This stimulated the search for other less dangerous pesticides, and led to the introduction of organic compounds, such as tar, petroleum oils, and dinitro-o-cresol. The latter compound eventually replaced tar oil for control of aphid eggs, and in 1933 was patented as a selective herbicide against weeds in cereal crops. Unfortunately this is a very poisonous substance which was first used as an insecticide in 1892 to control the Nun moth, an important pest of forest trees.[9]

The 1930s really represent the beginning of the modern era of synthetic organic pesticides—important examples included the introduction of alkyl thiocyanate insecticides (1930); salicylanilide (Shirlan) (1931), the first organic fungicide; dithiocarbamate fungicides (1934), valuable as foliar sprays for the control of a range of pathogenic fungi such as the scabs and rots of fruit and potato blight; 2,4-dinitro 6-(1′-methyl-*n*-heptyl)phenyl crotonate or dinocap (1946) and chloranil (tetrachloro-1,4-benzoquinone) (1938) were two other protectant fungicides, the former being especially valuable against powdery mildews. Other organic compounds used during this period were azobenzene, ethylene dibromide, ethylene oxide, methyl bromide, and carbon disulphide as fumigants; phenothiazine, *p*-dichlorobenzene, naphthalene, and thiodiphenylamine as insecticides.

In 1939 Dr Paul Müller discovered the powerful insecticidal properties of dichlorodiphenyltrichloroethane or DDT; following successful initial field tests in Switzerland against Colorado potato beetle, it was manufactured in 1943 and soon became the most widely used single insecticide in the world (see Chapter 5). Present fears about the long-term deleterious effects of DDT and other organochlorine insecticides on the ecosphere (see Chapter 15) must not allow us to forget our tremendous debt to DDT. This compound controls louse-borne typhus and is equally effective against malaria-carrying mosquitoes; its use certainly materially helped the Western powers to win World War II since it permitted military operations to be conducted in the tropics where otherwise the danger of epidemics would have been too great[9]—probably no chemical, even penicillin, has saved so many lives as DDT.[10] Following the success of DDT, several useful insecticidal analogues such as methoxychlor were discovered and a number of different types of organochlorine compounds were also found to be potent contact insecticides.[11]

Benzene hexachloride (or correctly hexachlorocyclohexane) was first prepared by the English chemist Michael Faraday in 1825, although its insecticidal properties were not recognized until 1942. From about 1945 several insecticidal chlorinated hydrocarbon cyclodiene compounds were introduced, though they did not come into widespread use until the middle 1950s. Common examples include aldrin, dieldrin, heptachlor, and endrin. However, in spite of their early promise, these organochlorine insecticides are now much less used due to their pollution of the environment (see Chapter 15).

The organophosphorus compounds[12] represent another extremely important class of organic insecticides. Their early development stemmed from wartime research on nerve gases for use in chemical warfare by Dr Gerhard Schrader and his team in Germany, Early examples included the powerful insecticides schradan (octamethylpyrophosphoramide), which acts as a systemic insecticide against aphids and red spider, and the contact insecticide parathion (*O,O′*-diethyl-*p*-nitrophenyl phosphorothionate) which is remarkably effective against aphids, red spider, and eelworms. Unfortunately both these compounds are highly poisonous to mammals and later research in this field has been increasingly directed towards the discovery of more selective and less poisonous insecticides. Malathion (1950) was the first example of a wide-spectrum organophosphorus insecticide combined with a very low mammalian toxicity and more recently other safe compounds such as the selective aphicide menazon (1961) have been developed (see Chapter 6). An important advantage of organophosphorus insecticides is that after application they are generally rapidly degraded to non-toxic materials; consequently they are not persistent, like organochlorine insecticides, and therefore do not tend to accumulate in the environment and along food chains (see Chapter 15).

A closely related group of insecticides are the carbamate esters first discovered by the Geigy company in Switzerland in 1947, although the most generally effective member of the group carbaryl (*N*-methyl-α-naphthylcarbamate) was not introduced until nearly a decade later. This is becoming of increasing importance as a replacement for DDT (see Chapter 6). The natural product pyrethrum has been used as an insecticide since 1850 but is unstable in light (see Chapter 4). Dr Elliott and his team at Rothamsted Experimental Station in England, however, demonstrated in 1973 that it was possible to prepare synthetic analogues of permethrin which were much more potent insecticides than natural pyrethrum and were also photochemically stable. Since then a large number of synthetic pyrethroids have been marketed as insecticides and they are currently the most important group of insecticides.

In 1943 Templeman and Sexton working for Imperial Chemical Industries in England independently discovered the herbicidal activity of the phenoxyacetic acids. Two well-known examples are 2-methyl-4-chloro-(MCPA) and 2,4-dichloro(2,4-D)phenoxyacetic acid. These compounds are translocated in plants and are extremely valuable for the selective control of broad-leaved weeds in cereal crops. They are very safe to use, and these two compounds in fact are the most widely used pesticides in Britain (see Chapter 8).

In 1951 Kittleson working for the Standard Oil Company in America introduced an important new fungicide called captan (or *N*-trichloromethylthiotetrahydrophthalimide). This had outstanding properties as a protectant fungicide against a wide spectrum of pathogenic fungi on fruit and vegetable crops. Subsequently a number of other *N*-trichloromethylthio compounds have been marketed as foliage fungicides (see Chapter 7).

The bipyridinium herbicides diquat and paraquat were introduced by Imperial Chemical Industries Ltd in 1958. These are very quick-acting herbicides which are absorbed by plants and translocated causing desiccation of the foliage. These herbicides are strongly adsorbed by clay constituents in soil, so they are effectively deactivated as soon as they come into contact with soil. They are useful total weed killers and will rapidly kill off all top growth. Paraquat is used in 'chemical ploughing' in which the weeds are killed by spraying with paraquat, followed by immediate reseeding—this method is specially valuable in areas where there is danger of soil erosion.

The idea of the internal treatment of plants with chemicals (plant chemotherapy) is not new and goes back at least to the twelfth century when various substances such as spices, colouring matters, and medicines were inserted into the boreholes of fruit trees to endeavour to improve the fruit.[13] Some rather macabre experiments were carried out by Leonardo de Vinci in the fifteenth century in which arsenic was injected into fruit trees to make the fruit poisonous.

The study of plant pathology developed rapidly during the eighteenth century and experiments on the movement of substances, such as dyes and mineral salts, were carried out. Certain plant diseases were shown to result from nutrient deficiency, such as chlorosis from lack of iron, and attempts were made to cure the plants by injection of mineral salts. In the early 1900s toxic compounds, such as potassium cyanide, were injected into plants in an effort to kill insect pests. An examination was also made of the effects of injecting dyes and disinfectants into plum trees infected with silver leaf disease, and later studies[13] showed that 8-quinolinol sulphate was an effective chemotherapeutant against the disease. In America in the 1920s injection of sweet chestnut trees was studied as a means of controlling blight; lithium salts had some inhibiting effect but injections of thymol were much more effective. Some early workers also tried root applications of chemicals for the control of phytopathogenic fungi; thus Massee (1903) claimed to reduce cucumber mildew by root treatment with aqueous copper sulphate, and Spinks (1913) found that lithium salts inhibited the development of powdery mildew on wheat and barley.

However, little significant progress in plant chemotherapy occurred until the late 1930s, because by this time the limitations of the conventional surface fungicides were obvious. Also many new organic compounds were becoming available, and outstanding success had been achieved in the field of human chemotherapy. In 1935 the important range of sulphonamide drugs was introduced, and in 1938 Hassebrauk demonstrated that root treatment with sulphanilamide protected wheat seedlings from attack by rust spores.

In 1940 Chain and Flory showed that penicillin was highly effective against bacterial infections in man. This stimulated the search for other medically useful antibiotics and chloramphenicol, 'Aureomycin', and streptomycin were soon discovered, and by 1952 streptomycin was being used for systemic control of certain fungal pathogens and bacterial diseases of plants (see Chapter 7).

World War II not only enhanced the development and commercial production of antibiotics, but it also provided the basis for Schrader's work on organophosphorus compounds, several of which proved highly efficient systemic insecticides.[12] The very valuable phenoxyacetic acid selective herbicides were also introduced, so that by the 1950s a range of commercial systemic insecticides and herbicides was available. However, it was not until the late 1960s that effective systemic fungicides appeared on the market,[14] and their development represents an important breakthrough in the field of plant chemotherapy.

The major classes of systemic fungicides developed from 1966 are oxathiins, benzimidazoles, thiophanates, and pyrimidines. Other effective systemic fungicides at present in use include antibiotics, morpholines, organophosphorus compounds,[14] and most recently the sterol biosynthesis inhibitors, e.g. triazoles (see Chapter 7).

Ever since man had a home it was invaded by rats and mice which also raided his food stores. The rat is one of man's most formidable enemies. It does significant damage to the fabric of buildings and is the carrier of some of man's dreaded diseases such as Weil's disease and bubonic plague, the black death of the Middle Ages which in 1348–9 killed a quarter of the population of Europe and between 1896 and 1917 was responsible for nearly 10 million deaths.

Chemicals that control rats are termed rodenticides (see Chapter 11). The first really effective compound, warfarin, was developed by the Wisconsin Alumni Research Foundation in 1944. It is an anticoagulant, which has been used in human medicine. Rats and mice are killed by internal haemorrhage and they will eat it in a suitable bait without becoming poison-shy.[15,16] However, in Britain a resistant strain of rats has appeared which are immune to normal doses of warfarin and these resistant or 'super' rats have increased considerably. Luckily, 'second generation' anticoagulants, like brodifacoum (1978), will control these resistant rats.[17]

The large-scale pesticide industry really dates from the end of World War II with the commercial introduction of the phenoxyacetic acid selective herbicides and the synthetic organochlorine and organophosphorus insecticides. It has been estimated[18] that the production of the major pesticide manufacturing countries in 1987 were worth the following amounts:

Western. Europe	$5670 million
Far East	$4835 million
United States	$4465 million

In 1984, the United States of America market accounted for 34% of the global total but declined through 1985 and 1986 to 30.5 and 26.4% respectively; this trend has continued with the share down to 24% in 1988.[17]

In the period from 1949 to 1965 the United Kingdom exports had expanded from £3 million to £10 million and by 1985 the figure had reached £511 million. The total value of all pesticides applied to world crops in 1949 was approximately £200 million, whereas the total exports of the major pesticide companies were ≃ £33 million; this clearly emphasizes the point that most of the pesticides are used in the manufacturing countries and not exported. The relatively low proportion exported may be partly due to the fact that many underdeveloped countries import chemical intermediates from industrialized countries and then carry out the final stages of production themselves in which case the chemical intermediates do not count as pesticide imports. Such relatively primitive countries like to have political control of pesticide manufacture. Thus in India hexachlorocyclohexane (Chapter 5) is the most widely used insecticide because it can be produced simply in crude form. The annual production of this compound in India in 1965 was some 30 000 tonnes which is about four times as much as was produced in the United States of America, though it represented only 20% of the total American production of all organochlorine insecticides.[16] The Pesticide Reviews of the American Department of Agriculture give the value of American pesticide production in 1987 as some $4465 million so that a reasonable estimate of total world pesticide production would be $21 500 million. British pesticide production was nearly £842 million, composed of £434 million of herbicides, £237 million of insecticides, £165 million of fungicides, and £6 million of miscellaneous agrochemicals.

In temperate countries herbicides are the major type of pesticide used: in Britain current sales to farmers are 52% herbicides, 30% fungicides, 10% insecticides, and 8% miscellaneous agrochemicals. In contrast, in tropical and subtropical areas a much larger relative amount is spent on insecticides.

The pesticides market tends to be dominated by relatively few products; thus in the United States of America of a total production of 440 000 tonnes (1985) six compounds accounted for nearly 70% of the synthetic herbicide market — alachlor (19%), atrazine (17%), butylate (12%), trifluralin and metolachlor (8% each) and 2,4-D (5%). For fungicides, the predominant product was sulphur with the major organics captan and the dithiocarbamates; with synthetic insecticides methyl parathion, carbofuran, terbufos, and chlorpyrifos account for some 50% of the total.

The rapid growth in the application of agrochemicals in industrial countries like those of Western Europe and the United States of America has been stimulated by the high cost and shortage of agricultural labour. A recent assessment (1989) of the current world usage of pesticides based on their user cash value suggested about 24% in the United States of America, 28% in Western Europe, 24% in the Far East, 9% in Eastern Europe and the Union of Soviet Socialist Republics, 8% in Latin America, and 7% in the rest of the world.[17] Relatively small amounts of pesticides are used in the underdeveloped countries. However, it is precisely in these countries that there is the greatest need for crop protection chemicals. They contain 49% of the world population and 46% of the total world cultivated land area and suffer the greatest losses of crops due to pests. On a world scale pests destroy about half of the annual crop during growth, harvesting, and storage, but in the underdeveloped countries, e.g. India, Africa, and Latin America, losses are some 70% of everything produced;[19] they need to double food production over the next 20 years to feed the increasing population.

In these countries, even after the harvest has been gathered, there are appreciable losses in storage due to insects, rats, mites, and fungi. Losses in the tropics are much greater than in temperate countries; in West Africa estimated losses in storage are 25% and in India some 8%. The major world crops are cereals, cotton, and soyabeans, with oilseed rape becoming of increasing significance. The agrochemical usage based on crops in 1988 were fruit/vegetables 26%; rice 12%, maize and cotton 11% each; wheat 10%; soyabean 9%; sugar beet 4%; others 17%.[17]

Properly constructed stores can keep out rats, and fungi and mites can be inhibited by controlling grain moisture content at < 13%. Insects are more difficult to control by physical methods and there is a need for the further development of safe insecticides for use on stored crops.

Clearly there is a tremendous potential for expansion in agrochemical sales to such countries, but their poor economic state clearly discourages increased expenditure on pesticides. Extrapolation of existing trends suggests a steady growth in the use of pesticides although it is unlikely that there will be any marked change in their geographical distribution; by the end of the 1990s perhaps some 16% of pesticides will be used in the developing countries.

The increasing use of pesticides in the underdeveloped countries is absolutely vital if such countries are to obtain the greater supplies of food necessary to feed their large populations adequately. Several examples highlight the value of pesticides in reducing crop losses. Thus in Ghana, which is the world's premier cocoa exporting country, the application of insecticides has almost trebled the yields by effectively controlling the damage to the crop by the capsid bug, and in Pakistan extensive use of insecticides on the sugar crop increased the yield by 30%. Rice is the staple food for the majority of mankind; in Japan the introduction of organophosphorus insecticides and systemic fungicides resulted in dramatic yield increases. In Asia, Japan produces the highest rice yield (62 decitonnes/ha) as compared with India (19.8) and some African countries (5). The United Nations Food and Agriculture Organization (FAO) has estimated[19] that without the use of pesticides some 50% of the total cotton production in developing countries would be destroyed by pests prior to harvest. It is becoming increasingly evident to the world food authorities that pesticides may be the single most important factor in improving food production in the underdeveloped countries.

Herbicides are likely to remain the major class of pesticides used in developed countries while insecticides take pride of place in developing countries. Current world consumption is 44.2% herbicides, 28.8% insecticides, and 20.9% fungicides, leaving 6.1% for growth regulators and miscellaneous agrochemicals.

The steady growth in agricultural production in the 1970s fuelled a substantial world growth in agrochemicals averaging some 6.3% per annum in real terms. However, in 1983, there was a 2.9% decline in the agrochemical market; sales recovered in 1985 but with a lower growth rate of 1.1% per annum. The global sales (1989) of $21 500 million represented a real demand increase of some 3.2%. This was the strongest rise since 1984; it is likely that the increase of approximately 3% will be maintained in 1991–2. The agrochemical market is approaching maturity which is characterized by a low growth rate in developed countries.

Rapid advances in biotechnology will lead to novel microbial products and new crop varieties; the former will become increasingly important crop protection agents and should gradually replace agrochemicals. However, these new methods are not likely to take more than some 5% of the crop protection market by the year 2000, though rapid progress can be expected early in the twenty-first century. It is concluded that the long-term growth for the agrochemical industry will be approximately 2% per annum in real terms, but higher (5%) in the less developed countries.[20]

Agrochemicals are becoming more potent in terms of the dose required (grams per hectare rather than kilograms per hectare) to control the pest.

In 1987, the European Economic Community agrochemical market was $5670 million and is predicted to rise to $6500 million (1990) (30% of the world market). The fastest growth will be shown by Spain and Portugal where accession to the EEC favours the development of larger, more efficient farms—in Portugal an annual growth rate of 6% is predicted. The United Kingdom market at 3% of world sales appears to have reached a plateau (1987–8).

In the United States of America, the agrochemicals market increased from $3504 million (1980) to an estimated $4465 million (1987).[17] Over the next five years it is predicted that the largest growth in pesticide usage will occur in China, Brazil, India, the Union of Soviet Socialist Republics, and the United States of America.[21] Crops are threatened by hundreds of disease organisms, have to compete with some 1800 problem weed species, and can be consumed by about 10 000 insects and 8000 nematodes.

Even today, when some $20 billion worth of agrochemicals are applied worldwide, almost half of the potential total agricultural production is lost—35% of the crop to weeds, pests, and diseases before harvest with a further 15% loss between harvest and sale. In the underdeveloped countries, the losses can be substantially greater, often 70% of the potential crop is lost to the consumer.

The estimated world population in 1987 was 5 billion compared with 2 billion at the end of the period 1830–1930. Currently some 700 million people are undernourished and 1.3 billion exist on an inadequate diet.[22] The world population, according to World Bank projections, is expected to increase to over 6 billion by the year 2000 and to over 8 billion by 2025. The population explosion is, however, not uniform; >90% of the increase has occurred in underdeveloped countries and hence the urgent need for much greater application of agrochemicals in these areas to increase food supply.

The efficiency of modern agrochemicals in controlling their target organisms and the resultant increase in crop yields is well illustrated by two further examples.[23] The yield of cotton in the United States of America, after treatment with cypermethrin against cotton bollworm, was 402 kg/ha (cf. the untreated crop yield of 67 kg/ha). The yield of wheat which had been treated with the herbicide diclofop-methyl against infestation by wild oats was 366 g/m^2 (cf. 143 g/m^2 for untreated wheat).

The greater efficiency of modern agricultural practice liberates land that can be used for recreational purposes; thus in the United States of America (1983) sufficient food was produced from 117 ha whereas in 1950 the production of the same quantity of food required 243 ha.

Currently on the world market there are approximately 1000 different active ingredients of pesticides which are formulated as at least 30 000 trade products.

In countries like the United States of America, the development of a pesticide from initial discovery in the laboratory to marketing takes at least 8 years. The costs of development have substantially increased; thus in 1964 the cost was $2.9

million but by 1987 it had risen to $50 million due to increasingly stringent environmental and toxicological tests demanded by the legislative bodies, such as the Environmental Protection Agency (EPA) in the United States of America.

It is also becoming increasingly difficult to discover a new agrochemical with significant advantages over existing products, and consequently the number of compounds which need to be screened to obtain one marketable product has substantially increased from one in 3600 (1964) to one in 16 000 (1985) and in 1989 was estimated to be one in 20 000. There has been an overall decline in the profitability of the agrochemicals industry from 11.5 % (1981) to 7.9 % (1986);[20] this is highlighted by the fact that all the ten major agrochemicals used in the United States of America (1987) were introduced prior to 1976, namely: glyphosate (1972), alachlor (1966), metribuzin (1971), carbaryl (1956), chlorpyrifos (1965), carbofuran (1967), chlorothalonil (1963), trifluralin (1963), bentazone (1975), and dicamba (1965).[20]

A new agrochemical will only be developed today if it is effective in protection of one or more of the following major world crops—maize, rice, soyabeans, cotton, wheat, or oilseed rape; this activity will ensure a sufficiently large potential market to justify the development costs.

Agrochemicals are today a very high risk business because the substantial sum (ca. $50 million) spent on development during the first 8 years must be recovered quickly; the life of the patent expires after 20 years and then other companies who have not had to bear the high development costs can manufacture the product and sell it, often at a lower price. The cumulative discounted cash flow for a successful pesticide is illustrated in Figure 1.1.

The agrochemicals industry today is much more complex; to protect the

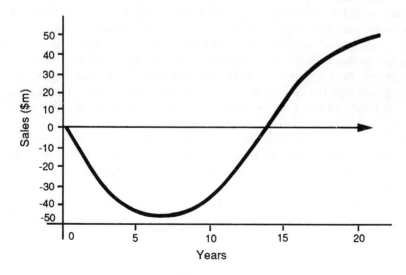

Figure 1.1

environment and consumers from dangerous agrochemicals, the standards demanded for approval and registration of products have become much more rigorous, and to satisfy the criteria may involve the company in an expenditure of some $5 million.

In 1980, the tests required by the American EPA involved the following experiments: assessment of acute toxicity based on the results of 90-day rat and dog feeding tests and similar ones extending over 2 years; rat reproduction studies over three generations and examination for teratogenicity in fish, shellfish, birds, and rodents; toxicity studies in birds and metabolic studies in rat, dog, and plants must be carried out; analysis must show that the residues in food crops are not more than 0.01 p.p.m., meat 0.1 p.p.m., and milk 0.005 p.p.m.

Long-term studies (up to 8 years) are made to determine the effects of the chemical on the ecosystem, including non-target organisms. These environmental studies are designed to identify any bioaccumulation problems, as occurred with organochlorine insecticides (see Chapter 15). The range of tests demanded by the various registration bodies are likely to be increased in the future. The present legislation contrasts sharply with the 1950s when only the 30–90 day rat feeding acute toxicity tests and residue analysis on food crops (1 p.p.m. maximum) were demanded.

In the United Kingdom all the legislation on pesticides is based on Part III of the Food and Environmental Protection Act 1985 and the Control of Pesticides Act 1986; the legislation has recently been reviewed.[24] The aims are to protect the health of humans, creatures, and plants; to safeguard the environment and to secure safe, efficient, and humane methods of pest control. It gives the Government the power to control the import, sale, supply, storage, use, and advertisement of pesticides. Information in connection with the control of pesticides is to be made available to the public and an advisory committee on pesticides will be established. No product may be used unless it has received provisional or full approval by ministers and they have consented to its use. Approved pesticides are listed by the Ministry of Agriculture, Fisheries and Food (MAFF).[18]

All sellers, storers, and users must take all reasonable precautions to protect the health of humans and the environment; users must take particular care to avoid pollution of water. From 1988, a recognized certificate of competence is required for anyone storing or selling pesticides.[25] Users must adhere to conditions set out in the approval of a given pesticide, e.g. the use of suitable protective clothing. From 1989, all contractors applying pesticides in agriculture, horticulture, and forestry must hold a recognized certificate of competence.[25]

In the United Kingdom, the control of substances hazardous to health (COSHH) regulations were introduced in 1988. By January 1990, industry has to carry out assessments of all substances hazardous to health within its working environment. Adequate control and protection measures must be devised; if some risk of exposure remains, staff must be equipped with protective clothing—this is particularly relevant to crop spraying operations. Employers must monitor exposure of staff by carrying out regular health surveillance as appropriate. COSHH regulations provide guidelines to good management practice to safeguard the health of workers.

Worldwide, even in developed countries, many of the pesticides discovered in the 1950s are still extensively used. There is urgent need for the introduction of more selective agrochemicals, particularly with different modes of action to combat the growing problems presented by resistant fungi and insects. There is a real danger that excessive emphasis on potential environmental hazards, especially in the United States of America, may result in the elimination of valuable agrochemicals and stifle the development of promising compounds due to overregulation. Such factors have caused a massive increase in the development costs of a new pesticide and for a time caused a reduction in the number of significant new products coming onto the market. The maximum number of new compounds were introduced in the 1950s and 1960s with some 18 per annum, but declined to 6 in the 1970s.[10]

Random screening has become less successful; consequently there has recently been more research targeting with greater resources being concentrated on areas of chemistry of proven biological activity. This approach, coupled with increasing use of computer graphics to provide a three-dimensional model of the active sites, has been quite successful and the number of new compounds coming onto the market has increased.

This approach has inevitably led to a clustering of new agrochemicals in certain areas, such as the triazole fungicides and synthetic pyrethroids which were launched in 1976 and 1977 respectively; in 1988 there were approximately 14 and 17 members of these groups on the market.[20]

The introduction of the synthetic pyrethroid insecticides, sulphonylurea herbicides, and triazole systemic fungicides in the 1970s and 1980s has set new standards of pesticidal potency; they are active in rates of grams rather than kilograms per hectare. The development of such compounds will be paralleled by improvements in application technology.[23] There will be increasing emphasis on the use of pesticides showing good selective toxicity to the target organism; such agrochemicals can be incorporated in integrated pest management (IPM) programmes which may include cultural and biological control measures as well as the use of behaviour-modifying chemicals.

IPM enables the use of smaller amounts of chemical pesticides and reduces the dangers of environmental pollution and the emergence of resistant strains of pests. In the developed countries in the 1990s there is likely to be a progressive shift from animal to crop production, resulting from increased use of vegetable proteins. Improvements in the technology of texturized vegetable meat substitutes will result in declining meat production and hence soyabeans and leguminous crops will become more important.

The move away from meat production has been stimulated by evidence relating an increased risk of heart disease to excessive consumption of saturated animal fats.

REFERENCES

1. Cremlyn, R. J. W., *Pest Articles and News Summaries*, **17**(3), 291 (1971).
2. Hassall, K. A., *The Biochemistry and Uses of Pesticides*, Macmillan, London, 1990.
3. Horsfall, J. G., *Principles of Fungicidal Action*, Chronica Botanica Co., Waltham, Mass., 1956.
4. Whitten, J. L., *That We May Live*, Van Nostrand, Princeton, N.J., 1966, p. 26.
5. De Ong, E. R., *Chemistry and Uses of Pesticides*, 2nd edn, Reinhold, New York, 1956.
6. Large, E. C., *The Advance of the Fungi*, H. Holt, New York, 1940, p. 13.
7. McCallan, S. E. A., 'History of fungicides', in *Fungicides, An Advanced Treatise* (Ed. Torgeson, D. C.), Academic Press, New York, 1967, p. 1.
8. Martin, H., and Woodcock, D., *The Scientific Principles of Crop Protection*, 7th edn, Arnold, London, 1983, p. 285.
9. Mellanby, K., *Pesticides and Pollution*, Collins, London, 1970, p. 111.
10. Boardman, R., *Pesticides in World Agriculture*, Macmillan, London, 1986.
11. O'Brian, R. D., *Insecticides: Action and Metabolism*, Academic Press, New York, 1967, p. 182.
12. Fest, C., and Schmidt, K.-J., *The Chemistry of Organophosphorus Insecticides*, Springer-Verlag, Berlin, 1973.
13. Wain, R. L., and Carter, G. A., 'Historical aspects', in *Systemic Fungicides* (Ed. Marsh, R. W.), 2nd edn, Longman, London, 1977, p. 6.
14. Cremlyn, R. J. W., *International Pest Control*, **15**(2), 8 (1973).
15. McMillen, W., *Bugs or People*, Appleton-Century, New York, 1965, p. 128.
16. Green, M. B., Hartley, G. S., and West, T. F., *Chemicals for Crop Improvement and Pest Management*, 3rd edn, Pergamon Press, 1987, p. 39.
17. British Agrochemicals Report 1988/89, British Agrochemical Association, Peterborough, 1989.
18. *Pesticides 1988*, Ministry of Agriculture, Fisheries and Food (MAFF), HMSO, London, 1989.
19. *Pesticides in the Modern World*, A symposium prepared by members of the Co-operative Programme of Agro-Allied Industries with FAO and other United Nations Organisations, Newgate Press, London, 1972.
20. Finney, J. R., *Proceedings of British Crop Protection Conference, Brighton*, **1**, 3 (1988).
21. *Farm Chemicals* (International edn), **85** (1985).
22. Büchel, E. H. (Ed.), *The Chemistry of Pesticides*, Wiley, New York, 1983, Ch. 1.
23. Reece, C. H., *Phil. Trans. Roy. Soc. London*, **B310**, 201 (1985).
24. Gilbert, D., and Macrory, R., *Pesticide Related Law*, British Crop Protection Council, Farnham, 1989.
25. *Revised Code of Practice for the Agricultural and Horticultural Use of Pesticides*, MAFF, HMSO, London, 1988.

Chapter 2
Physicochemical Factors

As mentioned in the introduction (Chapter 1), chemicals have been used for pest control for centuries but the large-scale pesticide industry really only dates from the end of World War II. The valuable new chemicals developed over the last 40 years or so have arisen as a result of effective collaboration between teams of chemists and biologists.[1]

Organic chemists first decide on some novel groups of compounds which they feel may have potential as pesticides; then they devise successful synthetic routes to obtain some candidate members of the selected groups. The chemicals, after purification and characterization by spectroscopic and analytical data, are submitted to the biologists for evaluation for potential pesticidal activity. In the initial primary screens a fairly high dosage of up to 500 p.p.m. may be used to pick out all compounds having some activity on the test organism. If activity is found the tests are repeated on that organism at lower dose rates, for instance at 200, 100, and 10 p.p.m. A series of further increasingly stringent tests are then applied, each stage causing candidate compounds to drop out. These screens are so good at eliminating compounds that the chances of a newly synthesized compound being a marketable pesticide are currently only about 1 in 20 000, but with effective screening techniques this random screening can give a reasonable return.[2] In fact, nearly all the present pesticides have been discovered as a result of more or less random screening of the chemicals available to the company.[3] The precise screening methods vary from company to company but generally the candidate chemicals are tested (a) against a range of insects of economic importance, both by spraying and orally by incorporating the chemical in the insects' food, (b) against a range of plant pathogenic fungi growing in nutrient agar plates (*in vitro* tests) and also against fungi growing on plants (*in vivo* tests), (c) against 10-15 different species of weeds growing in the greenhouse, and both soil and foliar applications are investigated, (d) against certain other pests, such as slugs, eelworms, rats, mice and mites, (e) for plant growth regulatory activity.

As an illustration the Plant Protection Division of Imperial Chemical Industries Ltd at Jealott's Hill Research Station in Berkshire, England, annually screen some 10 000 chemicals against disease organisms, pest insect species, weeds, and for possible useful effects on plant growth. If a chemical shows promise, the range of test organisms is widened and chemists use their skilled intuition to prepare structural modifications of the original chemical in an effort to enhance activity. The examination of a number of structural analogues hopefully will enable a structure–activity relationship to be elucidated, so leading both to a better pesticide and possibly throwing light on the mode of biocidal action. Only very few candidate chemicals survive the rigorous laboratory screening procedures. Those remaining are then formulated and applied in various ways to find out how to obtain the optimum results under field conditions. These experiments are initiated on a relatively small scale in the greenhouse and continue using small-scale plots and ultimately progress to large-scale field trials in Britain and at overseas field stations.

The chemicals must finally be tested under the actual conditions in which they are to be used in practice by Government Research Stations and farmers. During the various stages of development there is a steady decline in the number of chemicals under test as those which have proved unsuitable for various reasons are rejected. While development proceeds, efficient and economic methods of large-scale synthesis have to be devised by chemists; the production of large amounts of the material may present problems not apparent during the small-scale preparation.

Fairly early in the development of the pesticide, the toxicological properties have to be evaluated—this must cover not only determination of the acute toxicity but also any possible long-term effects on the environment in tests lasting at least five years (see Chapter 15). The acute toxicity is determined by testing the candidate chemical against various mammals, generally rats and mice. The toxicity is usually recorded as the LD_{50} value—this is the dose required to kill 50% of the population of test animals and is expressed as milligrams per kilogram of the body weight of the animal. There should be at least ten animals in the experiment and administration of the chemical can be oral (in the animal's food), by intravenous injection, or to the skin (dermal). Dermal toxicities are often slightly less than the oral values while intravenous toxicities are higher than the oral figures. In this book, the average values of the oral toxicities (LD_{50} values) as a single dose against rats are quoted to give some indication of the mammalian toxicities of the different types of pesticides. The smaller the LD_{50} value, the more toxic the chemical, so that toxicities of chemicals can be graded by the LD_{50} values as follows:[4]

	LD_{50} (mg/kg)
1. Supertoxic	<5
2. Extremely toxic	5–50
3. Very toxic	50–500
4. Moderately toxic	500–5000
5. Slightly toxic	5000–15 000
6. Relatively harmless	>15 000

In this respect it is vital to understand the mode of pesticidal action, and the chemical must have minimum ecological effects if it is to be a commercially viable pesticide in the modern pollution-conscious society. It usually takes a period of about ten years from the date of the initial discovery of useful biocidal properties of a chemical to the marketing stage and, with the increasing stringency of the environmental and toxicological tests which a candidate chemical has to pass, this gestation period is growing. Any candidate pesticide for the British market must be approved at several stages by the Pesticides Advisory Committee set up by the Ministry of Agriculture, Fisheries and Food (MAFF) which must be satisfied that it does the specified job before it may be included in the Ministry's annual list of approved products.[5] Most countries in the world have comparable requirements: the Environmental Protection Agency (EPA) in the United States of America is particularly stringent with reference to any residues left after using the chemical. If the mode of action and toxicological studies are satisfactory large-scale field trials are initiated.

At this stage the study of different types of formulation of the chemical have to be undertaken. Getting the right type of formulation is vital in obtaining the optimum results[3,4] for the pesticide; indeed it is true that a pesticide is only as good as its formulation. The majority of pesticides are applied as dusts or sprays, but they may also sometimes be used as granules, aerosols, fumigants, baits, or seed dressings.

The formulator tries to bring the active ingredient (a.i.) into a convenient form for application, either directly as powders, granules, or self-dispensing packs, e.g. aerosols, or sometimes after mixing with water or other readily available liquid diluent. The product must be formulated so that it is as effective as possible, but it must also be convenient to use, stable, and reasonably safe in storage and transport. The active ingredients of most pesticides are relatively insoluble in water, but are fairly soluble in lipophilic organic solvents like petroleum and xylene. These are, however, insoluble in water and so if the pesticide is dissolved in a suitable organic solvent and the solution diluted with water the organic layer quickly separates out from the water in the spray vessel.[6]

This problem can be overcome by addition of emulsifiers, which are surface active agents (surfactants), to the solution of the pesticide enabling it to form a stable emulsion of small globules $< 10 \mu g$ on mixture with water in the spray tank and the emulsion can then be sprayed onto the crop. Many water-insoluble pesticides are marketed in the form of thick self-emulsifiable concentrates (EC) consisting of a solution of the pesticide (ca. 0.25% w/v of a.i.) in an organic solvent, e.g. petroleum or other hydrocarbon oil containing oil-soluble surfactants. The formation of emulsifiable oils has been aided by the development of non-ionic emulsifying agents, e.g. polyglycol ethers, polyethylene oxides such as compound (**1**) (where $n \simeq 12$) and dimethyl silicone (**2**) are effective surfactants; these lower the surface tension of water and hence improve the spread of water on waxy surfaces.

$$C_8H_{17}-\!\!\!\bigcirc\!\!\!-OCH_2CH_2(OCH_2CH_2)_nOH$$

(1)

(2)

Such non-polar compounds have the advantage of being much more soluble in oils than ionic surfactants, and their effect is not greatly altered by the presence of saline impurities so that they do not form insoluble scums in hard water, which was a major trouble when soaps were used. An effective self-emulsifying oil then requires the use of several different surfactants, for instance two or more non-ionic agents with widely differing numbers of ethylene oxide groups in their chains, or more generally a mixture of anionic and non-polar surfactants. Cationic surfactants are also used in certain cases, e.g. with some fungicides which are themselves cationic surfactants. In this case, use of anionic surfactants would lead to loss of surfactant properties and formation of an insoluble gum—for this reason pesticides should not be mixed with each other until it is established that they are compatible.

Approximately 75% of all pesticides are used as sprays and the majority of these are applied as water emulsions prepared from ECs. If it is correctly formulated, the EC should remain suspended in the form of a milky emulsion for at least a day after dilution with water. If precipitation occurs prior to spraying, the sprayer nozzle will become blocked leading to uneven application of the pesticide.

The cost of petroleum solvents like cyclohexane or xylene has increased by >300% in the last six years and consequently the price of EC formulated products has escalated over recent years. Other cheaper methods of formulation for spraying are therefore being examined; for instance water-miscible liquids can be obtained if the active ingredient is soluble in water or alcohol. These formulations on dilution with water remain clear and are not milky like ECs.

When dilution has given the optimum concentration, the solution can be sprayed onto the crop. The chemical may also be used as a wettable powder (WP). The powder is prepared by grinding the material in a ballmill. However, few organic compounds produce a free-flowing powder without first mixing them with an inert inorganic mineral diluent, e.g. a talc or clay, and addition of dispersing and wetting agents, e.g. detergents, are also needed to give a fine suspension when the powder is mixed with water in the spray tank, which must not settle out on the bottom of the tank for at least half an hour, otherwise frequent agitation will be needed to prevent formation of sticky powders that tend to cake on storage. In certain instances, especially with inorganic compounds like copper fungicides, the mineral diluent is not necessary since these chemicals have no tendency to stickiness. Clay mineral diluents contain strongly acidic centres which may catalyse the decomposition of certain pesticides, whereas other chemicals may be sensitive to alkaline diluents. A knowledge of the chemical reactivity of the active ingredient, especially towards acidic or basic hydrolysis, is valuable information to aid selection of the mineral diluent to be used in the formulation of a wettable powder. The emulsion spray separates into two phases on impact, the water running off the plant surface while the pesticide solution remains on the surface. With non-systemic pesticides, it is essential to obtain a high coverage of the surface by the spray and this is often enhanced by the addition of various wetters and spreaders; further, the persistence of the dried deposit is sometimes improved by the use of chemical additives termed 'stickers'. In WPs good wettability and suspension stability are essential; this is assisted by having particles as small as possible, preferably $<25 \mu$m.

A liquid pesticide can be formulated as a WP by incorporation onto a suitable mineral adsorbent filler. Sometimes organic fillers, e.g. corn cobs or nut shells, are used. WPs often contain 50% a.i.

Another economical formulation is termed a flowable or suspension concentrate (SC); here the solid pesticide is suspended in a small quantity of a liquid medium, usually water or oil, and is blended with an inert dust diluent in the presence of ethylene glycol and suitable wetting and dispersing agents. A wet grinding technique is used commercially to produce a paste containing particles of the active ingredient (a.i.) of $<5 \mu$m diameter.

The choice of wetting agents used will depend on whether the flowable suspension is water or oil based. For this type of formulation, the pesticide must have a melting point $>60°$C and must not react with the dispersing medium (water or oil) in which it must be almost insoluble. In each case, on dilution with water the SC must form a stable aqueous dispersion of the a.i. in the spray tank.

A modification involves microencapsulation[7]—the insecticide is incorporated in small permeable spheres $\simeq 15 \mu$m in diameter composed of a suitable polymer; these are then blended with wetting agents and thickeners to provide the microencapsulated flowable concentrate. Some pesticides, particularly household and garden insecticide sprays, are formulated as oil solutions of the a.i.; these are contained in plastic bottles fitted with a spray atomizer. Pesticides may also be sold as oil concentrates to be further diluted with oil prior to spraying; such formulations are used to control flies, mosquitoes, and weeds in roadside verges.

Pesticides may also be applied as dusts, although this method of application is not generally as effective as spraying. The pesticide powders are made by adding the active ingredient to a finely ground carrier and grinding in a ballmill. The resultant dust concentrate may be further diluted with a solid diluent, often an inert clay. The particles must be the optimum size. If too small, they will cling together in the air blast and the coverage of the target surface will be poor. On the other hand, if too coarse the dust will not penetrate into the interior of plants and the active ingredient may separate from the particles of the diluent; this does not occur if the diameters of the particles are all less than 20 μm. The pesticide powder must not cake in humid climates or be so hard that damage is caused to the dusting machinery. The electrostatic charge is also important. This is acquired by friction as the dust passes through the blower; since the majority of leaf surfaces are negative, positively charged dusts adhere better. A dust must be more dilute than a wettable powder, generally about 2-5% of the active ingredient, because dusting machines usually cannot produce an even discharge at less than about 16.8 kg/ha. In spite of their ease in handling, formulation, and application, dusts are the least effective of the pesticide formulations because they have a relatively poor rate of deposition on the target. For instance in aerial dusting, which must be done on a calm day, some 10-40% of the pesticide reaches the crop while the remainder drifts away from the target. In similar conditions, aerial spraying using a water-emulsion formulation deposits 50-80% of the pesticide on the target.

Several insecticides are effective as seed dressings, especially against wheat bulb fly on cereals,[8] and in this context the stable synthetic pyrethroid permethrin (Chapter 4, page 59) appears very promising. With fungicides, application by seed treatment has become increasingly important with the development of a number of commercial systemic fungicides, e.g. carboxin (see Chapter 7, page 188). Various seed dressing techniques include the use of adhesive dust, wet slurries, and solutions.[9] These were originally devised to deal with broad-spectrum surface fungicides, particularly organomercurials, used as seed dressings on cereals to protect the crop against seed- and soil-borne diseases.[3] The techniques are, however, equally applicable to systemic fungicides and when used as seed sterilants, as well as protectants during germination, seed dressings provide a most economic and convenient method of application, since they avoid the high labour costs inherent in spraying or dusting operations. The majority of fungicides are, however, applied as foliar sprays or dusts; ultimately the efficiency depends on the amount of the active toxicant that reaches the site of action, which is sometimes not the formulated compound but a metabolite, e.g. benomyl (see Chapter 7, page 185). Gray showed[9] that the antifungal action of streptomycin was increased by the addition of glycerol, which acts as a humectant to the spray mixture. This slowed up the rate of drying on the foliage, allowing time for the compound to be absorbed by the foliage. Surfactants are often added to fungicides to improve their efficiency by increasing their wetting properties and the solubility of the active compound, as illustrated by the effect of the surfactant Tween 20 on triarimol.[8]

Small-scale application of pesticides can be conveniently carried out from aerosols—these give very fine spray drops by discharging a liquid contained in a pressure vessel above its boiling point through a very fine orifice. The droplets are further broken up by the propellant boiling when the liquid jet comes into contact with the air. This method of application is, however, expensive and is only used in small gardens, glasshouses, and in the home where convenience outweighs the cost of aerosol cans. Aerosols give a very fine spray with droplets in the 10-30 μm range. Currently we live in an aerosol culture—in 1984 some 2.4 billion aerosols were produced. They are used for dispensing hair sprays, body deodorants, furniture polish, window cleaners, water-repellents, de-icers for cars, and paints. Some 8% of these aerosols were insecticide sprays and insect repellents.

The majority of aerosols have, until recently, contained chlorofluorocarbons (CFCs), e.g. dichlorodifluoromethane (b.p. 28°C), as the propellants. However, there has been evidence that their use has caused serious damage to the ozone layer in the upper atmosphere. The resultant 'holes' in the ozone layers allow a greater amount of the sun's harmful UV radiation to reach the earth which would increase the incidence of skin cancers. For this reason, alternative propellants, e.g. carbon dioxide, are now being used in aerosols.[4] Some pesticides are sufficiently stable to be applied as smokes from specially manufactured smoke generators. The compound is mixed with an oxidant (e.g. sodium chlorate) and a combustible material (e.g. a carbohydrate), giving rise to a hot non-inflammable gas when it comes out from the orifice of the smoke generator.[3] Such generators are used in glasshouses, dwelling houses, and stores; the insecticides lindane HCH, DDT, and parathion and the fungicides karathane and captan are often applied in this way. A recent method of formulation is as granules; the pesticide is incorporated into porous clay pellets of 0.5–1.5 mm diameter which are coated with the pesticide by spraying them with a solution of the chemical to give the desired concentration ranging from 2 to 20% of the active ingredient. This type of formulation has the advantage that the pesticide can be applied in winds of up to 20 km/h without any appreciable drifting, in contrast to application as dusts or sprays; also the granules can often force their way through the foliage to reach the ground.

An alternative method of preparation of granules involves mixing sieved crushed nut shells or corn-cob chips with the powdered pesticide and a suitable adhesive in a rotar. Microgranules with particle sizes of 100–300 μm have recently been introduced; this formulation has, like controlled droplet application (see page 25) been made feasible by the development of spinning disc applicators.

Slow-release formulations are comparatively new; in 1963 Shell introduced No-Pest Strips[R], in which the volatile organophosphorus insecticide dichlorvos was incorporated into panels of polychlorovinyl resin which allowed the a.i. to slowly permeate into the atmosphere.[10] Other important examples include the formulation of trialkyltin antifouling agents in a synthetic acid resin; insect traps in which the pyrethroid insecticide is placed between two plastic strips. The herbicide 2,4-D may be bound to a lignin support, so that the a.i. is gradually released by hydrolysis; the molluscicide niclosamide is compounded in a neoprene support.[4,7]

Microencapsulation is an important slow-release technique in which the pesticide is wrapped in very small spheres (1–30 μm diameter) of permeable synthetic polymers or other materials that break down at different rates. The rate of diffusion of the pesticide is governed by the nature of the a.i. and the sphere. Microencapsulation is a valuable method of formulation of insecticides, e.g. chlorpyrifos, carbofuran, phorate, parathion, diazinon, and permethrin, and generally increases the effective life of the compound by 2–4 times.[11]

It has also been applied to fungicides, pheromones, plant growth regulators, repellents, herbicides, and rodenticides; in the latter case, microencapsulation can mask both odour and taste, so avoiding the problems of bait-shyness. Recent work[11] demonstrated that granules of chlorpyrifos in a thermoplastic granular matrix provide a viable alternative to the persistent organochlorines, e.g. aldrin, for control of soil insects. With a wide range of pesticides, by variation in the composition of the granule the a.i. can be released over periods varying from 150 days to 3 years.

Currently a new range of microencapsulated pesticides is being studied to give release over 40–100 days after application: fensulfothion, carbofuran, and terbufos have been obtained as controlled release granules which provide alternative nematicides in contrast to the normal volatile soil fumigants (see Chapter 10).

Pesticides may also be formulated as baits, generally in the form of palatable granules placed as doses in appropriate spots. The baits may incorporate an attractant; they are widely employed against domestic insects, e.g. flies, ants, cockroaches, and vertebrates, e.g. rats, mice, rabbits, birds, and slugs and snails.[2]

Formulation can be important in spray deposition, e.g. smaller deposits are obtained with wettable powders or flowables and the inclusion of oils appears to reduce the early loss of the pesticide by aiding penetration.[12]

The majority of modern agricultural pesticides are effective at doses of less than 1 kg/ha; consequently in order to achieve a reasonably uniform coverage the size of the particles used in dusting must be small (<20 μm diameter), and similarly when spraying small droplets must be produced. This demands for dusting, the use of an inert mineral diluent and for spraying dilution with water. Care must, however, be taken that too much water is avoided since it tends to increase run-off. Generally the amount of water is so arranged that run-off just occurs at the required applied dosage. A great variety of machinery has been developed for pesticide application, such as hydraulic sprays with either horizontal or vertical spray attachments and large air-blast machines used in orchards and plantations.[13] When spraying, the greater the volume of water employed and the lower the pressure applied the larger the spray droplets, and the coarser spray will be easier to direct to the target surface than a fine one and will be much less subject to drifting. Thus under these conditions weeds growing between rows of crops can be safely sprayed without drift onto the crop plants, whereas a fine spray can easily be blown hundreds of metres in a slight breeze and is obviously unsuitable for spraying herbicides in a restricted area.

Spraying can be performed at high volume (HV) >600 l/ha, medium volume (MV) (200–600 l/ha), low volume (LV) 50–220 l/ha, very low volume (VLV) (5–50 l/ha), or ultra low volume (ULV) <5 l/ha; the latter represents the minimum volume required to achieve economic control.[14] The precise volume applied depends on the target; however, the current trend is towards low-volume spraying, partly for economic reasons, as it reduces the need to transport large volumes of water, and partly because of the much lower application rates needed for modern agrochemicals.

Various types of spraying equipment are available; these contain different types of nozzles, which are devices through which the spray liquid emerges as a jet. Examples include hydraulic impact nozzles used for production of large droplets and hydraulic fan nozzles used for spraying flat surfaces, e.g. soil, water, and for aerial application (Plate 1). The centrifugal spinning disc nozzle, in which the spray jet impinges on a spinning disc, allows the application of minimum spray volumes and the droplet size can be regulated by altering the speed of rotation.[14]

A typical medium-volume, low-pressure farm sprayer tank has a capacity of 90–220 l and gives an output of some 1360 l/h. Such a sprayer operated from a tractor power take-off would spray at 225 l/ha when moving at some 5 km/h[13] (Plates 2 and 3). Water is used as the propellant and high pressures are needed to project the spray to the tops of standard trees in an orchard; this can be achieved by using powerful air blast sprayers in which the pesticide is mixed with a comparatively small volume of water. The solution or suspension is atomized by a spinning disc and the spray blown up to the trees. The same principle is used in modern knapsack sprayers, which can apply liquids at concentrations of a few kilograms per hectare, and are extensively used by small farmers. They have the advantage of being less heavy than the older sprayers and are consequently much more pleasant to use (Plate 4). In many instances, controlled droplet application (CDA) is more important in pesticide application than ultra-low-volume application (ULVA). With modern spray equipment the optimum droplet size can be selected for the particular target species, e.g. insect, fungus, or weed.

With herbicides, large droplets (>250 μm) tend to be used to reduce the risk of spray drift; however smaller droplets are often more effective. Thus sprays with the wild oat herbicide barban using 110 μm droplets gave better control of the weed than 200 or 400 μm droplets. For flying insects, the optimum droplet size was 10–50 μm; for insects on foliage 30–50 μm; for foliage 40–100 μm; and when spraying soil larger 200–500 μm droplets are favoured to avoid drift.[14]

In the 'Electrodyn' hand-held sprayer, introduced by ICI (1981), the pesticide is formulated in a special, non-aqueous system and is subjected to a 25 kV electric potential. The charged droplets repel one another and are attracted to the crop which is at earth potential. The droplets are formed in accurately controlled size which can be modified by variation of the flow rate and electric potential. This technique gives an excellent coverage of the crop with uniform distribution on the upper and lower surfaces of leaves. The electrodyn sprayer permits farmers to use very low spray volumes and little of the pesticide drifts away from the target; typically it can spray 1.5 ha at 0.5 l/ha.[15] Dramatically effective pest control can be achieved with ULV spraying (2 l/ha as compared with 200 l/ha using conventional spraying techniques). A range of pesticides are now available for application by the electrodyn sprayer, e.g. cypermethrin (30 g a.i./l/ha) to control bollworms on cotton and fluazifop butyl (300 g a.i./l/ha) to kill grass weeds in broad-leaved crops.[16] It is generally advantageous to achieve maximum retention; often this is favoured by small droplets and not too large a volume of spray liquid. Thus for cereals not more than 450 l/ha should be applied. Retention generally depends on the nature of the plant surface and also the droplet size, the spray volume rate, formulation, concentration, and the method of spray formation.[16] Rotary atomizers gave poor penetration and retention and were less efficient than hydraulic nozzles. Low-volume spraying may not always be beneficial, for instance control of apple mildew was better at 500–800 l/ha than at 50 l/ha, even when systemic fungicides were used.[15] After 50 years of application of synthetic pesticides our knowledge of the behaviour of droplets on impaction is still inadequate and consequently there may be advantages in multidroplet sized sprays.[17]

When aerosol droplets (diameter < 15 μm) fill a given volume of air so that visibility is reduced, a fog is produced. Some insecticides and fungicides are conveniently applied as fogs. In thermal fogging applicators, the pesticide, dissolved in a suitable oil, is vaporized by injection into a hot gas. The oil, when discharged into the atmosphere, condenses forming a dense fog; this method utilizes the fumigant action of the a.i.s and is valuable in the treatment of glasshouses, sheds, warehouses, and ships' holds because the fog penetrates cracks and crevices.[14]

Volatile pesticides are also employed as fumigants and fumigation is widely used for controlling insect pests in ships' holds, warehouses, silos, and rolling stock (see Chapter 10). The fumigant is usually released from pressure cylinders. Fumigation is also applied for soil sterilization and control of soil pests like nematodes (Chapter 12).

ADJUVANTS

These are ingredients that improve the properties of the pesticide formulation; they may be subdivided into activators or spray modifiers. Adjuvants may either be incorporated in the pesticide formulation or added to the diluted pesticide in the spray tank. Activators reduce the quantity of the a.i. required to achieve the desired result. They may modify the spectrum of activity, increase the aqueous solubility, or aid the penetration of the target. In some cases, the presence of the adjuvant is essential to make the chemical active.

Modifiers may function by improving the sticking or spreading properties of the pesticide on the target's surface; they may also increase the viscosity of the spray or promote foaming in order to avoid drift or reduce evaporation of the pesticide.

Physical factors are of considerable importance in determining the effectiveness of pesticides. Thus with surface-protectant fungicides like Bordeaux mixture (Chapter 7) the activity depends very much on the particle size used in the spray and the tenacity of the dried deposit on the leaves of the treated crop.[18] In many pesticides the achievement of the optimum balance between oil and water solubility is an important factor. Increasing the length of an alkyl chain aids oil or lipid solubility while depressing aqueous solubility. In a series of 2-alkylimidazolines it was found[18] that fungitoxicity increased with the number of carbon atoms in the alkyl side-chain up to seventeen carbon atoms and afterwards activity decreased. With N-dodecylguanidine (dodine), on the other hand, the optimum fungitoxicity occurred with a side-chain of twelve carbon atoms.[6]

Pesticides can be broadly classified into physical or chemical toxicants; in the former there is no clearly identifiable toxophore and the toxic effect depends on the physical properties of the whole molecule. For instance, oils such as kerosene are physical toxicants[6]—they kill plants by disorganizing the cells by causing them to lose water and they flood the breathing pores of insects causing rapid asphyxiation. Oils also wet insect cuticles and consequently they become trapped by the surface-tension forces of water. These suffocating and entangling effects are utilized in the method of killing mosquito larvae by spreading a thin film of oil over the surface of water in which mosquitoes breed.[2] Oil washes have been used for a long time to control insects and spider mites in orchards, and this method of controlling mites is becoming increasingly important as the mites are rapidly developing resistance to chemical acaricides (Chapter 5, page 86). The effectiveness of the treatment is enhanced by adding some polyisobutene to the oil which leaves a permanent sticky deposit on the trees and leaves which immobilizes the mites.[3,5]

Petroleum washes are often used on dormant fruit trees against insects and mites; the activity may be enhanced by inclusion of a suitable insecticide, e.g. DNOC.

Creosote oils are employed on a massive scale (800 million l/year in the United States alone) in the preservation of timber against rotting fungi and wood-boring beetles.[3,4] The oil is not very active, but can be formulated with fungicides, such as pentachlorophenol or triphenyl tin oxide, and organochlorine insecticides like aldrin or endrin. When petroleum oils are sprayed onto plant leaves, the air is displaced from the spaces in the mesophyll cells, dehydration occurs, and the leaf wilts. Carrots appear to be less sensitive to oils than other plant species, possibly due to their relatively high lipid content; oils have consequently been used for selective weed control in carrots.

If pesticides are to be active they must reach the ultimate site of action within the target organism. Thus even surface fungicides, like Bordeaux mixture, must be able to penetrate the fungal spore; similarly contact insecticides have to penetrate the insect cuticle and contact herbicides the plant cuticle when they impinge on it. The requirements if the pesticides are to be systemic in action are much more stringent, because in addition they must have the capacity to be absorbed by the roots or leaves or seeds of plants and be translocated to other parts of the plant. In this way the whole plant, including new growth, is protected from fungal attack or an established fungal infestation. With systemic insecticides, the total plant is rendered poisonous to any insect that eats or sucks it, but insects that just alight on the leaves are not hurt; thus systemic insecticides are likely to be inherently more selective in their toxicity than contact insecticides. Phytotoxicity is a much more difficult problem to overcome with systemic pesticides because they are brought into intimate contact with the host plant.

There have been many studies of the movement of xenobiotic substances in plants over the past fifteen years; the translocation of chemicals in plants may be considered in three stages.[19]

(a) *Entry into the free space within the tissues.* Water and solutes, after penetrating the leaf cuticle, pass into free space, defined as that part of the plant in contact by diffusion with the external environment. The cuticle controls the loss of water from the plant and aqueous solutions of many organic molecules can be transported across the cuticle, although the majority of chemicals enter by the root tips. Diffusion of an aqueous spray from the leaves into the free space will only occur while a liquid film remains on the leaf surface, so in cases where the speed of drying of the spray is a limiting factor uptake may be enhanced by addition of a humectant, e.g. glycerol.

(b) *Movement in the xylem.* The xylem vessels provide a system of water pathways communicating with the environment by free diffusion. The bulk of the movement of water and soluble minerals from the roots to the leaves is via the non-living xylem. This process does not involve the expenditure of metabolic energy and is driven by a pumping mechanism in the roots and the evaporation of water vapour from the leaf surface.

(c) *Movement in the phloem.* This is within the living parts of the cell and does require metabolic energy. Chemicals that have arrived at the leaves in the xylem are then distributed to the growing tissues of the plant via the phloem.

Movements (b) and (c) are distinct within the plant. The former occurs passively in the transpiration stream but requires evaporation at the surface and so can be reversed by immersion of the plant leaf in water, so checking evaporation at the immersed surface. The latter is dependent on the metabolic activity and can be prevented by treatments inhibiting metabolism or immobilizing the phloem.

Movement in the phloem has been demonstrated for some chemicals, e.g. 2,4-D and asulam, while other compounds like dalapon and maleic hydrazide move more freely and transfer from the phloem to the xylem, enabling them to be widely distributed in plants. Generally compounds possessing carboxyl, hydroxyl, or sulphamoyl groups are phloem mobile. The method of formulation of the pesticide influences uptake and has an indirect effect on the mode of transport.[13,19] Certain chemicals move from the leaves to the roots via the phloem and may be released into the surrounding soil. Such downward translocation may be valuable for control of soil fungi, as illustrated by pyroxychlor (Chapter 16, page 367). This effect was demonstrated with the hormone herbicides 2,3,6-trichloro- and 2,3,4,6-tetrachlorobenzoic acids (Chapter 8, p 225) when sufficient chemical was exuded into the soil to produce malformations in adjacent plants. Herbicides are distributed partly in the xylem and partly in the phloem.

Most of the early systemic fungicides showed typical xylem movement. These included several antibiotics, e.g. griseofulvin and streptomycin. Experiments with the first generation of commercial systemic fungicides (Chapter 7, page 185), such as benomyl, demonstrated that these are mainly transported in the xylem, but the pyrimidine systemic fungicides, e.g. ethirimol and dimethirimol (Chapter 7, page 191), appear to move unchanged in the phloem.

The majority of systemic organophosphorus insecticides are also transported in the xylem. If the compound is to move in the essentially aqueous plant sap, it must have sufficient aqueous solubility or be converted to such a compound after metabolism in plant tissue. On the other hand, if a compound is to achieve penetration via the foliage, it must be transported across the waxy leaf cuticle, demanding appreciable lipid solubility. To function as a systemic pesticide by foliar application the candidate compound must therefore have a reasonable lipophilic–hydrophilic balance.[20] If the compound is too lipophilic, it will remain held in the cuticular waxes and if too hydrophilic, will never penetrate the cuticle. Thus in the series of O,S-dimethyl-N-n-alkylphosphoramidates a parabolic relationship was established between systemic movement and the logarithm of the octanol–water partition coefficient (P) or relative hydrophobic constant (π), where π was defined as the difference between the logarithm of the partition coefficient of the parent compound (P_0) and the derivative (P) so that $\pi = \log P/P_0$. This implies the existence of an optimum lipophilic–hydrophilic value for maximum systemic translocation in plants, and in this series π was calculated to be 1.19. The N-n-propyl derivative, with a value of π of 1.31, nearest to the optimum value, showed the greatest systemic movement in a cotton leaf petiole. Some of the organophosphorus systemic insecticides (Chapter 6, page 139) are translocated unchanged (e.g. mevinphos), while in other cases it is the active metabolite that is translocated in the plant (e.g. schradan).

Soil has been formed by the gradual breaking down of the rocks of the earth's crust over many millions of years and different rocks produce soils with different characteristics.[21] When a pesticide is applied to the soil, the architecture of the soil becomes an important factor in its effectiveness. Soil is a dynamic system containing many inorganic and organic compounds which are constantly being chemically and microbiologically transformed. It is a very heterogeneous system containing solid, liquid, and gaseous components. The solid phase is mainly present in a finely divided form with a large surface area which is important for understanding the behaviour of chemicals in the soil system. The texture of the soil depends on the size of the particles it contains; sand particles have diameters of from 2 to 0.02 mm, while silt ranges from 0.02 to 0.002 mm diameter, and clay particles have diameters of less than 0.002 mm. They have a surface area corresponding to $2.3 \, m^2/g$ and play an important role in the dynamics of pesticides in soil.[21]

Clay minerals are composed of thin molecular layers held together by attractive forces forming an assemblage of layers. In the kaolinite type, the packing of the layers is so strong that no water molecules or ions can penetrate between them. On the other hand, in the montmorillonite type the individual layers are much more loosely packed, so that water molecules and ions can penetrate between the layers. The lattice of colloidal clay minerals is negatively charged, and the charge is neutralized by positively charged ions from the soil solution which are electrostatically attracted to the surface of the clay minerals. The cations are not, however, fixed and they can be exchanged by other cations, so montmorillonite has a much greater cation exchange capacity than kaolinite. Soil also contains organic colloids (humic substances) which are mainly negatively charged and possess a large surface area. All chemicals in soil are exposed to the physical and electrostatic attractive forces exerted by the soil colloids. Pesticides are adsorbed onto these surfaces, making it harder for them to be taken up by plants and microorganisms and partially protecting them from chemical and enzymic attack. Greenhouse experiments have shown a good correlation between pesticidal activity and such factors as the amount of clay and humus colloids present in the soil, soil humidity, and pH value. However, under field conditions, the correlations are not nearly as good, probably due to the effects of variation in climatic conditions, e.g. the intensity of sunlight and temperature, so great care is needed in the interpretation of the results of field tests. Certainly the activity of a pesticide is often markedly influenced by the moisture content of the soil at the time of application.

In conclusion, as the famous soil scientist Sir John Russell pointed out, 'in an apparently solid clod of earth only about half is usually solid matter, the other half is simply empty except for the air and water it contains'. A good silt loam is thus only half solid material by volume, because when the soil is in good physical condition the individual particles are packed together in aggregates and the pore spaces between the aggregates are partly filled with air and water.[21]

Soil generally contains some 1–5% by weight of organic matter, but in peat soils this may reach some 95% of the dry weight of the soil. Bacteria are by far the most important and numerous of the soil microorganisms—one spoonful of soil will contain billions of bacteria! Beneficial bacteria play an important role in breaking down organic matter to humus which is vital for the maintenance of good soil structure; other bacteria such as *Azotobacter* and *Clostridium* can convert atmospheric nitrogen to nitrates which subsequently can be utilized by plants. Other soil microorganisms are fungi; some are beneficial and some harmful to plants. The latter are the parasitic fungi which attack plant tissues, e.g. the species responsible for club root in brassicas. *Actinomycetes* attack beneficial bacteria and some are parasitic to plants, like that causing scab on potatoes.

The physicochemical properties of pesticides themselves are of great importance. For instance, water solubility may vary widely; thus the solubility of the herbicide simazine is only 5 p.p.m. whereas diuron is 42 p.p.m. and dalapon 500 000 p.p.m. This is an important factor in distribution processes in the soil: high solubility results in easier passage into the soil solution and penetration below the surface layers, but the material will also often be readily leached out from the soil by heavy rainfall.

The basic strength or alkalinity of a pesticide measures the ability of the compound to become positively charged by adsorption of hydrogen ions, as is illustrated by the herbicidal triazines (3):[22]

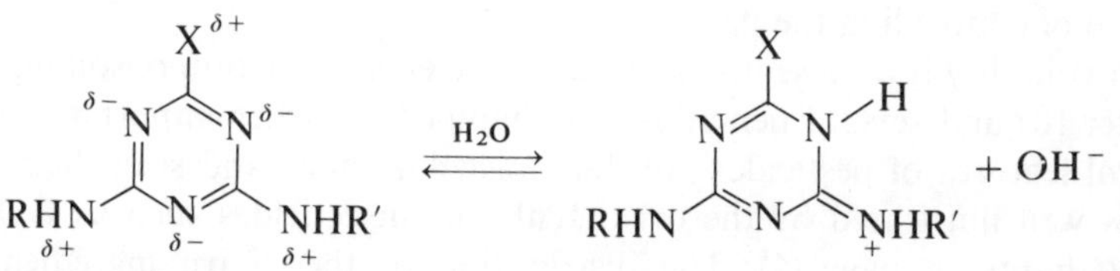

(3)   (4)

(where R,R′ = low alkyl (C_2—C_4) radicals and X = Cl, OCH_3, or SCH_3). The hydrogen ions in the soil are attracted by the negatively charged centres of the triazine (3), converting the originally neutral molecule into the cation (4). The greater the alkalinity of a pesticide, the larger its role in cation exchange reactions in the soil. Many other herbicides, e.g. ureas, contain nitrogen atoms which permit the adsorption of hydrogen ions with formation of cationic structures:

$$\geqslant N: + H_2O \rightleftharpoons \geqslant \overset{+}{N}-H + OH^-$$

Extreme cases are provided by compounds like paraquat (Chapter 8, page 253) that are themselves cations and so very readily exchange with cations in the soil colloids. These exchange reactions result in the herbicide becoming very strongly bound to the soil so that it is no longer available to be absorbed by the plant roots, which explains why paraquat and diquat are deactivated as soon as they come into contact with the soil. They act only through the leaves of plants, and when applied to soil they do not appreciably affect emergent seeds—hence the use of paraquat in 'chemical ploughing'. On the other hand, when bipyridinium herbicides are introduced into a nutrient solution without adsorptive colloids they are absorbed by the roots of plants growing in the solution and exert their full herbicidal activity.

Another extreme case is presented by herbicides that are strong acids (5), e.g. 2,4-D, MCPA, and dalapon, which evolve hydrogen ions and are converted to anions (6):

$$RCO_2H \rightleftharpoons RCOO^- + H^+$$

(5)   (6)

This causes them to be repelled by the negatively charged soil colloids and consequently they are not adsorbed onto their surfaces and so such compounds behave completely differently in soil from alkaline pesticides.

Another important physical property is volatility[23] related to vapour pressure, which is very considerable with such compounds as 2,4-D and carbamates, whereas ureas and triazines have only slight volatility. With certain pesticides the volatility is so high that they must be mixed immediately into the soil, otherwise they are vaporized into the surrounding atmosphere before they have a chance of controlling the pest.

High volatility is an essential feature for the effective action of soil fumigants (Chapter 10) and several nematicides (Chapter 12). Other important physico-chemical features of pesticides are the molecular shape and size; their importance is well illustrated by the organochlorine insecticides such as DDT and HCH (Chapter 5, page 84). The distribution of the active ingredient of a pesticide in the soil obviously depends on its chemical nature and mode of formulation. When a compound exists as a finely divided suspension with only comparatively little of the solid actually dissolved, quick distribution into the soil solution depends on the size of the suspended particles. Distribution continues into the soil colloids (soil particles), the biophase (plants and soil microorganisms), and the gaseous phase (soil air).

Application of a pesticide to the soil at an average dose of 5 kg/ha will not saturate the surface area of the soil colloids in the top 3 cm of soil. Consequently, pesticide molecules will be continually withdrawn from the soil solution via adsorption onto the soil colloids. This dynamic process results in further solid material from the applied suspension passing into the soil solution. The process of dissolution can therefore be accelerated by adsorption so even very insoluble materials can be dissolved in a reasonable time, and the compound becomes distributed between the adsorbed, solution, and gaseous phases. There is, of course, no final equilibrium attained since the soil system is not static but is in a continual state of flux. The pesticide concentration changes constantly as a result of plant absorption and biochemical and chemical degradation. With cationic materials, like paraquat, nearly all the compound is firmly adsorbed onto the soil colloids so that practically none of the chemical remains in the soil solution. On the other hand, with acidic pesticides, like 2,4-D, there is very little adsorption and the chemicals are concentrated in the soil solution and gaseous phases. Chemicals can finally be lost into the atmosphere by evaporation from the gaseous soil phase, and this process may represent a serious loss of the active ingredient with some pesticides, such as 2,4-D.

The selectivity of some herbicides, like the triazines, is a result of their low aqueous solubility combined with a fairly high degree of adsorption onto soil colloids.[22] Such chemicals do not therefore penetrate more than the top 15 cm of soil, so that deep-rooted plants, e.g. fruit bushes, are not affected by them although shallow-rooted weeds are killed.

Leaching of pesticides from the soil is obviously an important factor, which again will be determined partly by adsorption. This is also decisive in determining the distribution of the material into the biophase because plants and microorganisms cannot directly take up compounds adsorbed onto soil colloids. The process of adsorption onto soil colloids is in equilibrium with the removal (desorption) of the compound into the soil solution:

$$\text{Pesticide in soil solution} \quad \underset{\text{desorption}}{\overset{\text{adsorption}}{\rightleftarrows}} \quad \text{Pesticide adsorbed onto soil colloids}$$

If any element in the equilibrium changes, Le Chatelier's principle is obeyed and so the equilibrium shifts in such a way as to oppose the change. Thus if the water content in the soil increases as a result of precipitation, then the concentration of the chemical in the soil solution is maintained by increased desorption from the soil colloids. Similarly, when the concentration in the soil is reduced through absorption by plant roots or degradation, increased desorption maintains the equilibrium. On the other hand, if the pesticide concentration in the soil increases, the balance is now maintained by increased adsorption onto the soil colloids. Soil organic matter can absorb much more pesticide than the clay mineral of similar surface area, and with different clay minerals adsorption is proportional to the surface areas. In the case of organic substances the specific nature of the adsorbent is important; for instance, bog moss adsorbs very much more triazine herbicide than Wisconsin peat (Table 2.1).

Table 2.1

Soil components	Surface area (m^2/g)	$K_d = \dfrac{\text{adsorption (mg/kg)}}{\text{solution (mg/l)}}$	
		Simazine	Atrazine
Organic matter			
Bog moss	500–800	82.3	91.8
Wisconsin peat	500–800	21.5	21.5
Clay minerals			
Montmorillonite	600–800	12.2	5.3
Illite	65–100	8.5	4.3
Kaolinite	7–30	0	0

The titratable acidity of soil measures the quantity of hydrogen ions in the soil solution, and pesticides like triazines with basic centres can absorb positive hydrogen ions which then become firmly bound to the negatively charged soil colloids. Consequently, adsorption of basic compounds increases with the soil acidity, and hence the herbicidal activity of triazines is less in acidic than in alkaline soils.

In contrast, products with acidic groups like MCPA and dalapon are relatively easily leached down from the surface layers of the soil by water and are therefore valuable for controlling deeply rooted weeds and shrubs. Leaching is measured by applying the compound to a column of soil and noting the quantity of the pesticide eluted from the column with known volumes of water. The degree of leaching is dependent on the aqueous solubility of the compound, its chemical nature, and the pH value of the soil. Leaching will be favoured by a small adsorption capacity of the soil sample (i.e. one containing relatively few soil colloids such as clay and humus) and by high temperature and rainfall.

The adsorption of a pesticide or its distribution into the biophase (plants and microorganisms) depends on the absorption capacities of the biophase and on the nature of the soil. A soil of high adsorption capacity can lead to total inactivation of the pesticide because it never penetrates to the pest. Thus simazine applied to grasses at 10–20 kg/ha showed no herbicidal action when the grasses were growing in soil containing activated charcoal, whereas in similar experiments using soil without charcoal the grasses were completely killed.[22] Application rates must therefore be adjusted to take account of the particular soil conditions. Pesticides may be chemically and biologically degraded in soils, plants, and pests. The optimum life of a pesticide depends on the nature of both the crop and the pest; thus relatively short-acting herbicides are ideal for vegetables, otherwise subsequent crops may be injured by herbicidal residues in the treated soil. On the other hand, long-lasting herbicides, like simazine, are valuable for controlling weeds in sugar cane and fruit bushes. The majority of pesticides are biologically attacked, although the organochlorine soil insecticides, such as aldrin and dieldrin, tend to resist biodegradation and are therefore extremely persistent.[24] Comparatively few pesticides suffer purely chemical attack, although in some instances this can be catalysed by surface-active soil colloids. The hydrolysis of the chlorotriazines, e.g. simazine, is thus much faster in the presence of clay minerals, but in contrast biochemical degradation can be hindered by the compounds being adsorbed onto clay minerals. The relative active life of certain pesticides can be reversed by different soil and climatic conditions; among the herbicides methoxy- and chlorotri-azines, ureas, and picloram usually have comparatively long-term activity in soil, whereas carbamates, phenoxyalkanecarboxylic acids, and chloroaliphatic acids, like dalapon, have only short-term activity.

Soil colloids clearly play an important role in the fate of pesticides in soil; some organophosphorus systemic insecticides, e.g. mevinphos, phorate, and schradan (Chapter 6) gave better long-term control of aphids when applied to sandy soil than in clay loam. The amount of organic matter in the soil appears to be a major factor restricting the absorption of phorate from the soil by plant roots, and generally in affecting the persistence of organophosphorus insecticides; thus diazinon, though stable to neutral hydrolysis, nevertheless is easily hydrolysed in the presence of clay minerals.[20] Metallic ions in soils often interact with organophosphorus insecticides; the cupric ion is a very effective catalyst for the degradation of some organophosphorus esters, such as diazinon and chlorpyrifos.

Organophosphorus insecticides are generally much less persistent than organochlorines and are usually degraded in soil within 2–4 weeks of application, e.g. diazinon, dichlorvos, dimethoate, malathion, parathion, phorate, and mevinphos.

The amount of pesticide introduced into soil is relatively small and therefore will have little effect on the physicochemical state of the soil. In a French vineyard, where simazine had been used for weed control for ten years, a careful chemical, physical, and microbiological examination of the soil showed no recognizable changes; there was no accumulation of simazine and the yield and quality of the grapes were not impaired.

On the other hand, paraquat is adsorbed into the soil and may well accumulate over a long period of continuous application. The majority of recent agrochemicals, however, have such high activity that only very small doses (grams per hectare) are often required, so there appears to be little danger of any deleterious effect on the soil.

REFERENCES

1. *Pesticides in the Modern World*, A symposium by the Co-operative Programme of Agro-Allied Industries with FAO and other UN Organisations, Newgate Press, London, 1972, p. 30.
2. Fletcher, W. W., *The Pest War*, Blackwell, Oxford, 1974, p. 36.
3. Green, M. B., Hartley, G. S., and West, T. F., *Chemicals for Crop Improvement and Pest Management*, Pergamon Press, Oxford, 1987, p. 32.
4. Ware, G. W., *Fundamentals of Pesticides*, Thomson Publications, Fresno, Calif., 1986.
5. *Pesticides 1988*, Ministry of Agriculture, Fisheries and Food, HMSO, London, 1988.
6. Martin, H., and Woodcock, D., *The Scientific Principles of Crop Protection*, 7th edn, Arnold, London, 1983.
7. Williams, A., *Chem. in Britain*, **221** (1984).
8. Griffiths, D. C., Scott, G. C., Maskell, F. E., Roberts, P. F., and Jeffs, K. A., *Proceedings of British Insecticide and Fungicide Conference, Brighton*, **1**, 213 (1975).
9. Evans, E., 'Methods of application', in *Systemic Fungicides* (Ed. Marsh, R. W.), 2nd edn, Longman, London, 1977, p. 198.
10. Scher, H. B., and Comstock, M. J., *Advances in Pesticide Formulation Technology*, American Chemical Society Symposium Ser. 254 (1984).
11. McGuffog, D. R., *et. al.*, *Proceedings of British Crop Protection Conference, Brighton*, **2**, 429 (1984).
12. Cottrell, H. J. (Ed.), *Pesticides on Plant Surfaces*, Wiley, Chichester, 1987.

13. *Farm Sprayers and Their Use*, Ministry of Agriculture, Fisheries and Food, Bulletin No. 182, HMSO, London, 1961.
14. Matthews, G. A., *Pesticide Application Methods*, Longman, London and New York, 1979.
15. Durand, R. N., *et al.*, *Proceedings of British Crop Protection Conference, Brighton*, 3, 1083 (1984).
16. Southcombe, E. S. E. (Ed.), *Symposium on Application and Biology of Pesticides*, British Crop Protection Monograph No. 28, 1985.
17. Young, B. W., *Outlook in Agric.*, **15**(2), 80 (1986).
18. Cremlyn, R. J., *International Pest Control*, **5**, 10 (1963).
19. Jacob, F., and Newmann, S. T., 'General principles of the uptake and translocation of fungicides', in *Modern Selective Fungicides* (Ed. Lyr, H.), Longman, Harlow, 1987, p. 13.
20. Hartley, G. S., and Graham-Bryce, I. J., *Physical Principles of Pesticide Behaviour*, Vol. 2, Academic Press, London and New York, 1980.
21. Berger, K. C., *Sun, Soil and Survival*, University of Oklahoma Press, Cleveland, Ohio (1972).
22. Esser, H. O., *et al.* 's-Triazines', in *Herbicides—Chemistry, Degradation and Mode of Action* (Eds. Kearney, P. C., and Kaufmann, D. D.), 2nd edn, Vol. 1, Dekker, New York, 1975, p. 129.
23. Plimmer, J. R., 'Volatility', in *Herbicides—Chemistry, Degradation and Mode of Action* (Eds. Kearney, P. C., and Kaufmann, D. D.), 2nd edn, Vol. 2, Dekker, New York, 1975, p. 891.
24. *Organochlorine Insecticides* (Ed. Moriarty, F), Academic Press, New York, 1975.

Chapter 3

Important Biochemical Reactions in Pesticides

All living organisms are composed of one or more cells, which are therefore the basic units of life.[1,2] Many simple organisms contain just a single cell; such unicellular organisms include several species of algae. In contrast, more complex organisms, e.g. larger plants and animals, are made up of many cells (multicellular) in order to maintain adequate exchanges of energy and materials with the environment.

The cell is composed of a colloidal suspension of substances known as the *cytoplasm*; to preserve the structural identity of the cell it must be in a state of equilibrium with the surrounding environment. The energy needed for cellular activities comes from the respiration of glucose which requires oxygen; this reaches the cell by diffusion across the cell surface, but if the diameter of the cell exceeds approximately 1 mm oxygen will be unable to diffuse to the centre of the cell. This sets an upper limit to cell size, and in fact the majority of cells are considerably less than 1 mm in diameter.

The cell is separated from its environment by the *plasma* (or cell) membrane (Figure 3.1) which controls the exchange of materials between the interior of the cell and the external environment. The plasma membrane is essentially lipid in nature so that only those molecules possessing a reasonable lipid solubility will be able to permeate through the membrane. In the cytoplasm, cellular activities are located in membrane-bounded structures called *organelles* (small bodies), the most conspicuous of which is the *nucleus* which is usually spherical. The major proportion (>95%) of cellular DNA occurs in the nucleus and this directs the synthesis of more DNA and of the messenger RNA which controls protein synthesis. Nuclear DNA is bound up with protein molecules and is organized into strands folded into tubular structures known as chromosomes (coloured bodies)—the name deriving from the fact that these structures are readily stained with certain types of dyes. In cells that are on the point of dividing, the chromosomes are coiled into short thick segments, while at other times they unwind into longer thinner structures. The spherical *nucleolus*

37

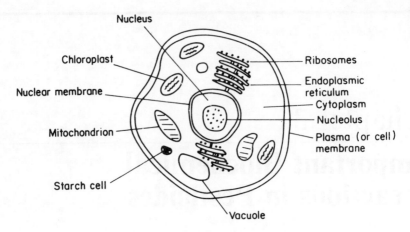

Figure 3.1   Diagram of a plant cell

(Figure 3.1) is often associated with a chromosome in the central portion of the nucleus and is composed of DNA, RNA, and protein; probably the DNA in the nucleolus consists of a large number of identical segments arranged end to end, and each segment probably directs the synthesis of one of the ribosomal RNA molecules used to form ribosomes, while other DNA segments in the nucleolus may direct the synthesis of ribosomal protein. The nucleolus is therefore almost certainly responsible for ribosome synthesis. The information for all cellular activities is carried in the DNA of the cells and is encoded in the precise sequences of the constituent bases adenine, guanine, and cytidine.

A *ribosome* provides a surface for protein synthesis and is a small particle found in the cytoplasm consisting of two spherical subunits each containing molecules of protein and ribosomal RNA (r-RNA), the latter probably being synthesized in the nucleolus. In some cells, ribosomes appear to be scattered throughout the cytoplasm while in others the ribosomes are attached to membranes within the cytoplasm, providing a large surface area for metabolic reactions, known as the *endoplasmic reticulum* (Figure 3.1).

Generally two organelles are involved in the production of cellular energy. *Mitochondria* contain different arrays of enzymes and electron carriers and are the site of the chemical oxidation (respiration) of glucose and the production of energy-rich phosphorus compounds, e.g. ATP. This is the powerhouse of the cell providing energy for the various cellular reactions, while *chloroplasts* are the site of photosynthesis, the conversion of light energy into chemical energy. These organelles (Figure 3.1) both contain membranes to which are bound enzymes and DNA, the latter enabling them to reproduce themselves within the cell. Chloroplasts also contain free ribosomes which are probably associated with protein synthesis; the chemical products of oxidation and photosynthesis are transported via the various membranes for utilization by the rest of the cell. These organelles, when isolated from the cell, can continue to function; thus isolated mitochondria will oxidize glucose and produce high-energy phosphates; likewise isolated chloroplasts will carry out photosynthesis.

Plant cells generally contain one or more large *vacuoles* (Figure 3.1) which are regions within the cytoplasm bounded by a single membrane containing a complex solution of inorganic and organic molecules, and also functioning as stores for enzymes. The cytoplasm of plant cells often contains starch granules and chloroplasts which are unique to plant cells. Some animals possess another kind of organelle known as a *lysosome*. This is a droplet bounded by a membrane containing several hydrolytic enzymes which can split the major cellular components, for instance nucleic acids, proteins, and polysaccharides. In higher animals lysosomes are important for the utilization of dead tissue and for degrading those cellular parts in need of frequent renewal.

The cell walls of plant cells are rigid, whereas an animal cell is bounded by a flexible cell wall. The cell walls or membranes control the exit and entry of substances between the cytoplasm and the external environment. The concentration of solute molecules within cells is usually higher than their external concentration, so that water tends to diffuse into the cell by the process termed osmosis, which causes the cell to expand and press against the cell wall so reducing the inflow of water; when these two pressures become equal water movement ceases and the cell no longer expands.

In large multicellular organisms, each individual cell need no longer have all the capacities necessary for the survival of the whole organism so the cells often become specialized and perform particular functions, e.g. a red blood cell is uniquely developed to transport oxygen. Such specialized cells are often assembled together to form a tissue and several different tissues may be further grouped together into large functional units known as organs, like the kidney, responsible for excretion and water balance in animals, and the leaf, which carries out photosynthesis and gas exchange in plants.

The major chemical constituents of cells are carbohydrates, lipids, proteins (some of which function as enzymes), and nucleic acids. The intact cell also contains simple inorganic cations, such as sodium, potassium, calcium, and magnesium, and anions like chloride and phosphate. In unicellular organisms the cell absorbs directly the nutrients it requires from the external environment, but in more complex animals and plants the food is first digested or metabolized to simpler low molecular weight compounds, e.g. amino acids (from proteins), fatty acids (from lipids), and glucose (from starch and sucrose), which are then carried to the cell in the bloodstream or the plant sap.

Living organisms, apart from certain bacteria, derive the energy needed for their vital life processes from organic compounds synthesized by another organism, or they utilize the energy from light.

RESPIRATION

Organisms obtaining their energy requirements from the respiration of preformed organic molecules obtained from their food include all animals, the majority of bacteria, fungi, and the non-green cells of plants.[3(a)] The energy derived from the respiration (biological oxidation) of foods such as carbohydrates, fats, and proteins is finally trapped in energy-rich phosphate bonds of adenosine-5'-triphosphate (ATP) which can then supply the energy needed for biochemical processes. The oxidation of glucose by burning in air can be expressed as:

$$C_6H_{12}O_6 + 6O_2 \longrightarrow 6H_2O + 6CO_2 + \text{heat energy}$$
Glucose

However, the biological oxidation process is much more complex and occurs in four main stages (Scheme 3.1):

(a) *Formation of acetyl coenzyme A (acetyl CoA) (glycolysis)*. This occurs in the cytoplasm and can be derived from breakdown of carbohydrates, fats, or proteins, but pesticides apparently only interfere directly with the route from glucose. Initially the foods are metabolized to simpler molecules: proteins are converted to α-amino acids; carbohydrates, like starch, to glucose; and lipids to fatty acids and glycerol.

The conversion of glucose into pyruvate involves the operation of a sequence of nine enzymic reactions occurring in the presence of oxygen. The net result of the process of respiration is the transformation of each molecule of glucose into approximately thirty molecules of ATP together with pyruvate. Finally, in the presence of oxygen, pyruvate is converted to acetyl CoA:

$$CH_3COCO_2H \quad + \quad NAD \quad + \quad CoASH \longrightarrow$$

Pyruvate   Nicotinamide   Coenzyme A
                      adenine
                      dinucleotide

$$CH_3COSCoA \quad + \quad NADH_2 \quad + \quad CO_2$$

Acetyl CoA   Reduced form
                    of NAD

The conversion of glucose to pyruvate is termed glycolysis, and the enzymes controlling the process are in the cytoplasm of the cell (Figure 3.1).

(b) *The tricarboxylic acid cycle*. This involves progressive oxidation of acetyl CoA and the transference of the reducing power chiefly associated with the production in the cycle of the reduced form of nicotinamide adenine dinucleotide ($NADH_2$) to the electron transport chain.

(c) *The electron transport chain*. This oxidizes reduced compounds, such as $NADH_2$, with liberation of energy. Oxidation may be defined as loss of electrons and as a consequence of the operation of the electron transport chain electrons pass from a relatively high level in $NADH_2$ to progressively lower levels in the electron carriers composing the chain until they combine with oxygen and protons to form water.

Reducing power is transferred from acetyl CoA to NAD via the tricarb-

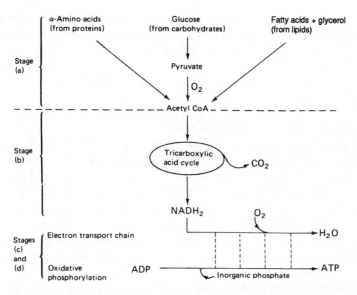

Scheme 3.1 An outline of the biochemical reactions involved in respiration

oxylic acid cycle. The electron transport chain is fed by the transfer of hydrogen atoms from acetyl CoA, the electrons from $NADH_2$, and from the reduced succinate dehydrogenase flavoprotein. Succinate dehydrogenase and cytochrome oxidase are two enzymes involved in the cycle; the enzymes and the electron carriers are bound to the mitochondrial membrane. Cytochrome oxidase occurs in very low concentration and the activity of uncouplers may be measured by reference to the concentration of this protein.

(d) *Oxidative phosphorylation* (Scheme 3.1). This couples the energy released in the electron transport chain to effect the phosphorylation of adenosine-5'-diphosphate (ADP) to adenosine-5'-triphosphate (ATP), the energy being trapped in the phosphate bonds. The precise nature of the coupling between electron transport and oxidative phosphorylation is not known. Oxidative phosphorylation may be measured by suspending mitochondria in an oxygen electrode which measures the concentration of oxygen in solution.

PHOTOSYNTHESIS

Plants also break down glucose to release water and carbon dioxide and the liberated energy is trapped in ATP by biochemical mechanisms similar to those already described for animals.

The important difference between plants and animals is the origin of their glucose: green plants and other photosynthetic organisms, such as algae and certain bacteria, have the ability to utilize light energy which is absorbed by the green pigment chlorophyll to synthesize glucose from carbon dioxide and water:[3(b),4]

$$6CO_2 + 6H_2O \xrightarrow[\text{light}]{\text{chlorophyll}} C_6H_{12}O_6 + 6O_2$$

Glucose

This process is known as photosynthesis.

The light energy is absorbed by the chlorophyll molecules in numerous chloroplasts and arranged in trapping units so that one chlorophyll molecule receives all the energy from the assembly. The energy is converted to chemical energy by reduction of an acceptor (A) in the plant, which in turn produces an oxidized chlorophyll molecule which must be reduced again before it can become the focus of the light absorption system. The net result is the reduction of A by a donor molecule, and the cyclic oxidation–reduction mechanism powered by light energy will result in an overall gain in chemical energy by the system as a whole.

The chemical energy is then used by the plant to reduce carbon dioxide to carbohydrates. The process is essentially the reverse of respiration and is accomplished in two stages via the so-called light and dark reactions. There are two light reactions in photosynthesis and both are of the general type shown in Figure 3.2. The light reactions I and II are coupled in series by the photosynthetic electron transport path as illustrated by Figure 3.2. This resembles the mitochondrial electron transport chain in respiration, and also involves quinone

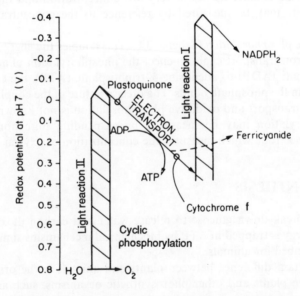

Figure 3.2   The scheme of reactions in photosynthesis

and cytochrome components; it is coupled to the photophosphorylation of $ADP \rightarrow ATP$.

Electrons are energized in light reaction II in which the donor molecule of Figure 3.2 is H_2O, while A is unknown. Removal of electrons from water results in the production of oxygen, and the passage of electrons down the photosynthetic electron transport chain is coupled to photophosphorylation, which can be inhibited and uncoupled by a number of compounds although these may not be identical to those that interfere with the mitochondrial electron system. After transfer down the photosynthetic electron transport chain, electrons have lost energy and are reactivated in light reaction I, which forms an unknown reduced acceptor that is more electronegative than NADP and can therefore reduce it to $NADPH_2$.[5] The electron transfer via light reactions I and II results in the production of oxygen, ATP, and $NADPH_2$ (Figure 3.2).

Photosynthesis is localized in subcellular organelles called chloroplasts. These are bonded by a double outer membrane containing chlorophyll-rich thylakoid membranes which carry out the light reactions whereby the sun's energy is harvested and trapped as ATP and $NADPH_2$. These final products of the light reactions are used to reduce carbon dioxide in the so-called dark reactions. Chloroplasts can be isolated by centrifugation of macerated leaves and have been extensively used in studies on the mode of action of herbicides. No herbicides are known to interfere with the dark reactions, but $> 50\%$ of current herbicides act on the light reactions of photosynthesis.

Herbicides that owe their activity to interference with photosynthesis generally inhibit photosynthetic electron transport by preventing light reaction II. Cell-free preparations of green plants or isolated chloroplasts can catalyse the photolysis of water in the presence of an electron acceptor (A) such as ferricyanide:

$$2A + 2H_2O \xrightarrow{\text{light, chloroplasts}} 2H_2A + O_2$$

This is known as the Hill reaction,[6(a)] catalysed by isolated chloroplasts, and may be measured by determination of the evolution of oxygen with an oxygen electrode or by a spectrophotometric method which determines the amount of H_2A formed. The Hill reaction with ferricyanide is apparently powered solely by light reaction II and the majority of herbicides, such as ureas and triazines, that act by inhibition of photosynthesis probably have light reaction II as their primary site of action. On the other hand, the bipyridinium herbicides function by utilizing photosynthesis for the production of stabilized free radicals which in the presence of oxygen give rise to toxic entities that kill the plant (see Chapter 8).

TRANSMISSION OF NERVOUS IMPULSES

The nervous system is characteristic of mammals and insects and carries signals from the various receptor sites (e.g. eyes, ears, and nose) to the brain.[4] Other specific receptor cells record senses such as temperature, taste, and pain. The external environment also transmits information to the nervous system by means of discrete electrical impulses along a long fibre of neurons (or nerve cells) known as the axon, eventually reaching the brain, so that the appropriate response can be made to the received stimuli. The brain and spinal cord together are termed the central nervous system (CNS) and this sends out signals and communicates with the rest of the body by the peripheral nervous system.[3(c),6(b)] Neurons utilize the electrical charges carried by ions and the activity of the nervous system ultimately depends on the neuron's capacity to maintain an unequal distribution of sodium and potassium ions on each side of the cell membrane. Under resting conditions, the electrical potential inside the membrane is negative with respect to the outside and the concentration of sodium ions (Na^+) inside the nerve cell is low, relative to the outside, whereas for the potassium ions (K^+) the reverse is true. This situation arises from the fact that Na^+ ions are actively transported out of the cell, while K^+ ions move into it. The unequal distribution of ions on the two sides of the cell membrane gives rise to an electrical potential.

The transmission of the nervous impulse is therefore an electrical process in which the current is carried by ions. When the axon meets another neuron there is a junction called a synapse generally some 20–30 nm wide. Nerve impulses are transmitted at the synapse by the release of a chemical transmitter, generally acetylcholine, although other neurotransmitters such as L-glutamate and γ-aminobutyric acid (GABA) are also involved in some synapses. Chemical transmitters are all small molecules (< 20 atoms).

When a nerve impulse arrives at the presynaptic membrane, acetylcholine is simultaneously released from the presynaptic cells and the chemical transmitter diffuses across the synaptic cleft to the postsynaptic membrane, where it binds to the acetylcholine receptor sites. The liberated acetylcholine must not persist in the synapse too long, otherwise there would be a continuous chain of nerve impulses. The transmitter is generally eliminated by combination with the enzyme acetylcholinesterase present in the postsynaptic membrane (Figure 3.3).

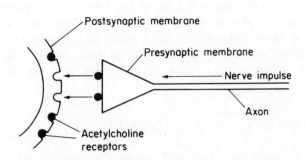

Figure 3.3 Schematic diagram of nerve synapse

Acetylcholinesterase catalyses the hydrolysis of acetylcholine to choline which does not act as a transmitter of nerve impulses:

$$(CH_3)_3\overset{+}{N}CH_2CH_2OCOCH_3 \xrightleftharpoons{\substack{\text{acetylcholinesterase} \\ (+H_2O)}}$$

Acetylcholine

$$(CH_3)_3\overset{+}{N}CH_2CH_2OH + CH_3CO_2H$$

Choline

The combination of acetylcholine with the receptor causes the postsynaptic cell to pass an impulse; subsequently acetylcholinesterase hydrolyses the acetylcholine so that stimulation of the receptor ceases, and the synapse is then available for release of a new transmitter.

The organophosphorus and carbamate insecticides (Chapter 6) owe their insecticidal properties to phosphorylation or carbamoylation of the enzyme acetylcholinesterase. This poisons the enzyme so that it cannot catalyse the hydrolysis of acetylcholine to choline; consequently there is an accumulation of acetylcholine at the synapse, which permits the continuous transmission of nerve impulses, and effective nervous coordination breaks down—the insect or mammal suffers convulsions and finally death.

Mammals contain several receptors having the capacity of binding to acetylcholine, and O'Brian has demonstrated[7] that the receptor isolated from fly heads has different binding properties from a vertebrate receptor. The receptor on the postsynaptic membrane may not be acetylcholinesterase since enzyme inhibitors do not block the receptor and may even enhance its activity.

Nicotine (Chapter 4, page 53) almost certainly owes its insecticidal properties to its ability to combine with the acetylcholine receptor in insects.[4]

A comparison of the acetylcholine receptor and the enzyme acetylcholinesterase as sites of action for insecticides shows that few compounds are considered to act on the receptor, while very many of them, e.g. organophosphates and carbamates, react with the enzyme. The observation[7] that insect receptors differ in their binding properties from those in mammals suggests that the receptors may offer a useful target for the development of compounds showing selective toxicity to insects.

DDT and other organochlorine insecticides probably act by interference with axonal transmission by binding to the nerve membrane and upsetting the sodium–potassium ion balance (Chapter 5, page 83). Pyrethroids affect both the peripheral and central nervous systems of insects, and the convulsion of the insect appears to be initiated by loss of potassium (Chapter 4, page 68).

TRANSPORT AND BIOLOGICAL PROCESSES IN PLANTS

In order to understand what happens when a pesticide is applied to a plant either directly by foliar spray, seed dressing, or via the soil, it is necessary to know the basis of plant biochemistry.

The plant root system absorbs water and minerals from the soil solution; the centre portion of the root contains two main types of conducting tissues, the xylem and phloem.[6(c)] The xylem is responsible for the movement of water throughout the plant since it is continuous from the root tip to the leaf veins, and contains an unbroken column of water. Water and minerals are carried up from the roots to the rest of the plant by the transpiration stream and the water movement is largely caused by the evaporation of water from the leaf surfaces which reduces the pressure in the leaf veins and consequently more water is sucked up from the roots. The suction pressure is substantial and may be 10–20 times the normal atmospheric pressure and so the water is easily transported upwards against the force of gravity.

The overall flow in the xylem is controlled by the opening and closing of small pores in the leaf surface (stomata), permitting movements of gases between the leaf cells and the surrounding air.[6(c)] In contrast, organic compounds produced by photosynthesis are transported in the phloem. This is a more complex conducting tissue than the xylem, since it also permits downward movement of certain chemicals from the leaves to the roots.[4] Such chemicals could prove valuable for the control of soil pathogens and there is evidence that some synthetic organic compounds, e.g. pyroxychlor (Chapter 16, page 367), act by downward translocation (see Chapter 2, page 28). Sugars move downwards and upwards in the phloem and the phloem sap contains approximately 25% of carbohydrates, chiefly sucrose, together with small amounts of amino acids.

Transport in the phloem system is governed by osmotic pressure; the leaf cells with high concentrations of sugars possess larger osmotic pressures than the non-photosynthetic leaf tissues so the pressure difference forces the sugars downwards towards the roots, thus providing nourishment for the whole plant. The plant skeleton is composed mainly of thick tough walls of xylem tissue which are necessary to withstand the suction forces in the xylem.

The characteristic shapes of different plants are controlled by hereditary instructions contained in the genetic code of the plant DNA and these are relatively independent of environmental factors. Hereditary instructions, however, also provide physiological responses that can modify plant growth, e.g. decide when lateral buds become active or how the plant responds to alterations in such external factors as temperature, light, and nutrients. The physiological changes in plants are controlled by chemical hormones, and by mimicking such compounds useful herbicides and growth regulators can be produced, for instance the phenoxyacetic acid selective herbicides (Chapter 8, page 219).

PROTEIN AND NUCLEIC ACID BIOSYNTHESIS

Ribonucleic acid (RNA) and deoxyribonucleic acid (DNA) have been recognized as the genetic material for some 30 years. The nature and distribution of RNA and DNA molecules defines the organism and direct and control the synthesis of all the compounds required by the organism, such as proteins.[6(d)] Proteinoid enzymes are involved in and direct cell growth and maintenance, including processing chemicals from the environment and oxidation reactions (respiration) that provide cellular energy.

The genetic material must be capable of reproduction or replication passing on the essential genetic information to daughter cells during replication so ensuring that the information in the progeny is identical to that of the parent. Chemically, nucleic acids consist of polymers made up of a phosphate group, a sugar (either D-ribose (RNA) or 2-deoxy-D-ribose (DNA)), and four nitrogenous bases (adenine, uracil, cytosine, and guanine (RNA) or adenine, thymine, cytosine, and guanine (DNA)).[2]

The genetic information contained in DNA is coded by a linear sequence of two purine (adenine and guanine) and two pyrimidine (cytosine and thymine) bases arranged in two interwoven strands forming part of a double helix. The precise linear structure of RNA or DNA is defined by the arrangement of the four bases; only certain pairs of bases easily form hydrogen bonds with each other, thus guanine (G) bonds with cytosine (C) and adenine (A) with thymine (T) in DNA or uracil (U) in RNA. These specific bonded base pairs C–G, A–T, and A–U are stable and are responsible for the double helical structure of the DNA molecule in which the two nucleic acid strands are held together by hydrogen bonding between the base pairs. The specificity of the base pairs implies that the two DNA strands of each double helix will be different but complementary.

Genetic information is transferred from the nucleic acid to the protein and protein synthesis may be summarized as shown (Scheme 3.2). The information

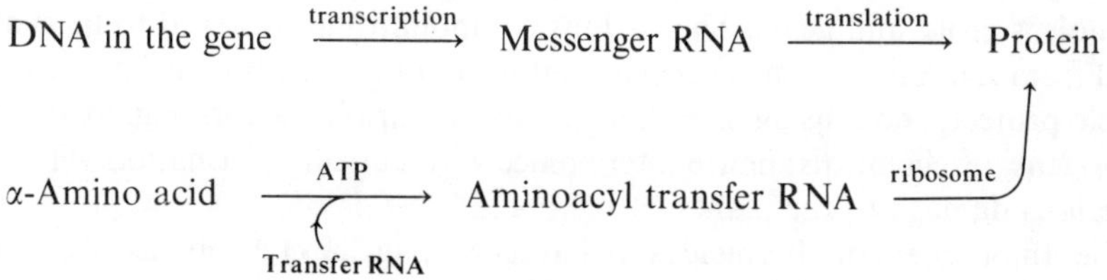

DNA in the gene $\xrightarrow{\text{transcription}}$ Messenger RNA $\xrightarrow{\text{translation}}$ Protein

α-Amino acid $\xrightarrow[\text{Transfer RNA}]{\text{ATP}}$ Aminoacyl transfer RNA $\xrightarrow{\text{ribosome}}$

Scheme 3.2 Basis of protein synthesis

contained in the DNA molecule is coded or transcribed onto a new RNA molecule known as messenger RNA, and is translated into a specific amino acid sequence defining a particular protein. This operation involves a specialized organelle called the ribosome whose sole function is protein synthesis (see page 38). The specific linear sequence of amino acids is dictated by the messenger RNA to which the ribosomes attach themselves. Protein synthesis from the free α-amino acids in the cytoplasm occurs in three steps:

(a) The α-amino acid reacts with ATP and a specific transfer RNA molecule to give an aminoacyl transfer RNA molecule. Each transfer RNA becomes attached to only one of the twenty α-amino acids usually found in proteins and a specific enzyme is needed to catalyse the formation of each aminoacyl transfer RNA.

(b) The aminoacyl transfer RNA is combined with the messenger RNA by base-pairing at a second specific binding site.

(c) Protein is synthesized by the ribosome in which the α-amino acid carried by the transfer RNA is joined onto the end of a growing protein chain through formation of a peptide linkage. In this way, the ribosome can be visualized as moving down the messenger RNA molecule and 'reading' its message. The synthesis, in spite of its complexity, is quite fast and the peptide chain may grow at the rate of some 25 amino acid units per second. The fungicides furalaxyl and metalaxyl (Chapter 7, page 208) apparently act by interference with RNA synthesis and the benzimidazole systemic fungicides were originally considered to owe their activity to inhibition of DNA synthesis. Later studies,[4] however, indicate that benzimidazoles prevent the assembly of tubulin into microtubules. Several antifungal antibiotics, e.g. cycloheximide, blasticidin S, and kasugamycin (Chapter 7, page 180), are probably fungicidal as a result of interference with protein synthesis.

SOME OTHER IMPORTANT BIOSYNTHETIC PROCESSES

Lipid Biosynthesis

Plant cuticles consist of a mixture of waxes, polymeric long-chain (C_{16} or C_{18}) hydroxycarboxylic acids, and other lipophilic components, e.g. long-chain alcohols, sterols, and acids. The various compounds are believed to be synthesized from malonic acid by successive addition of two-carbon atom units. The cuticle protects the cells against damage due to various factors, e.g. frost, wind, pathogens, or chemicals; hence interference with cuticular formation will result in serious damage to the plant.

The thiocarbamate herbicides (Chapter 8, page 229) probably owe their activity to disruption of the synthesis of long-chain fatty acids.[4]

Ergosterol is a major sterol in fungi, where it plays a vital role in membrane structure and function. Recent evidence indicates that the important group of azole and morpholine fungicides (Chapter 7, page 193) owe their activity to the inhibition of ergosterol biosynthesis. The formation of ergosterol from acetyl CoA involves many steps and the azole fungicides (e.g. triazoles) act by specifically blocking the step involving the 14α-demethylation of 24-methylene dihydrolanosterol.

The demethylation step is catalysed by a cytochrome P450 enzyme, and the azole fungicides appear to poison the enzyme by binding to the active site.

The morpholine fungicides, on the other hand, probably act at later stages of ergosterol biosynthesis by inhibition of $\Delta^8 \rightarrow \Delta^7$ isomerase or Δ^{14} reductase enzymes.[8]

Amino Acid Biosynthesis

Amino acids are important constituents of plants; phenylalanine is also a precursor of plant lignins which serve to protect and strengthen the cell walls of older (woody) plants.

Glyphosate (Chapter 8, page 257) inhibits the biosynthesis of the aromatic amino acids, phenylalanine, tyrosine, and tryptophan. Specifically it poisons the enzyme 5-enolpyruvyl shikimate-3-phosphate synthase (EPSP) which controls a vital step in the biosynthetic pathway. Plants treated with glyphosate slowly die due to lack of proteins arising from the blockage in the amino acid synthesis.

The sulphonylurea and imidazoline herbicides (Chapter 8, page 262) block the biosynthesis of essential branched-chain amino acids (valine, leucine, and isoleucine). These herbicides poison the enzyme acetolactate synthase (ALS) which catalyses the first step in the biosynthetic pathway; the blockage results in disruption of cell division in the plant.[9] The development of these highly potent herbicides has demonstrated that the biosynthesis of amino acids is a major target for herbicides.

Chitin Biosynthesis

Chitin, essentially a polymer of *N*-acetylglucosamine, is a structural feature of the insect cuticle and fungal cell wall (cf. cellulose in plants); hence its biosynthesis is an attractive target for selective fungicides. Chitin is synthesized from glucose via UDP-*N*-acetylglucosamine involving a reaction catalysed by the enzyme chitin synthetase, whereby *N*-acetylglucosamine units are transferred to the growing chitin polymer chain.

Polyoxin D, an antifungal antibiotic extensively used in Japan against rice blast disease, structurally resembles UDP-*N*-acetylglucosamine; it competitively inhibits chitin synthetase and interference with chitin biosynthesis is probably the primary mode of action of polyoxins.[4] Two other fungicides used to control rice blast, kitazin P and edifenphos (Chapter 7, page 207), may act by inhibition of chitin synthetase. Later studies,[4] however, indicate that these compounds inhibit the *N*-methyltransferase step in the biosynthesis of phosphatidyl choline, one of the most abundant phospholipids found in cellular membranes. The herbicides dichlobenil and chlorthiamid act by inhibition of cellulose biosynthesis; both compounds appear to prevent the incorporation of glucose to form cellulose, probably by action against the cellulose-synthesis complex in the cell membrane.

Melanin Biosynthesis

Tricyclazole (1) is a fungicide active *in vivo* against rice blast fungus at very low concentrations. The compound is appreciably less active *in vitro* and is suggested to block the biosynthesis of melanin which appears to be a factor in the pathogenicity of the fungus, because fungal mutants without melanin were not pathogenic.

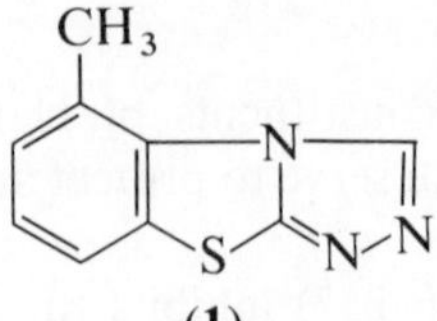

CH$_3$

(1)

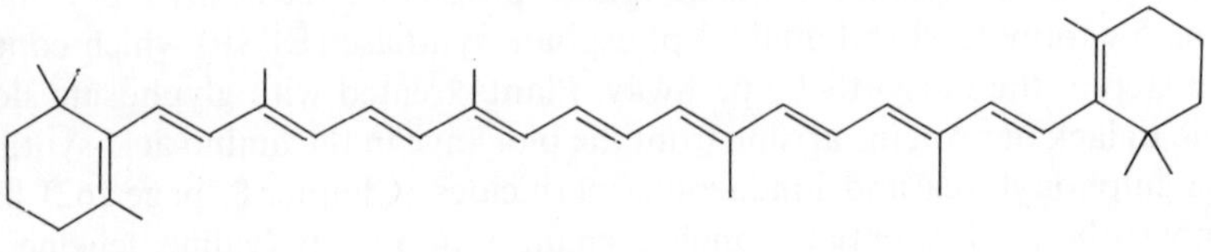

(2)

Carotenoid Biosynthesis

Carotenoids are yellow-orange pigments widely found in plant tissues; one of their functions is to protect the plant cell against photooxidations which would otherwise kill the plant.

Carotenoids, like β-carotene (2), are synthesized within the chloroplast from mevalonic acid via the isoprenoid pathway; the colour of β-carotene (2) arises from the long chain of conjugated double bonds in the molecule.

When plants are treated with pyridazinone herbicides, e.g. chloridazon (Chapter 8, page 248), the level of β-carotene is reduced. Inhibition of carotenoid biosynthesis is believed to be the primary biochemical mode of action of these herbicides. Aminotriazole, like the pyridazinones, is a bleaching herbicide which also probably interferes with carotenoid biosynthesis.[4]

METABOLISM

The various classes of pesticides may be regarded essentially as lipophilic molecules possessing certain functional groups which undergo well-defined biochemical reactions. The metabolic processes occurring in living organisms generally result in the formation of increasingly polar molecules which are more readily excreted.[10] The majority of metabolic processes experienced by pesticides involve detoxification, although sometimes the reverse is true, e.g. the *in vivo* transformation of parathion → paraoxon. The more toxic metabolites are fortunately generally further metabolized to non-toxic materials which are eliminated from the body.

Metabolism can be divided into two phases:

(a) *Primary metabolism*. This refers to the production of a *free* metabolite via biotransformation reactions, e.g. dehalogenation, dehydrohalogenation, desulphuration, epoxidation, hydrolysis, isomerization, oxidation, reduction, and nitrosation.

(b) *Secondary metabolism*. The conversion of polar metabolites containing free OH, CO_2H, or NH_2 groups into conjugates, e.g. glucoside formation (glucuronides) or condensation with amino acids or glutathione (GSH).

The biological action of a given pesticide is intimately related to its metabolism in the target organism and, in the case of systemic compounds, will also be affected by metabolism in the crop plants.[10]

REFERENCES

1. Steiner, R. F., *Life Chemistry*, Van Nostrand, New York, 1968, p. 1.
2. Rose, S., *The Chemistry of Life*, Penguin Books, 1970.
3. White, A., Handler, P., and Smith, E. L., *Principles of Biochemistry*, 5th edn, McGraw-Hill, New York, 1973: (a) p. 389, (b) p. 514, (c) p. 959.
4. Corbett, J. R., Wright, K., and Baillie, A. C., *The Biochemical Mode of Action of Pesticides*, Academic Press, London and New York, 1984.
5. Hill, R., *Essays in Biochemistry*, **1**, 121 (1965).
6. Stephens, G. C., and North, B. B., *Biology*, Wiley, New York, 1974: (a) p. 159, (b) p. 310, (c) p. 195, (d) p. 34.
7. O'Brian, R. D., in *Biochemical Toxicology of Insecticides* (Eds. O'Brian, R. D., and Yamamoto, I.), Academic Press, New York, 1970, p. 1.
8. Berg, D., 'Biochemical mode of action of fungicides', in *Human Welfare and the Environment*, Vol. 1., Pergamon Press, Oxford, 1983, p. 55.
9. La Rossa, R. A., and Falco, S. C., *Trends in Biotechnology*, **2**(6), 158 (1984).
10. Matsuma, F., and Murti, C. R. H. (Eds.), *The Biodegradation of Pesticides*, Plenum Press, London and New York, 1982.

Chapter 4
Naturally Occurring Insecticides

Plants have evolved over some 400 million years and to combat insect attack they have developed a number of protective mechanisms, such as repellency and insecticidal action. Thus a large number of different plant species contain natural insecticidal materials; some of these have been used by man as insecticides since very early times although many of them cannot profitably be extracted. However, several of these extracts have provided valuable contact insecticides which possess the advantage that their use does not appear to result in the emergence of resistant insect strains to the same degree as the application of synthetic insecticides (see Chapter 6, page 151).

Some botanical insecticides survive today; examples in ascending order of importance are nicotine, derris (rotenone), avermectins and pyrethrum.

NICOTINE

The tobacco plant was introduced into Europe about 1560, Sir Walter Raleigh began the practice of smoking tobacco in England in 1585, and as early as 1690 water extracts of tobacco leaves were being used to kill sucking insects on garden plants.[1] The active principle in tobacco extracts was later shown to be the alkaloid nicotine (1), first isolated in 1828 and the structure elucidated in 1893.[2,3] Nicotine occurs in tobacco plants as a salt with citric and malic acids to the extent of 1–8 % and may be extracted from the leaves and roots of the plants by treatment with aqueous alkali, followed by steam distillation.

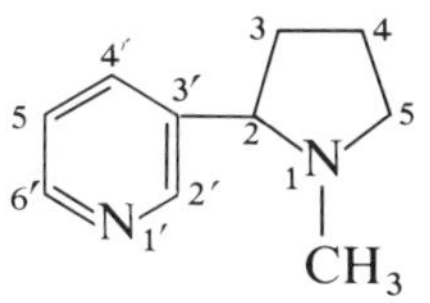

(1) 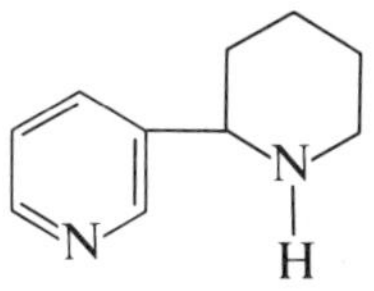   (2) 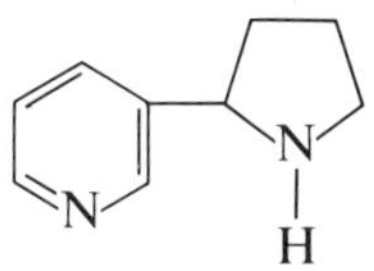   (3)

Natural nicotine is the laevorotatory (−) isomer $[\alpha]_D$ − 169°, the optical isomerism arising from the presence of the asymmetric $C_{(2)}$-carbon atom. The dextrorotatory (+) isomer is much less insecticidal so that synthetic nicotine in the racemic (±) form is only about 50% as active as the natural material.[3] Commercially nicotine is generally used as 'Black Leaf 40' which has been a popular garden spray for a long time. This is a concentrate containing 40% of nicotine sulphate.[2,4] Alkaline activators, like soap and calcium caseinate, are added to liberate the active free nicotine. Nicotine may also be applied as a dust.

Nicotine functions as a non-persistent contact insecticide against aphids, capsids, leaf miner, codling moth, and thrips on a wide variety of crops.[5] However, its use is rapidly declining and it is being replaced by synthetic insecticides, because of its comparatively high mammalian toxicity (LD_{50} (oral) to rats $\simeq 50$ mg/kg)[1] and its lack of effectiveness in cold weather. The compound is readily absorbed by the skin and any splashes must be washed off immediately.[4]

Smith and his coworkers in 1930 showed[2,3] that the bipyridyl derivative neonicotine or anabasine (2) possessed comparable aphicidal activity to that of nicotine (1). This compound, as the laevorotatory (−) isomer, was later isolated from a Turkestan weed *Anabasis apylla*, hence the name anabasine. Aphicidal activity was also shown by nornicotine (3). Subsequent examination of a range of nicotine analogues indicated that high insecticidal activity required the presence of a pyridine nucleus joined between the 2- and 3-positions to a saturated five- or six-membered ring. Other essential features included the 3-pyridylmethylamine residue with a highly basic side-chain nitrogen atom at least 4.2 Å away from the pyridine nucleus.[3]

Nicotine kills vertebrates because it mimics acetylcholine by combining with the acetylcholine receptor at the neuromuscular junction causing twitching, convulsions, and finally death.[4,6] There is evidence[6] that a similar mode of action accounts for the insecticidal activity where nicotine blocks synapses associated with motor nerves. For activity, it is essential that nicotine (1) and its active analogues, e.g. compounds (2) and (3), should have very similar molecular dimensions to acetylcholine (4) (Figure 4.1).

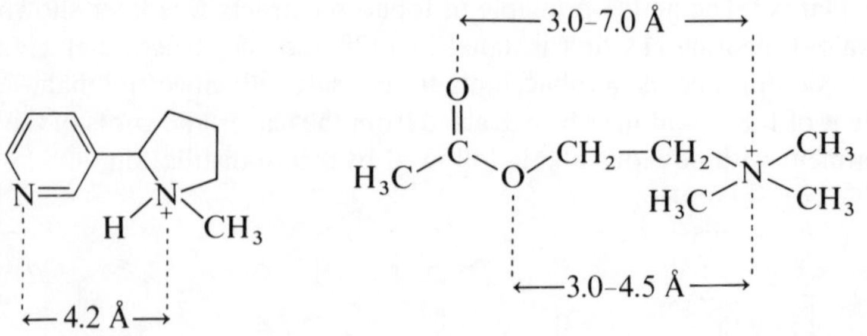

Figure 4.1 Molecular dimensions of nicotine and acetylcholine

Thus nicotine penetrates into the synapse and is then converted to the nicotinium ion containing a positive charge on the pyrrolidine nitrogen atom. This interacts with the acetylcholine receptor just like acetylcholine but, unlike acetylcholine, it is not susceptible to enzymic hydrolysis by acetylcholinesterase. This accounts for the observed toxic symptoms and there is considerable evidence[1,6] that the nicotinium ion is the active entity in nicotine.

ROTENOIDS

These are a group of insecticidal compounds occurring in the roots of *Derris elliptica* (from the East Indies and Malaya) and a species of *Lonchocarpus* (from South America). Derris has been used as an insecticide for a long time; thus Oxley (1848) recommended it for control of caterpillars.[1] Derris dust is manufactured by grinding up the roots and mixing the powder with a clay diluent.[4] Alternatively, the rotenoids can be extracted from the powdered roots with organic solvents.[2,3] The resultant resin by crystallization from ether or carbon tetrachloride gave rotenone (5), a white crystalline, laevorotatory solid whose structure was elucidated in 1932.[1,2]

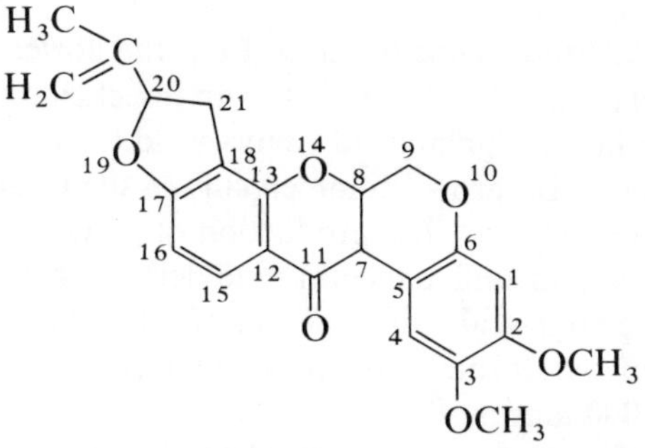

(5)

The mother liquors, after removal of rotenone, also afforded a number of analogous pentacyclic compounds (rotenoids), but these were not nearly as strongly insecticidal.

Attempts have been made to extract the active principles from derris and lonchocarpus roots to avoid the cost of transporting much inactive material contained in the roots. However, it was found that solvent extraction appeared to accelerate the decomposition to products of low insecticidal potency. This problem could be reduced by hydrogenation of the isopropenyl group in rotenone (5) to dihydrorotenone which, though highly active, was resistant to oxidative degradation.

Rotenoids are toxic to fish and many insects, but are almost harmless to most warm-blooded animals. In 1946 about ten million pounds of dried derris and lonchocarpus roots were imported into the United States of America, but this had declined to six million pounds by 1955. Derris was widely used in cattle and sheep dips for the control of ticks and other ectoparasites, but recently has been largely superseded by synthetic insecticides. Now it is primarily used in horticulture against aphids, caterpillars, sawflies, wasps, raspberry beetles, and red spider.[7] Rotenone is an extremely safe garden insecticide because it is degraded by light and air and does not leave residues, and has been widely used for more than 60 years (LD_{50} (oral) to rats $\simeq 135$ mg/kg).

The biochemical mode of insecticidal action appears to involve the inhibition of mitochondrial electron transport, and in isolated mitochondria rotenone (5) inhibits oxidation linked to $NADH_2$, although at low concentrations succinate oxidation was not affected. The inhibition of the electron transport chain appears to arise from the binding of rotenone to a component of the chain, but $NADH_2$ dehydrogenase is not inhibited.[6] The symptoms of insects poisoned by rotenone differ from those produced by insecticides acting on the nervous system, and are characterized by reduction in oxygen consumption, depressed respiration and heartbeat, and eventual paralysis.[3]

PYRETHROIDS

Pyrethrum is a contact insecticide obtained from the flower heads of *Chrysanthemum cinerariaefolium* and has been used as an insecticide since ancient times. The varieties grown in the highlands of Kenya yield the highest proportions of active ingredients; it is also grown commercially in the Caucasus, Iran, Japan, Ecuador, and New Guinea.[1,2,4] The production of pyrethrum as an insecticide dates from about 1850 and, unlike nicotine and derris, the use of pyrethrum has increased despite the large-scale introduction of synthetic insecticides. In 1965 the world output of pyrethrum was approximately 20 000 tons, with Kenya producing some 10 000 tons.

Pyrethrum owes its importance to the outstanding rapid knockdown action (a few seconds) on flying insects combined with a very low mammalian toxicity due to its ready metabolism to non-toxic products. Therefore, unlike DDT, pyrethrum is not persistent and leaves no toxic residues, which may be why this insecticide does not tend to induce the development of resistant insect populations. Pyrethrum is used to control pests in stored foods and against household and industrial pests. Pyrethrum aerosol sprays are excellent home insecticides because of their safety and rapid action.[7]

However, a major disadvantage of pyrethrum, especially for use against agricultural pests, lies in its lack of persistence due to its instability in the presence of air and light. Insects can often recover from exposure to sublethal doses of pyrethrum, which means that the compound must be mixed with small amounts of other insecticides to ensure that the treated insects do not recover.

Pyrethrum is obtained from the dried chrysanthemum flowers by extraction with kerosene or ethylene dichloride and the extract concentrated by vacuum distillation.[2] It contains four main insecticidal components which are collectively termed pyrethrins.[8] These are the esters of two cyclopentenolones (**6**; $R' = -CH=CH_2$ or CH_3) and two cyclopropanecarboxylic acids (**7**; $R = CH_3$ or CO_2CH_3). The structures of the main pyrethrins are therefore as shown in formula (**8**) in Figure 4.2.

The alcohols (**6**; $R' = -CH=CH_2$ and CH_3) are called pyrethrolone and cinerolone, while the carboxylic acids (**7**; $R = CH_3$ and CO_2CH_3) are chrysanthemic and pyrethric acids. Pyrethrin I is the most active of the natural pyrethrins. The acid components are capable of existing in *cis* and *trans* geometrical isomers due to the presence of the olefinic double bond, and each of these isomers can further exist in dextrorotatory $(+)$ and laevorotatory $(-)$ optical isomers. Similarly the alcohols can exist in four stereoisomeric forms. The stereochemistry of the pyrethroids has a vital influence on their insecticidal activity; thus the $(-)$-*trans*-chrysanthemates are practically inactive as compared with the $(+)$-*trans*-chrysanthemates, so it is fortunate that the natural

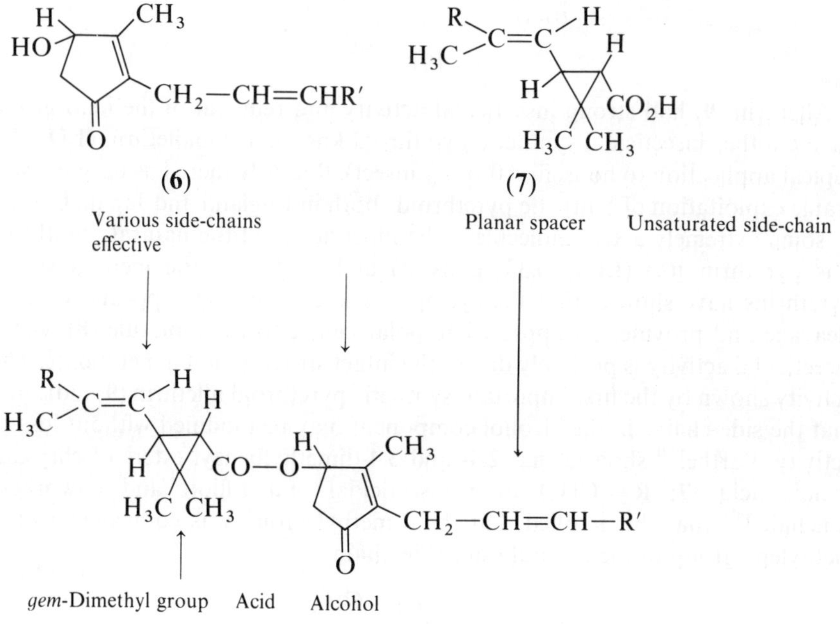

(6)

Various side-chains effective

(7)

Planar spacer   Unsaturated side-chain

(8)

gem-Dimethyl group   Acid   Alcohol

Compound	R'	R
Pyrethrin I	$-CH=CH_2$	$-CH_3$
Pyrethrin II	$-CH=CH_2$	$-CO_2CH_3$
Cinerin I	$-CH_3$	$-CH_3$
Cinerin II	$-CH_3$	$-CO_2CH_3$

Figure 4.2

(+)-*trans*-chrysanthemic acid can now be commercially synthesized by addition of diazoacetate to 2,5-dimethyl-2,4-hexadiene:

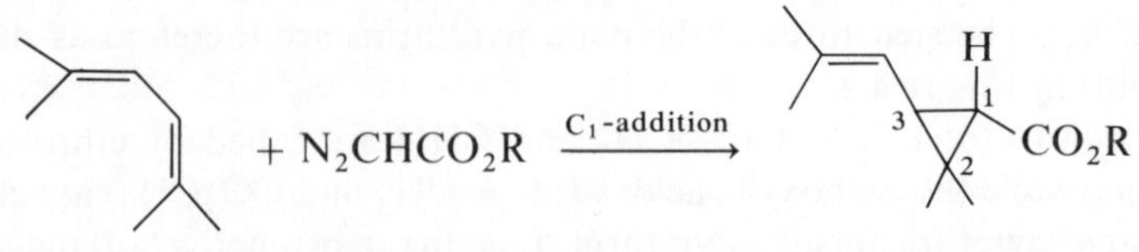

The synthesis of chrysanthemic acids and of cyclopentenolones[3] opened up the possibility of obtaining synthetic pyrethroids, the first of which was allethrin (**9**), prepared by esterification of synthetic ($\pm$)-chrysanthemic acid (**7**; R = CH$_3$) with the alcohol allethrolone (**6**; R' = H).

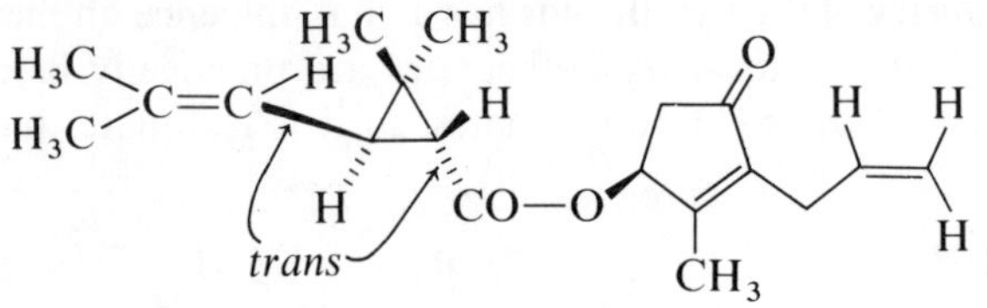

(**9**)

Allethrin (**9**) had strong insecticidal activity and removal of the keto group gave another insecticidal synthetic pyrethroid known as bioallethrin (LD$_{50}$ by topical application to houseflies 0.10 μg/insect). Recently there has been considerable exploitation of synthetic pyrethroids both in England and Japan, leading to some extremely active molecules. The most active of the natural pyrethrins was pyrethrin I(**8**) (LD$_{50}$ 0.33 μg/insect) and studies of the metabolism of pyrethrins have shown[9] that the cyclopropane ester linkage appears to resist cleavage and provides an appropriate polar centre to the molecule (**8**) whose insecticidal activity is probably due to the intact structure not a metabolite. The activity shown by the first important synthetic pyrethroid allethrin (**9**) indicated that the side-chains in the alcohol component may be modified without loss of activity. Barthel[10] showed that 2,4- and 3,4-dimethylbenzyl esters of chrysanthemic acid (**7**; R = CH$_3$) were insecticidal and Elliott and coworkers concluded[11] that this implied that the 4-methyl group was equivalent to the methylene group of the natural ester side-chain.

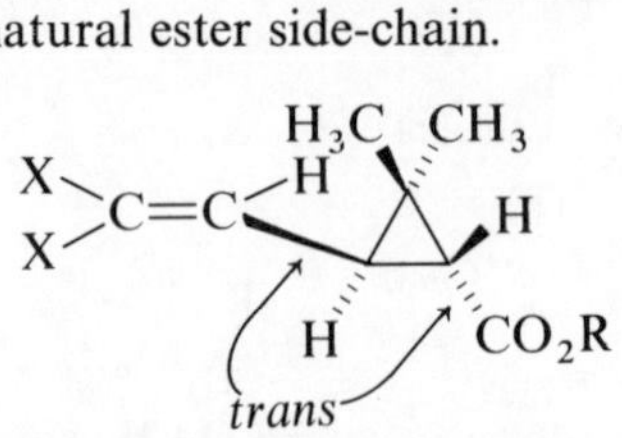

(**10**)

On this basis, 4-allylbenzyl chrysanthemate (**10**; X = CH$_3$, R = *p*-CH$_2$C$_6$H$_4$CH$_2$CH=CH$_2$), combining the structural features of allethrin (**9**) and the methylbenzyl chrysanthemates, was synthesized and was shown to be appreciably more active against houseflies (LD_{50} 0.02 μg/insect) than allethrin. Examination of a number of methylbenzyl chrysanthemates showed that 2,3,6- and 2,4,6-trimethyl substitution was particularly effective and so 4-allyl-2,6-dimethylbenzyl chrysanthemate was prepared which had a broader spectrum of insecticidal activity. This work demonstrated that the cyclopentenone ring of the natural pyrethrins could be replaced by a benzylic system without loss of activity.[11] The benzyl group can also be substituted by other aromatic systems and this approach led to the discovery of 5-benzyl-3-furylmethyl(+)-*trans*-chrysanthemate or bioresmethrin:

10: X = CH_3, R = CH_2—

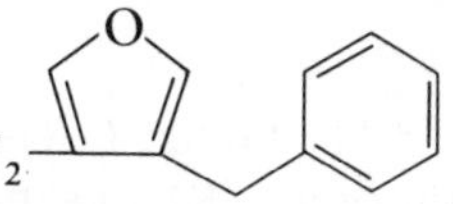

Bioresmethrin

This is an extremely potent insecticide (LD_{50} = 0.005 μg/insect),[9] but, although more active, bioresmethrin was also photosensitive and consequently was not persistent.

However, when the isobutenyl group of bioresmethrin was replaced by the dichlorovinyl group, the resultant compound NRDC 134:

10: X = Cl, R = CH_2—

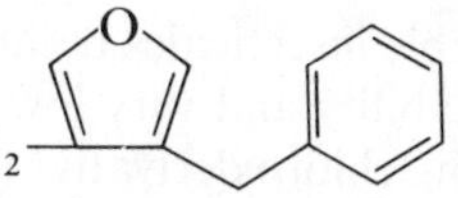

NRDC 134

was more toxic to houseflies and mustard beetles than most known insecticides.[12]

Chemical and spectroscopic evidence indicated that the furan ring in esters of 5-benzyl-3-furylmethyl alcohol was the probable site of photosensitized oxidative decomposition; therefore, in an attempt to discover more stable pyrethroids, other esters of 2,2-dichlorovinyl cyclopropanecarboxylic acid were synthesized. The racemic ester from 3-phenoxybenzyl alcohol, permethrin:

10: X = Cl, R = CH_2—

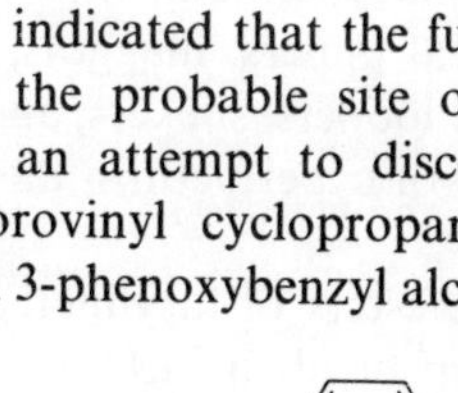

Permethrin

was as active against houseflies as bioresmethrin and 2.5 times as effective against mustard beetles. This compound showed much greater photostability and consequently was a moderately persistent insecticide (general dose required: 50–225 g a.i./ha); it was the first pyrethroid effective as a seed treatment against wheat bulb fly.[13] Replacement of the chlorine atoms by bromine and the introduction of the α-cyano group led to decamethrin (10a), the most active of eight possible stereoisomers,[14] which at the time of its discovery (1974) was the most potent insecticide known.

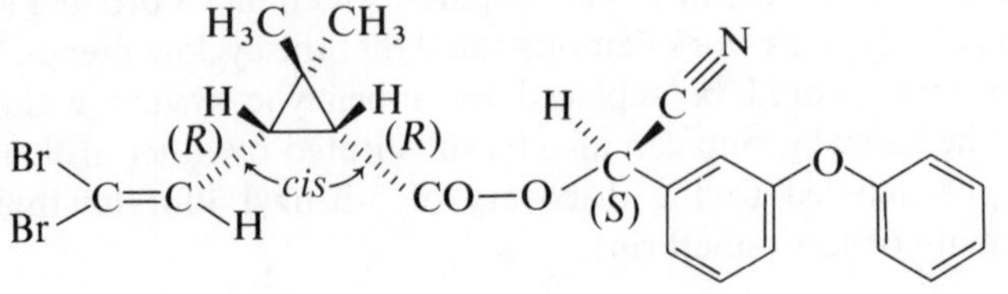

(10a)

Decamethrin, the crystalline (1R)-cis-(S)-α-cyano isomer (10a), was much more active than the liquid (R)-α-cyano isomer. The activity of decamethrin was quite exceptional, and again emphasizes the importance of stereochemistry in the activity of pyrethroids and suggests that the mode of action must involve a very specific interaction with a receptor site, probably in the central nervous system.[9,14]

The LD_{50} to houseflies and mustard beetles was 0.0003 μg/insect (≡ 0.03 mg/kg) which sets a new standard of insecticidal potency and is some 50 times more active than the natural pyrethrin I. These dihalovinyl pyrethroids are extremely valuable insecticides because of their high activity, combined with reasonable photostability and very low mammalian toxicity.[15]

The corresponding chloro derivative (10; X = Cl, R = CH(CN)C$_6$H$_4$OPh-*m*) as the (RS) cis-trans mixture, is known as cypermethrin which is a broad-spectrum insecticide (dose: 20–80 g a.i./ha) with good residual activity on plants. The more pure (1R cis) S and (1S cis) R enantiomeric pair of isomers called alphacypermethrin is a more potent insecticide which controls weevils, beetles, leaf miners, maggots, whiteflies, plant hoppers, sawflies, boll-worms, cutworms, loopers, suckers, and thrips at doses of 7–30 g a.i./ha. Plate 5 illustrates the use of cypermethrin and cyhalothrin to control insect pests on cotton.

A survey of the esters of furylmethanol led to the discovery of insecticidal activity in a group of phenylacetic acid esters and Japanese chemists at the Sumitomo Company introduced (1974) fenvalerate (11).[16] Fenvalerate is the (RS) mixture of four isomers and is used in doses of 20–150 g a.i./ha as an emulsifiable concentrate against a wide range of insect pests and is relatively stable in sunlight. The discovery of fenvalerate demonstrated that the cyclopropane ring was not essential for insecticidal activity. The (2S, 3S) stereoisomer of compound (11a), termed esfenvalerate and introduced by Shell, is active at 7–50 g a.i./ha against insects and is appreciably more potent than fenvalerate. Only esters of (S)-3-methyl-2-(4-chlorophenyl) butyric acid are active and this configuration can be readily related to the 1R configuration of the insecticidal cyclohexanecarboxylates (11b). Fenvalerate can therefore be regarded as an

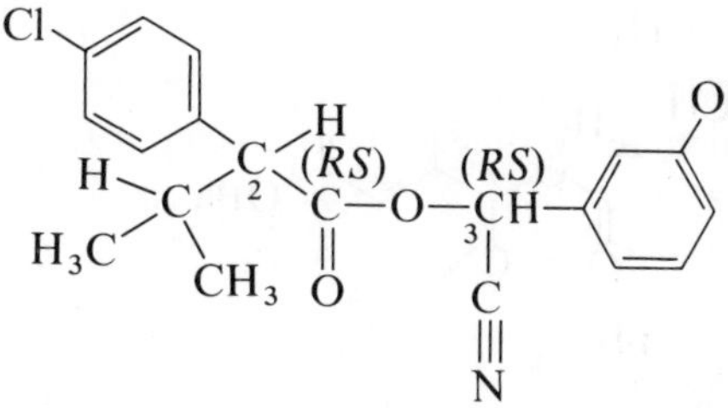

(11)

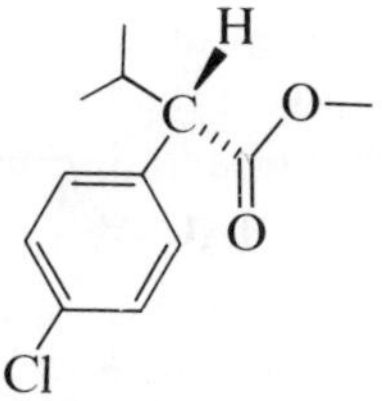

(11a) [2S]

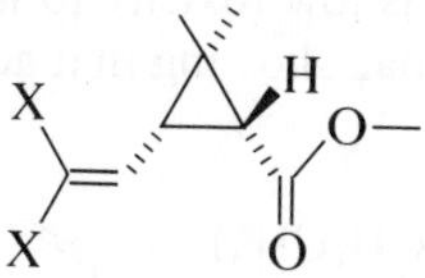

(11b) [1R cis]

analogue of the classical pyrethroid insecticides.[17] Esters of 1- or (3-)hydroxy-methylimidazolidine-2,4-dione (e.g. compound **(12)**) had the highest known knockdown activity against houseflies, mosquitoes, and cockroaches.[18]

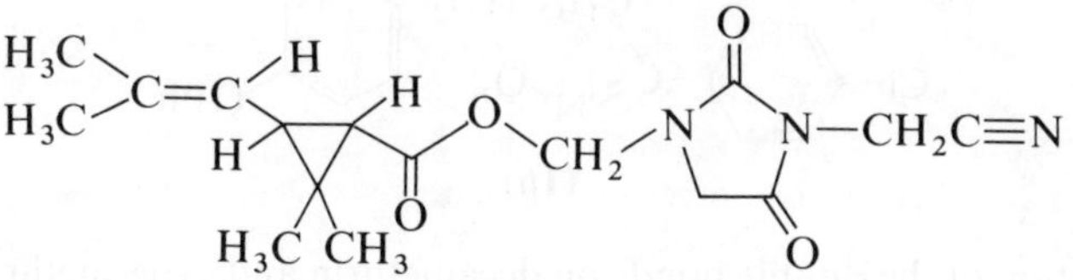

(12)

The American Cyanamid Company introduced a new series of active phenyl-acetic acid esters which were variations on the fenvalerate model; two examples are flucythrinate **(13)** [2S, αRS] and fluvalinate **(14)** [2R, αRS] (1981). Compound **(14)** is derived from the phenylacetate series by insertion of a nitrogen atom between the 2-carbon atom and the phenyl ring. In both series for maximum activity the preferred stereochemistry at the 2-carbon atom remains the same, although the substitution pattern is different.[17]

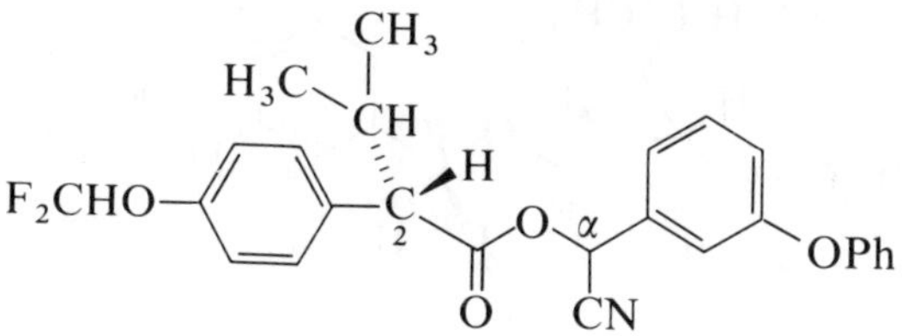

(13)

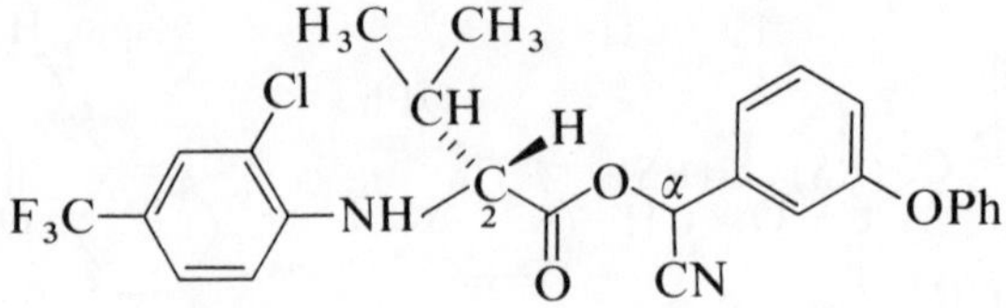

(14)

Several new substituted benzyl alcohols, e.g. (15) by Nippon Soda (1980), yield photostable pyrethroid insecticides. Compound (15) is very effective against rice hoppers and has low toxicity to fish. A group of analogous oxime esters (16) are also insecticidal, showing that active pyrethroids need not always be esters.[19]

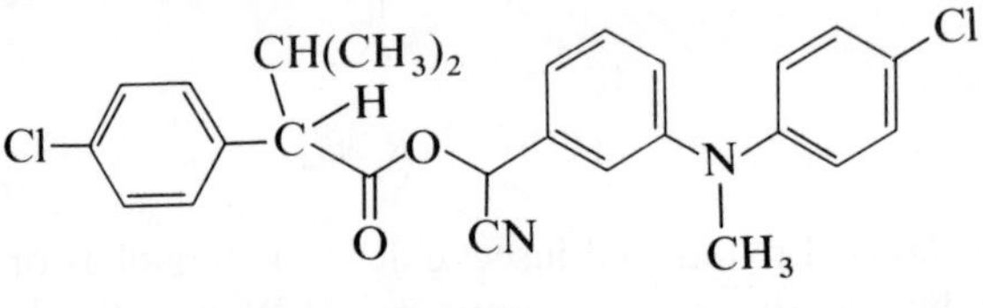

(15)

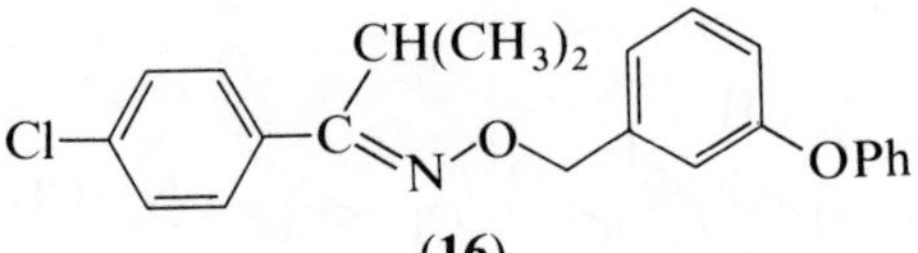

(16)

Bromination of the double bonds on decamethrin and cypermethrin gave the dibromo adducts tralomethrin (17) and tralocythrin (18). These adducts were both highly active, which suggested that a double bond in the side-chain was not essential for maximum potency, but other studies[20] indicated that the activity may be associated with *in vivo* conversion to the parent unsaturated compounds.

Later work[21-23] showed that one or both of the chlorine atoms in the dichlorovinyl group may be replaced by the trifluoromethyl group to give extremely active esters. The most potent compound was cyhalothrin, the (1R, αS) isomer (19). Cyhalothrin can exist as sixteen isomers but the synthesis

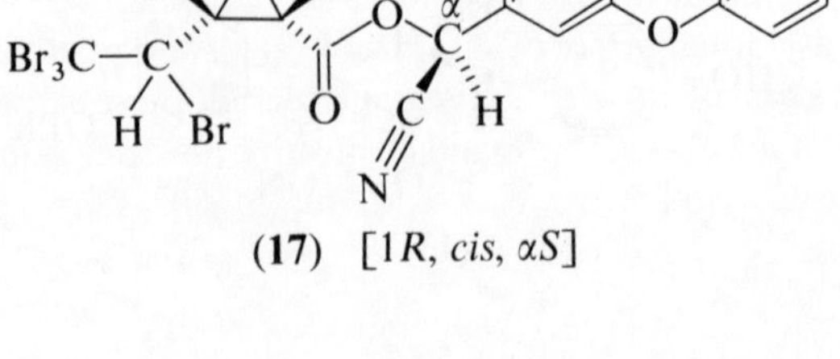

(17) [1R, *cis*, αS]

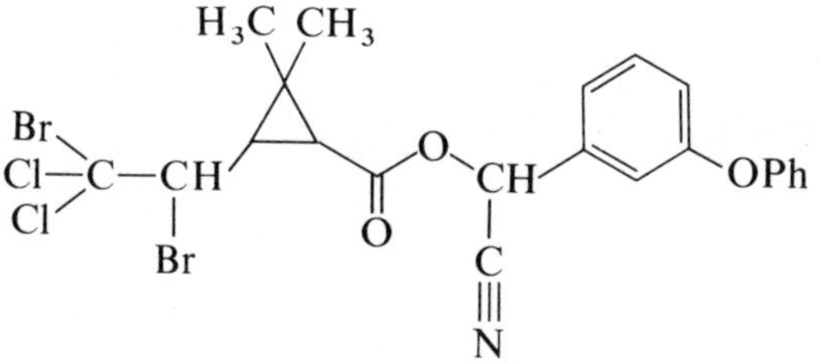

(18)

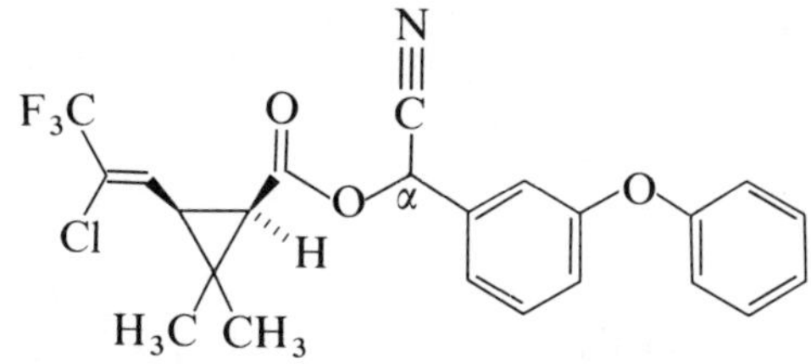

(19)

from 3-methyl-2-butenol **(20)** (Scheme 4.1) gives only the four *Z*(*cis*) isomers and virtually all the insecticidal activity resides in the (1*R*, α*S*) isomer which can be separated by high-performance liquid chromatography (HPLC) and fractional crystallization.

Cyhalothrin **(19)** has a comparatively high mammalian toxicity (LD_{50} oral to rat, 60 mg/kg), it is effective at low dose rates (5–30 g a.i./ha) against major insect pests on many crops, including oilseed rape and cotton (Plate 5). Although cyhalothrin **(19)** is more active against arthropod pests than permethrin and cypermethrin, it presents little hazard to honeybees. This represents an important advantage as compared with the widely used organophosphorus insecticides which are highly toxic to honeybees.

At normal rates cyhalothrin **(19)** shows low toxicity to birds, with no accumulation in eggs or tissues and no effect on earthworms. The half-life in soil is 3–12 weeks; in aerobic soil, it is degraded via hydrolytic and oxidative pathways to metabolites which finally undergo extensive mineralization to carbon dioxide (Scheme 4.2). In flooded soil degradation was slower and only hydrolysis products were detected.

Cyhalothrin **(19)** possesses contact, residual, stomach-acting, and repellant properties, but no fumigant or systemic action. It is marketed as an emulsifiable concentrate (EC) or an ultra low volume (ULV) formulation.

When applied as a foliar spray cyhalothrin **(19)** has no phytotoxicity towards major crops and controls a wide spectrum of lepidopterous pests, e.g. cotton bollworm, better than cypermethrin and is effective at lower rates. Plant viruses are a major cause of reduced yields and quality in many agricultural and horticultural crops.

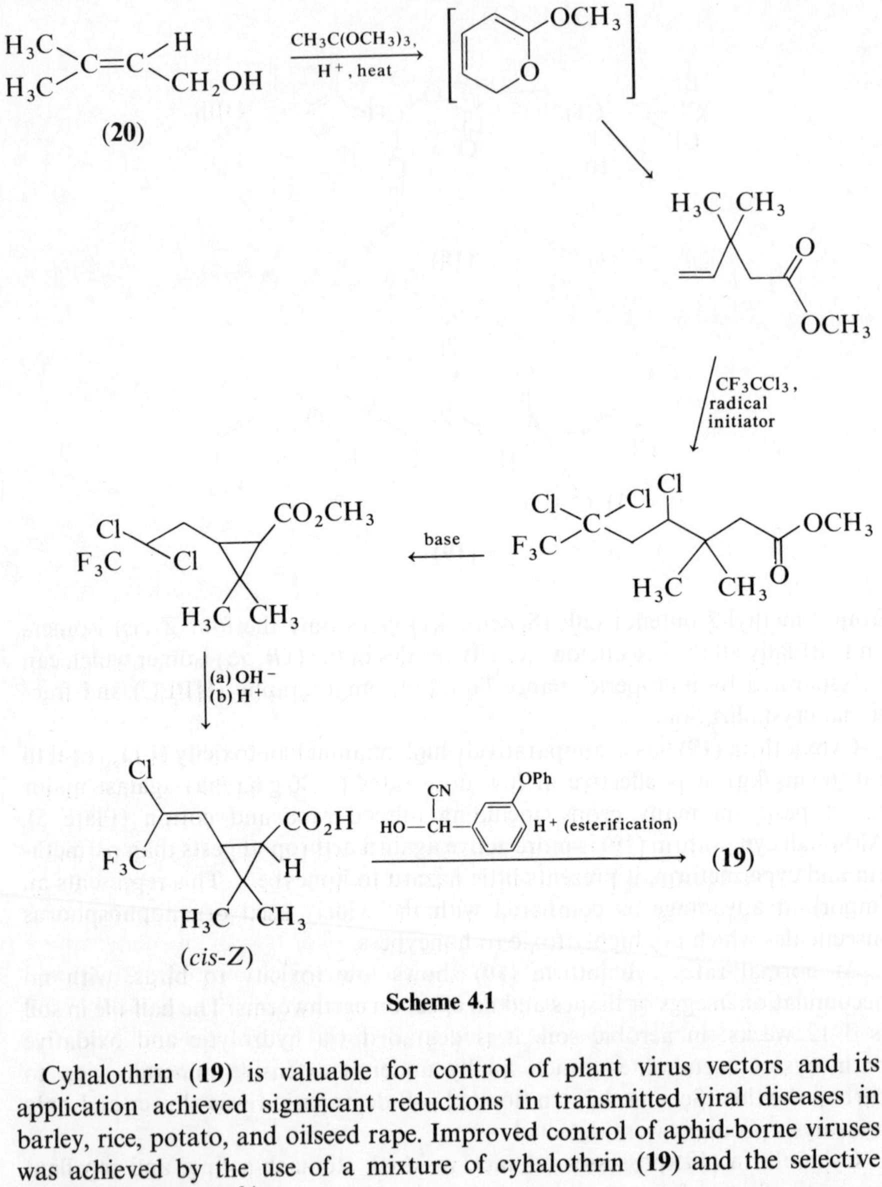

Scheme 4.1

Cyhalothrin (**19**) is valuable for control of plant virus vectors and its application achieved significant reductions in transmitted viral diseases in barley, rice, potato, and oilseed rape. Improved control of aphid-borne viruses was achieved by the use of a mixture of cyhalothrin (**19**) and the selective aphicide pirimicarb.[24]

Tefluthrin (**21**; X = CH₃) is the first pyrethroid that is really effective as a soil insecticide. Tefluthrin is a 1:1 mixture with the (1S, 3S) stereoisomer and is sufficiently stable in soil to provide good residual control of major soil insect pests at doses of 12–150 g a.i./ha. It is formulated for soil application as granules and may also be applied as a foliar spray or a seed dressing.[25] Tefluthrin (**21**) is especially valuable against corn rootworm and cutworm in the United States of

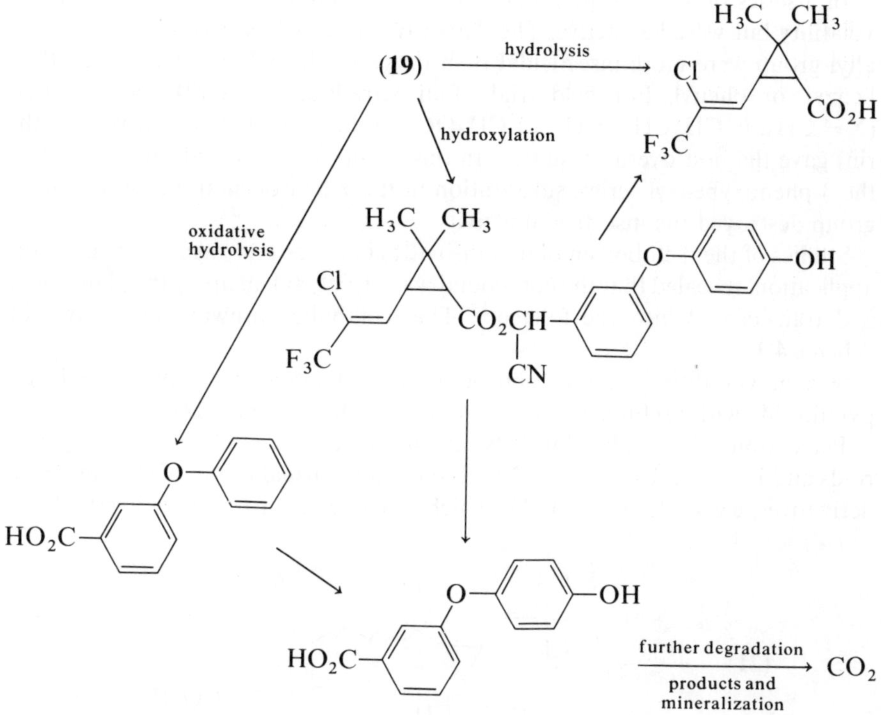

Scheme 4.2

America and to control a range of pests on maize and sugar beet in Western Europe. Tefluthrin will kill insects that have become resistant to the standard organophosphorus and carbamate insecticides and has a much lower acute mammalian toxicity (LD_{50} oral to rat $\simeq 1500$ mg/kg) than many conventional soil insecticides, which is accentuated by the markedly lower doses required. Tefluthrin (**21**) presents little hazard to earthworms and birds, especially when formulated as granules, but it is highly toxic to fish; the half-life in soil is 1–3 months and there is no danger of residue accumulation. Tefluthrin was designed to achieve the optimum combination of stability, volatility, and aqueous solubility.[26]

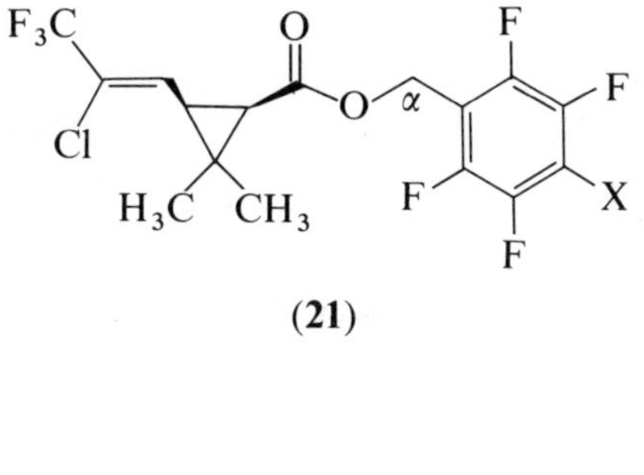

(**21**)

In a series of 4-substituted analogues, the pentafluoroesters had the highest volatility but were less active. The derivatives in which X was a lower alkyl or allyl group were more insecticidal than those in which X was trifluoromethyl, benzyl, or phenyl. For field trials, four tetrafluoroesters (21) were chosen (X = CH$_3$, F, CH$_2$CH=CH$_2$, or CH$_2$OCH$_3$); the 4-methyl derivative (tefluthrin) gave the best overall results.[26] In this group of pyrethroids, in contrast to the 3-phenoxybenzyl series, substitution of the α-hydrogen atom by the cyano group destroyed the insecticidal activity.

Studies of the metabolism of tefluthrin (21) in maize and sugar beet, after soil application, revealed that the compound was rapidly taken up by the plant roots and translocated into the foliage.[27] The metabolic pathways are shown in Scheme 4.3.

Several variations on the 3-phenoxybenzyl alcohol theme have produced pyrethroids with useful activity: an example is the miticide (22).

Perception of a relationship between the structural requirements of pyrethroids and DDT led Holan *et al.*[28] to develop an unusual group of cyclopropane derivatives, e.g. cycloprothrin (23) which is active against pests of livestock.

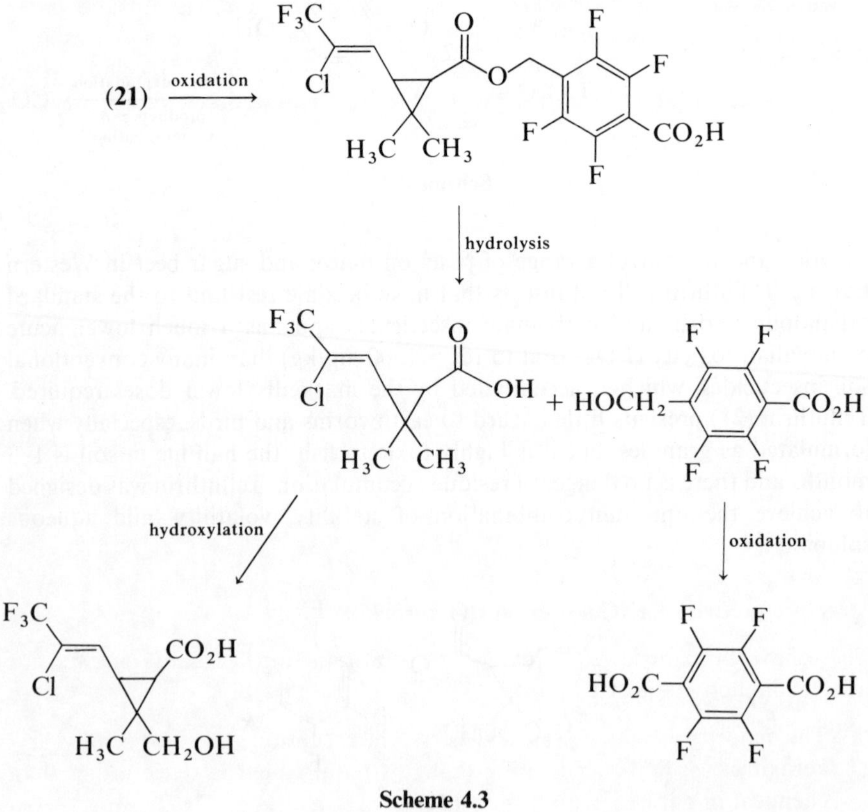

Scheme 4.3

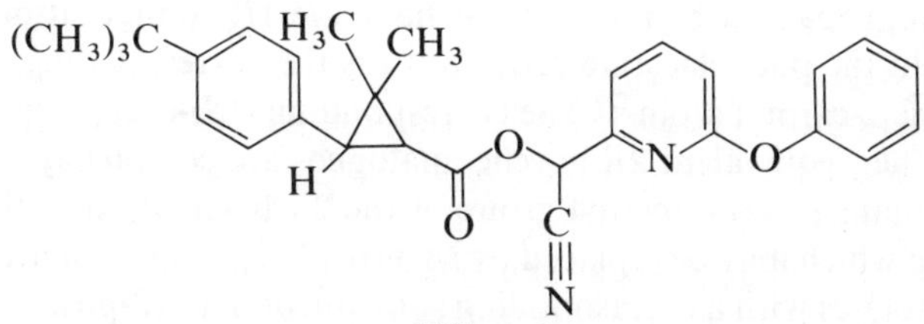

(22)

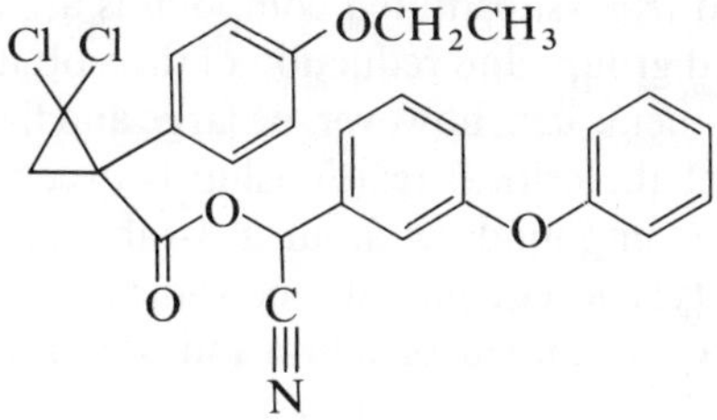

(23)

Insecticidal activity has been recently discovered in non-ester pyrethroids, e.g. etofenprox **(24)** and MTI 800 (R = R' = CH$_3$) **(25)**; activity in the latter compound was substantially enhanced in some fluorinated derivatives (e.g. R = H, R' = CF$_3$).[29] This area shows considerable scope, especially since analogues of MTI800 in which the C$_\alpha$ atom is replaced by silicon are also active.[29]

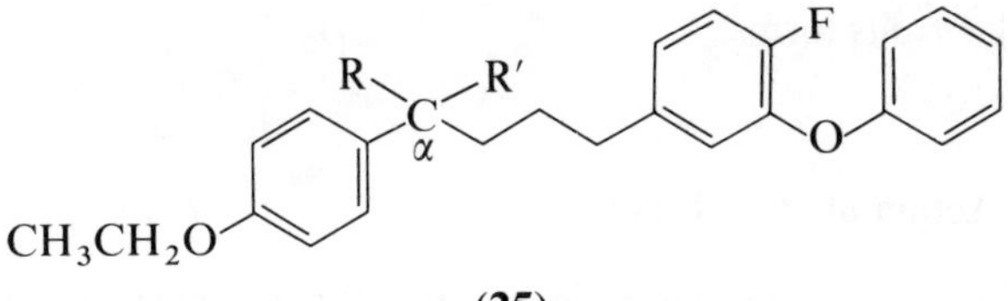

(24)

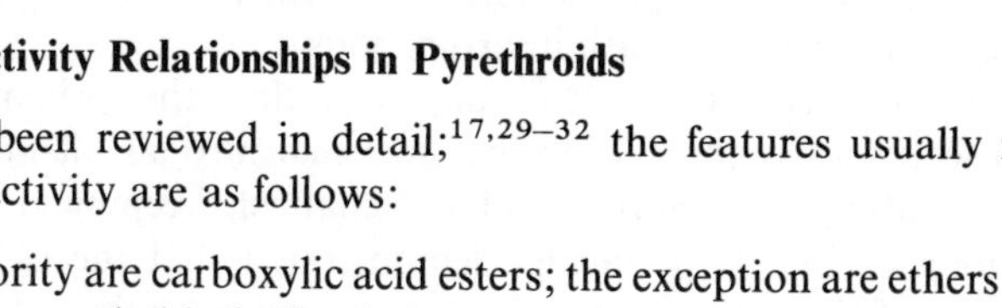

(25)

Structure–Activity Relationships in Pyrethroids

These have been reviewed in detail;[17,29–32] the features usually required for insecticidal activity are as follows:

(a) The majority are carboxylic acid esters; the exception are ethers of alkyl aryl ketoximes, probably indicating that this requirement is steric rather than chemical in nature.

(b) The cyclopropanecarboxylates must have the 1*R* configuration; the active isomers in the phenylacetate series possess the corresponding *S* configuration at the 2-carbon atom.[29] The corresponding (1*S*)-cyclopropanecarboxylates or their equivalent 2*R* acyclic analogues are completely inactive.

(c) They contain a *gem*-dimethyl group at the 2-carbon atom of the cyclopropane ring which may be replaced by a spiro alkyl group; similarly only those phenylacetates with a corresponding substituent in the 2-position are active.

The configuration at the 3-position of the cyclopropane ring is not critical and both *cis*- and *trans*-substituted compounds are active.

Bulky saturated groups and reduction of the isobutenyl group caused loss of activity. Substituents can, however, be large and flat, but the 3-substituent must not obstruct the critical relationship between the *gem*-dimethyl and the ester group. Compounds without a 3-substituent have relatively low insecticidal activity; consequently it is considered that a suitable substituent may assist transport to the site of action and/or reduce metabolic detoxification.

(d) Only esters of alkyl or aryl alkyl alcohols are insecticidal; the carbon atom attached to hydroxyl group must be sp^3 hybridized. Substitution of the α-carbon atom of 3-phenoxybenzyl alcohol by the cyano group enhances activity of the esters, which is invariably associated with the *S* configuration.

These criteria apply to the insecticidal action against a broad range of insect species and are therefore related to the shape and symmetry of the site of action and hence reflect the capacity of the molecule to produce physiological change at the receptor site.

Comparative studies indicate the vital importance of stereochemistry and the overall shape of the molecule in determining the insecticidal potency of pyrethroids.[14,29] In the majority of compounds, the maximum activity is associated with the 1*R trans* stereochemistry as shown by natural pyrethrin I. However, this generalization may be modified by other molecular and electronic effects, e.g. in decamethrin the 1*R cis* isomers were more active than the 1*R trans* forms.

The Mode of Action of Pyrethroids

The symptoms of insects poisoned by pyrethroids clearly shows that the chemical attacks the insects' nervous systems. Pyrethroids cause hyperexcitation followed by convulsions and death in arthropods.

The majority of evidence[6,17,33,34] indicates that the primary target of pyrethroid insecticides is the sodium channel in neuronal membranes which modulates the gating characteristics of the channel, keeping it open for an abnormally long period. Experiments demonstrated that the GABA-induced chloride current of the rat was unaffected by a concentration of decamethrin which resulted in an appreciable effect on the sodium gate mechanism.

The effect on the sodium channel alters the action potential, so that the depolarizing after potential is increased and prolonged; this results in repetitive firing of the neurons followed by convulsions and blockage of nerve conduction.[6] The predominance of particular symptoms can often be related to specific classes of pyrethroids.[34] This view of the primary mechanism of action is supported by *in vitro* studies of electrical impulses in the nervous system.

Examination of the effects of pyrethroids and DDT on the giant axons of crayfish and squid indicate[17] that both chemicals act on the same, or a very similar, site in the insects' nervous system. DDT and the pyrethroids are the only insecticides to exhibit a negative temperature coefficient of insecticidal action and produce similar lesions on the motor nerve terminals.

There are a large number of cases of cross-resistance between DDT and pyrethroids and genetic evidence suggests that pyrethroids have a single site of action. A good correlation occurs between the capacity of pyrethroids to cause hyperactivity and their rapid knockdown action on houseflies. The latter requires ready access to the site of action which is therefore more likely to be in peripheral sensory axons that are sited not very deeply within the insect.[6] Although the primary site of action of pyrethroids appears to be the nerve sodium channels, they also interact with potassium channels, GABA-activated chloride channels, and membrane-bound ATPases.[17]

METABOLISM

The rate and mechanism of metabolism has a major influence on the toxicology of a compound.[17] In mammals (e.g. rats), pyrethroids are very rapidly metabolized by three main processes: ester cleavage (by oxidases and esterases), oxidation (oxidases), and hydroxylation. These processes are illustrated by the metabolism of permethrin (**26**) in rats[17] (Scheme 4.4).

The conjugates are water soluble and are excreted in urine, so pyrethroids are not stored to any significant extent in fatty tissues. Metabolism in birds appears similar to that in mammals. In insects, the metabolism has been less extensively studied; however, in those compounds containing the isobutenyl side-chain, e.g. allethrin (**9**), the major metabolic pathway involves oxidation by mixed function oxidases (MFO) of the *trans* methyl group of the isobutenyl moiety, the rest of the structure remaining intact (see Scheme 4.5).

Pyrethroids suffer facile degradation in soil under aerobic conditions leading ultimately to carbon dioxide (see page 63). Under anaerobic soil conditions degradation to carbon dioxide is slower. Pyrethroids are virtually immobile in soils, as would be expected with highly lipophilic molecules. The metabolism in plants resembles that in mammals, except that oxidation of the primary alcohol metabolites is much less pronounced. The metabolites are usually rapidly conjugated with sugars and amino acids and there is no translocation of the pyrethroid from treated soil or leaves into the plant.

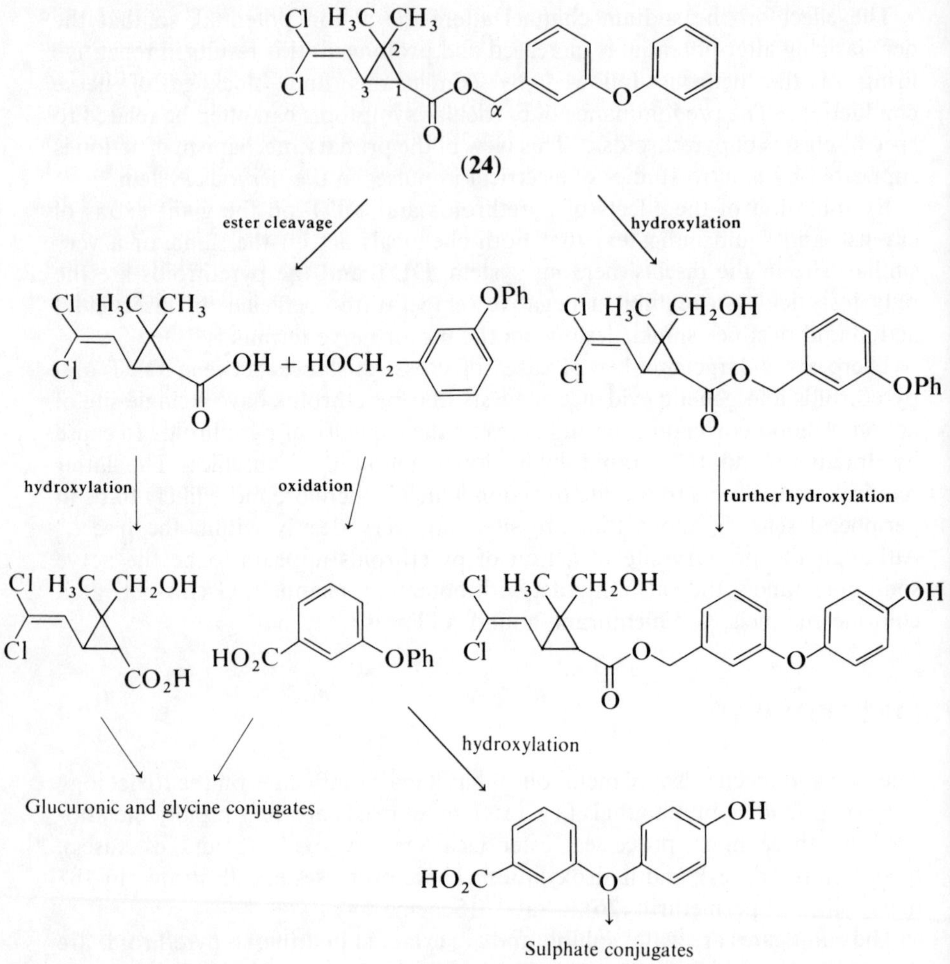

(24)
ester cleavage
hydroxylation
hydroxylation
oxidation
further hydroxylation
hydroxylation
Glucuronic and glycine conjugates
Sulphate conjugates

Scheme 4.4

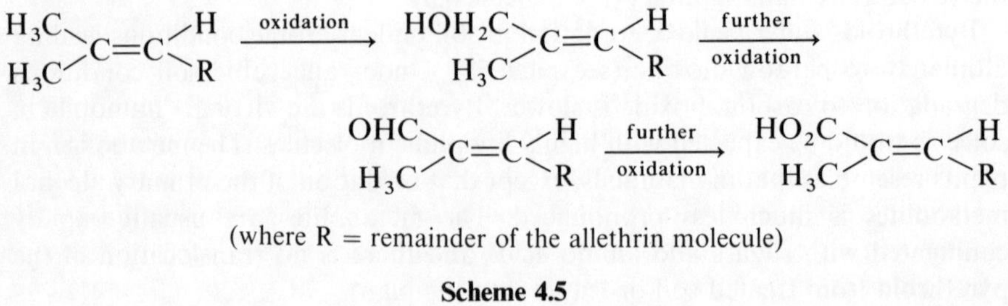

oxidation
further oxidation
further oxidation

(where R = remainder of the allethrin molecule)

Scheme 4.5

In cyclopropane-based pyrethroids, the rate of metabolism is sensitive to the stereochemistry about the 1,3-bond: the presence of a substituent *trans* to the ester group enhances the rate of ester cleavage, while the corresponding *cis* isomer is much less rapidly metabolized. In the pyrethroids derived from 3-phenoxybenzyl alcohols, the introduction of a cyano group on the α-carbon atom increases the stability of the molecule towards hydrolytic and oxidative metabolism. Decamethrin and *cis*-cypermethrin are the most metabolically stable pyrethroids; however, metabolism remains sufficiently rapid to permit efficient excretion by mammals. The non-cyclopropane pyrethroid, fenvalerate, possesses similar metabolic stability. In the synthetic pyrethroids, mammalian toxicity is generally moderate; however, the LD_{50} values are sensitive both to the stereochemistry (*cis/trans* ratio) and to the dosing vehicle.[17] *cis* Isomers have higher mammalian toxicity, especially in oil-based formulations; aqueous suspensions are usually the least toxic. For instance, with permethrin, these trends are illustrated in the Table 4.1. The figures demonstrate the need to record both the dosing medium and the isomer ratio when reporting acute toxicity data.

Table 4.1 LD_{50} values (mg/kg, oral to rat)

cis/trans ratio	H_2O	DMSO	Corn oil
80:20			220
60:40			440
40:60	2950	1500	1250
20:80			6000

The pyrethroids generally are some of the most potent insecticides known, but exhibit comparatively low mammalian toxicity. This combination of properties gives pyrethroids an almost unique degree of selectivity. Pyrethroids are now recognized as the third major class of synthetic insecticides and are expected to take approximately 15% of the world insecticide market by 1990.

In 1940 Eagleson discovered[3] that by formulating pyrethrum in sesame oil the insecticidal effectiveness was increased, although the oil was not itself insecticidal. The synergistic activity of the oil was shown to depend on the presence of compounds containing methylenedioxy groups, such as sesamin (**27**).[35] Later work resulted in the discovery of many other pyrethroid synergists which usually contained the methylenedioxyphenyl group. Three important commercial examples are piperonyl butoxide (**28**), sesamex (**29**), and sulfoxide (**30**).

The use of synergists with pyrethrum preparations permits the amounts of the active ingredients (a.i.) to be substantially reduced without loss of insecticidal potency; thus a mixture of one part of pyrethrum with two parts of piperonyl butoxide (**28**) is as effective against insects as seven parts of pyrethrum alone.

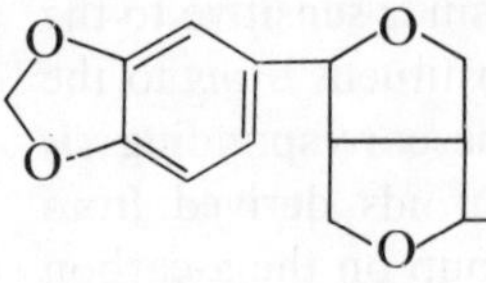

(27) 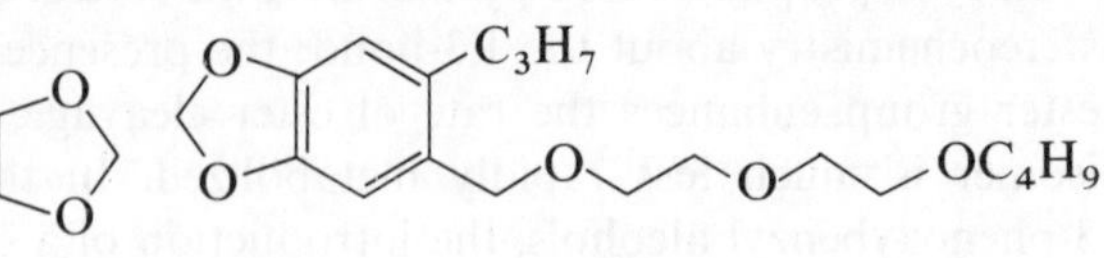   **(28)**

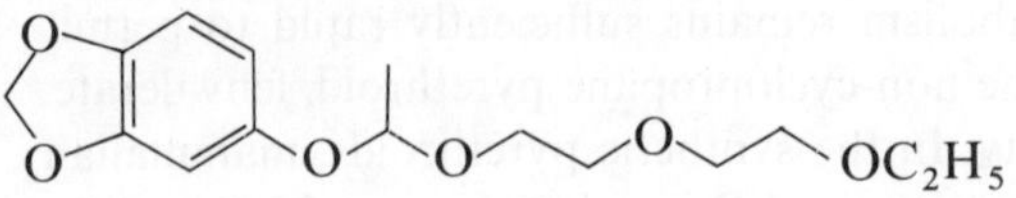

(29) 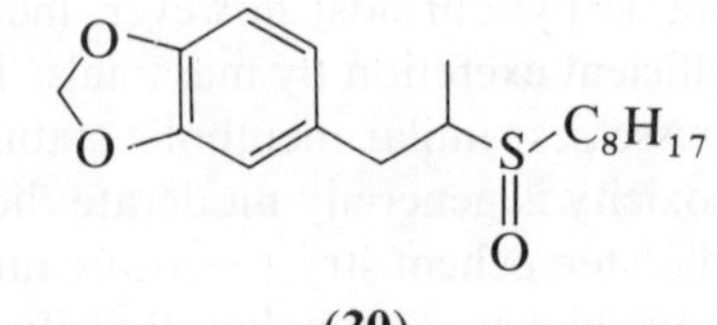   **(30)**

The synergists are expensive compounds, but their use is justified with pyrethroids which are both expensive and often not persistent. The synergist is most effective with those pyrethroids containing the isobutenyl side-chain which is particularly sensitive to oxidative detoxification (see page 69). The synergists act by interfering with the microsomal oxidation of the insecticide by mixed function oxidases (MFO) which are catalysed by cytochrome P-450. The methylenedioxyphenyl synergists apparently form a complex with cytochrome P-450 which inhibits the oxidases, so that the deactivation of the pyrethroid is reduced.[6,17]

Another group of naturally occurring insecticides are related to nereistoxin **(31)**, a toxin isolated from a species of marine worm (*Lumbriconereis heteropoda*). Synthetic modifications include cartap **(32)** and the 1,2,3-trithiole **(33)**.[2]

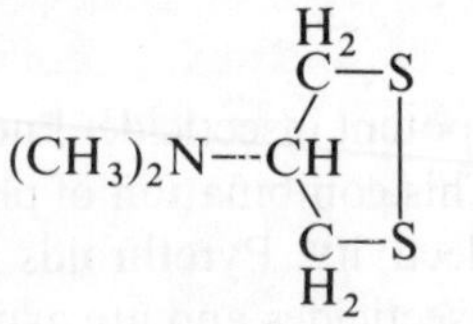

(31) 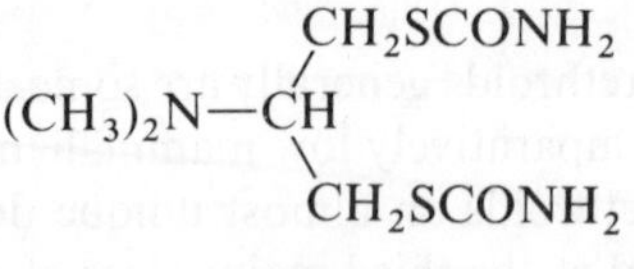   **(32)**

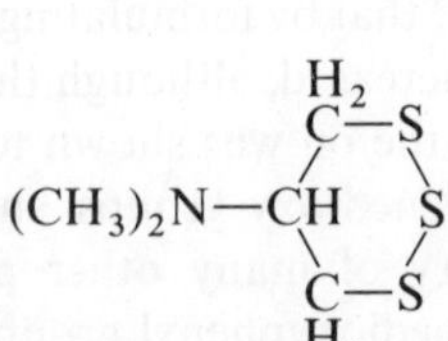

(33)

Cartap **(32)** is effective against lepidopterous and coleopterous insects and is important in the Japanese market.[2] It is generally used as the hydrochloride and is a slow-acting contact insecticide useful against rice stem borers, Colorado and

Mexican beetles, caterpillars, and weevils.[7] It has only moderate toxicity to mammals: LD_{50} (oral) to rats 380 mg/kg. Both cartap and nereistoxin, like nicotine (page 54), act by combination with the acetylcholine receptor in insects, and experiments with cockroaches show[6] that they block synaptic but not axonal nervous transmission.

1,2,3-Trithiole (33) was developed by Sandoz (1975) and is an efficient contact and stomach insecticide against several economically important pests.[15]

INSECTICIDAL BUTYLAMIDES

Several members of the *Compositae* and *Rutaceae* groups of plants contain insecticidal unsaturated butylamides, e.g. pellitorine (34) and fagaramide (35).

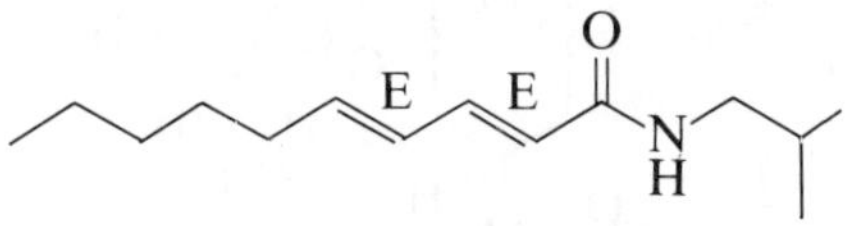

(34) 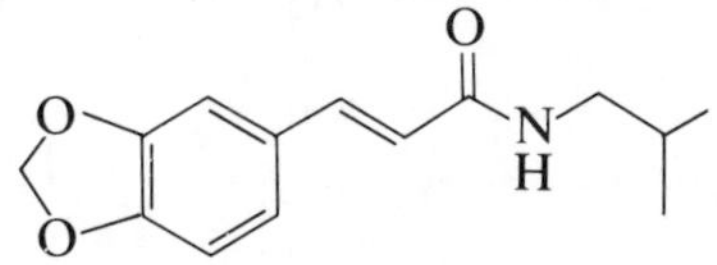   (35)

These compounds, like several of the pyrethrins, have a rapid knockdown effect on flying insects; their practical application as insecticides is, unfortunately, limited by chemical instability.[36,37] However, a number of synthetic analogues have been examined and several show promise as insecticides.[29] In particular studies at Rothamsted Experimental Station in England were based on a group of 159 naturally occurring isobutylamides from *Piperaceae* species. Only 28 showed insecticidal properties and many were unstable and mixtures of isomers; however, a stable well-characterized compound (36) (R = H) was discovered.

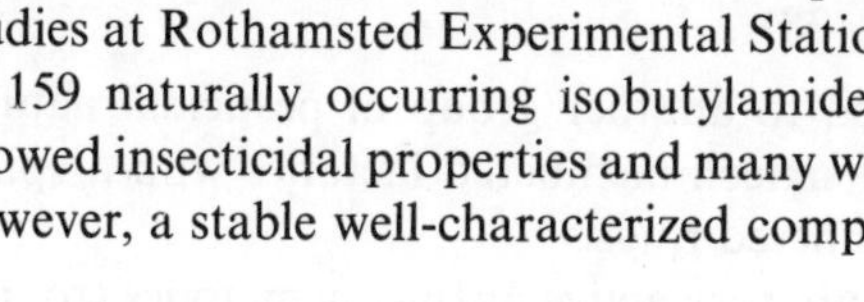

(36) 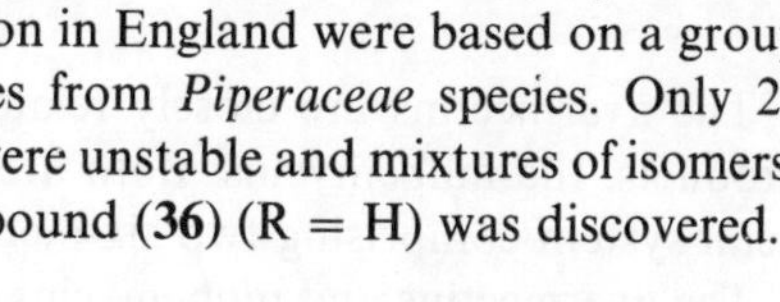   (37)

This showed significant insecticidal activity against houseflies and mustard beetles. In the synthetic analogues of compound (36), the compounds containing R = 4-bromo or 3,4-dibromo were the most active.[38] Further synthetic modification afforded the series (37) where the compounds in which Ar = C_6H_5, $3,5\text{-}F_2C_6H_3$, or dibenzofuran-3yl (R^1, R^2 = H or Me) were as active as other insecticides towards houseflies; these compounds were shown to be some four times more effective against houseflies that had become resistant to DDT and pyrethroids.

THE AVERMECTINS AND RELATED ANTIBIOTICS

In 1975 scientists at Merck Company Inc. discovered a high level of anthelmintic activity in a fermentation culture of *Streptomycetes avermitilis*, a Japanese soil microorganism.[29] Isolation and structural elucidation revealed a mixture of eight closely related natural products named avermectins, of which the most active member was avermectin B1a (**38**).[39]

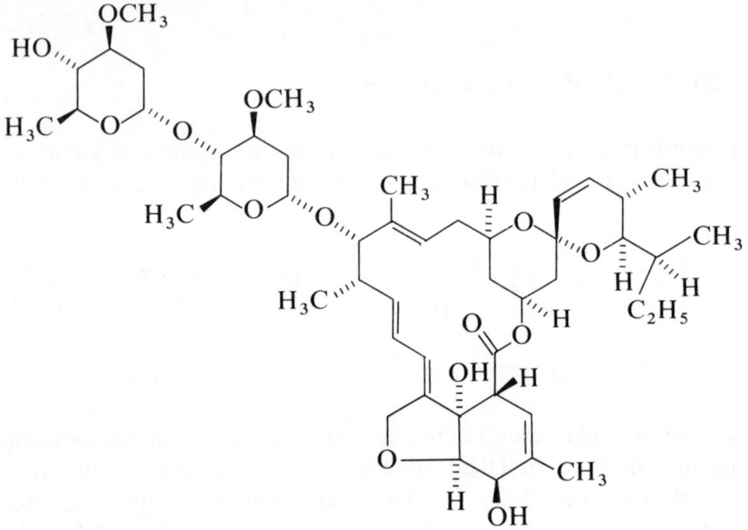

(**38**)

The avermectins are closely related to another group of pesticidal natural products, the milbemycins. Both are sixteen-membered lactones with a spiroketal system comprising two six-membered rings.

The avermectins and milbemycins are very active against nematodes (roundworms) and other arthropods (insects, ticks, lice, mites) but they are inactive against tapeworms or flukes. They are among the most potent insecticides known and their spectrum of insecticidal activity is in many ways complementary to that of pyrethroids which enhances their potential value.

Avermectin B1a interacts with GABA receptors and stimulates the binding of the inhibitory neurotransmitter which regulates the chloride ion channels in the neural membrane. The avermectins act as GABA antagonists, thereby causing muscular paralysis and death of the parasite.[36,39] Avermectins affect many components of the GABA system in the mammalian central nervous system (CNS). Studies with the American cockroach, however, indicate that avermectin B1a acts by activation of the chloride channels, keeping them in the open position allowing excess chloride ion to enter the postsynaptic region of the GABA-ergic cells, and it is concluded that the chloride ion channel is the primary biochemical target for the avermectins.[6] Attempts to obtain simpler synthetic analogues of these compounds have so far proved unsuccessful.[29]

MISCELLANEOUS COMPOUNDS

Picrotoxinin (PTX) (**39**) has been isolated from the seeds of *Anamirta cocculus* and is moderately toxic to insects, e.g. cockroaches, but is more toxic to mice. PTX, like avermectins, acts on GABA-regulated chloride ion channels and functions as a GABA receptor antagonist. Other compounds with a similar mode of action include the trioxabicyclooctanes, e.g. the phosphorus ester (**40**) and bicycloorthocarboxylates (**41**); all are nerve poisons which are not cholinesterase inhibitors.[36]

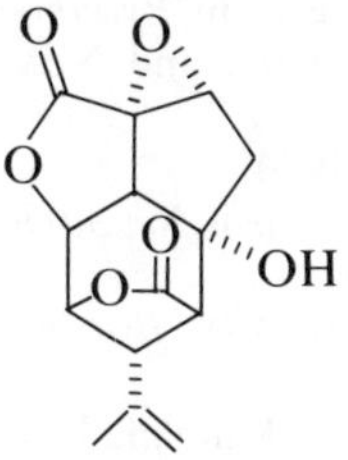

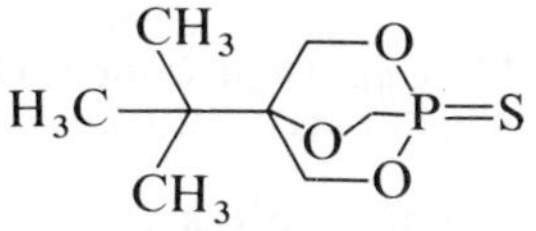

(**40**) 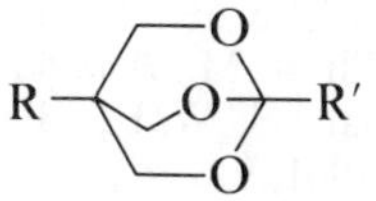   (**41**)

(**39**)

The compounds (**39**) to (**41**) were generally more toxic to mammals than to insects; the bicycloorthocarboxylates (**41**), however, can sometimes show selective toxicity to houseflies and cockroaches and, consequently, have potential for development as novel insecticides (see Chapter 5, page 101).

Insect neuropeptides (INPs) generally containing five to ten amino acid residues play vital roles as circulating neurohormones and neurotransmitters.[40] They control many aspects of the insect's growth, development, and reproduction together with important physiological and metabolic processes; 25 such processes are thought to be mediated via these neuropeptides. By 1988, some 27 INPs had been identified and their structures elucidated. For instance, in the locust, the adipokinetic hormones, AKH I and II, stimulate lipid metabolism during periods of sustained activity, e.g. migratory flight. Increased knowledge of the structures and physiology of INPs should provide new potential target sites for the design of novel insecticides. Peptide synthesis is one possible area of exploitation—if certain physiologically active peptides were introduced to the insect at the wrong time, or not expressed when required, severe disruptive effects would result.

If the genes coding for INPs could be introduced into crop plants, then insects attacking the crop could be killed. Another promising line of research would be the design of novel peptides and other molecules to block the INP receptors.

REFERENCES

1. Jacobson, M., and Crosby, D. G. (Eds.), *Naturally Occurring Insecticides*, Dekker, New York, 1971.

2. Fuchs, R. A., and Schroder, R., 'Agents for control of animal pests', in *The Chemistry of Pesticides* (Ed. Bucher, E. H.), Wiley, London and New York, 1983.

3. Martin, H., and Woodcock, D., *The Scientific Principles of Crop Protection*, 7th edn, Arnold, London, 1983.

4. Woods, A., *Pest Control*, McGraw-Hill, London, 1974, p. 82.

5. *Approved Products for Farmers and Growers*, Ministry of Agriculture, Fisheries and Food, London, 1989.

6. Corbett, J. R., Wright, K., and Baillie, A. C., *The Biochemical Mode of Action of Pesticides*, 2nd edn, Academic Press, London and New York, 1984.

7. Worthing, C. R., and Hance, R. J. (Eds.), *Pesticide Manual*, 9th edn, British Crop Protection Council, Croydon, 1991.

8. Elliott, M., and Janes, N. F., 'Chemistry of natural pyrethrins', in *Pyrethrum, the Natural Insecticide* (Ed. Casida, J. E.), Academic Press, London and New York, 1973, p. 56.

9. Elliott, M., *Bulletin of World Health Organization*, **44**, 315 (1970).

10. Barthel, W. F., 'Synthetic pyrethroids', *Advances in Pest Control Research*, **4**, 33 (1961).

11. Elliott, M., Janes, N. F., and Graham-Bryce, I. J., *Proceedings of 9th British Insecticide and Fungicide Conference, Brighton*, **2**, 373 (1975).

12. Elliott, M., Farnham, A. W., Janes, N. F., Needham, P. H., Pulman, D. A., and Stevenson, J. H., *Nature (London)*, **246**, 169 (1973).

13. Griffiths, D. C., Scott, G. C., Jeffs, K. A., Maskell, F. F., and Roberts, P. F., *Proceedings of 8th British Insecticide and Fungicide Conference, Brighton*, **1**, 213 (1975).

14. Elliott, M., Farnham, A. W., Janes, N. F., Needham, P. H., and Pulman, D. A. 'Insecticidally active conformations of pyrethroids', in *Mechanism of Pesticide Action*, ACS Symposium Series No. 2, 1974, p. 80.

15. Watkins, T. I., and Weighton, D. M., *Reports on Progress of Applied Chemistry*, **60**, 404 (1975).

16. Ohno, N., *et al.*, *Pesticide Science*, **7**, 241 (1976).

17. Leahey, J. P. (Ed.), *The Pyrethroid Insecticides*, Taylor and Francis, London, 1985.

18. Hirano M., *et al.*, *Pesticide Science*, **10**, 291 (1979).

19. Bull, M. J., *Pesticide Science*, **11**, 249 (1980).

20. Elliott, M., Janes, N. F., *et al.*, *Proceedings of British Crop Protection Conference, Brighton*, **3**, 849 (1984).

21. Briggs, C. G., Elliott, M., and Janes, N. F., 'Synthetic pyrethroids', in *Pesticide Chemistry: Human Welfare and the Environment* (Eds. Myamoto, J., and Kearney, P. C.), Vol. 2, Academic Press, London and New York, 1983, p. 157.

22. Bentley, P. D., *et al.*, *Pesticide Science*, **11**, 156 (1980).

23. Robson, M. J., *et al.*, *Proceedings of British Crop Protection Conference, Brighton*, **3**, 853 (1984).

24. Perrin, R. M., and Gibson, R. W., *International Pest Control*, **27** (6), 142 (1985).

25. Jutsum, A. R., Gordon, R. F. S., and Ruscoe, C. N. E., *Proceedings of British Crop Protection Conference, Brighton*, **1**, 97 (1986).

26. McDonald, E., Punja, N., and Jutsum, A. R., *Proceedings of British Crop Protection Conference, Brighton*, **1**, 199 (1986).

27. Amos, R., and Leahey, J. P., *Proceedings of British Crop Protection Conference, Brighton*, **2**, 821 (1986).

28. Holan, G. B., O'Keefe, D. F., Walser, R., and Virgana, C. T., *Nature (London)*, **272**, 734 (1978).

29. *Recent Advances in the Chemistry of Insect Control*, Vol. I (Ed. Janes, N. F.), Royal Society of Chemistry, London, 1985; Vol. II (Ed. Crombie, L.), Royal Society of Chemistry, Cambridge, 1990.

30 Elliott, M., and Janes, N. F., 'Recent structure–activity correlations in synthetic pyrethroids', in *Advances in Pesticide Chemistry* (Ed. Geissbuhler, H.), Pt. 2, Pergamon Press, Oxford, 1978, p. 166.

31. Elliott, M., and Janes, N. F., *Chemical Society Reviews*, 7, 473 (1978).

32. Elliott, M., and Janes, N. F., *Pesticide Science*, 11, 513 (1980).

33. Irving, S. R., *Proceedings of British Crop Protection Conference, Brighton*, 3, 859 (1984).

34. Hart, R. J., 'Mode of action of agents used against arthropod parasites', in *Chemotherapy of Parasitic Diseases* (Eds. Campbell, W. C., and Screw, R.), Plenum Press, New York, 1986.

35. Hodgson, E., and Tate, L. G. *Insecticide Biochemistry and Physiology*, Heyden, London, 1976, p. 115.

36. Casida, J. E., 'Insecticides acting as GABA receptor antagonists', in *Pesticide Science and Technology* (Eds. Greenhalgh, R., and Roberts, T. R.), Blackwell, Oxford, 1987.

37. Matolcsy, Gy, Nádasy, M., and Andriska, V., *Pesticide Chemistry*, Elsevier, Amsterdam, 1988.

38. Elliott, M., *et al.*, *Pesticide Science*, 18, 223 (1987).

39. Baker, R., and Swain, C. J., *Chemistry in Britain*, 25 (7), 692 (1989).

40. Evans, P. D., Robb, S., and Cuthbert, B. A., *Pesticide Science*, 25, 71 (1989).

Chapter 5

Synthetic Insecticides I: Miscellaneous and Organochlorine Compounds

Synthetic insecticides have recently been gaining ground at the expense of natural products. However, synthetic pyrethroids which were originally derived from natural pyrethrum are currently the most exciting group of insecticides (see Chapter 4, page 71).

The four main groups of insecticides are organophosphates, carbamates, pyrethroids, and organochlorines; in 1983 their estimated market values were $1640, 1100, 600 and 560 million respectively.

The organochlorine insecticides have caused environmental pollution (see Chapter 15, page 350). Consequently they have been banned or severely restricted in developed countries. The sales of this group of insecticides is therefore declining and this trend is expected to continue, although on economic grounds their use in Third World countries remains significant, especially for control of insect disease vectors, e.g. mosquitoes and lice. Miscellaneous other compounds used in insect control programmes ($380 m in 1983) include organotins, acylureas, formamidines, pheromones, and insect growth regulators (IGRs).[1]

MISCELLANEOUS COMPOUNDS

The earliest synthetic contact insecticides were inorganic materials.[2] The pigment Paris Green, a copper acetoarsenite of approximate composition $Cu_4(CH_3COO)_2(AsO_2)_2$, was successfully employed in the United States of America (1864) for the control of Colorado beetle on potatoes. Lead arsenate, $PbHAsO_4$, was also used in 1892 against the gipsy moth in forests in the eastern United States,. This compound was one of the most effective agents against codling moth caterpillars in orchards.[3] It is almost insoluble in water and is often formulated with organomercurial compounds, so providing a mixture that also controls apple and pear scab. In view of the high intrinsic mammalian toxicity of lead, preparations containing calcium arsenates are often preferred because environmental pollution by lead presents a serious problem since lead is a cumulative poison.

Such arsenical compounds are general stomach poisons and any selective toxicity to insects depends on the fact that the insect consumes a much larger amount of vegetable matter in relation to its body weight as compared with man. Moreover, the insect consumes the freshly contaminated crop whereas man only eats part of the crop after a suitable interval of time between application of the insecticide and harvesting.

Due to the highly poisonous nature of arsenic and the potential dangers of environmental contamination, sodium arsenite and lead arsenate have been banned, and the use of other inorganic arsenicals has markedly decreased. Lead arsenate is not now used in orchards because fruit growers rely mainly on organophosphorus and carbamate compounds (Chapter 6) for the control of insect pests.[4]

The action of arsenical insecticides is probably due to the production of water-soluble inorganic arsenic ions. Paris Green and sodium arsenite produce arsenite ions, while calcium and lead arsenates give arsenate ions. Both kill insects by inhibition of respiration: arsenate ions (ions of arsenic acid $O=As(OH_3)$ are known[5] to uncouple oxidative phosphorylation, probably because arsenate mimics the phosphate ion and becomes incorporated into key high-energy intermediates which quickly decompose if they contain arsenate rather than the normal phosphate.

Arsenite ions (ions of arsenious acid $As(OH)_3$) act by inhibition of the pyruvate or the α-ketoglutarate dehydrogenase systems, and probably interference with oxo-acid oxidation is the main site of toxic action.

Dinitrophenols

Dinitrophenols and their derivatives are very versatile pesticides and have been used as insecticides, fungicides, and herbicides. Potassium dinitro-o-cresylate was marketed in Germany in 1892 as a moth-proofing agent. In tests against the eggs of the purple thorn moth, phenol and cresols were found[2] to be toxic, and the activity was progressively increased by the introduction of two nitro groups in the molecule, but further nitration reduced activity. This led to the development of 4,6-dinitro-o-cresol (DNOC), or more correctly 2-methyl-4,6-dinitrophenol (1; R = CH_3), and the corresponding 2-cyclohexylphenol (1; R = cyclohexyl), formulated in petroleum oil as sprays against the eggs of aphids, winter moth, and red spider mites.[6]

DNOC (1; R = CH_3) is obtained by sulphonation of o-cresol followed by controlled nitration. Such dinitrophenols have high mammalian toxicities and their use has caused environmental damage; they are also phytotoxic so that

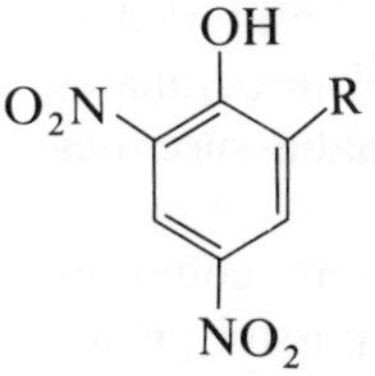

OH
O_2N —— R

NO$_2$
(1)

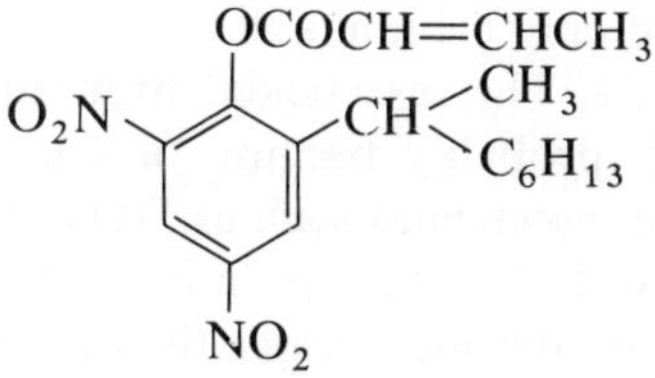

$OCOCH=CHCH_3$
O_2N —— $CH{\large<}^{CH_3}_{C_6H_{13}}$

NO$_2$
(2)

OCOR
O_2N —— $CH{\large<}^{CH_3}_{CH_2CH_3}$

NO$_2$
(3) 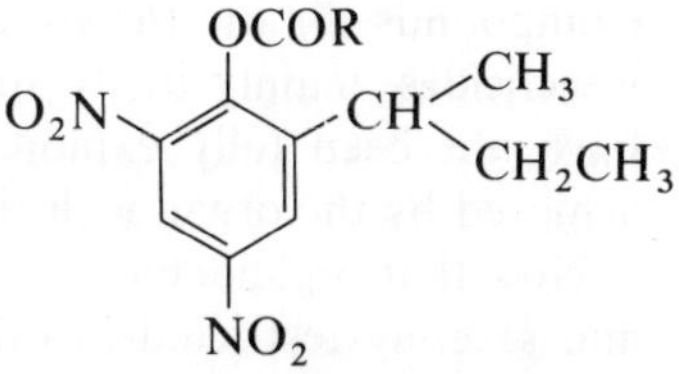

(3a) R = $CH=C(CH_3)_2$
(3b) R = $OCH(CH_3)_2$

they cannot be applied to foliage, and they can be used as rapid-acting herbicides[2] (Chapter 8, page 219).

Certain related dinitrophenols, such as dinocap (2) and binapacryl (3a), although less toxic to mammals and insects, are useful acaricides and fungicides (Chapter 7, page 170); the related carbonate, dinobuton (3b), is an excellent acaricide. Many of the dinitrophenols are highly toxic to all forms of life; they are contact poisons which have the ability to penetrate to the site of action within the cell. The toxicity arises from their interference with respiration through uncoupling oxidative phosphorylation in mitochondria;[2,5] in addition some of the herbicidal effects may be due to direct interference with photosynthesis (Chapter 3).

Various esters (e.g. compounds (2) and (3)) are used against powdery mildews and mites and these are apparently hydrolysed *in vivo* to the free 2,4-dinitrophenol which is the active toxicant.

Organic Thiocyanates

In a series of alkyl thiocyanates (4) obtained by reaction of alkyl halides with sodium thiocyanate, the insecticidal activity increased with the length of the alkyl chain up to the dodecyl derivative (4; R = $C_{12}H_{25}$) and then declined. The dodecyl compound was the most active because it possessed the optimum oil/water solubility balance for penetration of the insect cuticle (Chapter 2, page 29). A more useful compound was, however, 2-(2-butoxyethoxy) ethyl thiocyanate or lethane (4; R = $-CH_2CH_2OCH_2CH_2OC_4H_9$), discovered in 1936.[2,4]

Isobornyl thiocyanate or thanite (6), obtained from the terpene isoborneol (5) by treatment with chloroacetyl chloride and sodium thiocyanate, is also a useful derivative:

$R-S-C{\equiv}N$

(4)

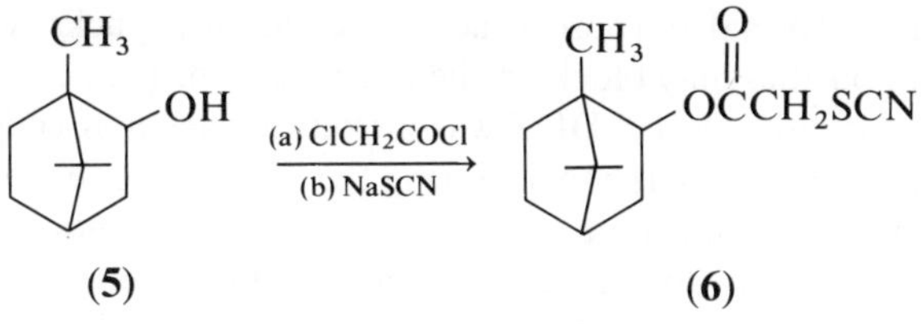

CH$_3$ —— OH

(5)

$\xrightarrow[\text{(b) NaSCN}]{\text{(a) ClCH}_2\text{COCl}}$

CH$_3$ —— $OCCH_2SCN$ (with $\overset{O}{\overset{\|}{C}}$)

(6)

Compounds (5) and (6) are the only members of this group currently used as insecticides, mainly in fly sprays. The insecticidal properties of thiocyanates have not been fully exploited, probably because of the dramatic successes achieved by the organochlorine insecticides such as DDT.

Now that organochlorine insecticides have been banned in several countries and severely restricted in others, the insecticidal thiocyanates possibly merit further investigation. They are remarkably rapid-acting compounds against flying insects—indeed their knockdown action is almost as quick as that of the pyrethroids (Chapter 4), and they also show[2] ovicidal activity against a number of insect eggs.

Thiocyanates may owe their insecticidal action to the *in vivo* liberation of the cyanide ion in the insect body, and there is evidence[5] that these compounds release cyanide in living mice and houseflies. However, despite this evidence, other factors may well operate in their mode of action, since the release of cyanide ion would not seem to be consistent with their rapid knockdown effect.

ORGANOCHLORINE INSECTICIDES AND VARIOUS ACARICIDES

The most important member of this group of insecticides is 1,1,1-trichloro-2,2-di-(*p*-chlorophenyl)ethane, also termed dichlorodiphenyltrichloroethane, or DDT (7). This compound was first prepared by Zeidler (1874) but its powerful insecticidal properties were not discovered until 1939 by Müller of the Swiss Geigy Company. DDT is manufactured by condensation of chloral and chlorobenzene in the presence of an excess of concentrated sulphuric acid:[2,4]

$$
\text{Cl}-\!\!\left\langle\bigcirc\right\rangle + \underset{\underset{\displaystyle CCl_3}{|}}{\overset{\overset{\displaystyle O}{\|}}{CH}} + \text{Cl}-\!\!\left\langle\bigcirc\right\rangle \xrightarrow[(-H_2O)]{H_2SO_4} \text{Cl}-\!\!\left\langle\bigcirc\right\rangle-\!\!\underset{\underset{\displaystyle {}_1CCl_3}{|}}{{}_2CH}-\!\!\left\langle\bigcirc\right\rangle-\!\!\text{Cl}
$$

Chlorobenzene   Chloral   (7)

The crude product consists of some 80% of the desired *p,p'* compound (7) together with approximately 20% of the *o,p'* isomer and a trace of the *o,o'* isomer. Only the *p,p'* isomer has significant insecticidal activity; pure DDT can be obtained as a white powder (m.p. 108°C) by recrystallization from ethanol. However, the increased cost involved in purification is only justified when DDT is used for special purposes, such as for aerosol packs, when the presence of sticky impurities may block up the fine nozzle through which the solution of the pesticide is discharged.[4] DDT was introduced as an insecticide in 1942 and was manufactured on a large scale during the War. Its application in Naples in 1944 checked, for the first time in medical history, a potentially serious outbreak of typhus in which, during one month, a million people were dusted with the insecticide; its subsequent use in India and other countries of the Far East has substantially decreased the death rate from malaria. DDT is now banned for agricultural uses in most developed countries but is still important in the control of insect vectors. The benefits to mankind from the use of organochlorine insecticides have been tremendous and DDT was once the most widely used insecticide in the world. The annual production was more than 100 000 tonnes in the late 1950s, although since that period the annual production has declined and in 1971 it was 20 000 tonnes. At the time of its discovery, the main advantages of DDT appeared to be its stability, persistence of insecticidal action, cheapness of manufacture, moderate mammalian toxicity (LD_{50} (oral) to rats, 150 mg/kg), and broad spectrum of insecticidal activity. DDT kills a wide variety of insects, including domestic insects and mosquitoes, but it is not very effective against mites and does not act nearly as rapidly on flying insects as pyrethrum or thiocyanates.[6,7]

The discovery of the insecticidal properties of DDT (7) stimulated the search for analogous organochlorine compounds; although many hundreds of compounds were synthesized only comparatively few have been sufficiently active and cheap for commercial exploitation. Two examples are methoxychlor (8) and 1,1-dichloro-2,2-di-(p-chlorophenyl)ethane, known as DDD (9).

(8)

(9)

(10)

None of these showed such a high general activity as DDT, although DDD has a slightly different spectrum of insecticidal activity. The compounds (8) and (9) have much lower mammalian toxicities (LD_{50} (oral) to rats, 3400 and 5000 mg/kg respectively) than DDT, and methoxychlor does not appear to be concentrated in animal fats—a most valuable asset, bearing in mind the widespread environmental pollution arising from the majority of organochlorine insecticides (Chapter 15, page 350).

DDT was formerly used to control flies in milking sheds, but as it was found in the milk it is now banned in most countries for this purpose. However, methoxychlor is permitted since no residues of the compound could be detected in the milk.[2] DDD (9) has found use on food crops since it is appreciably less toxic than DDT. Generally for insecticidal potency, a DDT-type molecule (10) must contain p-substituents X, which may be either halogens or short-chain alkyl or alkoxy groups, Y, which is almost always hydrogen, and Z, which may be CCl_3, $CHCl_2$, $CH(NO_2)CH_3$, or $C(CH_3)_3$. In a given series (10) with fixed X and Y substituents, successive substitution of the Z substituent by the groups from CCl_3 to $C(CH_3)_3$ is generally accompanied by a progressive decline in insecticidal potency.[1,2] The insecticidal activity of DDT and its analogues is greatly influenced by molecular shape and size, and various hypotheses have been proposed to account for the influence of molecular geometry. Thus it has been proposed that for activity, a DDT analogue (10) requires a Z group of sufficient steric size, e.g. trichloromethyl, to inhibit the free rotation of the two planar phenyl rings so that they will be constrained to positions of minimum steric crowding, termed a trihedral configuration. The idea was supported by examination of DDT analogues containing differently sized Z groups; thus $Z = t$-butyl afforded active compounds such as the non-chlorinated p,p'-dimethoxydiphenyl derivative (10; $X = OCH_3$, $Y = H$).

Another suggestion emphasized[2] the importance of free rotation of the phenyl rings in DDT analogues. If such rotation was impossible, the compound would be inactive, as was the case with the o,o' isomer of DDT. The third hypothesis argued that the active DDT-type molecules functioned essentially as physical toxicants. They had the optimum molecular size and shape to penetrate the interstices of the cylindrical lipoprotein molecules, so orientated to permit ions and small molecules to pass through but to exclude larger DDT-type molecules. When the penetrating molecule possessed regions with strong attractive forces, e.g. chlorine atoms, interaction with the membrane might affect its permeability. However, there were many exceptions to these hypotheses and although such a large amount of DDT has been used since World War II, the precise mechanism of its toxic action remains uncertain. The general symptoms of DDT poisoning in insects and vertebrates are violent tremors, loss of movement followed by convulsions, and death, clearly indicating that DDT acts on the nervous system and produces damage in nervous tissue at much lower concentrations than induce toxic effects in other tissues and enzyme systems.[5]

Structure–activity studies of DDT analogues (10) emphasize the importance of molecular topography. Many theories have been proposed to explain the influence of molecular geometry.[8]

DDT appears to alter the permeability of the axonial membrane by delaying the closing of some of the sodium channels. It therefore acts as a nerve poison by upsetting the sodium balance in nerve membranes and so the mode of action of DDT resembles that of the pyrethroids (Chapter 4, page 63), although the kinetics of the closing of the channels are different.[5,7] Holan (1969) explained the insecticidal activity of a series of diaryl halocyclopropane DDT analogues on the basis of their capacity to bind to the lipoprotein interface of the axonal membrane.[1,7]

All the active molecules were regarded as molecular wedges, the base of which was represented by the two phenyl rings, while the trichloromethyl groups functioned as the apex of the wedge. Holan postulated that the base of the wedge interacts with a protein region of the axial membrane and the apex fits into a pore in the lipid part of the membranes. The two-point attachment of the wedge locks the DDT onto the axonal membrane so that the apex opens a gate for the passage of sodium ions, which alters the permeability of the axon and disrupts normal axonal nerve conduction.

The complex of the insecticide with the membrane probably dissociates more readily at higher temperatures, possibly explaining why the insecticidal activity of DDT displays a negative temperature coefficient. Pyrethroids are also more insecticidal at lower temperatures and probably have a similar mode of action. DDT causes repetitive discharge in insect nerves as well as complex changes in the permeability of the nerve membrane to sodium and potassium ions (see Chapter 3, page 44). The latter effect appears to be the basis of the toxicity of DDT and related organochlorine compounds; the specificity to insects arises from the high permeability of the insect cuticle to the passage of the toxicant while the mammalian skin constitutes an effective barrier.

By 1950, a number of examples of DDT-resistant strains of insects (e.g. houseflies and cabbage root fly) had been reported and the problems raised by them had become serious.[9] Resistance to a particular insecticide will be more widespread if the toxicant is persistent and the insect has a short life cycle, which explains why, given the very large-scale use of DDT and other organochlorine insecticides, there was a rapid emergence of resistant insects (see Chapter 6, page 151).

Several factors have been proposed[9] to account for the phenomenon of insect resistance (Chapter 6, page 153), such as morphological characteristics (e.g. the development of thick impermeable insect cuticles), slow rate of uptake of DDT,[4] behaviouristic tendencies, and increased detoxification of DDT and other organochlorine insecticides. The biochemical defence mechanisms are almost certainly the most significant.[5,10]

The metabolism of DDT occurs by a number of different pathways, but the most important appears to be the dehydrochlorination of DDT (7) to give the dichloroethylene DDE (11)[10,11] by the non-microsomal enzyme DDT-dehydrochlorinase (Scheme 5.1). This enzyme has been isolated in a pure state from DDT-resistant houseflies. In this connection, it is interesting that dehydrochlorination by boiling alcoholic sodium hydroxide is one of the few chemical reactions of DDT.

DDE (11) is a highly persistent metabolite which is therefore a major environmental pollutant (Chapter 15, page 355) but has only slight insecticidal activity; in birds and mammals it is further slowly metabolized to the carboxylic acid DDA (12) (Scheme 5.1) and this is sufficiently water soluble to be excreted as amino acid conjugates.[11]

The proposal that metabolic conversion to DDE (11) is the major detoxification mechanism in DDT-resistant insects is supported by experiments in which DDT-resistant houseflies (*Musca domestica*) are shown to become susceptible to

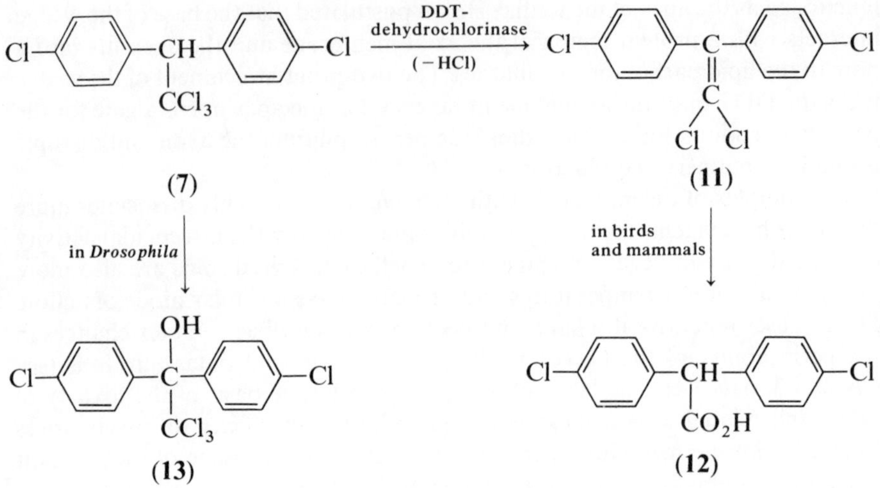

Scheme 5.1

a mixture of DDT and certain non-insecticidal structural analogues, such as 1,1-di(p-chlorophenyl)ethanol or 1,1-di-(p-chlorophenyl)ethane. Such compounds function as DDT synergists (cf. pyrethrum synergists, Chapter 4, page 71) because they inhibit the *in vivo* activity of the dehydrochlorinase.[2]

Sometimes compounds not structurally related to DDT may also show synergistic activity, e.g. N,N-di-n-butyl-p-chlorobenzenesulphonamide. Another enzymic detoxification pathway appears to operate in DDT-tolerant strains of fruit flies (*Drosophila melanogaster*) and probably in the cockroach. This involves an oxidase-catalysed conversion of the central α-hydrogen atom of DDT to the hydroxyl group, giving dicofol (13), a compound used commercially as an acaricide, but without general insecticidal activity; chlorfenethol with the —C(CH₃)OH— bridge is a closely related acaricide.

Mites and ticks (acari) are sufficiently different from insects in their biochemistry for certain compounds to be highly toxic to them (acaricidal) while they are relatively innocuous to true insects (Plate 6). Mites belong to the same class as spiders and differ from insects in having eight legs rather than six and unsegmented bodies. The low insecticidal activity of many acaricides is also an advantage because most of the natural predators of mites are insects.

Indeed, it has been discovered that several bridged diphenyl derivatives, some of which are closely related to DDT, are valuable acaricides against phytophagous mites and ticks, although they are almost devoid of toxicity to insects.[4] One example, already mentioned, is dicofol (13), which was introduced as an acaricide in 1952. Another related compound is chlorobenzilate (14), prepared from p-chlorobenzaldehyde via the benzoin condensation (Scheme 5.2). The acaricidal activity of chlorobenzilate (14) may be related to its ability to bind to muscarinic acetylcholine.[5]

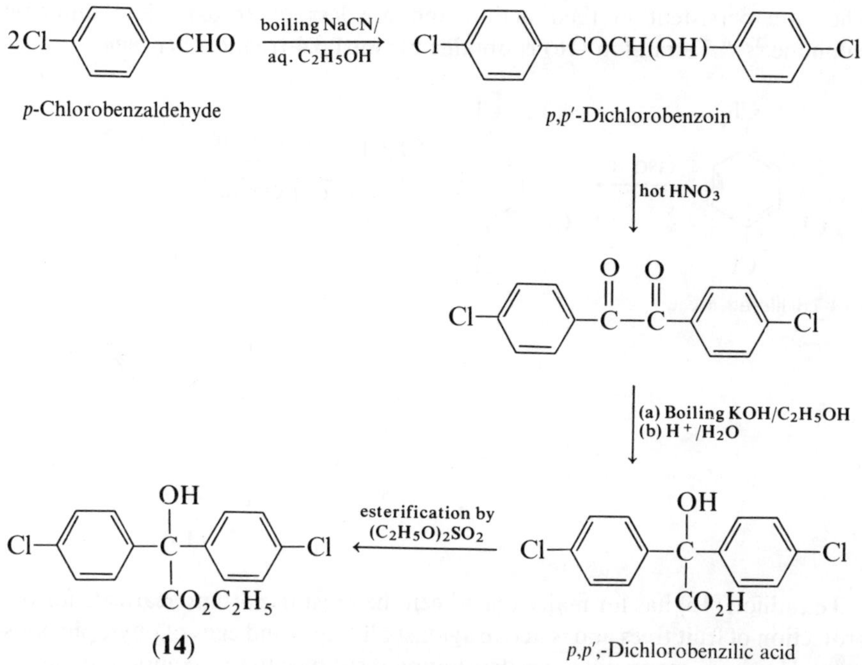

$$2\,Cl-\!\!\!\bigcirc\!\!\!-CHO \xrightarrow[\text{aq. C}_2\text{H}_5\text{OH}]{\text{boiling NaCN/}} Cl-\!\!\!\bigcirc\!\!\!-COCH(OH)-\!\!\!\bigcirc\!\!\!-Cl$$

p-Chlorobenzaldehyde   *p,p′*-Dichlorobenzoin

hot HNO$_3$ ↓

$$Cl-\!\!\!\bigcirc\!\!\!-\overset{\overset{O}{\|}}{C}-\overset{\overset{O}{\|}}{C}-\!\!\!\bigcirc\!\!\!-Cl$$

(a) Boiling KOH/C$_2$H$_5$OH
(b) H$^+$/H$_2$O ↓

$$Cl-\!\!\!\bigcirc\!\!\!-\underset{CO_2C_2H_5}{\overset{OH}{\underset{|}{\overset{|}{C}}}}-\!\!\!\bigcirc\!\!\!-Cl \xleftarrow[\text{(C}_2\text{H}_5\text{O)}_2\text{SO}_2]{\text{esterification by}} Cl-\!\!\!\bigcirc\!\!\!-\underset{CO_2H}{\overset{OH}{\underset{|}{\overset{|}{C}}}}-\!\!\!\bigcirc\!\!\!-Cl$$

(14)   *p,p′*,-Dichlorobenzilic acid

Scheme 5.2

Another acaricide of this type is bromopropylate or isopropyl 4,4′-dibromo-benzilate. The earliest acaricide was azobenzene (1945) which may be used as an aerosol or smoke against red spider in greenhouses. Another early compound was benzyl benzoate, whose acaricidal properties were substantially enhanced by the introduction of two chlorine atoms in the *p* positions, giving dichlorodi-phenoxymethane (DCPM) (15), obtained from methylene chloride and sodium *p*-chlorophenate:

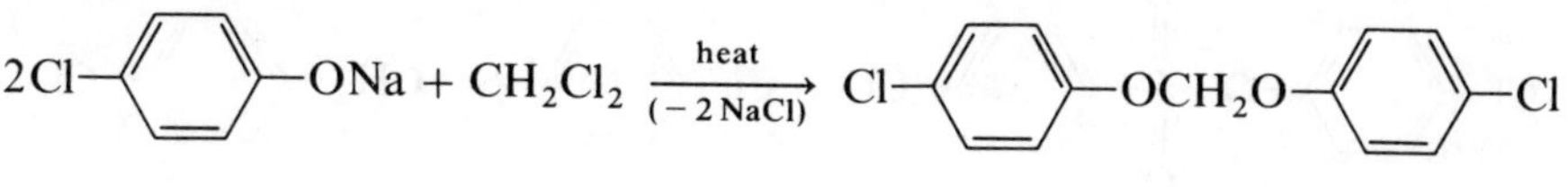

$$2\,Cl-\!\!\!\bigcirc\!\!\!-ONa + CH_2Cl_2 \xrightarrow[(-2\,NaCl)]{\text{heat}} Cl-\!\!\!\bigcirc\!\!\!-OCH_2O-\!\!\!\bigcirc\!\!\!-Cl$$

(15)

Diphenyl sulphone is also acaricidal and is specially toxic to the eggs of the fruit tree spider mite. As has been observed with azobenzene, the introduction of one chlorine atom did not reduce the ovicidal activity of diphenyl sulphone, but 4,4′-dichloroazobenzene and 4,4′-dichlorodiphenyl sulphone both had reduced toxicity. On the other hand, further chlorination of 4,4′-diphenyl sulphone restored the acaricidal properties;[2] thus both 2,4,5-trichlorodiphenyl sulphone and the 2,4,5,4′-tetrachloro derivative (16) are valuable agricultural acaricides.

They are persistent in their action and are less phytotoxic than diphenyl sulphone.[2,4,6] Tetradifon (**16**) is obtained from 1,3,4-trichlorobenzene:

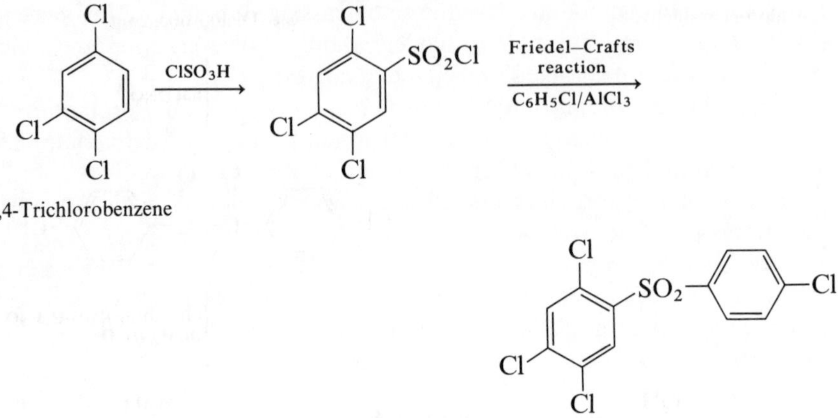

1,3,4-Trichlorobenzene

(**16**)

Tetradifon (**16**) has for many years been the most important acaricide for the protection of fruit trees and is active against all stages and eggs of phytophagous mites. However, many mites are developing resistance to this compound and so it is now generally formulated as mixtures with other acaricides such as cyhexatin.

Substituted phenyl benzenesulphonates show acaricidal and insecticidal properties; the most effective ovicide was the 4,4'-dichloro derivative known as chlorfenson (**17**). This is very useful since, with rapidly breeding mites, control against the egg stage is often more effective than killing the mites themselves. Benzyl phenyl sulphides are also acaricidal and again the best-known example is the 4,4'-dichloro derivative or chlorbenside (**18**) introduced in 1953; this showed good systemic acaricidal activity in treated plants and was quite persistent.

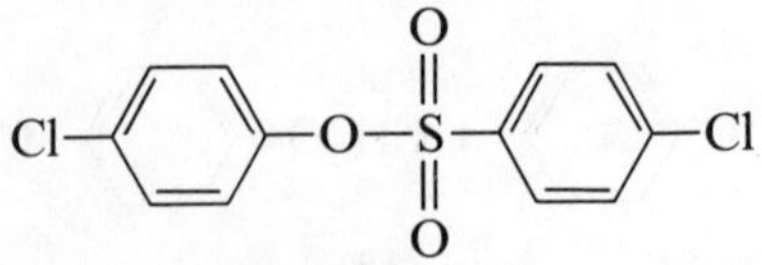

(**17**) 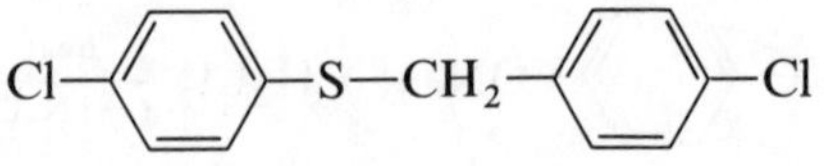   (**18**)

Many of the most potent acaricides in the group of bridged diphenyl compounds, like DDT itself, contain 4,4'-dichloro atoms. Nothing appears to be known about the biochemical mode of action of most of these acaricides,[5] although this is not really surprising since the study of the physiology of mites is much less developed than that of insects. The wide variety of different possible bridging groups between the two phenyl rings makes it difficult to formulate precise structure-activity relationships.[2] The chemistry of acaricides has been reviewed.[12] Most of the compounds have selective toxicity to mites; some of them in addition show selectivity towards specific species of mites. None of the bridged diphenyl compounds have the embarrassing stability of DDT since the bridging groups are susceptible to biodegradation; however, mites are rapidly becoming resistant to bridged diphenyl derivatives.

A number of organophosphorus insecticides show acaricidal properties, e.g. parathion, malathion, and diazinon (Chapter 6), and also carbamates (e.g. aldicarb) and dinitrophenols (e.g. DNOC and dinocap).

Some important acaricides include formamidines, like chlordimeform (**19**) and amitraz (**20**); these act as monoamine oxidase inhibitors. However, recent evidence[1,5] indicates that the primary biochemical action is probably interference with the octopamine receptor. Chlordimeform (**19**) suffers *in vivo* N-demethylation in plants, animals, and soil to a compound that binds to the receptor and hence disrupts nervous co-ordination.

Chlordimeform (**19**) (1972) acts as a broad-spectrum acaricide active against adult mites, eggs, and larvae.[6] It is also insecticidal, effective against cotton bollworm and budworm (some 500 tonnes were used against cotton insects in the United States of America in 1982), cockroaches, and many lepidopterous pests.[1]

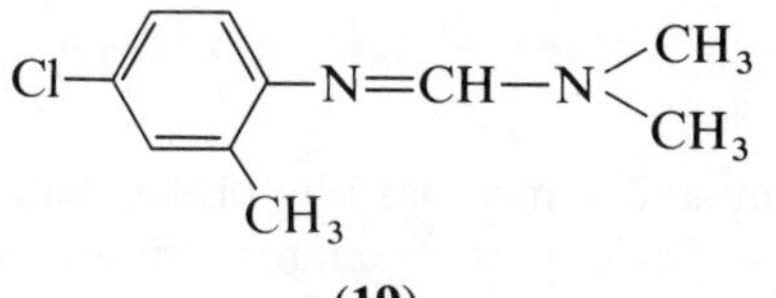

$$Cl-\bigotimes\text{(}CH_3\text{)}-N{=}CH-N\begin{smallmatrix}CH_3\\CH_3\end{smallmatrix}$$

(19)

Another useful amidine is amitraz (**20**), developed by Boots Limited (1973) and prepared from 2,4-dimethylaniline, ethyl orthoformate, and methylamine:

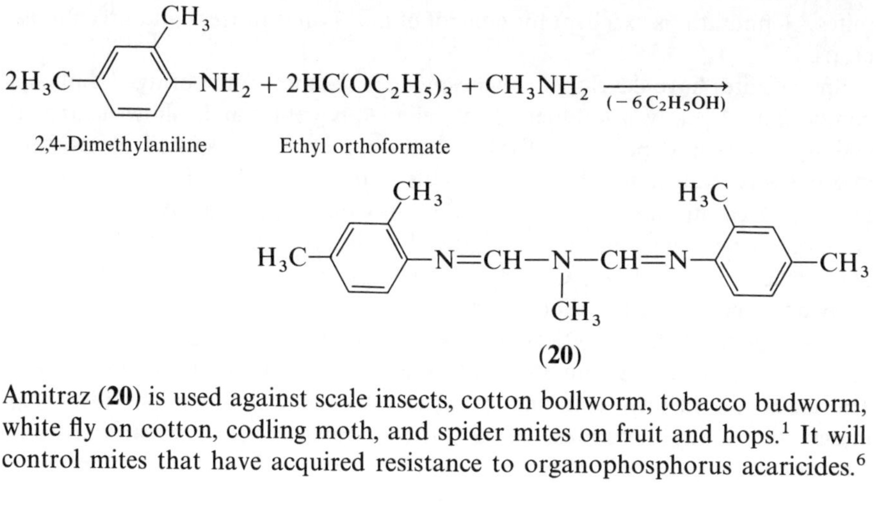

$$2H_3C-\bigotimes(CH_3){-}NH_2 + 2HC(OC_2H_5)_3 + CH_3NH_2 \xrightarrow[(-6C_2H_5OH)]{}$$

2,4-Dimethylaniline   Ethyl orthoformate

$$H_3C-\bigotimes(CH_3){-}N{=}CH-\underset{CH_3}{N}-CH{=}N-\bigotimes(H_3C){-}CH_3$$

(20)

Amitraz (**20**) is used against scale insects, cotton bollworm, tobacco budworm, white fly on cotton, codling moth, and spider mites on fruit and hops.[1] It will control mites that have acquired resistance to organophosphorus acaricides.[6]

Both formamidines **(19)** and **(20)** have low mammalian toxicities and do not accumulate in the environment.

Several other compounds possessing the formamidine moiety, $RN{=}CH{-}NR_2$, are insecticidal; there appears to be considerable potential in this group for further development which may lead to new insecticides with high activity needing very low dose rates of $\simeq 15$ g/ha, comparable to some of the synthetic pyrethroids.

The mode of action, as an octopamine agonist, is novel and results in interference with nerve transmission at octopamine-sensitive nerve functions. Such octopamine agonists may prove a valuable source of new insecticides.[4]

Some organotin compounds are useful acaricides; the best-known example is cyhexatin **(21)** while fenbutatin **(22)** is one of the leading acaricides. In spite of

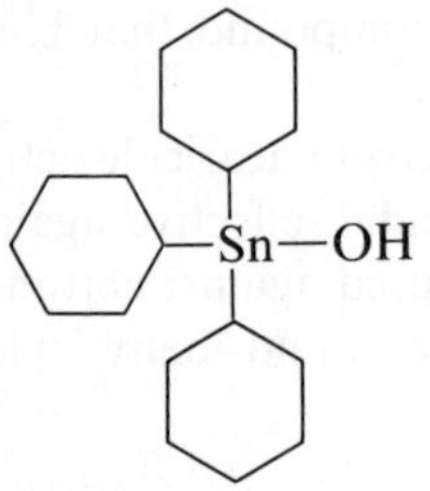

(21) 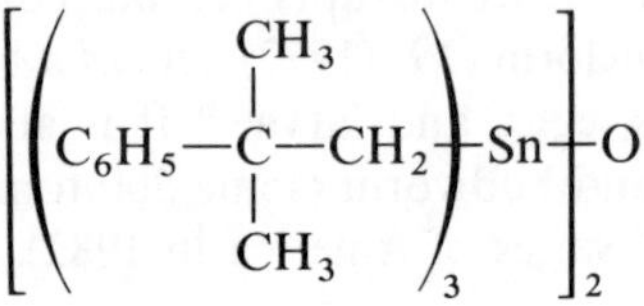   **(22)**

being rather slow acting and sometimes phytotoxic, fenbutatin has the important advantage of low toxicity to predatory mites and insects.[3,12] These organotin acaricides probably owe their activity to inhibition of oxidative phosphorylation.[5]

Cyhexatin has a fairly low mammalian toxicity: LD_{50} (oral) to rats is 540 mg/kg. Both compounds **(21)** and **(22)** have good residual activity against mites;[1] fenbutatin is excellent for control of mites on fruit trees and greenhouse crops.

Spider mites have developed a widespread resistance to the bridged diphenyl compounds. Luckily a number of miscellaneous compounds show acaricidal activity which is depicted in Table 5.1. To minimize the spread of resistant mites, it is recommended that the available acaricides are listed in the following groups, based on known or expected cross-resistance patterns: A—organotins; B—clofentezine, hexythiazox; C—bridged diphenyls; D—pyrethroids; E—flubenzimine; F—tetradifon; G—amitraz; H—propargite; I—quinomethionate; J—benzoximate; and K—dinobuton.

It is important that a variety of products are used with not more than one compound from each group applied to the same crop in a season. The chosen acaricides should have a minimum effect on beneficial predatory mites and insects and finally there must be regular monitoring to identify early signs of resistance.[13]

Table 5.1 Miscellaneous acaricides

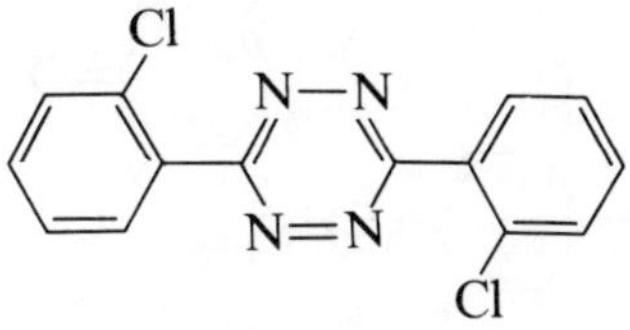

Clofentezine

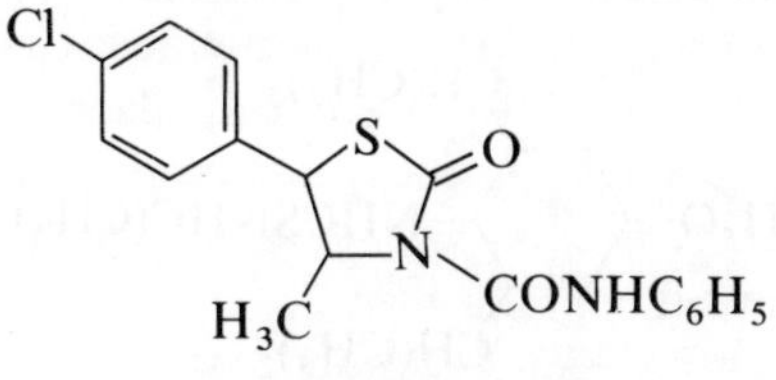

Hexythiazox

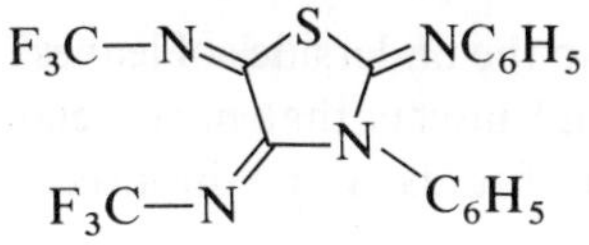

Flubenzimine

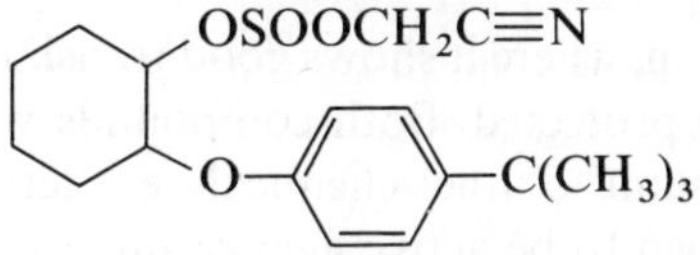

Propargite

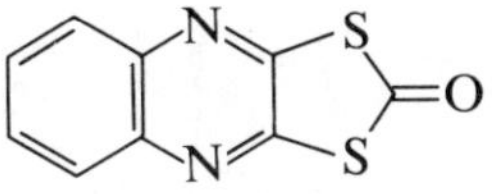

Quinomethionate

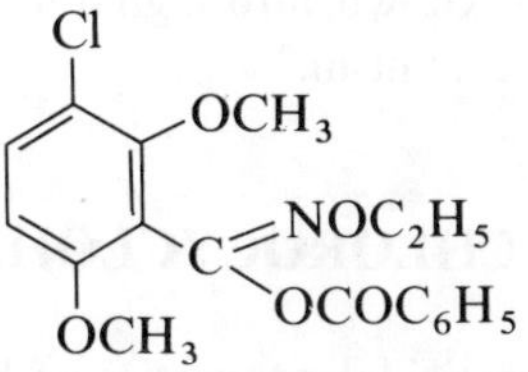

Benzoximate

The majority of the compounds are heterocyclic. Clofentezine (Fisons–Boots, 1981) is a specific acaricide, chiefly acting as an ovicide, which is used on top fruit, citrus, cotton, vegetables, and ornamentals with good residual activity. Hexythiazox (Nippon Soda, 1984) is a non-systemic acaricide used at 3–5 g a.i./ 100 l of spray which is toxic to the eggs, larvae, and nymphs of phytophagous mites on cotton, fruit, and vegetables. Flubenzimine (Bayer AG, 1979), a closely related 1,3-thiazolidine derivative, is effective against mites at 500 mg/l. It inhibits the development of larvae hatching out from eggs and retards the growth of all stages of mites; the mode of action implies that early application is essential for effective control in the field. Quinomethionate (Bayer AG, 1960) is a selective non-systemic acaricide and fungicide for use against mites and powdery mildew on fruits and vegetables. Propargite (Uniroyal Inc., 1973) controls phytophagous mites on a wide range of crops; it is moderately toxic (LD_{50} (oral) to rats, 220 mg/kg). Benzoximate (Nippon Soda) is effective at 10–13 g a.i./100 l of spray against mites.[1,6] Apart from propargite, these acaricides all have very low mammalian toxicities ($LD_{50} > 2000$ mg/kg).

Recent acaricides include CGA 106630 (**23**) and NC 129 (**24**),[14] both of which control phytophagous mites and other insect pests on cotton.[15] The former is also used on vegetables and ornamentals, whereas the latter is only applied to

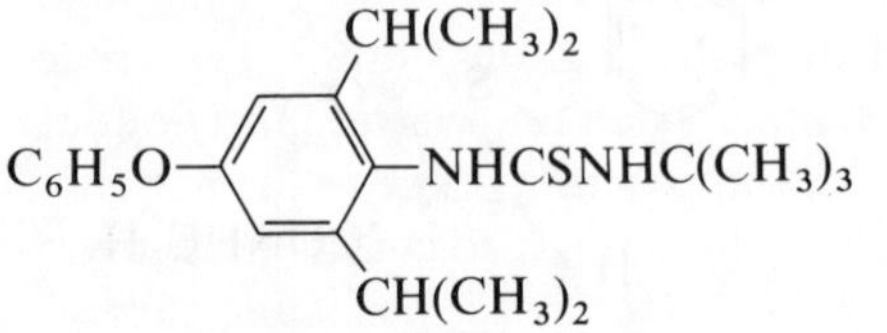

$$C_6H_5O-\underset{\underset{CH(CH_3)_2}{|}}{\overset{\overset{CH(CH_3)_2}{|}}{\bigcirc}}-NHCSNHC(CH_3)_3$$

(23)

$$(CH_3)_3C-\bigcirc-CH_2S\underset{\underset{Cl}{\overset{N-N}{\nearrow}}}{}\overset{C(CH_3)_3}{\underset{}{}}=O$$

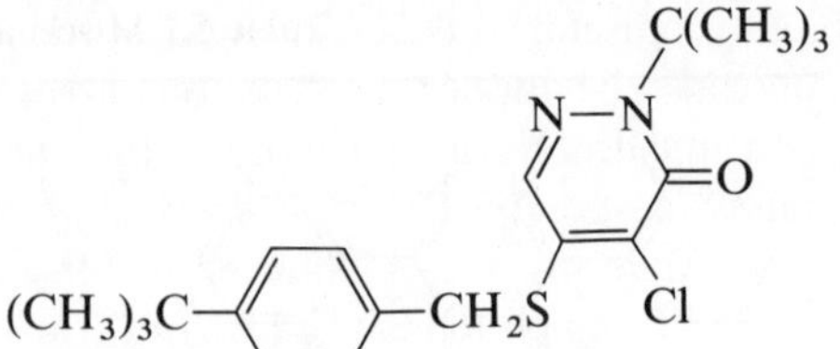

(24)

cotton, where it shows good translaminar activity so the undersides of leaves are also protected. Both compounds will kill mites and insects that have become resistant to other chemicals, e.g. tetradifon and pyrethroids. Some thioureas are known to be active against mites (see page 102).

The modes of biological action of these miscellaneous acaricides are apparently not known, although it has been suggested that several of them may have a novel mechanism.[5]

HEXACHLOROCYCLOHEXANE (HCH)

The insecticidal properties of hexachlorocyclohexane (HCH) were discovered independently in 1942 by Imperial Chemical Industries in Britain and in France. HCH is prepared by treatment of benzene with chlorine under the influence of ultraviolet light without catalysts. Hexachlorocyclohexane theoretically can exist as eight different stereoisomers of which five are actually found in the crude product. Only the γ isomer or lindane (25) (shown in Scheme 5.3) has powerful insecticidal properties.[8] The cyclohexane ring exists as the non-planar chair conformation in which the bonds attached to the six carbon atoms of the ring are either axial (a) or equatorial (e). The axial bonds are perpendicular to the general plane of the ring system while the equatorial bonds lie in the general plane of the ring. On this basis, the structure of the γ isomer is depicted as (25) containing three adjacent axial and equatorial chlorine atoms; this structure may therefore be termed the a,a,a,e,e,e isomer.

Many attempts have been made to increase the proportion (13%) of the γ isomer (25) in the mixture of isomers obtained, but without success. The crude product is sometimes used, but it can be purified by extraction with hot methanol followed by fractional recrystallization to give substantially pure ($>99\%$) γ-HCH or lindane, the latter name arising from its original isolation by Van der Linden (1912). Lindane has a similar insecticidal spectrum to DDT, but its physical properties are more suitable than those of DDT for use as a soil insecticide because of its greater volatility and water solubility.[1] Lindane, like DDT, rapidly penetrates the insect cuticle and can exert a significant fumigant action in a dry atmosphere. It is stable to heat and can be volatilized unchanged and is a useful soil dressing against the attack of soil insects. The crude mixture may also be applied as dusts to control various soil pests, as well as flea beetles and mushroom flies. As sprays, lindane is valuable against many sucking and biting pests, and as smokes for control of pests in grain stores.[6] The crude material, however, suffers from the disadvantage of an unpleasant musty odour and taste which tends to taint foodstuffs. This apparently is due to the presence of the other isomers as pure γ-HCH has no smell, but is also considerably more expensive than the crude mixture. The symptoms of insect poisoning superficially resemble those of DDT, and γ-HCH is known to be a neurotoxicant; thus a concentration of 10 μm increases the frequency of spontaneous discharges in the cockroach nerve cord and extends the synaptic cleft after discharge.[5] Lindane rapidly penetrates the cuticle of cockroaches and accumulates in the peripheral regions of the central nervous system, quickly causing tremors, loss of bodily coordination, convulsions, and prostration. Like DDT, lindane probably kills insects by bringing about a sodium–potassium imbalance in nerve membranes. However, it is possible that its biochemical site of action is somewhat different from that of DDT because it has been observed[2] that houseflies exposed to either γ-HCH or the cyclodiene insecticides exhibited a characteristic wing 'fanning' effect which was not observed with DDT. Also DDT-resistant houseflies were generally susceptible to both lindane and the cyclodienes (see Chapter 5, page 99).[2] In weakly alkaline media and in dogs, rabbits, and rats,[10] lindane (**25**) suffers dehydrochlorination to give mainly 1,2,4-trichlorobenzene. The metabolism in insects also involves dehydrochlorination; thus one of the initial products of metabolism of lindane in houseflies was the monodehydrochlorinated compound, pentachlorocyclohexene. This was the major metabolite

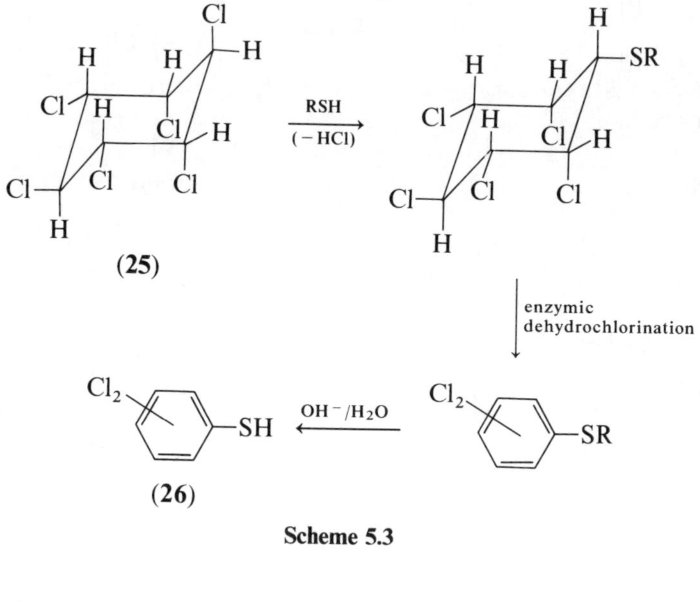

Scheme 5.3

isolated from lindane-resistant houseflies one hour after treatment with the insecticide (25), although pentachlorocyclohexene was not obtained after similar experiments with susceptible flies. Pentachlorocyclohexene is an intermediate metabolite since the amount of this compound isolated from treated flies decreased with time.

Most of the lindane is metabolized in flies to water-soluble compounds which with aqueous alkali gave isomeric dichlorothiophenols (26), probably formed via reaction with a sulphydryl compound (RSH), related to glutathione[10] (Scheme 5.3). The resistance to γ-HCH observed in houseflies seems chiefly due to their enhanced ability to metabolize the insecticide to such non-toxic materials.

THE CYCLODIENE GROUP

The insecticidal properties of chlordane (27) were reported in 1945—this was the first member of a remarkable new group of organochlorine insecticides.[2,4,7] These compounds are prepared from hexachlorocyclopentadiene (28) by the Diels–Alder reaction; for instance with cyclopentadiene the product is chlordene (29). This is only slightly toxic to insects, but subsequent addition of chlorine gave the highly active compounds chlordane (27) and heptachlor (30) (Scheme 5.4).

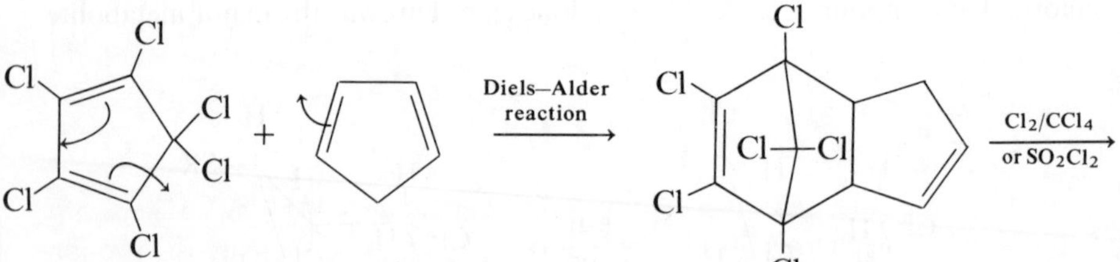

Scheme 5.4

An important feature of the synthesis is that cyclopentadiene is so reactive that it can undergo the Diels–Alder reaction with itself. On the other hand, hexachlorocyclopentadiene (**28**) cannot undergo a self-Diels–Alder and so may be used for the synthesis of a large number of other ring systems.

The stereochemistry of the cyclodienes is complex; for instance chlordene (**29**) could exist in two possible configurations known as *exo* and *endo* isomers— actually chlordene is solely the *endo* isomer:

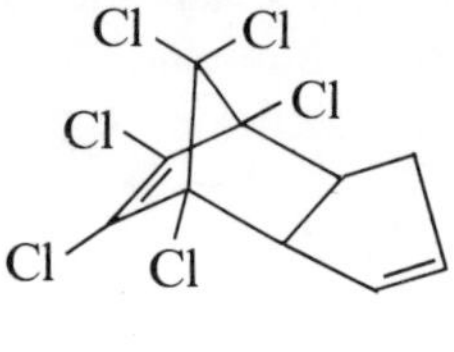

exo- 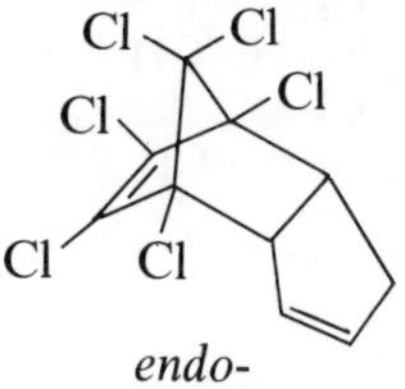   *endo-*

(**29**)

The further addition of chlorine across the double bond gives chlordane (**27**), which is a mixture of the *cis* and *trans* isomers:

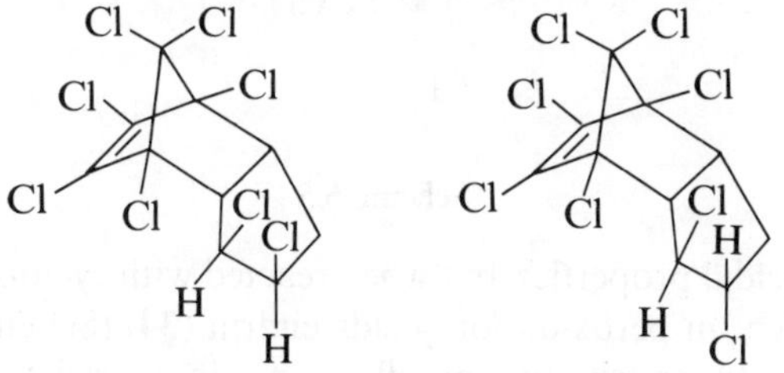

cis   *trans*

(**27**)

Commercial chlordane also contains some heptachlor (**30**), which is generally more insecticidal than chlordane, so efforts were made to increase the proportion of heptachlor in production. Experiments showed that when the chlorination of chlordene (**29**) was carried out by sulphuryl chloride in the presence of benzoyl peroxide catalyst, heptachlor (**30**) was the major compound obtained. In animal, plant, and insect tissue heptachlor is converted to the epoxide (**31**), which is slightly more insecticidal than the parent compound.

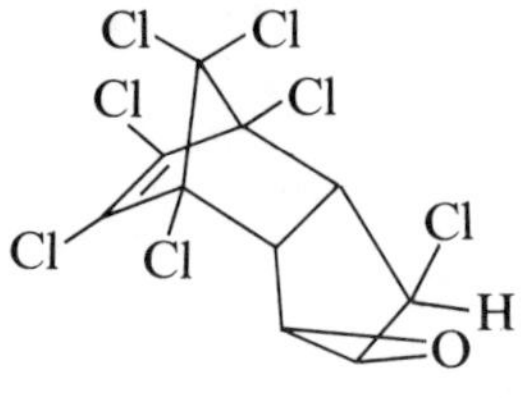

(**31**)

The most important cyclodiene insecticides are those containing four fused five-membered rings and are prepared by the addition of various dienophiles to hexachlorocyclopentadiene (28). The simplest Diels–Alder adduct of hexachlorocyclopentadiene is obtained with vinyl chloride followed by treatment with alcoholic potassium hydroxide (Scheme 5.5). The resultant compound (32) has

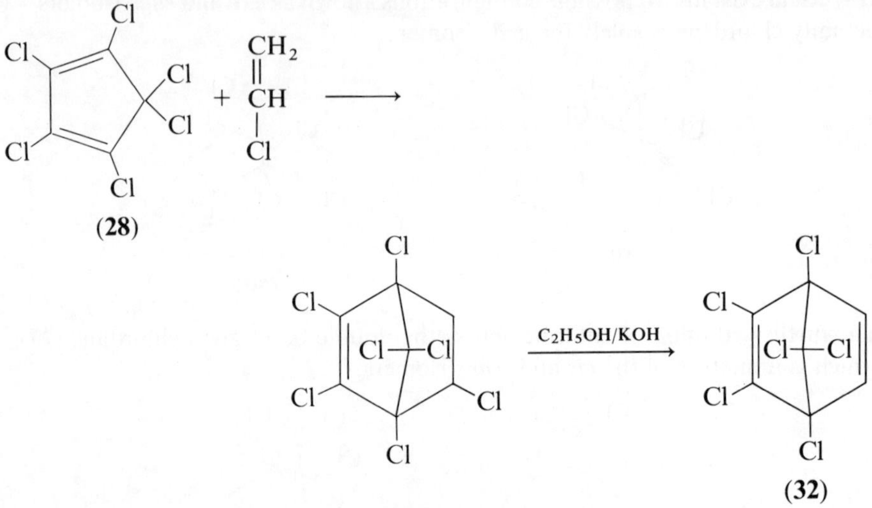

(28)

C₂H₅OH/KOH

(32)

Scheme 5.5

almost no insecticidal properties, but when reacted with cyclopentadiene it gives isodrin (33), which on peroxidation yields endrin (34) (Scheme 5.6). Isodrin is also metabolized to endrin in houseflies and in a variety of animals, e.g. rabbits.[10]

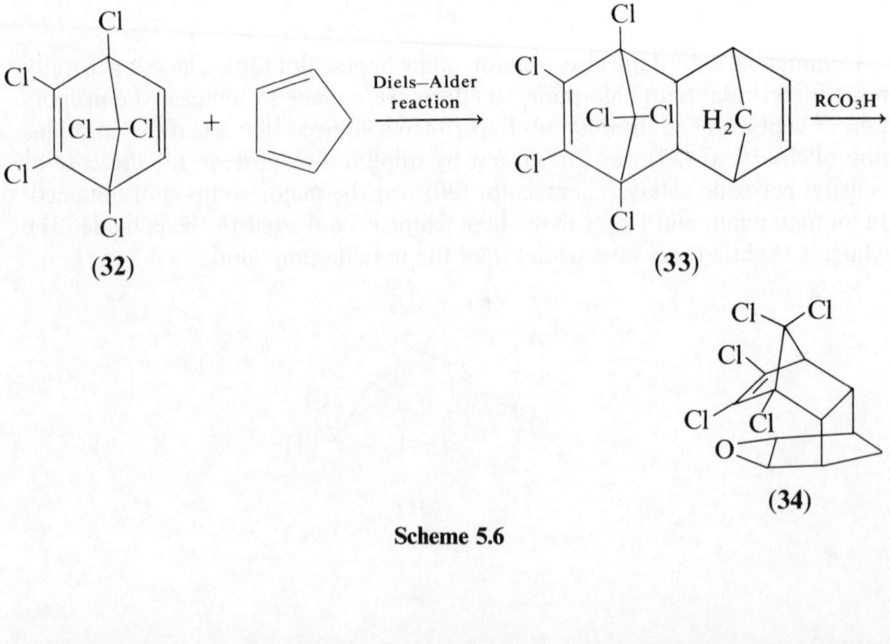

(32)

Diels–Alder reaction

(33)

RCO₃H

(34)

Scheme 5.6

Both isodrin (33) and endrin (34) are insecticidal, but only endrin has been commercially exploited. When the Diels–Alder adduct of vinyl chloride and cyclopentadiene is dehydrochlorinated and the resultant product subjected to a second Diels–Alder reaction with hexachlorocyclopentadiene, the product is not isodrin but the isomer known as aldrin, which on epoxidation affords dieldrin (35) (Scheme 5.7).

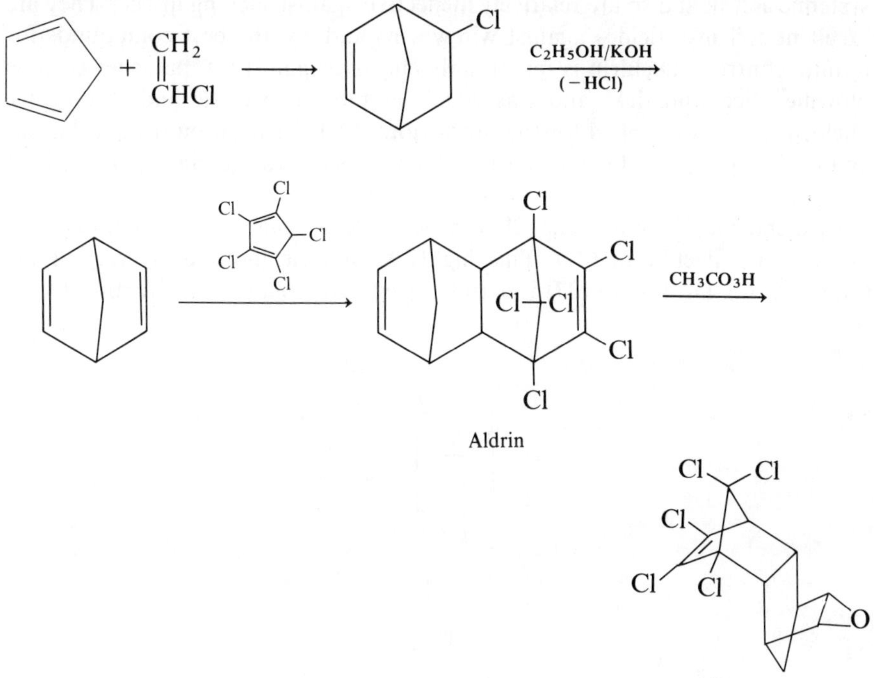

Aldrin

(35)

Scheme 5.7

Considering the isomers isodrin and aldrin, only the former, by sunlight or treatment with hydrogen bromide, is isomerized to the high-melting caged structure (36). The formation of this derivative (36) demands that the two double bonds must lie in close proximity; therefore isodrin must have the *endo–endo* configuration while aldrin is the *endo–exo* isomer[2] (Scheme 5.8).

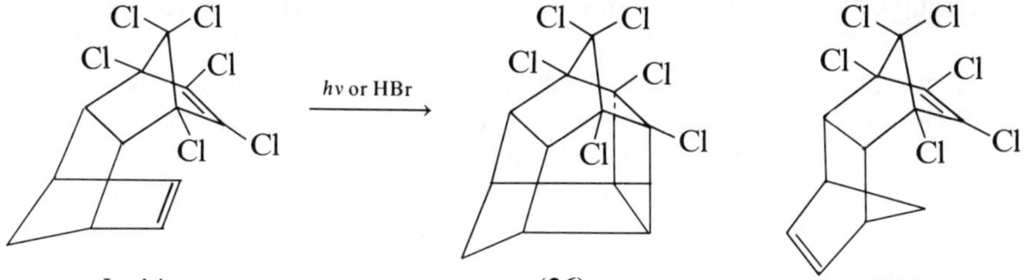

Isodrin   (36)   Aldrin

Scheme 5.8

Aldrin and dieldrin are the best-known members of the cyclodiene group of insecticides and are named after Diels and Alder, the discoverers of the diene synthesis. Both are chemically very stable and do not react even with caustic soda solution. Aldrin, dieldrin, and endrin (34) were the most active general contact insecticides until the introduction of the synthetic pyrethroids (Chapter 4). Like DDT they are highly lipophilic and persistent, but have little systemic action and so are relatively ineffective against sucking insects. They are excellent soil insecticides against wireworms and are the best compounds for termite control.[2] Dieldrin is remarkably effective against ectoparasites such as blowflies, lice, and ticks and was widely employed in cattle and sheep dips. Dieldrin was also applied for the protection of fabrics from moths and beetles and against carrot and cabbage root flies and as a seed dressing against wheat bulb fly.[7]

Hexachlorocyclopentadiene (28) dimerizes on heating with aluminium chloride, to form dechloran (37). This highly symmetric molecule is very stable (m.p. 485 °C). Dechloran (37) is extremely effective against the fire ant at very

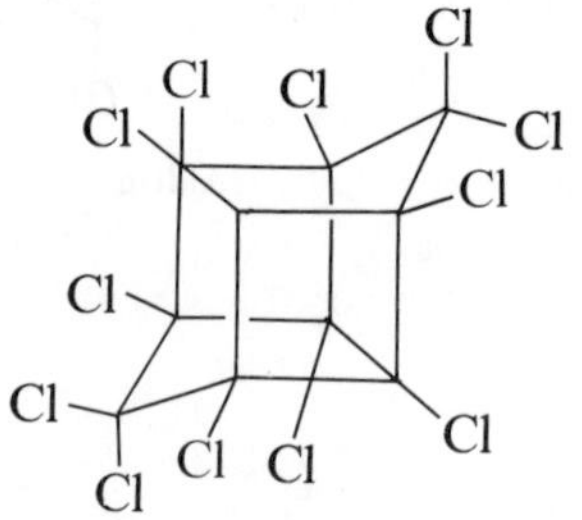

(37)

low doses (10 g/ha) (LD$_{50}$ (oral) to rats, 350 mg/kg). However, the compound showed carcinogenic and teratogenic effects and was banned by the Environmental Protection Agency in the United States of America (1978).[8]

Other related cyclodienes include the cyclic sulphite esters, e.g. endosulfan (38), obtained as shown in Scheme 5.9. Endosulfan has a similar spectrum of insecticidal activity to aldrin, except that it is also acaricidal.[6] Endosulfan (38), unlike most organochlorines, is degraded in the environment and does not accumulate. It is the only organochlorine insecticide permitted for use in the United States of America.[1] In common with the majority of organochlorine insecticides, most cyclodienes are persistent lipophilic molecules which are not readily biodegraded and tend to accumulate in the environment (see Chapter 15, page 354). The cyclodiene insecticides possess significantly higher acute mammalian toxicities as compared with DDT or lindane; thus the LD$_{50}$ (oral) values to rats are: aldrin, dieldrin, and heptachlor 40 mg/kg; endosulfan 35 mg/kg; and endrin 4 mg/kg.

Soloway[16] concluded that active cyclodienes needed to contain two correctly

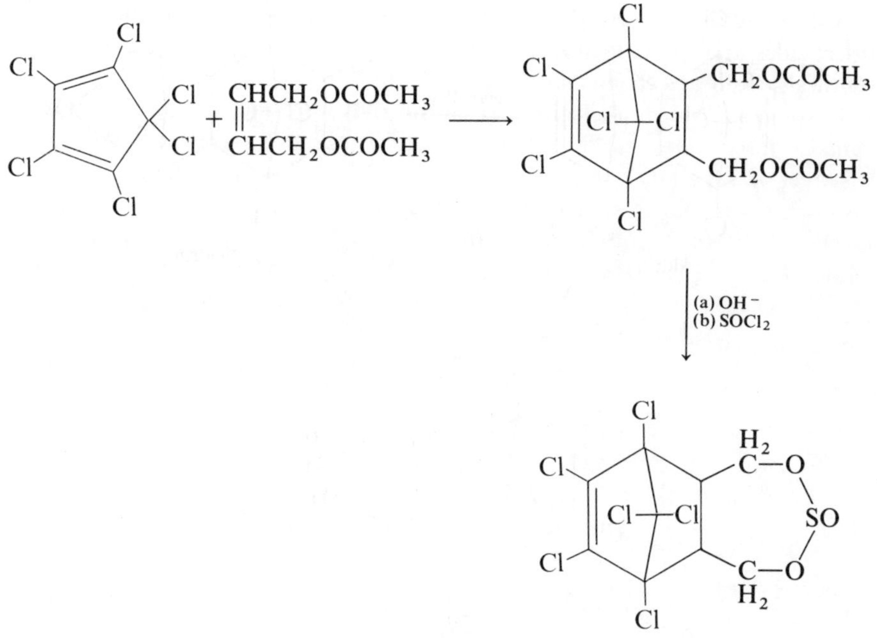

Scheme 5.9

separated electronegative centres which possibly become attached to the biological site of action. The overall molecular topography is also vital; the active compounds have similar molecular shapes and in the epoxides (e.g. dieldrin) the orientation of the oxygen atom appears decisive. Comparatively little is known about the *in vivo* metabolic degradation of the cyclodienes; aldrin is metabolized in insects, soil, and plants to dieldrin via aldrin *trans*-diol (**39**)[10,11] (Scheme 5.10).

It may therefore be postulated that *trans*-diol (**39**) is one of the active forms of dieldrin; thus when dieldrin was applied to a cockroach ganglion there was an inhibition period before an electrophysiological response was observed. However, treatment with *trans*-diol (**39**) gave a more rapid response.[2] HCH and the cyclodienes clearly act on the nervous system; treated insects exhibit violent trembling, convulsions, and paralysis. There are many cases of cross-resistance between cyclodienes and γ-HCH which suggests a similar mode of action but appears to be different from that of DDT; flies resistant to DDT are generally readily controlled by application of cyclodienes or HCH.

The mode of action of HCH and the cyclodienes appears to be due to these compounds causing an excessive release of acetylcholine which results in loss of nerve co-ordination.[5]

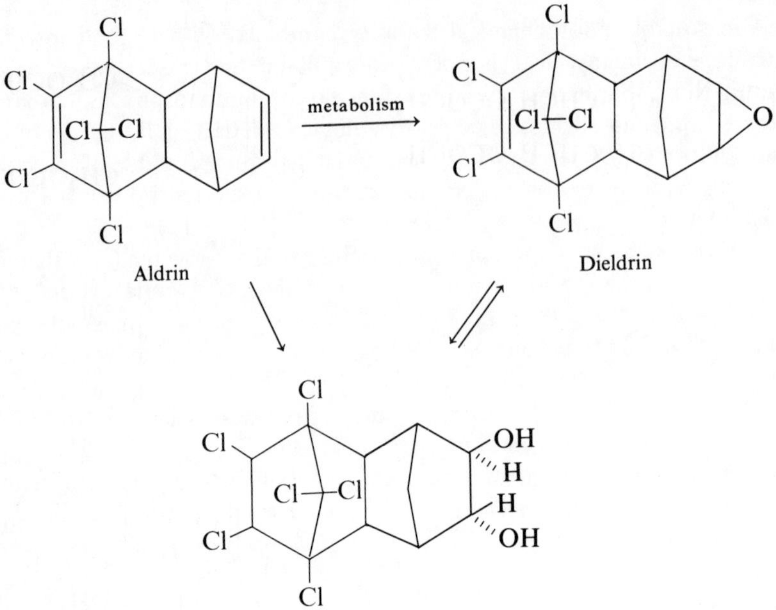
Cl
Cl
Cl
Cl
Cl
Cl
Aldrin
metabolism
Dieldrin
O
OH
H
H
OH

(39)

Scheme 5.10

MISCELLANEOUS INSECTICIDES

The development of widespread resistance to the main classes of insecticides—organochlorines, organophosphorus, carbamates, and pyrethroids—has emphasized the need to introduce new types of insecticides with different biochemical sites of action.

Many different chemical structures have shown insecticidal properties. Examples include *nitromethylene heterocycles*, which were examined by Soloway *et al.*[16a] The most potent member of this class was 2-nitromethylene-1,3-thiazinane; this showed good insecticidal activity, rapid action, low mammalian toxicity, and persistence.[17] However, it was too unstable in sunlight for use as an agricultural insecticide. To overcome photoinstability, many derivatives were synthesized and eventually the formyl derivative (**40**) was selected. This pos-

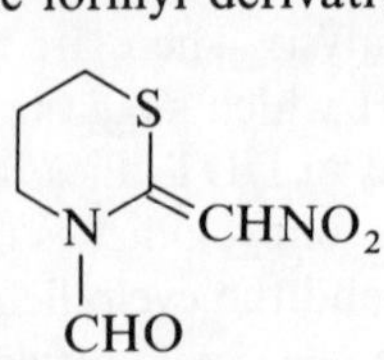
S
N
CHNO2
CHO

(**40**)

sessed reasonable photochemical stability combined with a good insecticide spectrum, e.g. against leaf hoppers, caterpillars, and insect pests on rice, including those that had acquired resistance to organophosphorus, carbamate, and pyrethroid insecticides. Electrophysiological studies of the effects of compound (40) on the American cockroach indicated a novel mode of insecticidal action—the activation of acetylcholine receptors in the central nervous system of the insect.

Benzophenone hydrazones represent another group of insecticides with a novel mode of action; the most active members were the ethoxycarbonyl derivatives (41) and (42) which can be prepared from the appropriate benzophenone[18] (Scheme 5.11). The compounds (41) and (42) were moderately toxic (LD_{50} (oral) to rats, 250 mg/kg) and had a broad spectrum of insecticidal activity; they controlled caterpillars, flies, termites, ants, stored product pests, and locusts. Benzophenone hydrazones were particularly effective against phytophagous caterpillars: spraying at rates of 250–400 g a.i./ha controlled all instars of the caterpillars. The caterpillars were killed 48–72 hours after ingestion of the toxicant; both the *E* and *Z* isomers of compounds (41) and (42) showed similar biological activity.

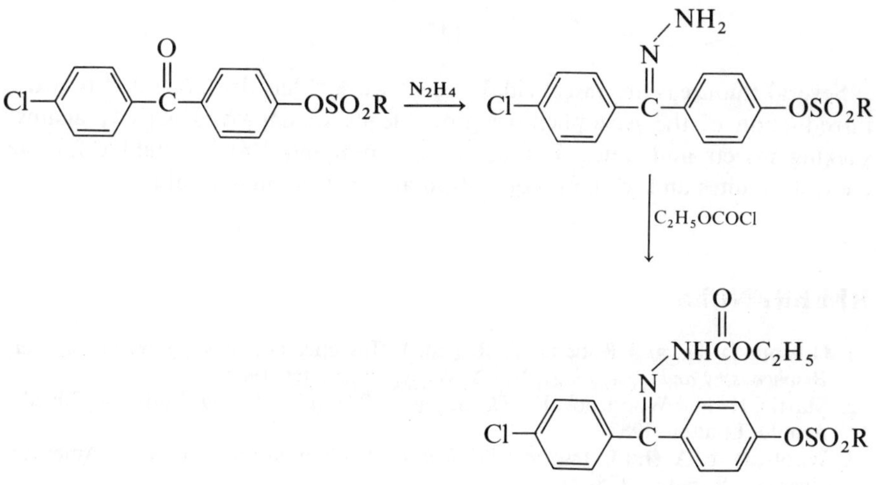

(41) R = CH_3
(42) R = CH_2Br

Scheme 5.11

Bicycloorthocarboxylates (43) are a new class of potent insecticide[19,20] (see Chapter 4, page 75). Derivatives in which R,R' were certain alkyl or aryl substituents (e.g. propyl or phenyl) showed high toxicity to houseflies and cockroaches. The activity was increased by formulation with piperonyl butoxide, which reduced the rate of oxidative detoxification.

Bicycloorthocarboxylates (**43**) showed a positive temperature–activity coefficient in poisoning houseflies, and acted on the neuromuscular function in cockroaches to inhibit GABA-mediated synaptic transmission, probably by closing off the chloride channels. The potency of *t*-butylbicycloorthobenzoates (**44**) as a GABA-gated chloride channel blocker and as a toxicant for houseflies or mice was increased twenty-fivefold by the introduction of cyano groups in the 3′,4′- or 3 and 4′-positions or an ethynyl group in the 3- or 4′-positions.[20]

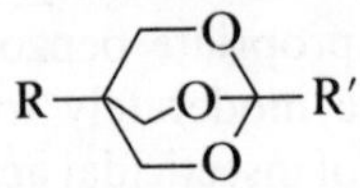

(**43**) 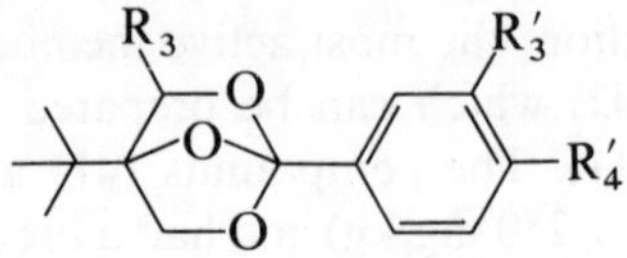   (**44**)

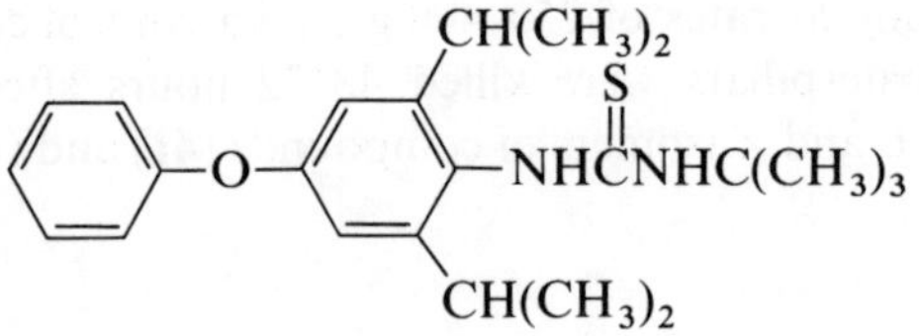

(**45**)

Several thioureas are insecticidal; recent work[20] has demonstrated that the introduction of the *para*-phenoxy group leads to improved activity against sucking insects and mites. For instance, compound (**45**) is valuable for the control of mites and whiteflies on cotton and *Plutella* in vegetables.

REFERENCES

1. Hutson, D. H., and Roberts, T. R. (Eds.), 'Insecticides', in *Progress in Pesticide Biochemistry and Technology*, Vol. 5, Wiley, Chichester, 1985.
2. Martin, H., and Woodcock, D., *The Scientific Principles of Crop Protection*, 7th edn, Arnold, London, 1983.
3. Woolston, E. A. (Ed.), *Arsenical Pesticides*, ACS Symposium Ser. No. 7, American Chemical Society, 1975.
4. Green, M. B., Hartley, G. S., and West, T. F., *Chemicals for Crop Improvement and Pest Management*, 3rd edn, Pergamon, Oxford, 1987.
5. Corbett, J. R., Wright, K., and Baillie, A. C., *The Biochemical Mode of Action of Pesticides*, 2nd edn, Academic Press, London and New York, 1984.
6. *The Pesticide Manual*, 9th edn, British Crop Protection Council, Thornton Heath, 1991.
7. Moriarty, F. (Ed.), *Organochlorine Insecticides*, Academic Press, London and New York, 1975.
8. Matolcsy, Gy, Nádasy, M., and Andriska, V., *Pesticide Chemistry*, Elsevier, Amsterdam, 1988.
9. Brown, A. W. A., and Pal, R., *Insecticide Resistance in Arthropods*, World Health Organisation, Geneva, 1971.

10. Matsumura, F., and Murti, C. R. M. (Eds.), *The Biodegradation of Pesticides*, Plenum Press, London and New York, 1982.
11. Miyamoto, J., *et al.*, *Pesticide Metabolism*, Blackwell, Oxford, 1988.
12. Knoles, C. O., Armad, S., and Shrivastava, S. P., 'Chemistry and selectivity of Acaricides', in *Insecticides* (Ed. Tahori, A. S.), Vol. 1, Gordon and Breach, New York, 1972, p. 77.
13. Lemon, R. W., *Proceedings of British Crop Protection Conference, Brighton*, **3**, 1089 (1988).
14. Strabert, H., Drabet, J., and Rindlisbacher, A., *Proceedings of British Crop Protection Conference, Brighton*, **1**, 25 (1988).
15. Hirata, K., *Proceedings of British Crop Protection Conference, Brighton*, **1**, 41 (1988).
16. Soloway, S. B., *Advances in Pest Control Research*, **6**, 85 (1965).
16a. Soloway, S. B., *et al.*, 'Nitromethylene insecticides', in *Advances in Pesticide Science* (Ed. Geissbühler, H.), Part 2, Pergamon Press, Oxford, 1979, p. 206.
17. Harris, M., Price, R. N., Robinson, J., and May, T. E., *Proceedings of British Crop Protection Conference, Brighton*, **1**, 115 (1986).
18. Giles, D. P., Copping, L. G., and Willis, R. J., *Proceedings of British Crop Protection Conference, Brighton*, **2**, 405 (1984).
19. Palmer, C. J., and Casida, J. E., *J. Agricultural Food Chemistry*, **33**(5), 976 (1985); *Chemical Abstracts*, **108**, 200061 f (1988).
20. Crombie, L. (Ed.), *Recent Advances in the Chemistry of Insect Control*, Vol. II, Royal Society of Chemistry, Cambridge, 1990, p. 212.

Chapter 6
Synthetic Insecticides II: Organophosphorus and Carbamate Compounds

The organic chemistry of phosphorus goes back to 1820 when Lassaigne first studied the reactions of alcohol with phosphoric acid.[1,2] In 1854 Clermont prepared tetraethyl pyrophosphate (TEPP) by heating the silver salt of pyrophosphoric acid with ethyl chloride, although the powerful insecticidal properties of this compound were not discovered until some 80 years later.

Serious investigations into the synthesis of toxic organophosphorus compounds as potential nerve gases began during World War II. At Cambridge Saunders and his colleagues[3] studied alkyl fluorophosphates such as tetramethylphosphorodiamidic fluoride or dimefox (1), while in Germany Schrader made the highly active nerve gases tabun (2) and sarin (3). All these compounds are powerful insecticides, but on account of their extremely high mammalian toxicities, they have never been extensively used as insecticides.[1]

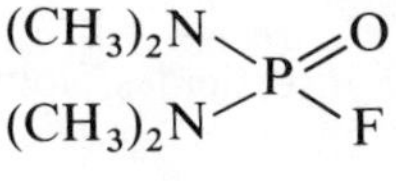

$$
\begin{array}{ccc}
\underset{(CH_3)_2N}{\overset{(CH_3)_2N}{>}}P\underset{F}{\overset{O}{<}} &
\underset{C_2H_5O}{\overset{(CH_3)_2N}{>}}P\underset{CN}{\overset{O}{<}} &
\underset{(CH_3)_2CHO}{\overset{H_3C}{>}}P\underset{F}{\overset{O}{<}} \\
(1) & (2) & (3)
\end{array}
$$

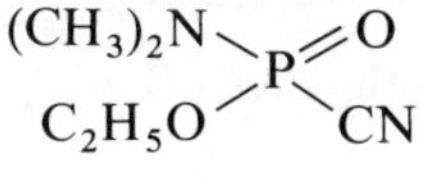

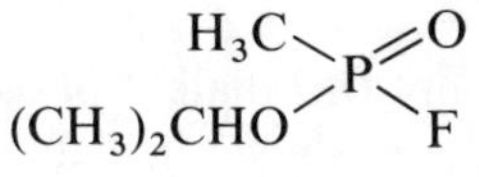

In 1941 Schrader prepared octamethylpyrophosphoramide, known as schradan (4), by controlled hydrolysis of the phosphorochloridate (5) (Scheme 6.1). Schradan can be manufactured by a one-stage process from phosphorus oxychloride and dimethylamine without isolating the intermediate chloridate. Historically schradan was the first organophosphorus compound recognized to be a potent systemic insecticide, though dimefox (1) is also systemically active. However, schradan has a high mammalian toxicity (LD_{50} (oral) to rats $\simeq 8$ mg/kg) and it has been replaced by the less toxic dimeton series.

$$\text{-}_3)_2NH + POCl_3 \longrightarrow [(CH_3)_2N]_2POCl + 2(CH_3)_2\overset{+}{N}H_2Cl^-$$

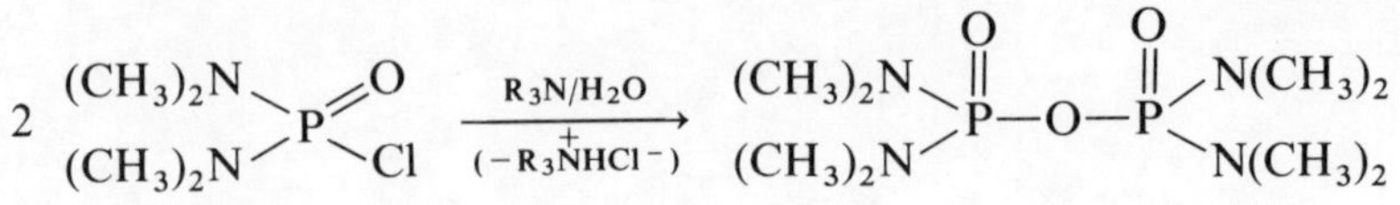

$$2\ \underset{(CH_3)_2N}{\overset{(CH_3)_2N}{\diagdown}}P\overset{\diagup O}{\underset{\diagdown Cl}{}} \xrightarrow[(-R_3\overset{+}{N}HCl^-)]{R_3N/H_2O} \underset{(CH_3)_2N}{\overset{(CH_3)_2N}{\diagdown}}P\overset{\overset{O}{\|}}{\underset{}{}}-O-P\overset{\overset{O}{\|}}{\underset{}{}}\underset{\diagdown N(CH_3)_2}{\overset{\diagup N(CH_3)_2}{}}$$

$$(5) \hspace{4cm} (4)$$

Scheme 6.1

Schrader[2,4] synthesized O,O-diethyl-O-p-nitrophenyl phosphate or paraoxon (**6**; X = O) by reaction of O,O-diethyl phosphorochloridate with sodium p-nitrophenate (Scheme 6.2). Paraoxon is a powerful contact insecticide with moderate persistency and very high mammalian toxicity (LD_{50} (oral) to rats $\simeq 2$ mg/kg). Another important organophosphorus contact insecticide was parathion-ethyl (**6**; X = S) which is prepared by a similar preparative route using thiophosphoryl chloride; this has a broad spectrum of insecticidal activity but again very high mammalian toxicity (LD_{50} (oral) to rats, 3 mg/kg).[5] The analogue, parathion-methyl, is now preferred since it has similar insecticidal activity, but is slightly less toxic (LD_{50} (oral) to rats, 15 mg/kg). In the United States of America, some 6000 tonnes of parathion-methyl was used in 1982, although it is being gradually superseded by less toxic compounds.

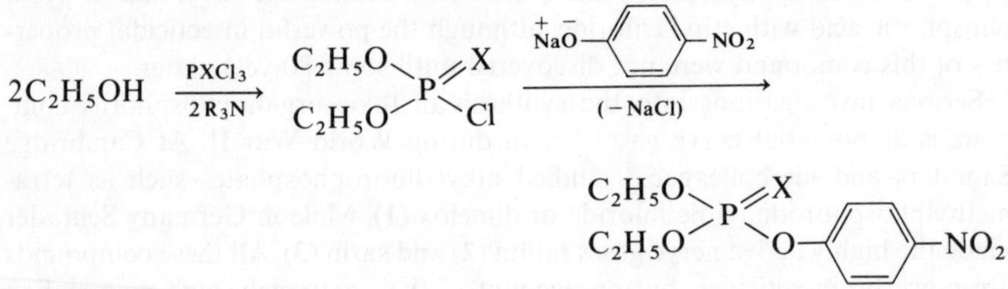

$$2C_2H_5OH \xrightarrow[2R_3N]{PXCl_3} \underset{C_2H_5O}{\overset{C_2H_5O}{\diagdown}}P\overset{\diagup X}{\underset{\diagdown Cl}{}} \xrightarrow[(-NaCl)]{NaO-\!\!\langle\bigcirc\rangle\!\!-NO_2}$$

$$\underset{C_2H_5O}{\overset{C_2H_5O}{\diagdown}}P\overset{\diagup X}{\underset{\diagdown O-\langle\bigcirc\rangle-NO_2}{}}$$

$$(6)$$

Scheme 6.2

Many O,O-dialkyl phosphorothioates (**7**) are valuable insecticides, some of which are shown in Table 6.1.

Pure phosphorothioates are, however, poor inhibitors of acetylcholinesterase; their insecticidal activity *in vivo* is the result of oxidative desulphuration which converts the thiono (P=S) into the phosphoryl group (P=O). The corresponding phosphoryl derivatives are often powerful inhibitors of acetylcholinesterase and the oxidative desulphuration has been shown to occur in plants, insects, and mammals. The P=S → P=O conversion has been demonstrated with a wide variety of organophosphorus insecticides, e.g. parathion, fenitrothion, diazinon, malathion, dimethoate, fonofos, etc. The phosphorothioates (P=S) can therefore be regarded as proinsecticides because they need *in vivo* activation before they are insecticidal.[6]

Table 6.1 Some important O,O'-dialkyl phosphorothioates (7)

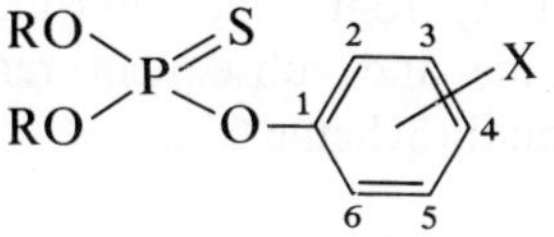

$$RO-\underset{RO}{\overset{S}{P}}-O-\underset{6\quad 5}{\overset{2\quad 3}{\bigcirc}}-X_4$$

(7)

Name	R	X	LD_{50} (oral) to rats (mg/kg)
Bromophos	CH_3	4-Br 2,5Cl_2	3500
Bromophos-ethyl	C_2H_5	4-Br 2,5Cl_2	50
Cyanophos	CH_3	4-CN	600
Fenitrothion	CH_3	3-CH_3 4-NO_2	800
Fensulfothion	C_2H_5	4-$SOCH_3$	5
Fenthion	CH_3	3-CH_3 4-SMe	200
Iodofenphos	CH_3	2,5Cl_2 4-I	2100
Parathion	C_2H_5	4-NO_2	3
Parathion-methyl	CH_3	4-NO_2	15
Heptenophos	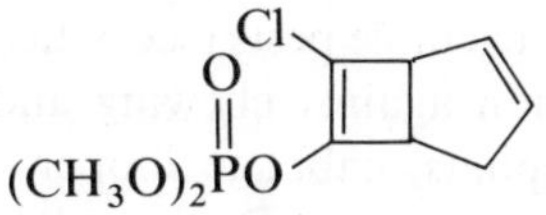 $(CH_3O)_2PO$ (structure)		100

The introduction of substituents, e.g. halogen into the phenyl ring of parathion (**6**; X = S) were found to drastically reduce the mammalian toxicity with little effect on the insecticidal potency. Good examples of this are seen in bromophos and iodofenphos (Table 6.1). Bromophos is a contact and stomach-acting insecticide effective against sucking and chewing insects, e.g. flies and mosquitoes; it has a remarkably low mammalian toxicity.[5] Cyanophos controls lipidopterous pests and sucking insects on fruit and vegetables; it is useful in locust-control programmes and as a domestic insecticide. Fenitrothion is a broad-spectrum contact insecticide, effective for the control of many chewing and sucking pests, e.g. locusts, aphids, caterpillars and leafhoppers. It is also used as a domestic insecticide and against mosquitoes. Fensulfothion is a persistent soil insecticide and nematicide which is used in many crops, e.g. bananas, potatoes, cotton, citrus, strawberries, and coffee. Fenthion is a persistent contact insecticide, valuable against fruit flies, leafhoppers, cereal bugs, and weaver birds in the tropics. It is successively metabolized to the sulphoxide and the sulphone, both of which are insecticidal. Iodofenphos, a contact insecticide and acaricide, is used in public hygiene to control flies, mosquitoes, ants, bedbugs, cockroaches, fleas, and warehouse pests. Parathion is a contact insecticide and acaricide with some fumigant action, very effective against soil insect pests, but suffers from very high mammalian toxicity. Heptenophos is a systemic insecticide, valuable against sucking pests, e.g. aphids, and against ectoparasites (fleas, lice, mites, and ticks) on animals.[5]

The general synthesis of these phosphorothioates may be illustrated by the preparation of fenitrothion (**8**) from 3-methyl-4-nitrophenol by condensation with thiophosphoryl chloride and subsequent reaction of the intermediate dichlorothioate with methanol (Scheme 6.3).

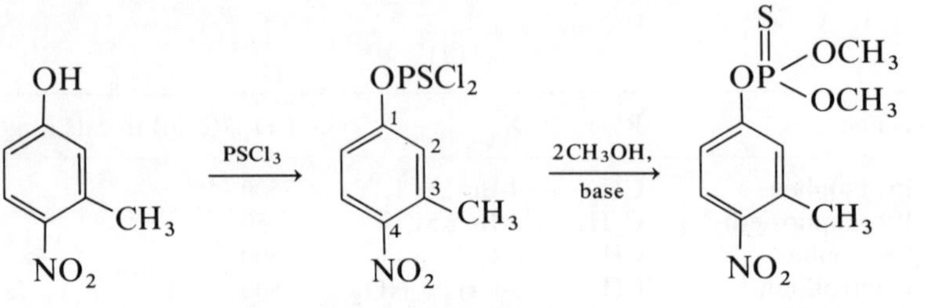

3-Methyl-4-nitrophenyl   (**8**)
phosphorodichloridothioate

Scheme 6.3

Profenofos (**9**) was introduced by Ciba-Geigy (1975) as an insecticide for control of important cotton and vegetable pests (see Scheme 6.4). Profenofos (**9**) is very active by spray application against chewing and sucking insects and mites, e.g. cotton bollworms, aphids, cabbage looper, and thrips. It has a comparatively low mammalian toxicity (LD_{50} (oral) to rats, 360 mg/kg).[5] Such *O*-alkyl *O*-aryl *S*-n-propylphosphorothiolates show enhanced insecticidal activity against lepidopterous insects as compared with *O,O*-dialkyl phosphorothionates.[6]

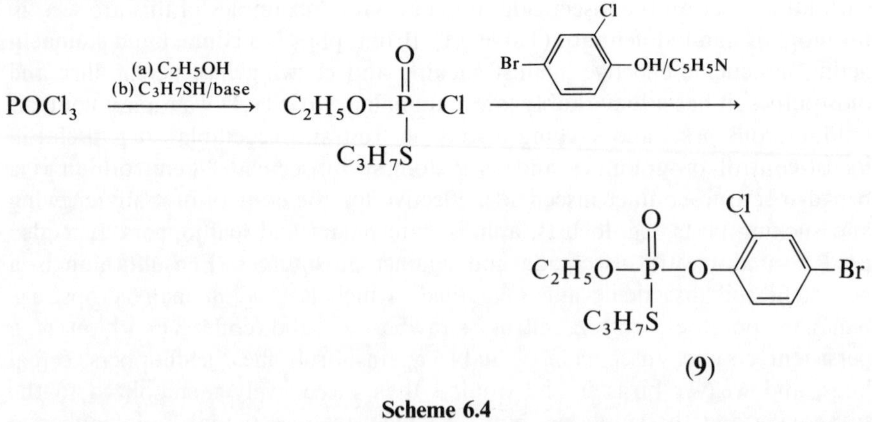

(**9**)

Scheme 6.4

Demeton (**10**; R = C_2H_5) and demeton-methyl (**10**; R = CH_3) were first synthesized by Schrader in 1950. The commercial products are mixtures; thus demeton-methyl contains 70% of the thionate (demeton-*O*-methyl, **10**; R = CH_3) and 30% of the thiolate (dimeton-*S*-methyl, **11**; R = CH_3). Demeton-methyl is manufactured by condensation of *O,O*-dimethylphosphorochloridothionate and 2-ethylthioethanol[1,2] (Scheme 6.5).

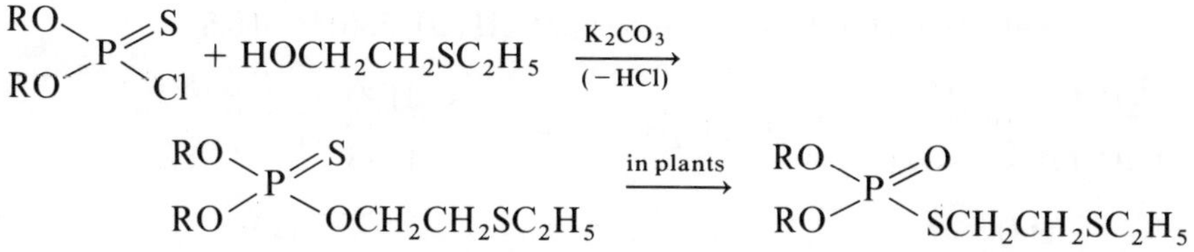

$$\text{RO}\diagdown\text{P}\diagup^{\text{S}}_{\text{Cl}} + \text{HOCH}_2\text{CH}_2\text{SC}_2\text{H}_5 \xrightarrow[(-\text{HCl})]{\text{K}_2\text{CO}_3}$$

$$\underset{\textbf{(10)}}{\text{RO}\diagdown\text{P}\diagup^{\text{S}}_{\text{OCH}_2\text{CH}_2\text{SC}_2\text{H}_5}} \xrightarrow{\text{in plants}} \underset{\textbf{(11)}}{\text{RO}\diagdown\text{P}\diagup^{\text{O}}_{\text{SCH}_2\text{CH}_2\text{SC}_2\text{H}_5}}$$

Scheme 6.5

Demeton-methyl functions both as a contact and a systemic insecticide and acaricide, with an LD_{50} value (oral) to rats $\simeq 180$ mg/kg. In plants, it rearranges to the thiolate form (**11**; R = CH_3) which is largely responsible for the systemic activity. The pure thiolate (demeton-S-methyl, **11**; R = CH_3) may be prepared by alkylation of the corresponding phosphate (Scheme 6.6).

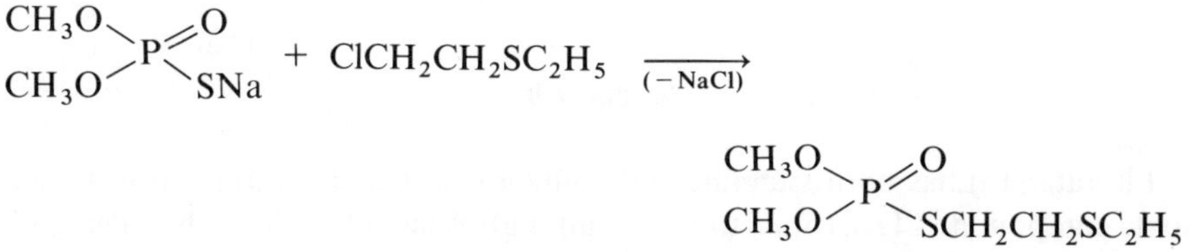

$$\underset{\text{CH}_3\text{O}}{\text{CH}_3\text{O}}\diagdown\text{P}\diagup^{\text{O}}_{\text{SNa}} + \text{ClCH}_2\text{CH}_2\text{SC}_2\text{H}_5 \xrightarrow[(-\text{NaCl})]{}$$

$$\underset{\textbf{(11)}}{\text{CH}_3\text{O}\diagdown\text{P}\diagup^{\text{O}}_{\text{SCH}_2\text{CH}_2\text{SC}_2\text{H}_5}}$$

Scheme 6.6

The oral toxicity (LD_{50}) of demeton-S-methyl to rats is 50 mg/kg and it has a fairly persistent action probably due to *in vivo* oxidation in plants to other insecticidal compounds which are more resistant to hydrolysis. Demeton-S-methyl (**11**; R = CH_3) by oxidation with hydrogen peroxide is converted into dimeton-S-methylsulphone (**12**; R = CH_3) (Scheme 6.7). Demeton-S-methyl sulphone (**12**; R = CH_3) has specific systemic activity against aphids, red spider mites, and leafhoppers on most agricultural and horticultural crops. In general, this group of organophosphorus insecticides is quite rapidly translocated in plants chiefly via the transpiration stream in the xylem.

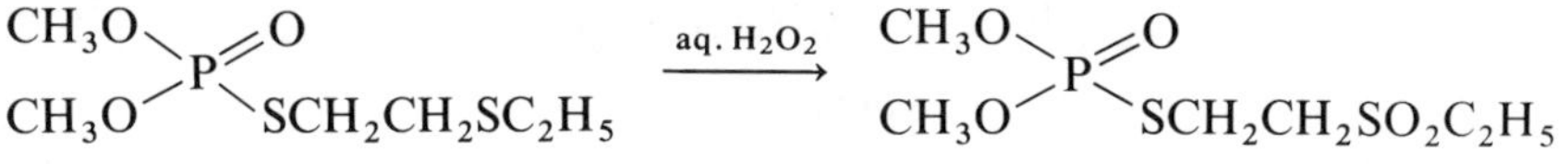

$$\underset{\textbf{(11)}}{\text{CH}_3\text{O}\diagdown\text{P}\diagup^{\text{O}}_{\text{SCH}_2\text{CH}_2\text{SC}_2\text{H}_5}} \xrightarrow{\text{aq. H}_2\text{O}_2} \underset{\textbf{(12)}}{\text{CH}_3\text{O}\diagdown\text{P}\diagup^{\text{O}}_{\text{SCH}_2\text{CH}_2\text{SO}_2\text{C}_2\text{H}_5}}$$

Scheme 6.7

Phorate (O,O-diethyl-S-2-ethylthiomethylphosphorodithioate) (**13**) is prepared[1,2] by condensation of chloromethylethyl sulphide with sodium O,O-diethylphosphorodithioate (**14**) (Scheme 6.8). An alternative preparative route

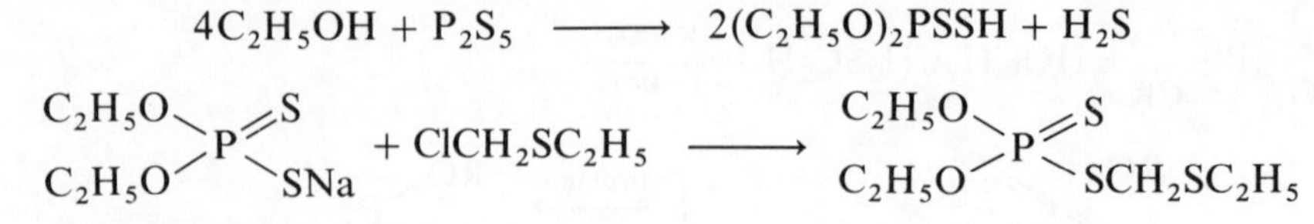

$$4C_2H_5OH + P_2S_5 \longrightarrow 2(C_2H_5O)_2PSSH + H_2S$$

$$\begin{array}{c} C_2H_5O \\ C_2H_5O \end{array} P \begin{array}{c} S \\ SNa \end{array} + ClCH_2SC_2H_5 \longrightarrow \begin{array}{c} C_2H_5O \\ C_2H_5O \end{array} P \begin{array}{c} S \\ SCH_2SC_2H_5 \end{array}$$

(14)   (13)

Scheme 6.8

involves reaction of O,O-diethylphosphorodithioic acid with formaldehyde, followed by condensation with ethyl mercaptan (Scheme 6.9).

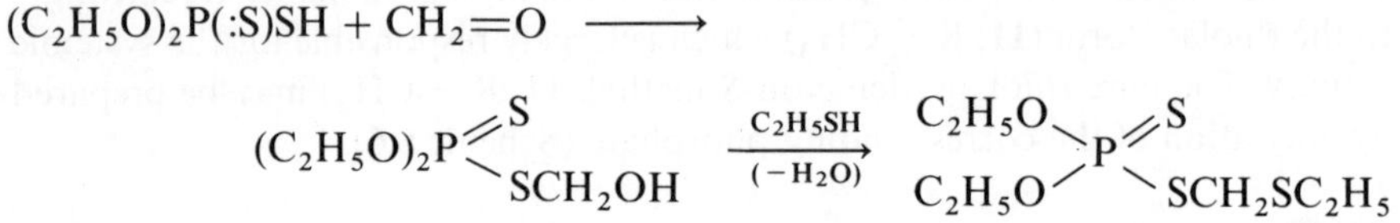

$$(C_2H_5O)_2P(:S)SH + CH_2{=}O \longrightarrow$$

$$(C_2H_5O)_2P \begin{array}{c} S \\ SCH_2OH \end{array} \xrightarrow[(-H_2O)]{C_2H_5SH} \begin{array}{c} C_2H_5O \\ C_2H_5O \end{array} P \begin{array}{c} S \\ SCH_2SC_2H_5 \end{array}$$

(13)

Scheme 6.9

Phorate (13) has both systemic and contact insecticidal action and is a very toxic compound (LD_{50} (oral) to rats, 2 mg/kg). When phorate is absorbed and translocated in plants, it is oxidatively metabolized (see page 133). Phorate is employed for the control of aphids, carrot fly, fruit fly, and wireworm in potatoes.[5] It protects plants for a fairly long period due to the greater persistency of the sulphoxide metabolite in plants. A minimum interval of six weeks must be observed between the last application and harvesting of edible crops.

Disulfoton (15; $R = C_2H_5$) and thiometon (15; $R = CH_3$) are synthesized by analogous reactions using 2-chloroethyl ethyl sulphide. Disulfoton is formulated as granules for control of aphids and other insects in many crops, while thiometon is valuable as a systemic insecticide and aphicide for vegetables, fruit and cereals.[5,6]

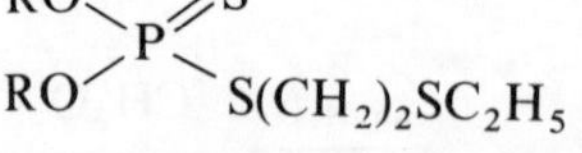

$$\begin{array}{c} RO \\ RO \end{array} P \begin{array}{c} S \\ S(CH_2)_2SC_2H_5 \end{array}$$

(15)

Thionoethers, like demeton, disulfoton, and thiometon, are generally oxidized *in vivo* rapidly to the sulphoxide and then slowly to the sulphone; consequently there is gradual accumulation of the sulphoxide which is probably the principal toxicant. Thioether oxidation increases the anticholinesterase activity, but not as much as oxidative desulphuration (see page 128). The metabolic processes are illustrated for disulfoton (15; $R = C_2H_5$) (Scheme 6.10).

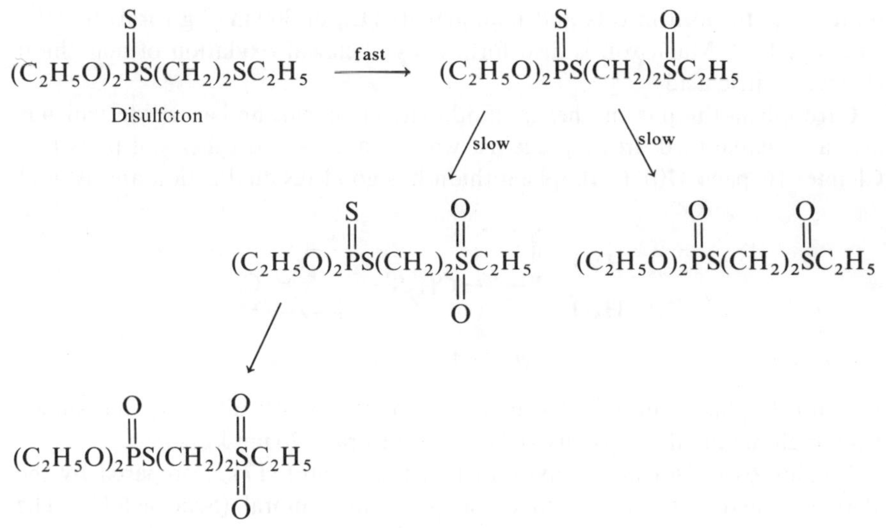

$$
\text{(C}_2\text{H}_5\text{O)}_2\overset{\overset{\text{S}}{\|}}{\text{P}}\text{S(CH}_2)_2\text{SC}_2\text{H}_5 \quad \xrightarrow{\text{fast}} \quad \text{(C}_2\text{H}_5\text{O)}_2\overset{\overset{\text{S}}{\|}}{\text{P}}\text{S(CH}_2)_2\overset{\overset{\text{O}}{\|}}{\text{S}}\text{C}_2\text{H}_5
$$

Disulfoton

slow   slow

$$
\text{(C}_2\text{H}_5\text{O)}_2\overset{\overset{\text{S}}{\|}}{\text{P}}\text{S(CH}_2)_2\underset{\underset{\text{O}}{\|}}{\overset{\overset{\text{O}}{\|}}{\text{S}}}\text{C}_2\text{H}_5 \qquad \text{(C}_2\text{H}_5\text{O)}_2\overset{\overset{\text{O}}{\|}}{\text{P}}\text{S(CH}_2)_2\overset{\overset{\text{O}}{\|}}{\text{S}}\text{C}_2\text{H}_5
$$

$$
\text{(C}_2\text{H}_5\text{O)}_2\overset{\overset{\text{O}}{\|}}{\text{P}}\text{S(CH}_2)_2\underset{\underset{\text{O}}{\|}}{\overset{\overset{\text{O}}{\|}}{\text{S}}}\text{C}_2\text{H}_5
$$

Scheme 6.10

Malathion (**16**), introduced in 1950 by the American Cyanamid Company, is synthesized by addition of *O,O*-dimethylphosphorodithioic acid to diethyl maleate (Scheme 6.11). Commercially the preparation is carried out as a one-stage process in which the dithioic acid (obtained from the reaction of methanol and phosphorus pentasulphide) is added to diethyl maleate in the presence of catalytic quantities of base and hydroquinone to prevent polymerization of the maleate. Malathion (**16**) is an important and widely used contact insecticide and acaricide for the control of aphids, red spider mites, leafhoppers, and thrips on a very wide range of vegetable and other crops.[5,6] It was important in the history of the development of organophosphorus insecticides since it was the first member with a broad spectrum of contact insecticidal activity combined with a remarkably low mammalian toxicity: LD_{50} (oral) to rats $\simeq 1300$ mg/kg. Malathion (**16**) is also valuable to control insect vectors, e.g. mosquitoes, and can be used as a substitute for organochlorine insecticides. The selectivity arises from metabolic activation in insects to the phosphoryl analogue, malaoxon, which is

$$
4\text{CH}_3\text{OH} + \text{P}_2\text{S}_5 \longrightarrow 2(\text{CH}_3\text{O})_2\text{P}(=\text{S})\text{SH} + \text{H}_2\text{S}
$$

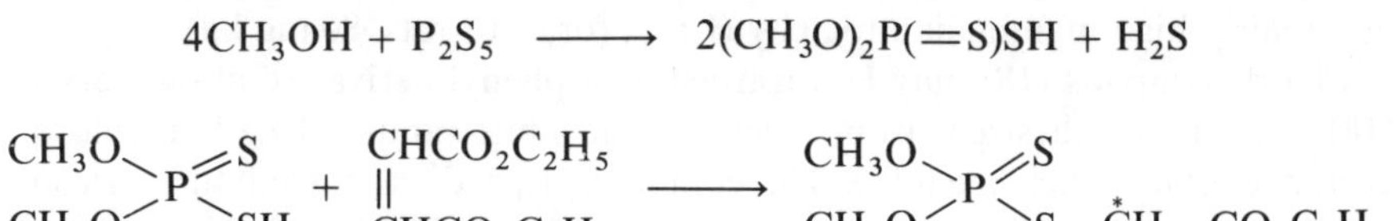

$$
\begin{matrix} \text{CH}_3\text{O} \\ \text{CH}_3\text{O} \end{matrix}\!\!\underset{}{\overset{}{\text{P}}}\!\!\begin{matrix} \text{S} \\ \text{SH} \end{matrix} \quad + \quad \begin{matrix} \text{CHCO}_2\text{C}_2\text{H}_5 \\ \| \\ \text{CHCO}_2\text{C}_2\text{H}_5 \end{matrix} \quad \longrightarrow \quad \begin{matrix} \text{CH}_3\text{O} \\ \text{CH}_3\text{O} \end{matrix}\!\!\underset{}{\overset{}{\text{P}}}\!\!\begin{matrix} \text{S} \\ \text{S}-\overset{*}{\text{C}}\text{H}-\text{CO}_2\text{C}_2\text{H}_5 \\ \quad\quad | \\ \quad\quad\text{CH}_2\text{CO}_2\text{C}_2\text{H}_5 \end{matrix}
$$

(**16**)

(C^* = asymmetric carbon atom)

Scheme 6.11

more toxic to both insects and mammals (LD_{50} of 300 mg/kg (oral) to rats) (see page 129). Malaoxon is also formed by chemical oxidation of malathion (16) with nitric acid.

Carbophenothion is another thionothioate insecticide and acaricide, which is also a valuable seed dressing against wheat bulb fly and other soil pests (see Chapter 16, page 370). Carbophenothion has good residual action and is used

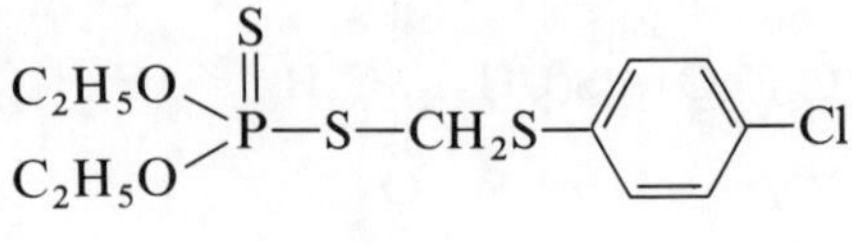

Carbophenothion

to control aphids, mites, and scale insects on dormant fruit trees,[5] but suffers from high mammalian toxicity (LD_{50} (oral) to rats, 20 mg/kg).

Dichlorvos (2,2-dichlorovinyl dimethyl phosphate) (17) is prepared by the Perkow reaction from trimethyl phosphite and chloral (Scheme 6.12). The Perkow reaction (see page 122) is very useful for the synthesis of vinyl phosphates, e.g. chlorfenvinphos, tetrachlorvinphos, and mevinphos.

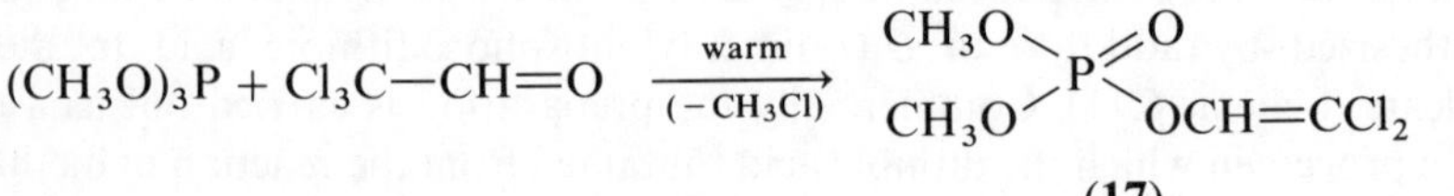

(17)

Scheme 6.12

Dichlorvos is a volatile contact and fumigant insecticide of short persistence, and is some thousand times more volatile than the majority of organophosphorus insecticides—hence the fumigant action. Dichlorvos is used[5] as a domestic insecticide against flies since it is very rapidly degraded in mammals to non-toxic products (see page 132). Dichlorvos (17) as a slow-release formulation is used as a specific anthelmintic for dogs, horses, and pigs. Dichlorvos, a very active insecticide, is probably the active ingredient of the proinsecticides butonate and naled[6] (Scheme 6.13). Dichlorvos is also useful against flies, caterpillars, and thrips, in glasshouses. Care must be taken in its application as it has a fairly high mammalian toxicity (LD_{50} (oral) to rats, 80 mg/kg).

Chlorfenvinphos (18) may be regarded as a phenyl derivative of dichlorvos (18) and is synthesized from triethyl phosphite and 2,4,α,α-tetrachloro-acetophenone[1,2] (Scheme 6.14). The commercial product is a mixture of geometric isomers, mainly the Z isomer as shown above. The E and Z isomers are distinct compounds because there is no rotation about the double bond. It is relatively stable towards hydrolysis and is useful as an insecticide against, for instance, cabbage root fly, carrot fly, citrus fly, mushroom fly, and Colorado beetle. It is specially toxic to rats (LD_{50} (oral), 15 mg/kg) but is less toxic to other mammals (LD_{50} (oral) to mice, 117–200 mg/kg). Chlorfenvinphos (18) is

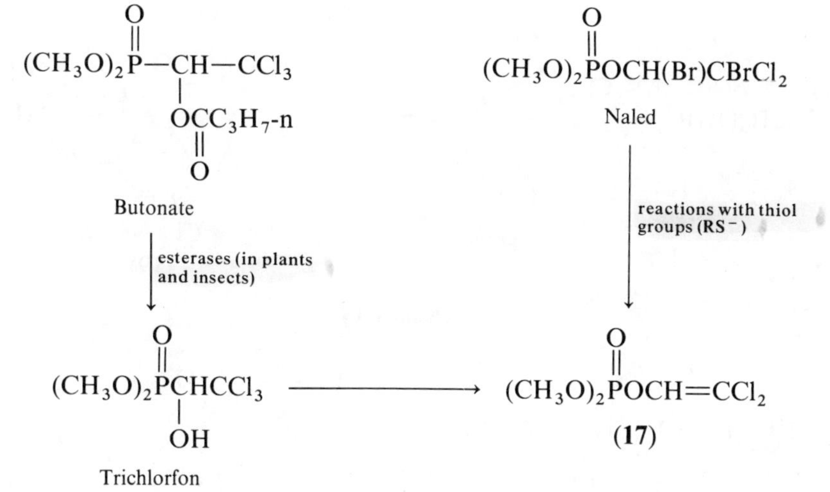

$$(CH_3O)_2P\overset{\overset{O}{\|}}{}-\underset{\underset{\overset{\|}{O}}{OCC_3H_7\text{-}n}}{CH}-CCl_3$$

Butonate

$$(CH_3O)_2\overset{\overset{O}{\|}}{P}OCH(Br)CBrCl_2$$

Naled

| |
esterases (in plants
and insects)

reactions with thiol
groups (RS^-)

$$(CH_3O)_2\overset{\overset{O}{\|}}{P}\underset{OH}{CHCCl_3} \qquad\longrightarrow\qquad (CH_3O)_2\overset{\overset{O}{\|}}{P}OCH=CCl_2$$

Trichlorfon   **(17)**

Scheme 6.13

$$(C_2H_5O)_3P \;+$$

(18)

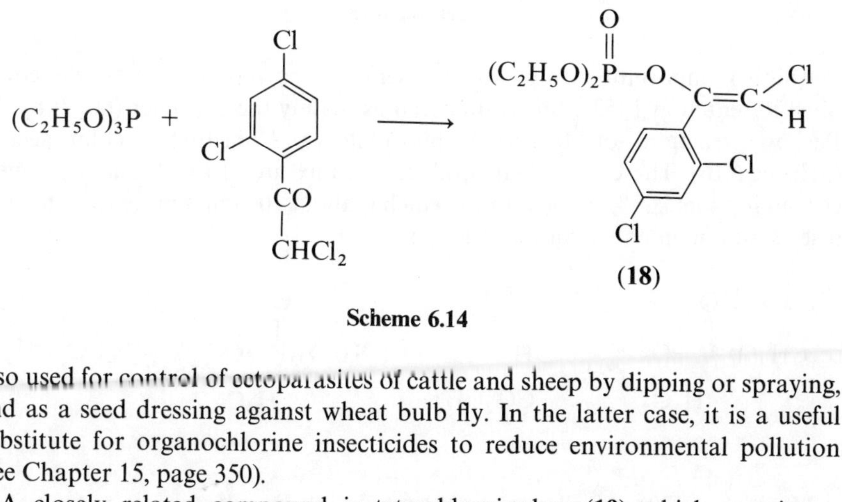

Scheme 6.14

also used for control of ectoparasites of cattle and sheep by dipping or spraying, and as a seed dressing against wheat bulb fly. In the latter case, it is a useful substitute for organochlorine insecticides to reduce environmental pollution (see Chapter 15, page 350).

A closely related compound is tetrachlorvinphos **(19)** which was introduced by Shell Research Limited (1966) and is synthesized by the Perkow reaction from trimethyl phosphite and 2,4,5,α,α-pentachloroacetophenone[2] (Scheme 6.15). The crude product is a mixture of the geometric isomers; the Z isomer is obtained (98%) by recrystallization.[2] Tetrachlorvinphos **(19)** is an extremely safe insecticide: LD_{50} (oral) to rats $\simeq 5000$ mg/kg and, like malathion, is used for the control of domestic and garden pests, for instance flies in dairies and livestock barns (Plate 7). Generally it is effective against codling and winter moths, caterpillars, flea beetles, and weevils.[7] Tetrachlorvinphos **(19)** is good against livestock ectoparasites and is rapidly metabolized and eliminated from animals in the urine within a few days.

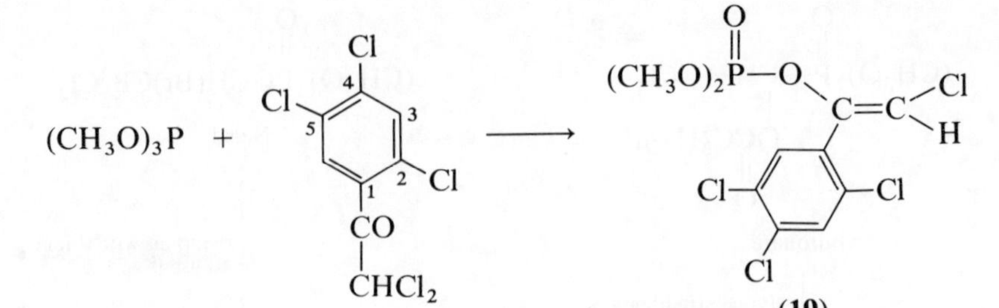

Scheme 6.15

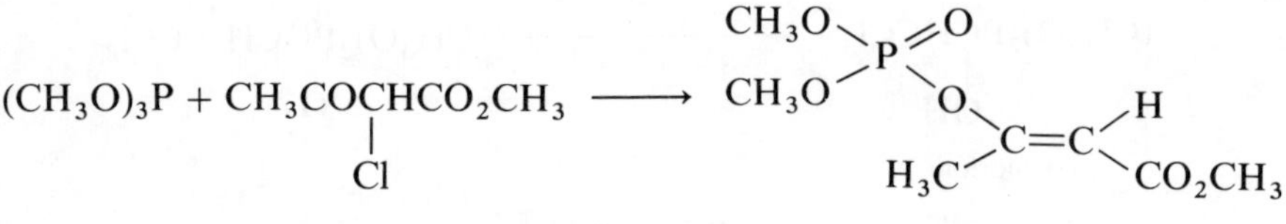

Scheme 6.16

Another important member of this series is mevinphos (**20**) developed by Shell Research in 1953;[5] this is prepared as mainly the E isomer (60%) by the Perkow reaction of trimethyl phosphite and methyl α-chloroacetate (Scheme 6.16). The commercial product is a mixture of the E and Z isomers containing some 60% of the E form, which is about 100 times more toxic to both insects and mammals than the Z form:

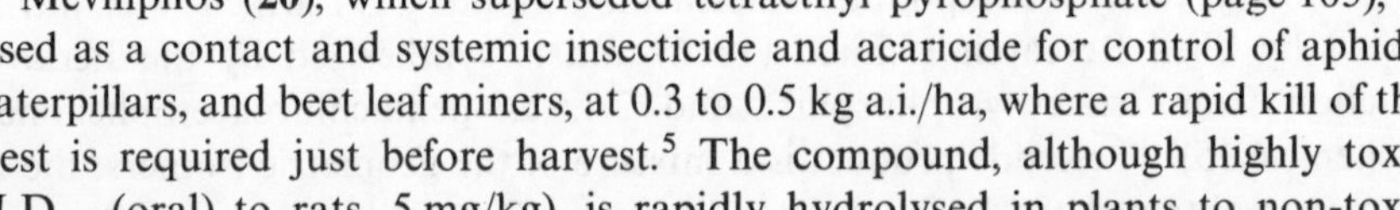

E isomer   *Z* isomer

Mevinphos (**20**), which superseded tetraethyl pyrophosphate (page 105), is used as a contact and systemic insecticide and acaricide for control of aphids, caterpillars, and beet leaf miners, at 0.3 to 0.5 kg a.i./ha, where a rapid kill of the pest is required just before harvest.[5] The compound, although highly toxic (LD$_{50}$ (oral) to rats, 5 mg/kg), is rapidly hydrolysed in plants to non-toxic materials. It is therefore a short-acting insecticide and after 24 hours the compound has been almost completely degraded (see Scheme 6.17). It is recommended that a minimum period of three days should be observed between the last application of mevinphos and the harvesting of edible crops.

Dicrotophos and monocrotophos (LD$_{50}$ (oral) to rats, 20 and 15 mg/kg respectively) are powerful contact and systemic insecticides and acaricides with

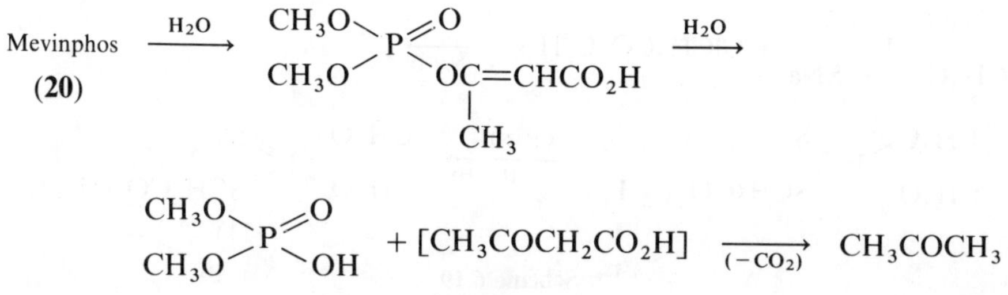

Mevinphos $\xrightarrow{\text{H}_2\text{O}}$

$$\begin{array}{c} CH_3O \\ CH_3O \end{array} P \begin{array}{c} \diagup O \\ \diagdown OC=CHCO_2H \\ \quad\quad | \\ \quad\quad CH_3 \end{array} \xrightarrow{\text{H}_2\text{O}}$$

(20)

$$\begin{array}{c} CH_3O \\ CH_3O \end{array} P \begin{array}{c} \diagup O \\ \diagdown OH \end{array} + [CH_3COCH_2CO_2H] \xrightarrow[(-CO_2)]{} CH_3COCH_3$$

Scheme 6.17

a broad spectrum of activity, used to control pests on such crops as cotton, rice, soyabeans, maize, coffee, citrus, and potatoes.[5]

Phosphorus amides, like schradan (page 135) and dicrotophos, are activated by *in vivo* oxidation of the amido moiety. In its metabolism, the N-methyl group of dicrotophos is converted initially to the N-methylol group, which subsequently suffers either conjugation or elimination of formaldehyde. In this way, dicrotophos is converted to the more insecticidal monocrotophos (Scheme 6.18); the latter can be further metabolized to the unsubstituted amide which is also active.[6]

A useful insecticide is *O,O*-dimethyl-*S*-methylcarbamoylmethyl phosphorodithioate, known as dimethoate (**21**). This can be prepared from sodium *O,O'*-dimethylphosphorodithioate and ethyl chloroacetate (Scheme 6.19).

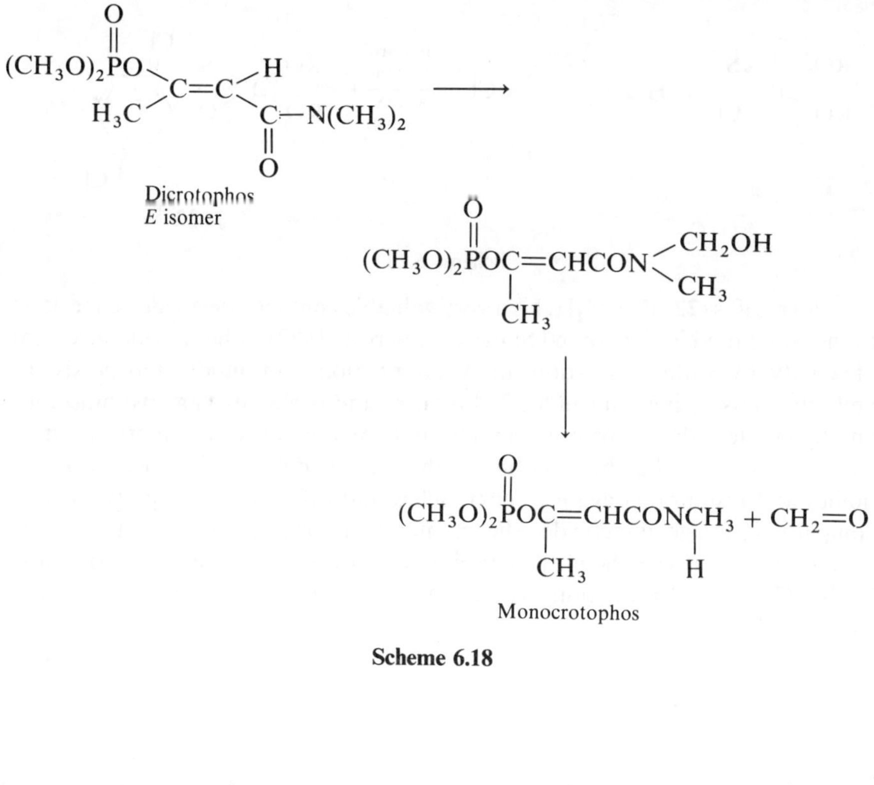

$$\begin{array}{c} \quad\quad O \\ \quad\quad \| \\ (CH_3O)_2PO \\ \quad\quad\quad\quad C=C \\ H_3C \quad\quad\quad\quad C-N(CH_3)_2 \\ \quad\quad\quad\quad\quad\quad \| \\ \quad\quad\quad\quad\quad\quad O \end{array} \longrightarrow$$

Dicrotophos
E isomer

$$\begin{array}{c} \quad\quad O \\ \quad\quad \| \\ (CH_3O)_2POC=CHCON \diagup CH_2OH \\ \quad\quad\quad | \quad\quad\quad\quad \diagdown CH_3 \\ \quad\quad\quad CH_3 \end{array}$$

$$\begin{array}{c} \quad\quad O \\ \quad\quad \| \\ (CH_3O)_2POC=CHCONCH_3 + CH_2=O \\ \quad\quad\quad | \quad\quad\quad\quad\quad | \\ \quad\quad\quad CH_3 \quad\quad\quad\quad H \end{array}$$

Monocrotophos

Scheme 6.18

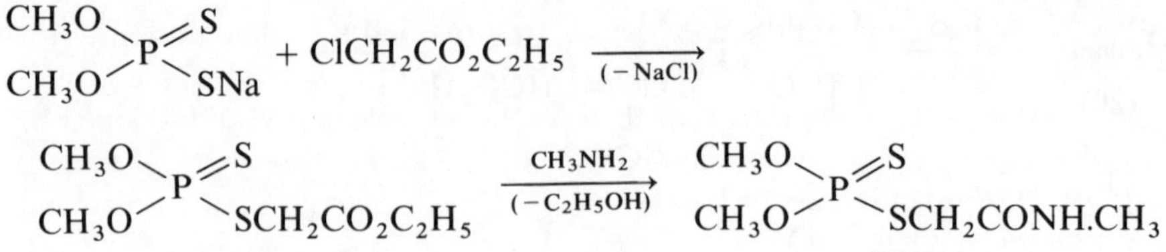

$$\underset{CH_3O}{\overset{CH_3O}{\diagdown}}P\overset{S}{\underset{SNa}{\diagup}} + ClCH_2CO_2C_2H_5 \xrightarrow[(-NaCl)]{}$$

$$\underset{CH_3O}{\overset{CH_3O}{\diagdown}}P\overset{S}{\underset{SCH_2CO_2C_2H_5}{\diagup}} \xrightarrow[(-C_2H_5OH)]{CH_3NH_2} \underset{CH_3O}{\overset{CH_3O}{\diagdown}}P\overset{S}{\underset{SCH_2CONH.CH_3}{\diagup}}$$

$$(21)$$

Scheme 6.19

Dimethoate (21) functions as a systemic and contact insecticide and acaricide, effective against red spider mites and thrips on most agricultural and horticultural crops and also plum and apple sawflies and olive and wheat bulb flies.[5] Dimethoate (21) has a moderate mammalian toxicity (LD_{50} (oral) to rats $\simeq 230$ mg/kg) and, unlike the majority of organophosphorus insecticides, dimethoate is not absorbed by the lipid phase and hence has good residual properties. In plants and animals dimethoate (21) is metabolized to the rather more toxic phosphoryl analogue known as O-methoate which has been used for the control of aphids on hops.

Many organophosphorus insecticides contain heterocyclic moieties with nitrogen heterocycles well represented.[3] The only important pyridine derivatives are chlorpyrifos (22; $R = C_2H_5$) and chlorpyrifos-methyl (22; $R = CH_3$) (see Scheme 6.20). In the condensation, the temperature must be carefully controlled to avoid isomerization to the corresponding S-methyl esters.

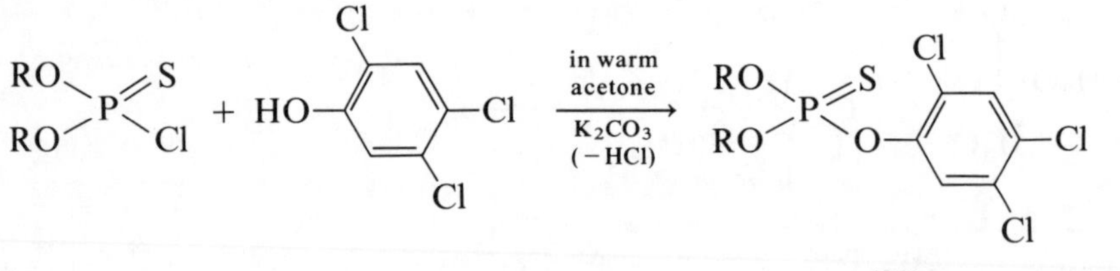

$$\underset{RO}{\overset{RO}{\diagdown}}P\overset{S}{\underset{Cl}{\diagup}} + HO\text{—}\underset{}{\bigcirc}\text{—}Cl \xrightarrow[\substack{K_2CO_3 \\ (-HCl)}]{\substack{in\ warm \\ acetone}} \underset{RO}{\overset{RO}{\diagdown}}P\overset{S}{\underset{O}{\diagup}}\text{—}\underset{}{\bigcirc}\text{—}Cl$$

$$(22)$$

Scheme 6.20

Chlorpyrifos (22; $R = C_2H_5$) is a very valuable contact insecticide; some 3500 tonnes were used in the United States of America (1982). It has a wide spectrum of activity, by contact, ingestion, and vapour action.[5] It is moderately persistent and retains its activity in soil for 2–4 months and is valuable against mosquito and fly larvae, cabbage root fly, aphids, and codling and winter moths on fruit trees. Chlorpyrifos has become one of the most widely applied insecticides in homes and restaurants against cockroaches and other domestic pests.[7] It is a comparatively safe insecticide; the mammalian toxicity LD_{50} (oral) to rats $\simeq 160$ mg/kg and the chemical is rapidly detoxified in animals. Chlorpyrifos-methyl (22; $R = CH_3$) is quite volatile and is used to control insects in grain stores.[6]

Several organophosphorus insecticides are obtained by condensation of the alkali or ammonium salt of *O,O*-dimethylphosphorodithioic acid with a heterocyclic chloromethyl compound.[8] One example is menazon (22a) developed by Imperial Chemical Industries Limited in 1961[4,8,9] and prepared by reaction of sodium *O,O*-dimethylphosphorodithioate with 2-chloromethyl-4,6-diamino-1,3,5-triazine (Scheme 6.21).

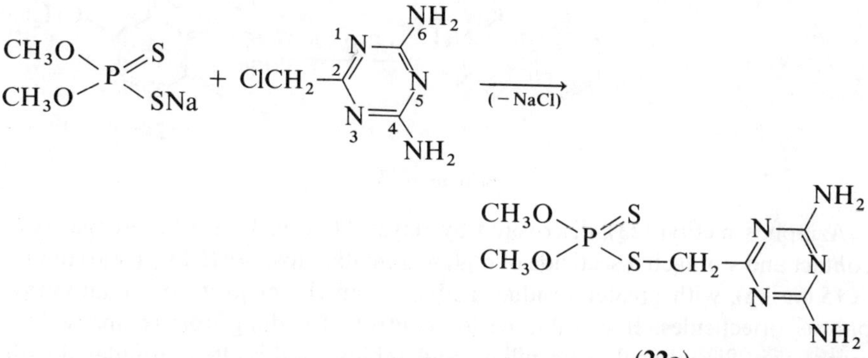

Scheme 6.21

Menazon (22a) is a selective systemic aphicide and acaricide with a persistent residual action[3] for control of aphids on a wide range of crops. Menazon is a remarkably safe insecticide (LD_{50} (oral) to rats $\simeq 1900$ mg/kg) and is a poor cholinesterase inhibitor *in vitro*, but is selectively activated *in vivo* in aphids. Other examples include phosphorylated derivatives of the heterocycles benzotriazine, pyrimidine, pyridine, coumarin, and quinazoline. One of the most active members is azinphos-methyl (23), obtained from *N*-chloromethylbenzazimide (24) (Scheme 6.22). *N*-Chloromethylbenzazimide (24) is obtained from anthranilic acid (Scheme 6.23).

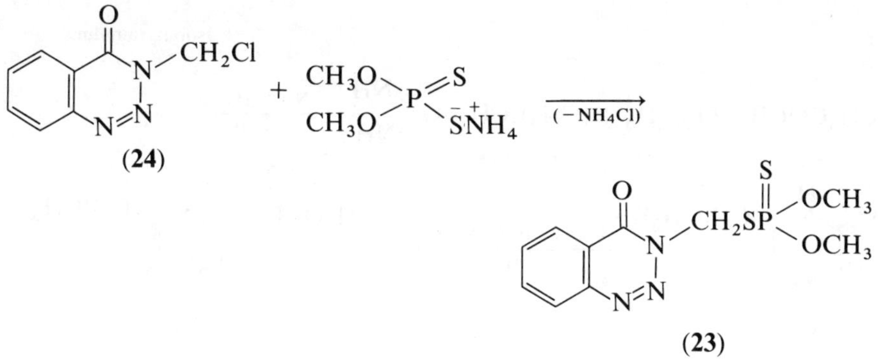

Scheme 6.22

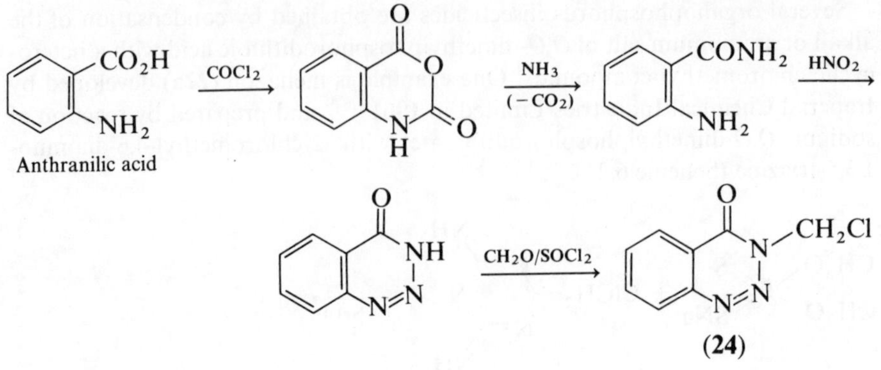

Scheme 6.23

Azinphos-methyl (**23**), discovered by Bayer AG (1953), is a broad-spectrum contact and stomach insecticide of high mammalian toxicity (LD_{50} (oral) to rats $\simeq 15$ mg/kg), with greater residual activity than the majority of organophosphorus insecticides. It is valuable for control of codling, tortrix and winter moths, fruit flies, weevils, caterpillars, and aphids,[5] and is often formulated with demeton-*S*-methyl sulphone. In insects azinphos-methyl (**23**) is metabolized to the more toxic phosphoryl analogue. Azinphos-ethyl is also used as an insecticide and is more effective against red spider mites.

Diazinon (**25**) was introduced by the Geigy Company in 1952 and incorporates the pyrimidine nucleus; it is obtained by condensation of ethyl acetoacetate and isobutyramidine.[2] Isobutyramidine is obtained as shown in Scheme 6.24.

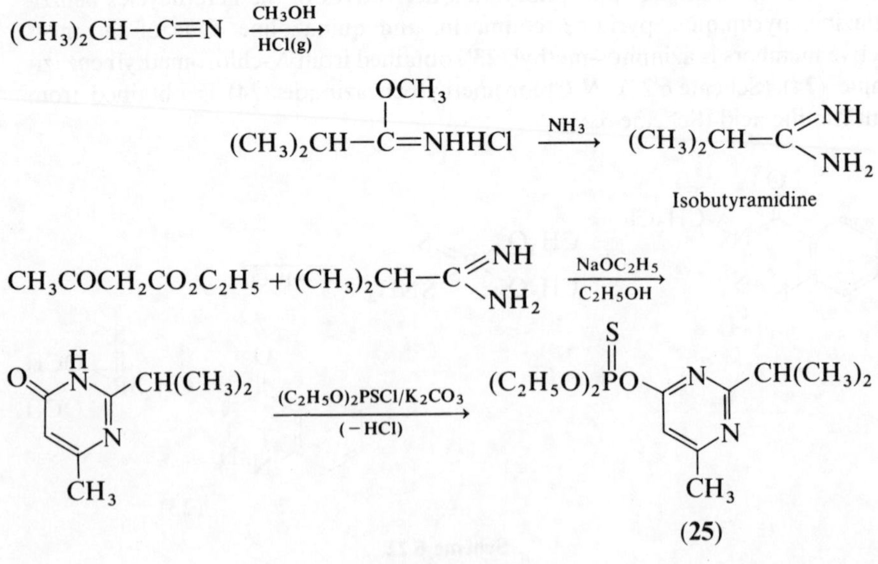

Isobutyramidine

(**25**)

Scheme 6.24

Diazinon (25) is a contact insecticide for soil and foliar treatment with fairly good residual activity; it is sufficiently volatile to be active against flies.[8] It is effective against a number of soil, fruit, vegetable, and rice pests, e.g. cabbage root, carrot and mushroom flies, aphids, spider mites, thrips, and scale insects together with domestic and livestock pests.[2,5,9,10] A minimum period of two weeks is needed between the last treatment and harvesting of edible crops.[5] Diazinon has fairly low mammalian toxicity:[2] LD_{50} (oral) to rats $\simeq 108$ mg/kg. Pirimphos-methyl (26) is similarly prepared by condensation of *N,N*-diethylguanidine, ethyl acetoacetate, and *O,O*-dimethylphosphorochloridothioate (Scheme 6.25).

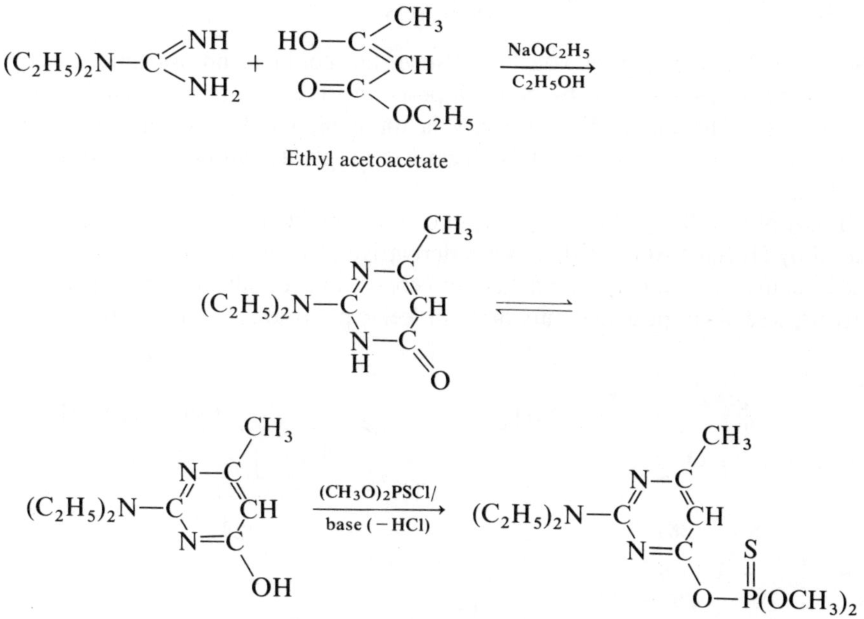

$$(C_2H_5)_2N-C{\overset{NH}{\underset{NH_2}{\diagup}}} \;+\; \underset{\text{Ethyl acetoacetate}}{HO-C{\overset{CH_3}{\underset{\underset{OC_2H_5}{O=C}}{\diagup\!\!\!\diagup CH}}}} \xrightarrow[\text{C}_2\text{H}_5\text{OH}]{\text{NaOC}_2\text{H}_5}$$

$$(C_2H_5)_2N-C{\overset{N-C\diagdown CH_3}{\underset{\underset{H}{N-C}\diagdown O}{\diagup CH}}} \;\rightleftharpoons$$

$$(C_2H_5)_2N-C{\overset{N-C\diagdown CH_3}{\underset{\underset{OH}{N=C}}{\diagup CH}}} \xrightarrow[\text{base}\,(-\text{HCl})]{(CH_3O)_2PSCl/} (C_2H_5)_2N-C{\overset{N-C\diagdown CH_3}{\underset{\underset{O-\overset{\overset{S}{\parallel}}{P}(OCH_3)_2}{N=C}}{\diagup CH}}}$$

(26)

Scheme 6.25

Pirimphos-methyl (26) was introduced by Imperial Chemical Industries in 1970 as a broad-spectrum insecticide effective against pests in stored products, e.g. beetles, weevils, mites, and moths, and against insects affecting public health, e.g. flies, cockroaches, mosquitoes, lice, bed bugs, and fleas.[5] It has very low mammalian toxicity (LD_{50} (oral) to rats, 2000 mg/kg) and will control insects that have become resistant to organochlorine insecticides; it is fast acting and has both contact and fumigant action. The ethyl ester, pirimphos-ethyl, finds use as an insecticide in mushroom farms and greenhouses.[6]

Quinalphos (27) developed by Bayer AG (1969) is prepared by reaction of *o*-phenylenediamine, chloroacetic acid, and *O,O*-diethylphosphorochloridothioate

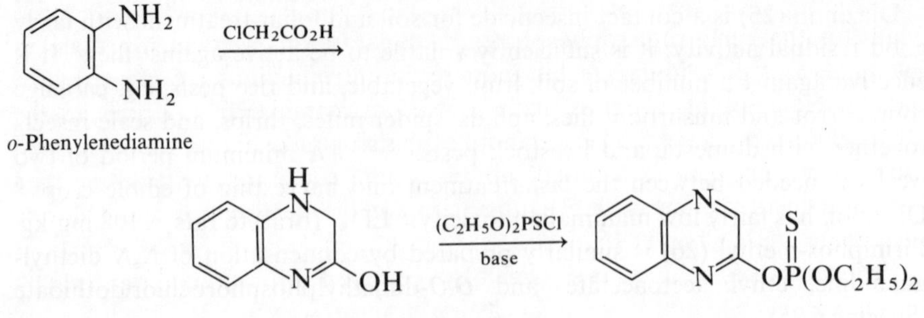

o-Phenylenediamine

(27)

Scheme 6.26

(Scheme 6.26). Quinalphos is a broad-spectrum contact and systemic insecticide, applied as a spray to control pests in cereals, brassicas, and other vegetables.[6] The mammalian toxicity is quite high (LD_{50} (oral) to rats is 70 mg/kg), but the compound is degraded in plants within a few days of application.

Triazophos (**28**) is a broad-spectrum contact insecticide and acaricide introduced by Hoechst AG (1970) and is a derivative of 1,2,4-triazole. Triazophos is used against flies and leatherjackets in cereals, maize, oilseed rape, brassicas, carrots, weevils in peas and cutworms in leeks, potatoes, and other crops.[5]

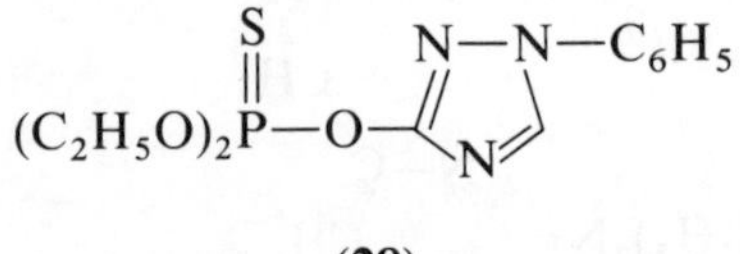

(28) 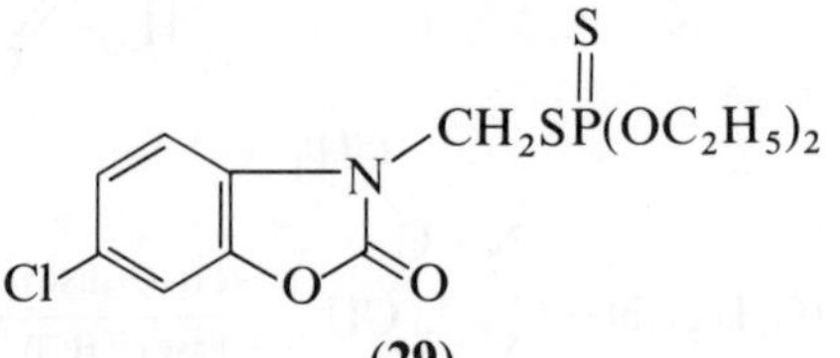   (29)

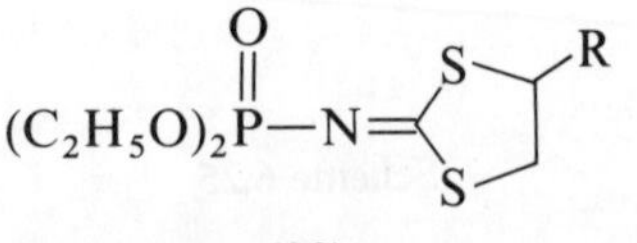

(30)

Several organophosphorus insecticides contain other heteroatoms. Examples include phosalone (**29**), used to control aphids in cereals, oilseed rape, brassicas, and fruit trees; it is rapidly metabolized via the phosphorothioate.

Phosfolan (**30**; R = H) and mephosfolan (**30**; R = CH$_3$) are examples of systemic insecticides containing heterosulphur atoms.[9,11]

Some 250 organophosphorus compounds are currently registered for use as insecticides. In 1983 their estimated market value was $1640 million out of a total value for insecticides of $4280[6] (cf. market values of $1100 million for carbamates, $560 million for organochlorines, and $600 million for pyrethroids).

Many of the early organophosphates, like parathion, were highly toxic to mammals as well as to insects, but more recent introductions, e.g. chlorpyrifosmethyl and tetrachlorvinphos, have low acute mammalian toxicities in the 2000–5000 mg/kg range. The organophosphates, unlike the organochlorines, are relatively non-persistent and do not bioaccumulate in the environment; they are consequently often valuable substitutes for the persistent organochlorine insecticides[6] (see Chapter 15, page 350).

Elementary phosphorus is obtained by reduction of phosphate minerals by fusion with carbon and silica in an electric furnace:

$$Ca_3(PO_4)_2 + 3SiO_2 + 5C \xrightarrow{1400°C} 3CaSiO_3 + 5CO + 2P$$

Key phosphorus compounds for synthesis of organophosphates are obtained as follows: phosphorus pentasulphide by heating a mixture of phosphorus and sulphur at 350 °C; phosphorus trichloride by passing chlorine over molten phosphorus and subsequent distillation from the residual solid phosphorus pentachloride; phosphoryl chloride or phosphorus oxychloride by oxidation of phosphorus trichloride or by controlled hydrolysis of phosphorus pentachloride. Thiophosphoryl chloride is prepared by passing phosphorus trichloride vapour over molten sulphur in the presence of a suitable catalyst, e.g. charcoal. The various organophosphorus insecticides are generally synthesized via four main types of reaction:[2,3,11]

(a) Condensation of phosphorus oxychloride, or thiophosphoryl chloride with an alcohol, thioalcohol, or amine in the presence of a suitable tertiary base or other acid-binding agent such as sodium carbonate (e.g. synthesis of parathion, page 106). Careful temperature control is vital to avoid isomerization into the corresponding S-esters (Scheme 6.27). Such reactions are all essentially bimolecular nucleophilic substitutions at an electrophilic phosphorus centre and are designated $S_N2(P)$.

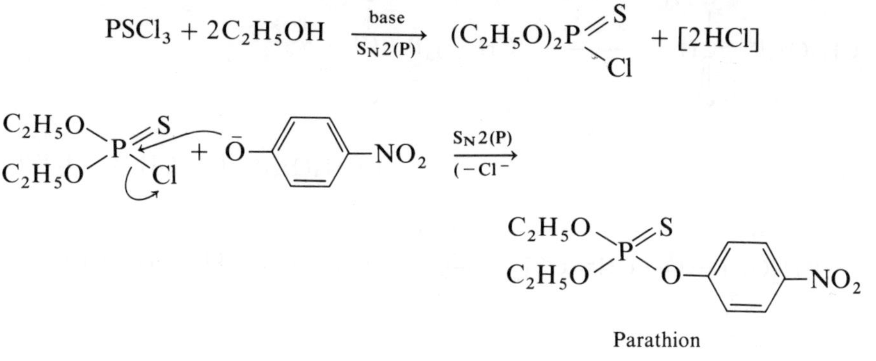

$$PSCl_3 + 2C_2H_5OH \xrightarrow[S_N2(P)]{base} (C_2H_5O)_2P{\overset{S}{\underset{Cl}{\diagup}}} + [2HCl]$$

Parathion

Scheme 6.27

(b) Reaction of the appropriate alcohol with phosphorus pentasulphide gives the corresponding O,O-dialkylphosphorodithioic acid:

$$P_2S_5 + 4ROH \xrightarrow{\text{warm}} 2(RO)_2P(\!=\!S)SH + H_2S$$

Examples are the synthesis of phorate (page 109) and malathion (page 111).

(c) Pyrophosphates are prepared by partial hydrolysis of the appropriate phosphorochloridate in the presence of a small amount of water (half a molar equivalent) and a tertiary amine:

$$2(RO)_2P(\!=\!O)Cl \xrightarrow[(-HCl)]{H_2O/R_3N} (RO)_2\overset{\overset{O}{\|}}{P}\!-\!O\!-\!\overset{\overset{O}{\|}}{P}(OR)_2$$

An example is the synthesis of schradan (page 105). Pyrophosphates may also be obtained by reaction of the phosphorochloridate and the sodium salt of a phosphate:

$$(RO)_2P(\!=\!O)Cl + (RO)_2P(\!=\!O)ONa \xrightarrow[(-NaCl)]{} (RO)_2\overset{\overset{O}{\|}}{P}\!-\!O\!-\!\overset{\overset{O}{\|}}{P}(OR)_2$$

(d) Phosphites can be obtained by reaction of phosphorus trichloride with alcohols in presence of a tertiary amine:

$$3ROH + PCl_3 \xrightarrow{3R_3N} (RO)_3P + [3HCl]$$

An example of the use of phosphites in pesticide synthesis is provided by dichlorvos (page 112). This is obtained from trimethyl phosphite and chloral by the Perkow reaction which involves attack of the nucleophilic P^{III} atom on the electrophilic carbonyl carbon atom followed by rearrangement—the final stage being an Arbusov reaction[4] (Scheme 6.28). A wide

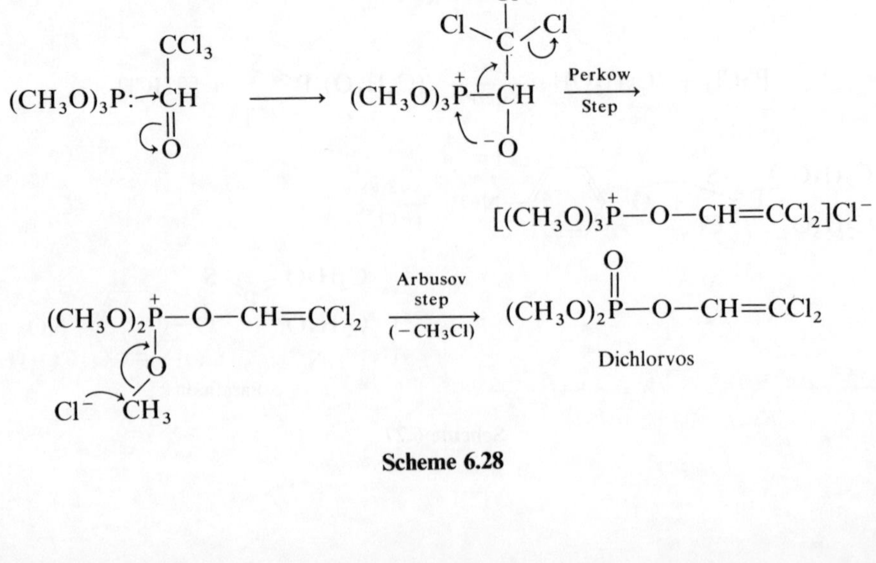

$$[(CH_3O)_3\overset{+}{P}\!-\!O\!-\!CH\!=\!CCl_2]Cl^-$$

Scheme 6.28

variety of P^{III} compounds will react with α-halogenocarbonyl compounds to give vinyl phosphates.

MODE OF ACTION OF ORGANOPHOSPHORUS INSECTICIDES

The insecticidal organophosphorus compounds inhibit the action of several enzymes, but the major action *in vivo* is against the enzyme acetylcholinesterase.[2,3,11-14] This controls the hydrolysis of the acetylcholine (31) generated at nerve junctions into choline (32). In the absence of effective acetylcholinesterase, the liberated acetylcholine accumulates and prevents the smooth transmission of nervous impulses across the synaptic gap at nerve junctions. This causes loss of muscular coordination, convulsions, and ultimately death (see Chapter 3, page 45). Acetylcholinesterase is an essential component of the nervous system of both insects and mammals so the basic mechanism of toxic action of the organophosphorus compounds is considered to be essentially the same in insects and mammals. The active centre of the enzyme acetylcholinesterase contains two main reactive sites: an 'anionic site' which is negatively charged and binds onto the cationic part of the substrate (acetylcholine, 31) and the 'esteratic site' containing the primary alcoholic group of the amino acid serine which attacks the electrophilic carbonyl carbon atom of the substrate. The normal enzymic hydrolysis of acetylcholine (31) to choline (32) may therefore be illustrated as shown (Figure 6.1).

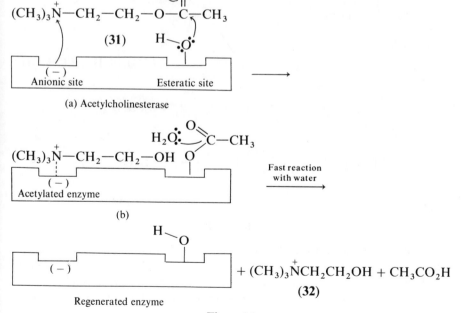

Figure 6.1

Figure 6.1(a) depicts the formation of the initial enzyme–substrate complex by the orientation of the active centres of acetylcholinesterase to the substrate (acetylcholine). Figure 6.1(b) shows formation of the acetylated enzyme, which is subsequently rapidly hydrolysed to choline (32) and acetic acid, leaving the enzyme with both its active sites intact, so permitting it to repeat the enzymic hydrolytic process on further substrate molecules releasing several thousand choline molecules per second.[7]

The majority of active organophosphorus compounds have the general structure (33), where R,R' are generally lower alkyl, alkoxy, alkylthio, or

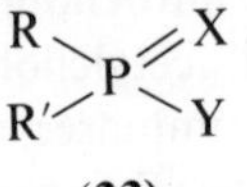

$$\begin{array}{c} R \\ \diagdown \\ R' \diagup \end{array} P \diagup^{\textstyle X}_{\textstyle \diagdown Y}$$

(33)

substituted amino groups; X is oxygen or sulphur; and Y is a good leaving group or one capable of being metabolized into such a group.

The organophosphorus compound (33) mimics the natural substrate acetylcholine (31) by binding itself to the esteric site of acetylcholinesterase (Figure 6.1). The subsequent reaction between the enzyme (ECH_2OH) and the organophosphorus compound is of the bimolecular S_N2 type and mirrors the normal three-stage reaction between the enzyme and acetylcholine (Scheme 6.29).

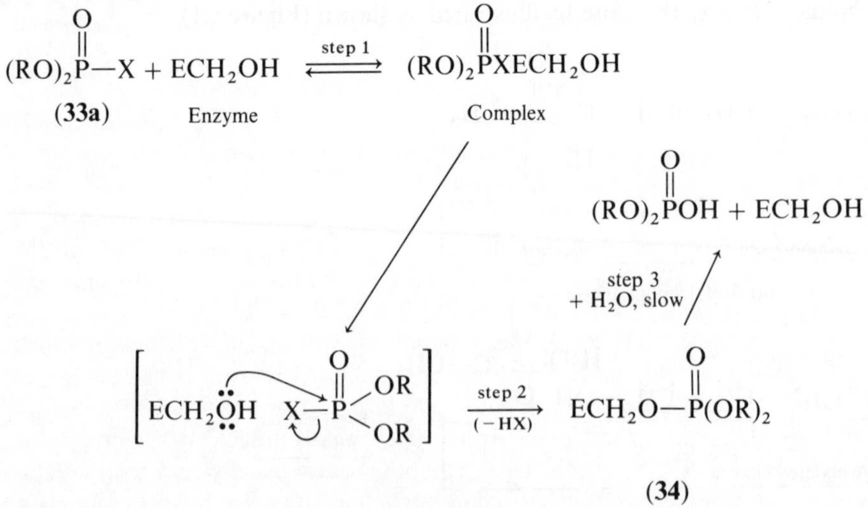

$$\begin{array}{ccc} & O & & O \\ & \| & & \| \\ (RO)_2P{-}X + ECH_2OH & \overset{\text{step 1}}{\rightleftharpoons} & (RO)_2PXECH_2OH \end{array}$$

(33a)   Enzyme   Complex

$$\begin{array}{c} O \\ \| \\ (RO)_2POH + ECH_2OH \end{array}$$

step 3
+ H_2O, slow

$$\left[ECH_2\overset{..}{\underset{..}{O}}H \quad X{-}\overset{\displaystyle O}{\overset{\|}{P}}\diagup^{\textstyle OR}_{\textstyle \diagdown OR} \right] \xrightarrow[(-HX)]{\text{step 2}} \begin{array}{c} O \\ \| \\ ECH_2O{-}P(OR)_2 \end{array}$$

(34)

Scheme 6.29

Initially a complex is formed between the enzyme and the phosphate (step 1) which subsequently gives the phosphorylated enzyme (34) (step 2); the latter is slowly hydrolysed to the free enzyme (step 3). In contrast, the normal reaction between acetylcholinesterase and acetylcholine (31) is shown in Scheme 6.30.

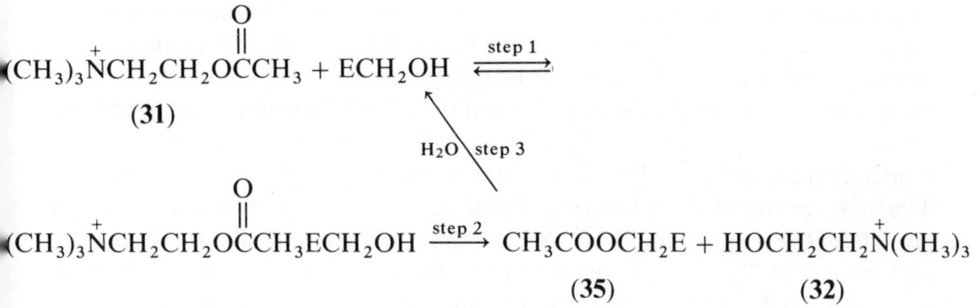

$$(CH_3)_3\overset{+}{N}CH_2CH_2O\overset{\overset{\displaystyle O}{\|}}{C}CH_3 + ECH_2OH \underset{}{\overset{\text{step 1}}{\rightleftharpoons}}$$

(31)

$$H_2O \diagdown \text{step 3}$$

$$(CH_3)_3\overset{+}{N}CH_2CH_2O\overset{\overset{\displaystyle O}{\|}}{C}CH_3ECH_2OH \xrightarrow{\text{step 2}} CH_3COOCH_2E + HOCH_2CH_2\overset{+}{N}(CH_3)_3$$

(35)   (32)

Scheme 6.30

The acetylated enzyme (35) is very rapidly hydrolysed by water (step 3) so that the active enzyme is quickly regenerated, enabling the hydrolysis of acetylcholine to choline to be effectively catalysed by the enzyme. On the other hand, when an organophosphate is present the inactive phosphorylated enzyme (34) is only very slowly hydrolysed to the active enzyme because the phosphorus–oxygen bond is much stronger than the carbon–oxygen bond of the acetylated enzyme (35). Therefore the organophosphate effectively poisons the enzyme by phosphorylation and thus blocks efficient hydrolysis of acetylcholine into choline.[4,8,14] The resultant build-up of acetylcholinesterase causes abnormally high levels of activation of the synapse and blockage of nerve function.[13]

Although the majority of organophosphorus compounds active as phosphorylating agents show some insecticidal activity, it is difficult to predict how effective they will be in practice, so a wide range of analogous compounds were obtained for biological screening to identify the optimum insecticide. Sometimes closely related compounds display markedly different types and degrees of activity; thus specific 2,4-dichlorophenyl- and 2,4,5-trichlorophenyl-phosphorothioates function as nematicides, acaricides, or soil insecticides.[1] The introduction of 3-methyl and chloro groups into parathion (page 107) substantially reduced the mammalian toxicity, but it would be hard to predict these effects since the substituents do not greatly alter the physiochemical properties.[7]

Generally there is evidence[2] that compounds with *m*-substituents are much more effective as insecticides than would be predicted from consideration of electronic effects, and this may be a reflection of the steric properties of *m*-substituents. Many commercial organophosphorus insecticides contain the thiophosphoryl (P=S) group. These compounds are usually very weak inhibitors of acetylcholinesterase *in vitro*, but are activated *in vivo* by mixed-function oxidases to the corresponding phosphates (P=O); another type of important *in vivo* oxidation is the conversion of strongly electron donor groups like C_2H_5S (in, for example, phorate, page 133) and $(CH_3)_2N$ (in, for example, schradan, page 105) into electron-withdrawing groups[2,15] (see page 135).

Several enol phosphates are powerful cholinesterase inhibitors and have been developed as insecticides. The phosphorylating ability of such compounds is

enhanced by protonation and may be due to interaction of the carboxylic group
with the esteratic site of the enzyme (see Figure 6.1, page 123). Such interaction
would probably be sensitive to steric factors—thus *E*-mevinphos is substantially
more active as an anticholinesterase agent than the *Z* isomer, probably because
interaction between the carboxylic group and the enzyme would be sterically
hindered in the *Z* isomer by the presence of the *cis*-diethylphosphoryl moiety.[2]
The high activity of the *E* isomer is therefore explicable in terms of obtaining a
good fit of the toxicant molecule to the anionic and esteratic sites of the enzyme
(see page 123). The distance between the active sites of insect acetylcholinesterase is 4.5–5.9 Å and may be related to the distance between the phosphorus
atom and the carbonyl carbon atom in the geometric isomers of mevinphos as
follows: *E* isomer P–C* distance = 4.3–5.2 Å (good fit) whereas for the *Z* isomer
the corresponding distance is 2.2–4.4 Å (bad fit) (see page 114).

Generally it is found[1,2] that phosphorylating ability is reduced by the
presence of bulky groups attached to phosphorus, and for this reason most
insecticides contain lower alkoxy groups. Another example of the importance of
obtaining a good fit of the toxicant molecule to the anionic and esteratic sites of
acetylcholinesterase is provided by amiton (**36**), prepared from *0,0*-diethyl-
phosphorochloridothioate and sodium 2-diethylaminoethoxide by a route
involving the thiono–thiolo rearrangement[2] (Scheme 6.31).

$$(C_2H_5O)_2PSCl + NaOCH_2CH_2N(C_2H_5)_2 \xrightarrow[(-NaCl)]{}$$

$$\overset{\overset{\textstyle S}{\|}}{(C_2H_5O)_2P}-OCH_2CH_2N(C_2H_5)_2 \xrightarrow{70-80°C}$$

$$\overset{\overset{\textstyle O}{\|}}{(C_2H_5O)_2P}-SCH_2CH_2N(C_2H_5)_2$$

(**36**)

Scheme 6.31

Amiton (**36**) is a water-soluble, persistent systemic toxicant formerly used
against aphids, scale insects, and mites,[2] but it suffers from a very high
mammalian toxicity: LD_{50} (oral) to rats $\simeq 3$ mg/kg. At physiological pH values
amiton is ionized and such ionic compounds tend to be selectively toxic to
mammals because their ionic character appears to restrict their entry through
the sheath protecting the insect nervous system, whereas this ion-impermeable
barrier is absent in mammals.[2,8] Amiton (**36**) is a phosphoryl analogue of
acetylcholine and the high anticholinesterase activity of the compound arises
from the excellent affinity of the quaternary nitrogen atom for the anionic site of
the enzyme (see page 123).

Chirality in an insecticidal molecule may also influence its activity, and this is
to be expected since enzymes are themselves chiral molecules. The valuable

properties of malathion (16) (page 111) stimulated extensive investigation of analogous compounds, and it was discovered that the asymmetry in the succinate group due to the presence of the chiral carbon atom (C*) was associated with differences in insecticidal potency: *dextro*-malathion was appreciably more toxic to both houseflies and mice as compared with the *laevo* isomer and was also a more active inhibitor of acetylcholinesterase and liver carboxyesterase.[2,8]

Studies of the metabolism of selected types of organophosphorus insecticides in plants, animals, and insects have proved invaluable in elucidating the mode of action, predicting probable metabolites for related organophosphates, and accounting for selective toxicity. Many organophosphorus insecticides show little *in vitro* anticholinesterase activity and the *in vivo* activity is the result of net metabolic activation. Examples of activation processes are enzymic oxidations of phosphorothioates, like parathion diazinon, and malathion, involving conversion of P=S to P=O, of sulphides such as phorate, involving the transformation: S → SO → SO$_2$, and of amides, such as schradan, to the *N*-oxide or methylol:

$$(CH_3)_2N-\overset{\overset{O}{\|}}{P}- \longrightarrow (CH_3)_2\overset{\downarrow}{\underset{O}{N}}-\overset{\overset{O}{\|}}{P}- \longrightarrow \overset{HOH_2C}{\underset{H_3C}{>}}N-\overset{\overset{O}{\|}}{P}-$$

Such *in vivo* activation processes occur principally in the insect gut, fat body tissues, and in the mammalian liver.[15,16] The effectiveness of a given insecticide to a specific insect will depend on the balance of biochemical activation and detoxification processes occurring in the insect species. It is important to realize that the metabolic pathways will often differ in different organisms, e.g. insects, mites, soil microorganisms, plants, or vertebrates.[17]

Enzymes known as mixed function oxidases (MFO) occur in animal, fish, and insect cell microsomes and, in the presence of molecular oxygen and reduced nicotinamide adenine dinucleotide phosphate (NADPH$_2$) or reduced nicotinamide adenine dinucleotide (NADH$_2$), they will oxidize a variety of lipophilic substrates such as steroids, lipids, and foreign organic compounds.[1,2] One of the oxygen atoms is incorporated into the substrate (RH) while the other is reduced to water:

$$RH + NADPH_2 + H^+ + O_2 \xrightarrow{MFO} ROH + NADP^+ + H_2O$$

These enzymes are not specific and only require a substrate of high lipophilicity. However, the microsomal hydroxylation of foreign compounds requires the presence of a special microsomal pigment, cytochrome P 450, while liver microsomes have a different electron transport system requiring cytochrome b_5.[2] Examples of processes effected by MFO are the following:

(a) Hydroxylation:

$$RH \rightarrow ROH$$

(b) *O*- and *N*-dealkylation:

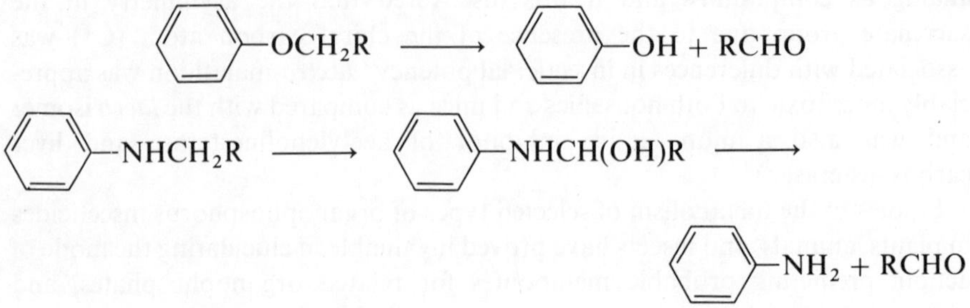

$$\text{C}_6\text{H}_5\text{—OCH}_2\text{R} \longrightarrow \text{C}_6\text{H}_5\text{—OH} + \text{RCHO}$$

$$\text{C}_6\text{H}_5\text{—NHCH}_2\text{R} \longrightarrow \text{C}_6\text{H}_5\text{—NHCH(OH)R} \longrightarrow$$

$$\text{C}_6\text{H}_5\text{—NH}_2 + \text{RCHO}$$

(c) Oxidative desulphuration:

$$\overset{|}{\underset{|}{>}}\!P{=}S \longrightarrow \overset{|}{\underset{|}{>}}\!P{=}O; \quad {>}C{=}S \longrightarrow {>}C{=}O$$

(d) Oxidation of sulphides:

$$\text{RR'S} \longrightarrow \text{RR'SO} \longrightarrow \text{RR'SO}_2$$

(e) Deesterification:

$$(\text{RO})_2\overset{\text{S}}{\overset{\|}{\text{P}}}\text{—OAr} \longrightarrow (\text{RO})_2\overset{\text{S}}{\overset{\|}{\text{P}}}\text{OH} + \text{ArOH}$$

(f) Epoxidation:

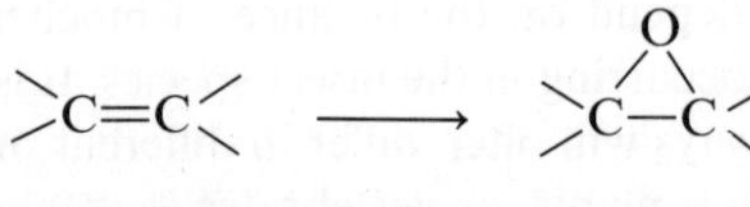

$$>\!C{=}C\!< \longrightarrow >\!\overset{\text{O}}{\overset{\diagup\diagdown}{C{-}C}}\!<$$

(g) Oxidation of tertiary amines:

$$\text{RN(CH}_3)_2 \longrightarrow \underset{\downarrow}{\text{RN(CH}_3)_2} \longrightarrow \text{RN}\!\!<\!\!\overset{\text{CH}_2\text{OH}}{\underset{\text{CH}_3}{}} \xrightarrow{(-\text{CH}_2\text{O})} \text{RNHCH}_3$$
$$\text{O}$$

The following are illustrative examples of the metabolism of some important organophosphorus insecticides. The metabolism of parathion (see page 106) is shown in Scheme 6.32.

Phosphorothioates, like parathion, are examples of proinsecticides.[6] They are poor inhibitors of acetylcholinesterase, whereas their oxo analogues are very active anticholinesterase agents. The toxicity of such compounds as parathion is therefore due to their *in vivo* oxidative desulphuration (Scheme 6.32, path c) by microsomal MFO in animals and insects; in plants the oxidation is probably achieved by peroxidases.[2,16,18] Paraoxon is subsequently deactivated by esterase-catalysed hydrolysis, and microsomal MFO can also effect some detoxification of parathion by dealkylation (Scheme 6.32, path b). In some cases, organophosphorus compounds may show differential toxicity as a result of

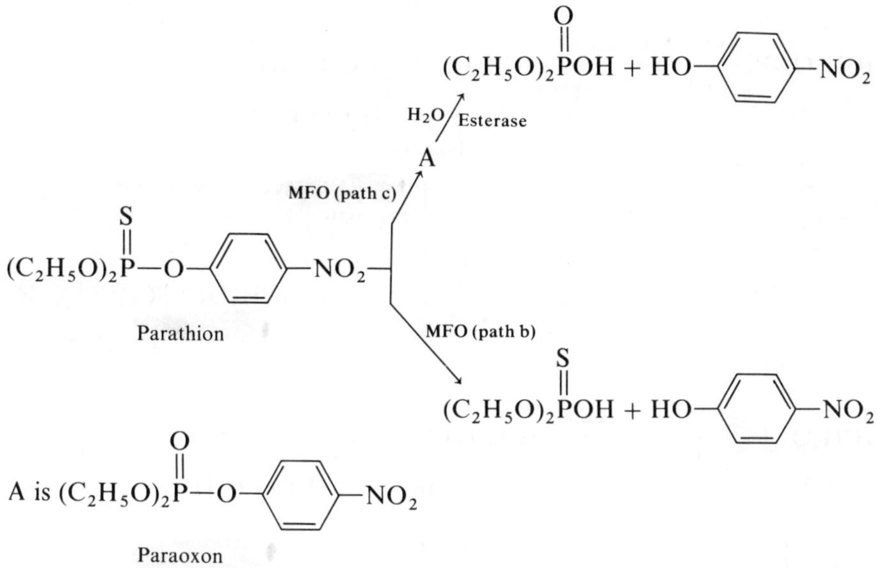

Scheme 6.32

different metabolism in mammals and insects. One of the first examples was malathion which, although an effective contact insecticide, showed remarkably low mammalian toxicity (page 111). Malathion is rapidly activated to malaoxon by oxidative desulphuration in insects and mammals[17,18] (Scheme 6.33, path c). Malaoxon is a highly potent anticholinesterase agent which is very toxic to insects and mammals. Malathion and its toxic metabolite, malaoxon, are detoxified by carboxyesterases which hydrolyse the carboethoxy moiety leading to polar, water-soluble, compounds that are excreted (Scheme 6.33). Vertebrates show a greater carboxyesterase activity, as compared with insects, so that the toxic agent malaoxon builds up more in insects than in mammals, accounting for the selective toxicity of malathion (**16**) towards insects. The major metabolic pathways for malathion (**16**) are shown in Scheme 6.33.

The metabolites shown (Scheme 6.33) agree with the main compounds isolated from treated mammals and insects. The relative proportion of metabolites obtained indicate a greater PS → PO conversion (path c) in insects than in mammals followed by hydrolysis of the P—S—C linkage, especially in houseflies. On the other hand, in mammals the major degradation is via hydrolysis of the carboethoxy linkage. The selectivity of malathion (**16**) towards insects is thus explained by the differences in the rates and routes by which it is metabolized in insects and mammals.

Dimethoate (**23**; page 115) has been extensively applied for insect control owing to its good insecticidal activity combined with comparatively low mammalian toxicity. The metabolism of dimethoate has been studied in plants, insects, and vertebrates.[1,2,17] The major features of the metabolism are similar

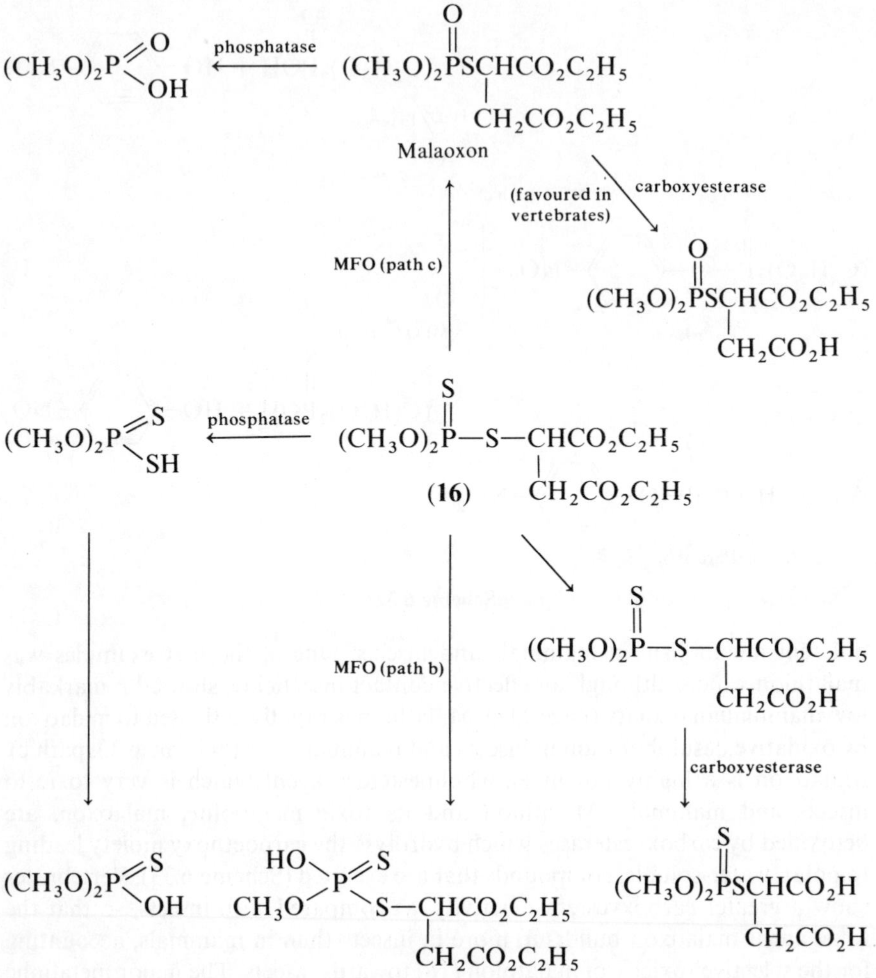

$$\text{(CH}_3\text{O)}_2\text{P}\!\!\underset{\text{OH}}{\overset{\text{O}}{<}} \quad \xleftarrow{\text{phosphatase}} \quad \text{(CH}_3\text{O)}_2\overset{\overset{\text{O}}{\|}}{\text{P}}\text{SCHCO}_2\text{C}_2\text{H}_5$$

$$\underset{\text{Malaoxon}}{\text{CH}_2\text{CO}_2\text{C}_2\text{H}_5}$$

(favoured in vertebrates)   carboxyesterase

MFO (path c)

$$\text{(CH}_3\text{O)}_2\overset{\overset{\text{O}}{\|}}{\text{P}}\text{SCHCO}_2\text{C}_2\text{H}_5$$
$$\text{CH}_2\text{CO}_2\text{H}$$

$$\text{(CH}_3\text{O)}_2\text{P}\!\!\underset{\text{SH}}{\overset{\text{S}}{<}} \quad \xleftarrow{\text{phosphatase}} \quad \text{(CH}_3\text{O)}_2\overset{\overset{\text{S}}{\|}}{\text{P}}\text{—S—CHCO}_2\text{C}_2\text{H}_5$$

$$\textbf{(16)} \qquad \text{CH}_2\text{CO}_2\text{C}_2\text{H}_5$$

MFO (path b)

$$\text{(CH}_3\text{O)}_2\overset{\overset{\text{S}}{\|}}{\text{P}}\text{—S—CHCO}_2\text{C}_2\text{H}_5$$
$$\text{CH}_2\text{CO}_2\text{H}$$

carboxyesterase

$$\text{(CH}_3\text{O)}_2\text{P}\!\!\underset{\text{OH}}{\overset{\text{S}}{<}} \qquad \underset{\text{CH}_3\text{O}}{\overset{\text{HO}}{>}}\text{P}\!\!\underset{\text{S—CHCO}_2\text{C}_2\text{H}_5}{\overset{\text{S}}{<}} \qquad \text{(CH}_3\text{O)}_2\overset{\overset{\text{S}}{\|}}{\text{P}}\text{SCHCO}_2\text{H}$$

$$\text{CH}_2\text{CO}_2\text{C}_2\text{H}_5 \qquad\qquad \text{CH}_2\text{CO}_2\text{H}$$

Scheme 6.33

in all organisms; the metabolic reactions occurring include *O*- and *N*-dealkylation by MFO (path b, page 128); hydrolysis of P—O and P—S bonds by phosphatases; activation by oxidative desulphuration by oxidases (path c, page 128) resulting in conversion of P=S → P=O; and deamination by amidases. The metabolism is summarized in Scheme 6.34. Dimethoate is generally more rapidly degraded in mammals and eliminated in the urine as water-soluble hydrolysis products; for instance, in sheep amidases give the metabolite (37), but in rats and mice both (37) and (38) were isolated while in guinea pigs the major metabolite was the dithioic acid (38). In mammals, the metabolism of dimethoate takes place mainly in the liver.[16]

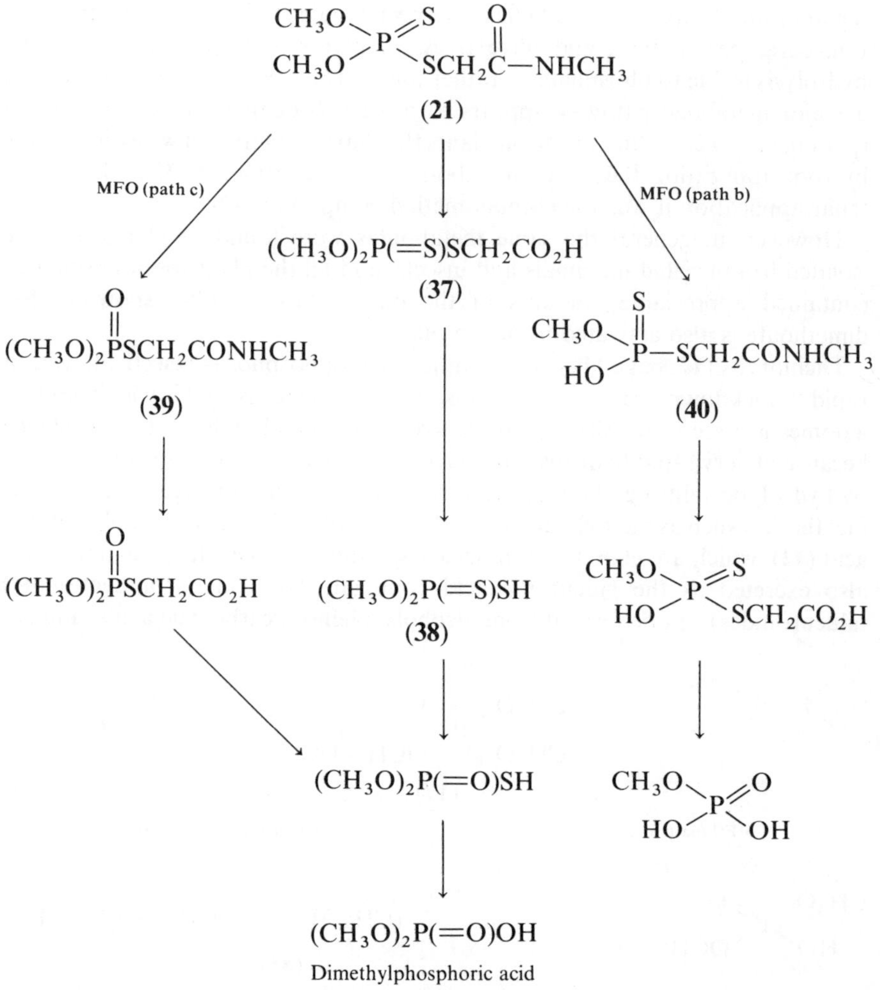

$$CH_3O \diagdown P \diagup^S \quad O$$
$$CH_3O \diagup \diagdown SCH_2C{-}NHCH_3$$

(21)

MFO (path c)

MFO (path b)

$$(CH_3O)_2P(=S)SCH_2CO_2H$$

(37)

$$O$$
$$\|$$
$$(CH_3O)_2PSCH_2CONHCH_3$$

(39)

$$CH_3O \diagdown \overset{S}{\underset{}{\|}}$$
$$HO \diagup P{-}SCH_2CONHCH_3$$

(40)

$$O$$
$$\|$$
$$(CH_3O)_2PSCH_2CO_2H$$

$$(CH_3O)_2P(=S)SH$$

(38)

$$CH_3O \diagdown P \diagup^S$$
$$HO \diagup \diagdown SCH_2CO_2H$$

$$(CH_3O)_2P(=O)SH$$

$$CH_3O \diagdown P \diagup^O$$
$$HO \diagup \diagdown OH$$

$$(CH_3O)_2P(=O)OH$$

Dimethylphosphoric acid

Scheme 6.34

In insects, both houseflies and cockroaches metabolized dimethoate **(21)** slowly, accounting for the selective toxicity to these insects. In several insect species attack by amidases is of less importance than that by phosphatases, but this does not apply to the boll weevil where, as in mammalian metabolism, the first degradation is amidase attack. The special toxicity of dimethoate **(21)** towards houseflies depends upon its rapid penetration into the fly and the substantial conversion to the more toxic oxo analogue **(39)** by oxidase desulphuration coupled with the sensitivity of fly acetylcholinesterase to phosphorylation. In the olive fruit fly, dimethoate **(21)** is also converted to the oxo compound **(39)** which is ultimately degraded to dimethylphosphoric acid. The metabolism of dimethoate (**21**) has also been studied[17] in various plants—cotton, corn, pea, potato, and olive trees. In plants, it is probable that total hydrolysis is due to phosphatases rather than carboxyesterases or amidases and a major metabolic pathway appears to involve *O*-demethylation by oxidases (path b, page 128). Thus, in cotton plants that have been treated with dimethoate by root application the major metabolite was the dithioate (**37**), whereas by foliar application it was the *O*-monomethyl compound (**40**).

However, in general the same metabolites were found in plants as were isolated from treated mammals and insects, and all the plant species examined contained appreciable amounts of the oxo metabolite (**39**), showing that dimethoate is also activated *in vivo* in plants.

Dichlorvos (**17**; page 112) is a volatile organophosphorus insecticide with a rapid 'knockdown' action on flies, mosquitoes, and moths, and is widely used as a domestic insecticide. Although quite toxic to mammals, it has little persistency because of very rapid hydrolysis to inactive compounds—thus it is metabolized by hydrolytic splitting of either the vinyl or methyl ester linkages to give polar metabolites such as methyl phosphate, dimethyl phosphate (**41**), and phosphoric acid (**42**), which are eliminated in urine (Scheme 6.35). Dichloroethanol (**43**) is also excreted as the glucuronic acid conjugate (**44**). Such conjugates (**44**) (glucuronides) can be formed from alcohols, phenols, carboxylic acids, amines,

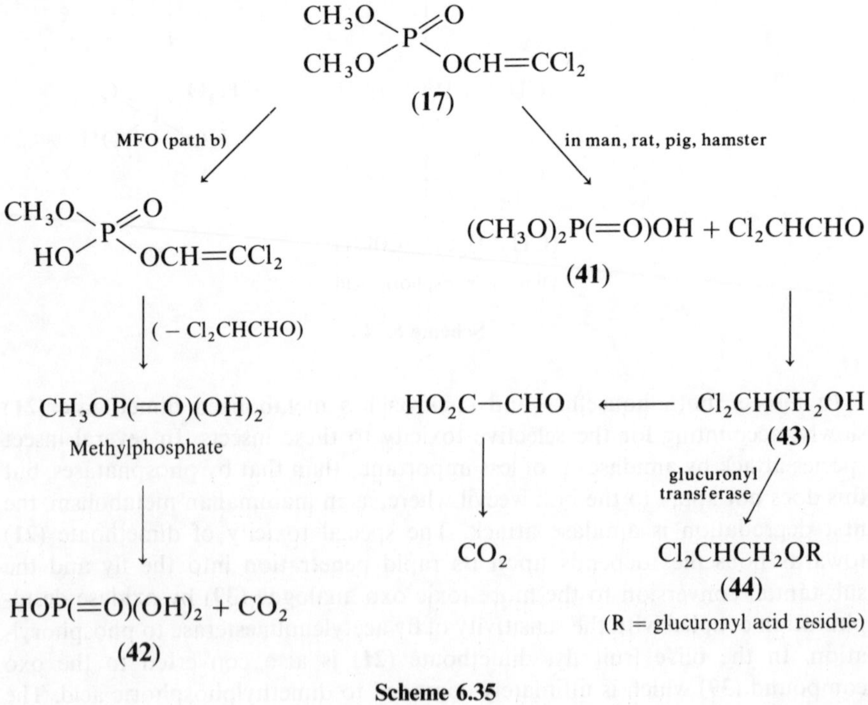

$$CH_3O\diagdown P\diagup O$$
$$CH_3O\diagup P\diagdown OCH{=}CCl_2$$
(**17**)

MFO (path b)   in man, rat, pig, hamster

$$CH_3O\diagdown P\diagup O$$
$$HO\diagup P\diagdown OCH{=}CCl_2$$

$(CH_3O)_2P({=}O)OH + Cl_2CHCHO$
(**41**)

$(- Cl_2CHCHO)$

$CH_3OP({=}O)(OH)_2$
Methylphosphate

$HO_2C{-}CHO \longleftarrow Cl_2CHCH_2OH$
(**43**)

glucuronyl
transferase

CO_2 Cl_2CHCH_2OR
(**44**)

$HOP({=}O)(OH)_2 + CO_2$
(**42**)

(R = glucuronyl acid residue)

Scheme 6.35

and thiols by the action of glucuronyl transferases and they have been observed in the *in vivo* metabolism of several organophosphorus insecticides, e.g. dichlorvos and parathion.[2] The metabolism of dichlorvos is summarized in Scheme 6.35.[17]

The systemic insecticide phorate (13; page 109) provides an example of the metabolism of a sulphide group. The various metabolites are shown in Scheme 6.36;[2,17] in plants, animals, and insects the sulphide group is oxidized to the corresponding sulphoxide (45) and sulphone (46) (path d, page 128). In plants (e.g. cotton and pea) there was also oxidative desulphuration in which P=S was converted to P=O (path c, page 128) by oxidases. On the other hand, in insects (e.g. cockroaches and bugs) oxidative desulphuration does not occur.

Hydrolysis by phosphatase cleavage of the P—S bond was especially significant in bollworms and weevils, where the principal metabolites were diethylphosphoric acid (47) and diethylphosphorothioic acid (48).[17] The initial

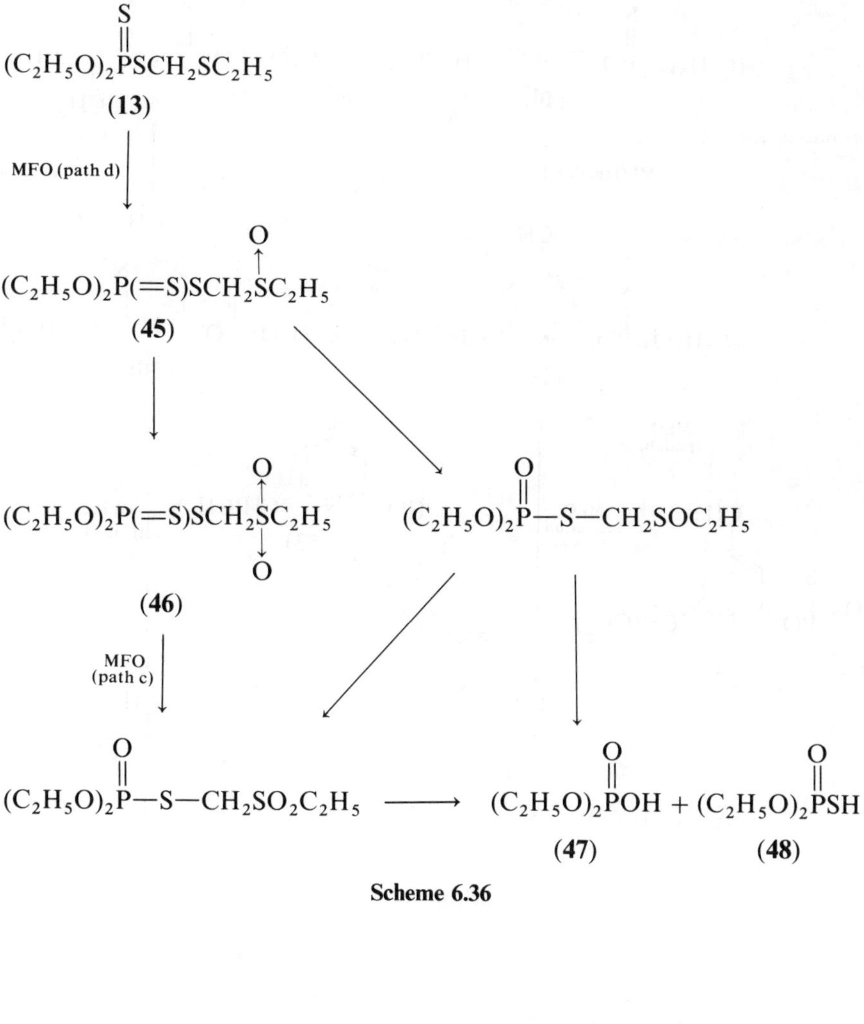

$$\underset{(13)}{(C_2H_5O)_2\overset{\overset{\displaystyle S}{\|}}{P}SCH_2SC_2H_5}$$

MFO (path d)

$$\underset{(45)}{(C_2H_5O)_2P(=S)S\overset{\overset{\displaystyle O}{\uparrow}}{C}H_2SC_2H_5}$$

$$\underset{(46)}{(C_2H_5O)_2P(=S)S\overset{\overset{\displaystyle O}{\uparrow}}{\underset{\underset{\displaystyle O}{\downarrow}}{C}}H_2SC_2H_5} \qquad \underset{}{(C_2H_5O)_2\overset{\overset{\displaystyle O}{\|}}{P}-S-CH_2SOC_2H_5}$$

MFO (path c)

$$(C_2H_5O)_2\overset{\overset{\displaystyle O}{\|}}{P}-S-CH_2SO_2C_2H_5 \longrightarrow (C_2H_5O)_2\overset{\overset{\displaystyle O}{\|}}{P}OH + (C_2H_5O)_2\overset{\overset{\displaystyle O}{\|}}{P}SH$$

$$\underset{(47)}{} \qquad\qquad \underset{(48)}{}$$

Scheme 6.36

oxidation of the sulphide to the sulphoxide occurs rapidly, whereas the subsequent oxidation to the sulphone is slow in plants, although it is rapid in mammals. This oxidation somewhat increases the activity of the original compound and enables plants to be protected rather longer due to the toxicity of the sulphoxide and sulphone metabolites. However, the activation is not as great as that effected by the process of oxidative desulphuration, i.e. the conversion of P=S to P=O.

The metabolism of diazinon (**25**, page 118) provides an example of an organophosphorus insecticide containing a heterocyclic moiety, and has been studied in mammals, insects, and plants. The major metabolites are shown in Scheme 6.37; those marked * are insecticidal. The selectivity of diazinon (**25**)

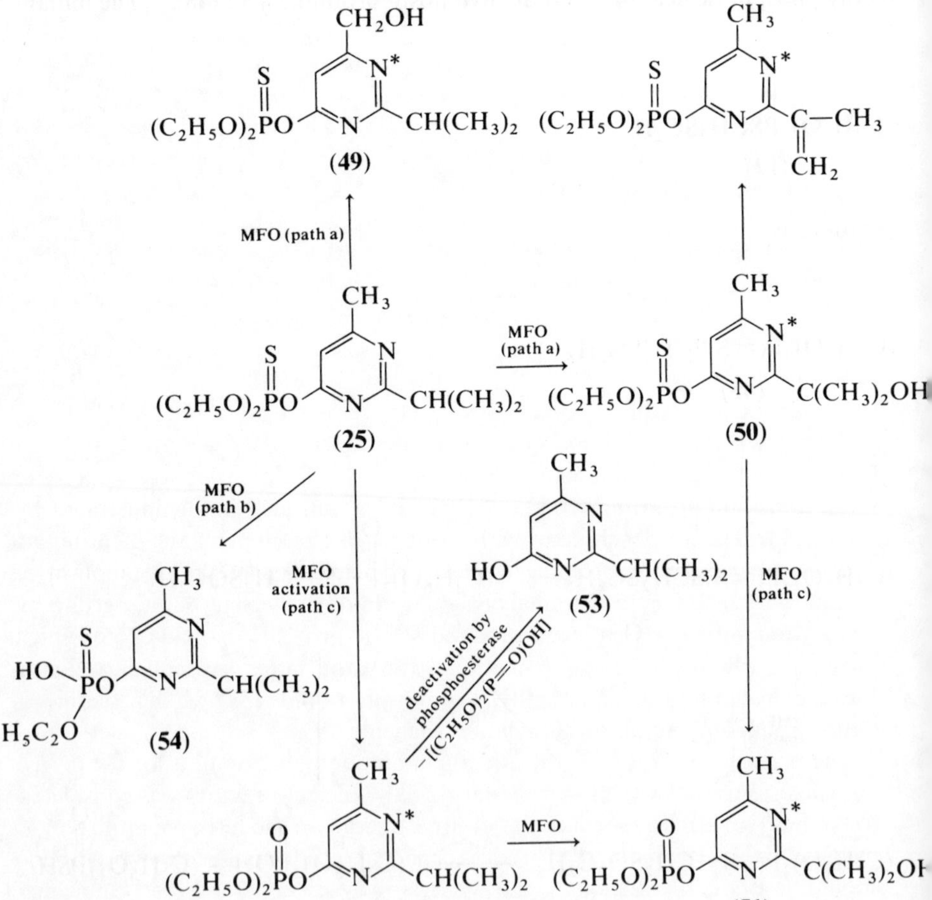

Scheme 6.37

towards insects is probably due to the relatively lower level of the oxo analogue found in mammals. Hydroxylation of the methyl and isopropyl side-chains by mixed function oxidases (path a, page 127) in rat liver, sheep, and cockroaches gives a series of active metabolites (**49** to **51**, Scheme 6.37).

Diazinon (**25**) was detoxified in rat liver and cockroaches by P—O-aryl cleavage promoted by glutathione (GSH) to give diethylphosphorothioic acid (**48**) and the pyrimidinyl glutathione (**52**) (Scheme 6.38).

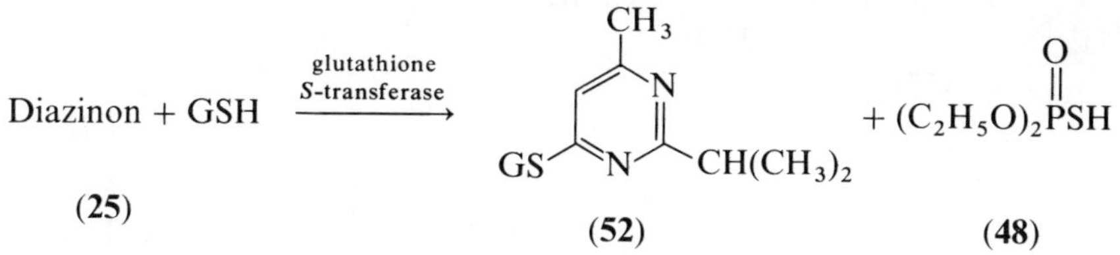

$$\text{Diazinon} + \text{GSH} \xrightarrow[\text{S-transferase}]{\text{glutathione}}$$

(**25**)

(**52**)   (**48**)

$+ (C_2H_5O)_2\overset{\displaystyle O}{\overset{\displaystyle \|}{P}}SH$

Scheme 6.38

Insects generally have poor transferase activity, although a diazinon-resistant strain of houseflies owed their tolerance to the presence of glutathione S-transferase.[2] Rat liver, but not insects, contains a phosphotriesterase with high specificity towards diazinon (**25**) which it hydrolyses to the inactive compounds diethylphosphoric acid and the 6-pyrimidinol (**53**)[1,2] (Scheme 6.37). The presence of this enzyme in mammals probably accounts for the selective toxicity of diazinon towards insects. The phosphoramidate, schradan (**4**, page 105), was one of the first systemic insecticides known and was active against aphids and mites on citrus, apple, and hops, but it suffers from the disadvantage of very high mammalian toxicity and is no longer used.

Schradan (**4**) itself is not an anticholinesterase agent and owes its high mammalian toxicity and insecticidal activity to *in vivo* activation by oxidation of the tertiary amino group. Oxidative activation can also be demonstrated by chemical treatment of schradan with oxidants like permanganate, dichromate, hydrogen peroxide, or bromine water.[17] The first step in the metabolism of schradan by insects and mammals is oxidation of the tertiary amino group by mixed function oxidases (path g, page 128) to give the N-oxide (**55**) which probably rearranges to the N-methylol (**56**), the latter subsequently giving formaldehyde and the heptamethylpyrophosphoramide (**57**), so that the bioxidation results in overall demethylation (Scheme 6.39).

The presence of oxygen in the dimethylamido moiety accounts for the *in vivo* anticholesterase activity of schradan; originally the active entity was considered to be the N-oxide (**55**) because the positive charge on the nitrogen atom would increase the electrophilicity of the phosphorus atom and hence the N-oxide should be a potent anticholesterase agent. However, later studies[1,2,17] indicated that the principal active metabolite was the methylol derivative (**56**), since the presence of this group permits hydrogen bonding (**58**) which will result in

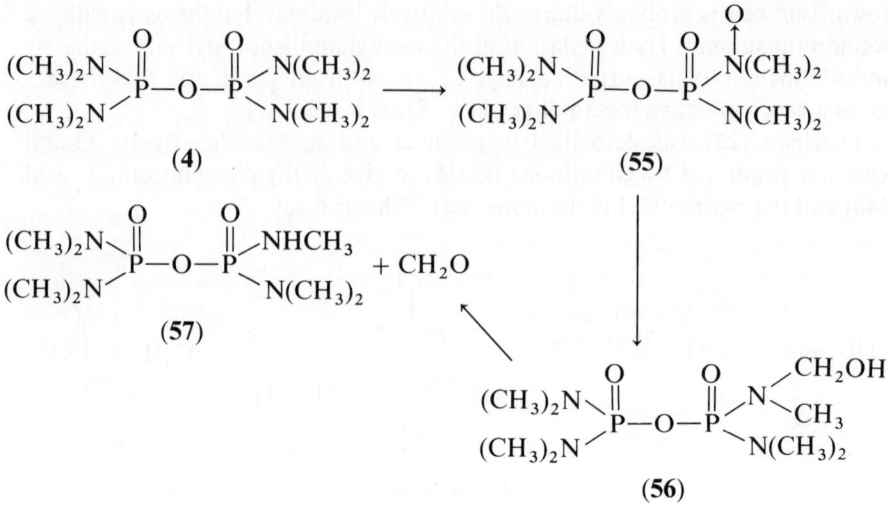

$$(CH_3)_2N\diagdown \underset{\underset{(CH_3)_2N\diagup}{}}{\overset{\overset{O}{\parallel}}{P}}-O-\overset{\overset{O}{\parallel}}{P}\diagdown \underset{N(CH_3)_2}{\overset{N(CH_3)_2}{}} \longrightarrow (CH_3)_2N\diagdown \underset{\underset{(CH_3)_2N\diagup}{}}{\overset{\overset{O}{\parallel}}{P}}-O-\overset{\overset{O}{\uparrow}}{\underset{}{P}}\diagdown \underset{N(CH_3)_2}{\overset{N(CH_3)_2}{}}$$

(4)   (55)

$$(CH_3)_2N\diagdown \underset{\underset{(CH_3)_2N\diagup}{}}{\overset{\overset{O}{\parallel}}{P}}-O-\overset{\overset{O}{\parallel}}{\underset{}{P}}\diagdown \underset{N(CH_3)_2}{\overset{NHCH_3}{}} + CH_2O$$

(57)

$$(CH_3)_2N\diagdown \underset{\underset{(CH_3)_2N\diagup}{}}{\overset{\overset{O}{\parallel}}{P}}-O-\overset{\overset{O}{\parallel}}{\underset{}{P}}\diagdown \underset{N(CH_3)_2}{\overset{N\diagdown^{CH_2OH}_{CH_3}}{}}$$

(56)

Scheme 6.39

increased electron drift from the phosphorus atom and should enhance its phosphorylating ability.

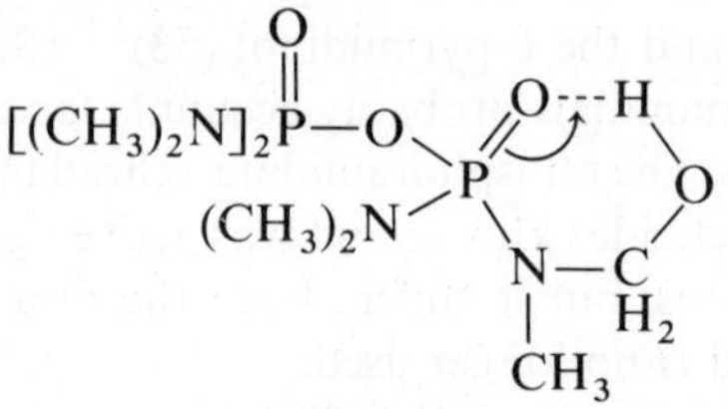

$$[(CH_3)_2N]_2\overset{\overset{O}{\parallel}}{P}-O\diagdown \underset{(CH_3)_2N\diagup}{\overset{}{P}}\diagdown \underset{N-C}{\overset{O\cdots H}{\diagdown_{O}}}$$
$$\underset{CH_3}{\overset{|}{}} \quad \overset{}{H_2}$$

(58)

The different metabolic pathways and the resultant balance of activation and detoxification processes, in many instances, account for the selective toxicity exhibited by several organophosphorus compounds. The most generally important metabolic activation process is microsomal oxidative desulphuration (path c, page 128), effected by oxidases whereby phosphorothioates are converted into the much more active phosphates (i.e. $P=S \rightarrow P=O$). The activation has been noted in the previous discussion of the metabolism of such insecticides as parathion (page 128), malathion (page 129), dimethoate (page 129), and diazinon (page 134).

Selectivity can sometimes be attributed to the differing activities of the oxidases from various insect species towards a given organophosphorus substrate. For instance, isopropyl parathion easily undergoes oxidative desulphuration by housefly oxidase but not by the bee enzyme, and this together with the greater sensitivity of housefly cholinesterase to isopropyl paraoxon accounts for the selective toxicity of isopropyl parathion to houseflies as compared to bees.

Steric effects may also be significant in some cases; for instance in the thiono analogues of mevinphos (page 114), the *E* isomer was selectively oxidized (activated) by mouse liver oxidase as compared to the *Z* isomer.

Metabolic oxidation of sulphides by MFO leading to formation of the corresponding sulphoxides and sulphones also leads to some activation of the original insecticide since oxidation enhances electron removal from the phosphorus atom. Examples are phorate (page 133) and demeton-methyl (page 108); such sulphide oxidation is valuable because its occurrence in plants and soils increases the persistency of action of systemic insecticides.

The metabolic oxidation of amide groups by MFO (path g, page 128) also results in activation. The classic example of this effect is schradan where overall oxidative demethylation occurs (page 135). Dimefox (**1**; page 105) is similarly activated. Another interesting example is dicrotophos (**59**) (page 115).

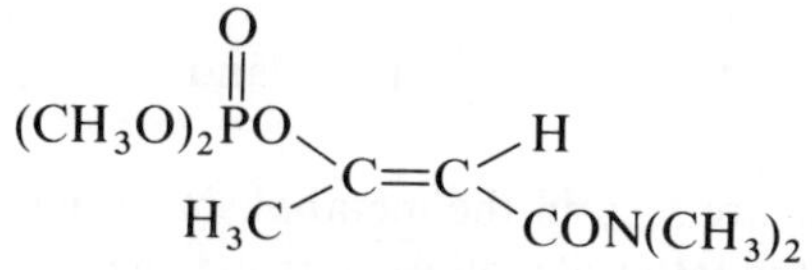

$$(CH_3O)_2\overset{\overset{\displaystyle O}{\|}}{P}O \underset{H_3C}{\diagdown} C = C \underset{CON(CH_3)_2}{\overset{H}{\diagup}}$$

(59)

This compound is mainly the *E*-crotonamide (**59**), which is more active than the *Z* isomer and was introduced by Ciba-Geigy in 1963 and by Shell in 1965. It acts as a systemic insecticide and acaricide of moderate persistence which is very effective against sap-feeding insect pests.[5,6] The metabolism of (**59**) involves oxidative demethylation to another insecticide monocrotophos (**60**) developed particularly for control of caterpillars and bollworms on cotton[5] (Scheme 6.40).

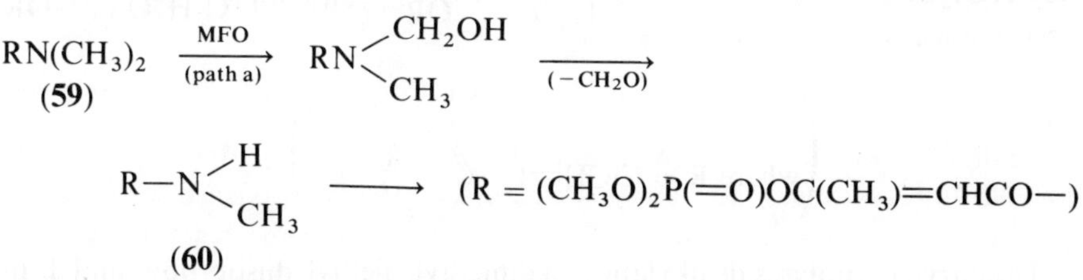

$$RN(CH_3)_2 \xrightarrow[\text{(path a)}]{MFO} RN \underset{CH_3}{\overset{CH_2OH}{\diagdown}} \xrightarrow[(-CH_2O)]{}$$

(59)

$$R-N \underset{CH_3}{\overset{H}{\diagdown}} \longrightarrow (R = (CH_3O)_2P(=O)OC(CH_3)=CHCO-)$$

(60)

Scheme 6.40

This is finally degraded to the unsubstituted amide (**61**) with the production of at least four active metabolites which contribute to the persistency of action of (**59**):

$$R-N \underset{CH_2OH}{\overset{H}{\diagdown}} \longrightarrow RNH_2 + CH_2O$$

(61)

The side-chains of arylalkanes may be oxidized by MFO (path a, page 127) to give alcohols which may be further metabolized by elimination or conjugation, e.g. the metabolism of diazinon (page 134).

Various metabolic processes also result in detoxification of organophosphorus insecticides. Such processes generally involve phosphorus ester bond cleavage which introduces a negative charge into the molecule so destroying activity as a phosphorylating agent. The products are also much more water soluble and so are readily excreted in the urine. Phosphorus ester bonds are cleaved by hydrolysis catalysed by phosphoroesterases which occur widely in mammalian tissues, insects, and microorganisms. These esterases also hydrolyse some carbamic and carboxylic esters. Phosphatases only hydrolyse partial esters of orthophosphoric acid but other esterases hydrolyse a variety of neutral phosphorus compounds, cleaving the bond between the phosphorus atom and the most acidic group or the so-called leaving group. As well as phosphorus ester bonds, bonds like

$$
\underset{}{>}\!\!\overset{\overset{\displaystyle O}{\|}}{P}\!\!-\!O\!-\!\overset{\overset{\displaystyle O}{\|}}{P}\!\!<, \qquad \overset{\overset{\displaystyle O}{\|}}{>\!P}\!\!-\!F \quad \text{and} \quad \overset{\overset{\displaystyle O}{\|}}{>\!P}\!\!-\!C\!\equiv\!N
$$

are cleaved. Examples are seen in the metabolism of parathion (page 128) and diazinon (page 134); the latter also illustrates the enzymic cleavage of a P—O-aryl bond in the presence of glutathione. Liver MFO often detoxify insecticides by oxidative dealkylation (path b, page 128); for instance, chlorfenvinphos (page 112) is deethylated:[15]

$$
(C_2H_5O)_2\overset{\overset{\displaystyle O}{\|}}{P}OR \xrightarrow[\substack{\text{NADPH}_2/O_2}]{\text{MFO}} \overset{\overset{\displaystyle OH}{|}}{CH_2CHO}\underset{C_2H_5O}{\overset{}{}}\!\!P\!\!\overset{\diagup O}{\diagdown OR} \xrightarrow{(-CH_3CHO)} \overset{HO}{\underset{C_2H_5O}{}}\!\!P\!\!\overset{\diagup O}{\diagdown OR}
$$

Chlorfenvinphos

$$
\left(\text{where } R = ClCH\!=\!\overset{\overset{\displaystyle Cl}{|}}{C}\!-\!\!\bigcirc\!\!-\!\!Cl \right)
$$

The enzyme catalyses dealkylation of dimethyl, diethyl, diisopropyl, and di-n-butyl esters, although generally ethyl esters appear to be the favoured substrates. Demethylation is more often effected by glutathione S-alkyltransferase; thus the demethylation of methyl parathion by rat liver MFO is greatly enhanced by the presence of glutathione (GSH) and similar results were noted with mevinphos (page 114) and tetrachlorvinphos (page 113). In each case, the products were the monodemethylated pesticide and S-methylglutathione:

$$
\overset{CH_3O}{\underset{CH_3O}{}}\!\!\overset{\overset{\displaystyle X}{\|}}{P}\!\!-\!Y \xrightarrow[\substack{\text{transferase}\\(\text{detoxification})}]{\text{glutathione } S\text{-alkyl}} \overset{HO}{\underset{CH_3O}{}}\!\!\overset{\overset{\displaystyle X}{\|}}{P}\!\!-\!Y + GS\!-\!CH_3
$$

(X = O or S)

Generally there is much greater alkyltransferase activity in mammalian liver than in insects and no transferase activity appears in sucking insects and mites, which probably accounts for the selective toxicity shown by the dimethyl phosphorus esters towards these species.

The metabolic cleavage of S-alkylphosphorothiolate bonds occurs by an oxidative mechanism for thiolothionates or by a hydrolytic mechanism for thiolate esters, although the detoxification of malathion (16) is chiefly due to hydrolysis by carboxyesterase (page 129).

Menazon (22a; page 117) in rats is principally degraded at the P—S bond and the main metabolite excreted is the 2-methylsulphone (62) (Scheme 6.41).

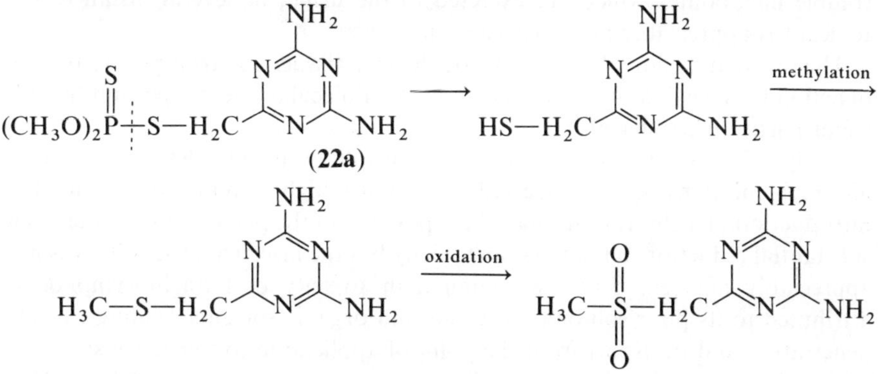

Scheme 6.41   (62)

The early organophosphorus insecticides like parathion, schradan, and tetraethyl pyrophosphate (TEPP) were highly active compounds but were also extremely toxic to mammals, and were the most dangerous chemicals ever to be used in agriculture. In applying such compounds, operators had to wear full protective clothing and a respirator;[5] they have caused several human fatalities and any birds or small mammals covered by the spray are killed. However, most organophosphorus compounds are comparatively rapidly biodegraded to non-toxic, water-soluble compounds which are quickly excreted by animals; consequently, unlike the organochlorine insecticides, they do not accumulate in the environment (see Chapter 15, page 350).[18]

Many organophosphorus compounds function as systemic insecticides which enable smaller amounts of the active ingredient to be used more effectively and reduce the harmful effects on natural predators. The widespread increase of insect strains resistant to organophosphorus insecticides has reduced the scope for the introduction of new compounds in this group. The enzymic hydrolysis of the neurotransmitter acetylcholine appears essential for the proper functioning of the nervous system of insects and mammals.[13] At first sight, the selective toxicity of many organophosphorus compounds, like malathion and tetrachlorvinphos, seems surprising; however, out of many thousands of organophosphorus compounds which have been synthesized there are now hundreds of insecticides with very favourable insect/mammalian toxicity ratios. Selectivity depends on several factors such as differential penetration, transport, sensitivity of the target site, and metabolism.[6,11]

The balance of metabolic activation and deactivation processes occurring in a given insect species is often critical. Various insect species may have different enzyme systems exhibiting different levels of activity towards the organophosphorus substrate; thus metabolism may be responsible for selectivity, e.g. malathion (page 129). In some cases, the organophosphorus compound may penetrate the insect cuticle more readily than the mammalian skin. Certainly, organophosphorus insecticides are rapidly biodegraded in mammals to water-soluble metabolites which are excreted in the urine; in several instances the toxicant is appreciably more persistent in insects.

The rate of diffusion of the insecticide through tissues and transport to the site of action is dependent on such factors as the molecular size, shape, and the oil/water partition coefficient.[6]

Study of a series of 2-chlorovinyl phosphate insecticides showed that although chlorfenvinphos (page 112) has a fairly high mammalian toxicity, the introduction of a chlorine atom at the 5-position of the phenyl ring resulted in a substantial reduction in mammalian toxicity but the insecticidal activity was not appreciably affected. The low mammalian toxicity of tetrachlorvinphos is attributed to its poor solubility in water and organic solvents, leading to slow penetration and transport from the point of application to the active site.

Acetylcholine and ionized anticholinesterase agents such as amiton (page 126) are extremely toxic to mammals and to certain insects, for instance aphids, mites, and scale insects, but are not effective against other species. Schradan (page 105) shows a similar spectrum of activity, and the selectivity arises from differences in the composition of the nervous system of insects and vertebrates. In insects, the nerve junctions are protected from ionic materials by a lipid nerve sheath and the thickness of this lipid barrier appears to be the decisive factor determining the resistance of insects towards schradan. Tolerant insects like American cockroaches and houseflies have a thick sheath whereas susceptible insects, such as rice bugs and green leafhoppers, have only a thin membrane.

The actual receptor site may also vary in different organisms so that, in spite of the specifically similar mechanism of toxicity of organophosphorus compounds (see page 123) in animals and insects, there are still a number of factors that can be exploited to obtain compounds showing selective toxicity towards a given target insect pest.

CARBAMATES

Carbamates are an economically important group of insecticides. Since 1958 they have increased their share of the market; in 1983, the estimated market value was $1100 million. They have the advantages of rapid action and a reasonable rate of biodegradation. Metabolism in the target organism limits toxicity and can provide the basis for selective toxicity; they are metabolized in plants which prevents the accumulation of potentially harmful residues. Some forty carbamates, mostly phenylmethylcarbamates, are currently used, and many are systemic in plants which enhances their value.[6]

The successful development of organophosphorus insecticides stimulated examination of other compounds known to possess anticholinesterase activity. One such compound is the alkaloid physostigmine (63), the active ingredient in calaban beans which has been used for trial by ordeal in West Africa.[9,19] The physiological properties of this alkaloid were supposed to be based on the phenylmethylcarbamate part of the structure and led to the discovery of a number of parasympathomimetic drugs like neostigmine (64).

The compounds being quite strong bases are ionized in aqueous solution and therefore have very low lipid solubility. Consequently they are unable to penetrate the ion-impermeable sheath surrounding the insect nervous system (cf. page 140). Therefore, efforts were made to synthesize compounds in which the N-substituted carbamate part of the molecule was attached to a less basic, more lipophilic moiety, since such compounds should show greater insecticidal activity. In 1951, the Geigy Company introduced 1-isopropyl-3-methylpyrazolyl-5-dimethylcarbamate (65). This water-soluble compound was a most effective systemic aphicide and was also active against houseflies, but showed very high mammalian toxicity (LD_{50} (oral) to rats $\simeq 12$ mg/kg) so the compound was not extensively developed.

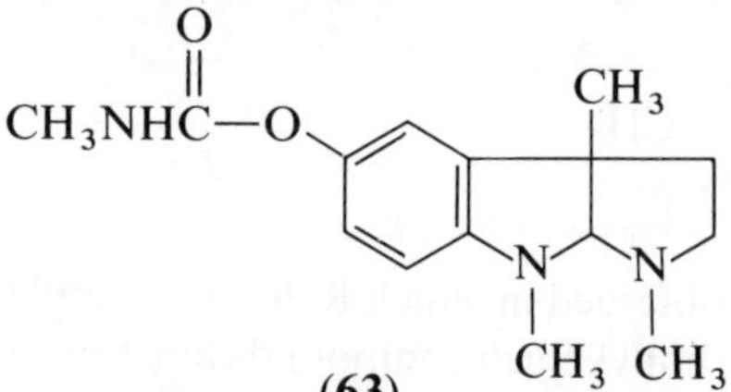

(63)

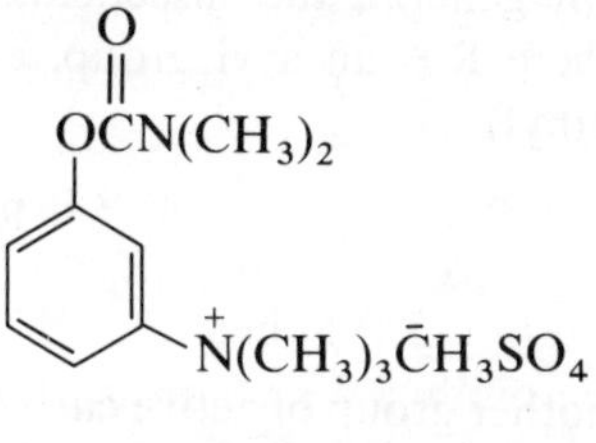

(64)

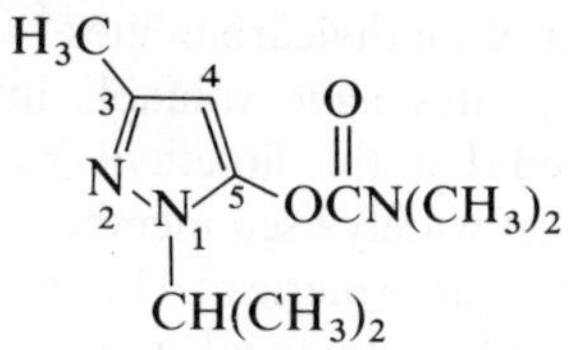

(65)

Later work showed[3,9] that in the dimethylcarbamate series heteroaromatic derivatives, especially those containing the pyrimidine nucleus, had much lower mammalian toxicities and in 1968 Imperial Chemical Industries Limited introduced pirimicarb (66), synthesized from ethyl α-methyl acetoacetate and dimethylaminoacetamidine (Scheme 6.42). Pirimicarb is a fast-acting specific

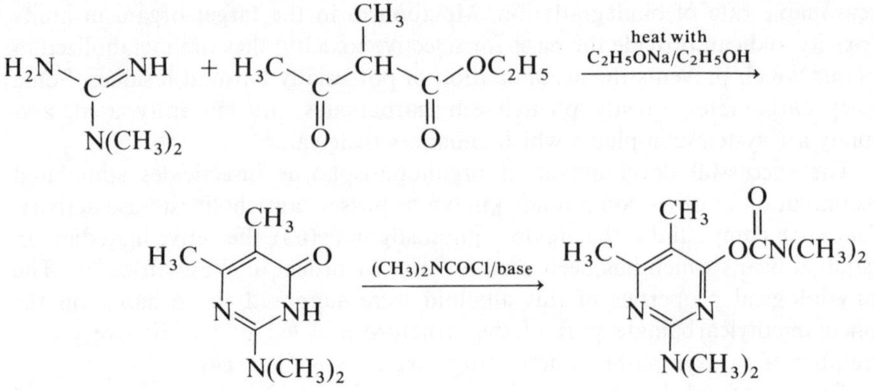

(66)

Scheme 6.42

systemic aphicide and is also effective against aphids that have developed resistance to organophosphorus insecticides[5,6] (Plate 8). The chemical is generally applied as a foliar spray to control aphids in farm and horticultural crops. It is also taken up by plant roots and translocated in the xylem and has only moderate mammalian toxicity (LD_{50} (oral) to rats $\simeq 147$ mg/kg). Pirimicarb has little toxicity towards bees, ladybirds, and other insects; hence it is valuable in integrated control programmes.

In general, the insecticidal carbamates have the following structure (**70**) (where R = an aryl group, e.g. phenyl, naphthyl, or heterocyclic; R' = H or methyl).

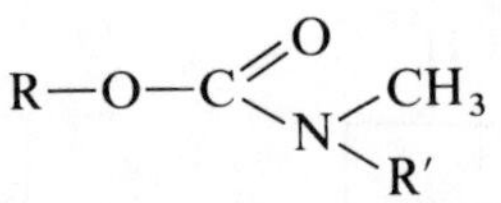

Another group of active carbamates can be obtained in which R' is, for example, acyl, sulphenyl, phosphorothio, or aminocarbonyl. Such compounds function as proinsecticides since they can be readily metabolized by reaction with nucleophiles to the corresponding N-methylcarbamates.[6]

Many monomethylcarbamates have valuable insecticidal activity and are much more easily hydrolysed than the dimethylcarbamates. Phenol carbamates are especially useful; the most widely used member is the α-naphthyl derivative, carbaryl (**67**), introduced by the American Union Carbide Company in 1956 which was the first successful commercial carbamate insecticide. Carbaryl is manufactured from α-naphthol[3] (Scheme 6.43). Carbaryl is a contact insecticide and fruit thinner with a broad spectrum of activity—effective against many pests of fruit, vegetables, and cotton. It can be applied to control earthworms and leatherjackets in turf[5,6] and can sometimes be used instead of DDT to reduce environmental pollution since it biodegrades and hence does not accumulate in the ecosystem.

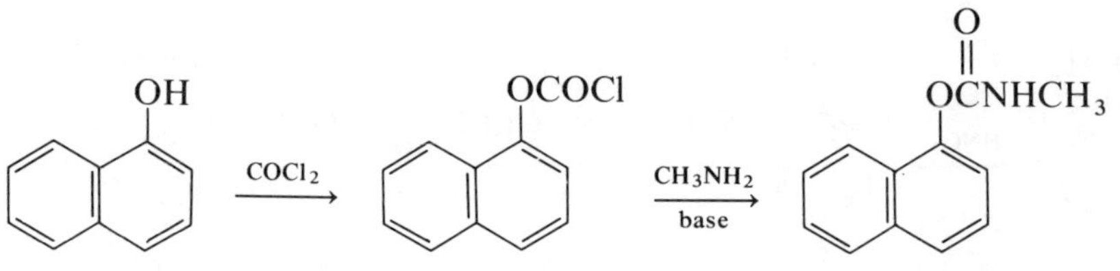

α-Naphthol   (67)

Scheme 6.43

Several other methylcarbamates **(68)** to **(71)** are useful insecticides. The *N*-methylcarbamates are conveniently obtained by reaction of the appropriate hydroxy compound with methylisocyanate (Scheme 6.44). This preparative route may be illustrated by the synthesis of the widely used insecticide, carbofuran (Scheme 6.45).

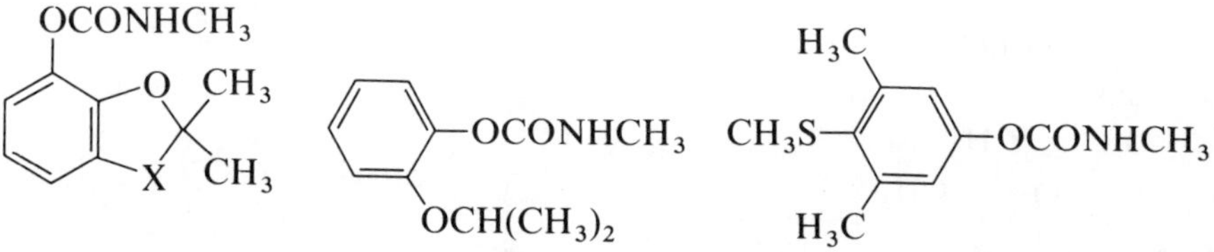

(68)   **(69)**   **(70)**

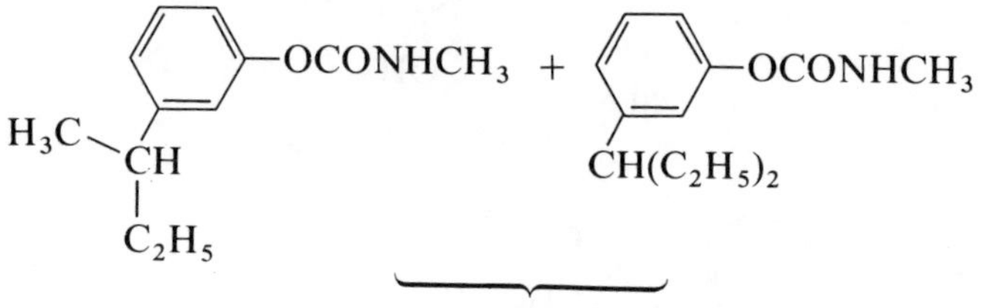

(71)

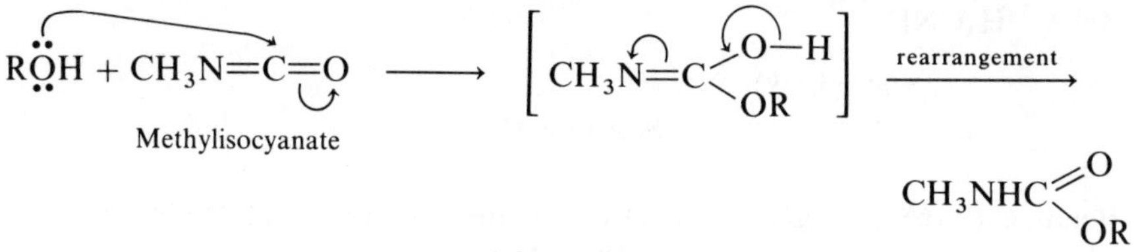

Methylisocyanate

$$CH_3NHC\!\!\begin{array}{c} O \\ \diagdown OR \end{array}$$

Scheme 6.44

Carbofuran (**68**; X = CH) is a broad-spectrum systemic insecticide, acaricide, and nematicide used on foliage at 0.5–4 kg/ha against insects and mites or incorporated in soil at 6–10 kg/ha for control of soil insects and nematodes.[5,6] The compound has high mammalian toxicity (LD_{50} (oral) to rats 8–14 mg/kg) but is rapidly metabolized to non-toxic products in plants and animals.[9]

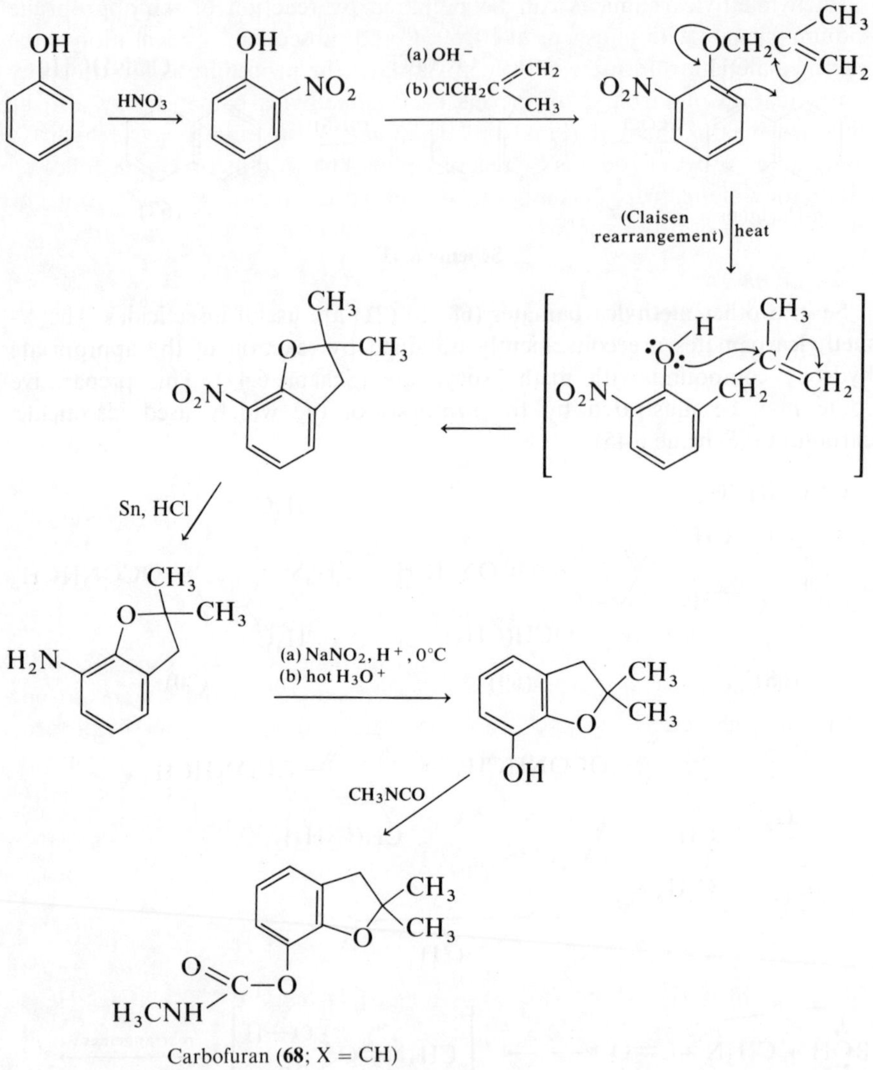

Carbofuran (**68**; X = CH)

Scheme 6.45

Bendiocarb (**68**; X = O), a contact insecticide with limited systemic action, is used to control wireworms and beetles in beet and maize.[6]

Propoxur (**69**) is used as a fumigant in greenhouses against whitefly and aphids[5] with moderate mammalian toxicity of LD_{50} (oral) to rats $\simeq 150$ mg/kg. It has residual and knockdown action on flies and mosquitoes and can be used as a domestic insecticide[9] (Plate 9).

Methiocarb (**70**) is a contact insecticide, which is also a useful molluscicide and bird repellent. Benfencarb (**71**) is a valuable soil insecticide.

Phenylmethylcarbamates can be prepared by reaction of the appropriate sodium phenate with phosgene at 10–50 °C and subsequent condensation of the intermediate chloroformate with methylamine; the procedure is illustrated by the formation of carbaryl (67) (page 142). Dimethylcarbamates may also be obtained by this method using dimethylamine. With more reactive hydroxy compounds, a better route is by reacting phosgene with methylamine followed by addition of the hydroxy compound to methylisocyanate. This is illustrated by the formation of propoxur (69) from catechol (Scheme 6.46).

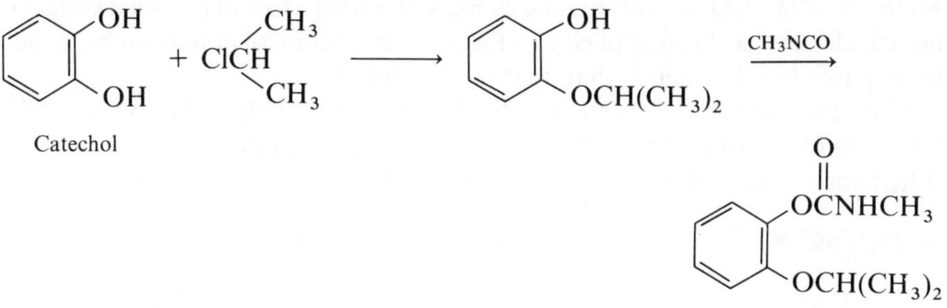

Scheme 6.46

Another group of insecticidal carbamates are the carbamoyloximes; the best-known example is aldicarb (72) which was introduced by the Union Carbide Company in 1965[3,5,6] and can be prepared from 2-methylpropene (Scheme 6.47). Aldicarb (72) is a systemic insecticide, acaricide, and nematicide which is formulated as granules for soil incorporation. It is effective for control

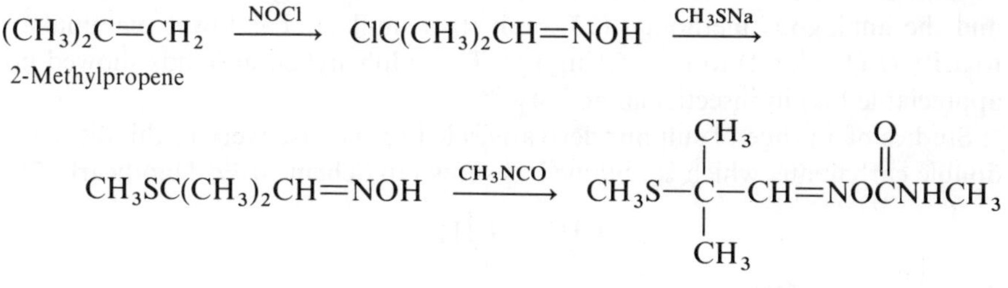

$$(CH_3)_2C{=}CH_2 \xrightarrow{\text{NOCl}} ClC(CH_3)_2CH{=}NOH \xrightarrow{\text{CH}_3\text{SNa}}$$

2-Methylpropene

(72)

Scheme 6.47

of aphids, nematodes, flea beetles, leaf miners, thrips and whiteflies on a wide range of crops. Aldicarb is readily translocated in plants after soil application where it is metabolized to the sulphoxide and the sulphone which are also active. The disadvantage of aldicarb (72) is the very high mammalian toxicity (LD_{50} (oral) to rats $\simeq 1$ mg/kg); hence it is only marketed as granules.[5]

Another example of an insecticidal oxime carbamate is provided by methomyl (73) (1968) which is obtained from acetaldehyde as shown in Scheme 6.48.

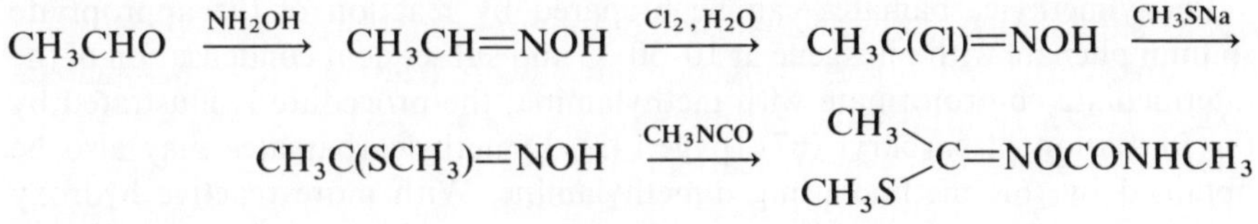

$$CH_3CHO \xrightarrow{NH_2OH} CH_3CH{=}NOH \xrightarrow{Cl_2,H_2O} CH_3C(Cl){=}NOH \xrightarrow{CH_3SNa}$$

$$CH_3C(SCH_3){=}NOH \xrightarrow{CH_3NCO} \underset{CH_3S}{\overset{CH_3}{>}}C{=}NOCONHCH_3$$

(73)

Scheme 6.48

Methomyl (**73**) (LD_{50} (oral) to rats, 20 mg/kg) is used as a soil and seed systemic insecticide and has been applied as a foliar spray to control aphids on hops, but its sulphoxide is less active than that from aldicarb.

Two other carbamoyl oxime insecticides are oxamyl (**74**) and thiofanox (**75**). Oxamyl (**74**), introduced by Du Pont (1969), is a soil-applied systemic insecticide, nematicide, and acaricide (LD_{50} (oral) to rats, 5 mg/kg) with a similar

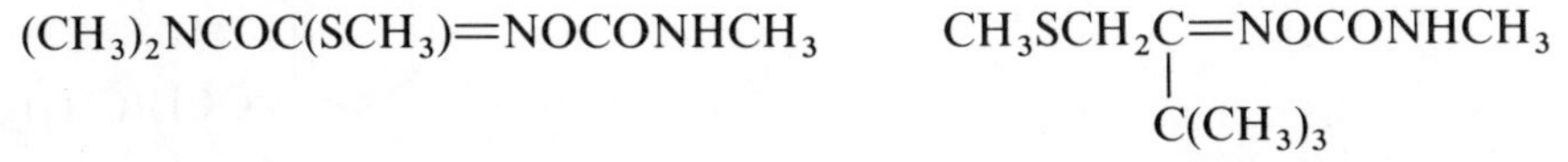

$$(CH_3)_2NCOC(SCH_3){=}NOCONHCH_3 \qquad\qquad CH_3SCH_2\underset{\underset{C(CH_3)_3}{|}}{C}{=}NOCONHCH_3$$

(74)   (75)

spectrum of activity to aldicarb.[6] Thiofanox (**75**) (Shell, 1973) is a soil-applied systemic insecticide used chiefly against aphids, capsids, and leafhoppers in sugar beet, potatoes, and some other crops.[6]

The problem with the methylcarbamoyl oxime insecticides is their high mammalian toxicity. Many efforts were made to alter the spectrum, but these were unsuccessful until the discovery of *N*-sulphonation (1974). Aldicarb (**72**) is safened 60 times by formulation as the *N*-trichloromethylsulphenyl derivative and the analogous methomyl (**73**) derivative similarly has lower mammalian toxicity (LD_{50} (oral) to rats, 50 mg/kg). The sulphenyl compounds showed no appreciable loss in insecticidal activity.[20]

Studies of a range of sulphur derivatives led to the discovery of thiodicarb, a double carbamate, which is obtained as shown in Scheme 6.49. Thiodicarb (**79**)

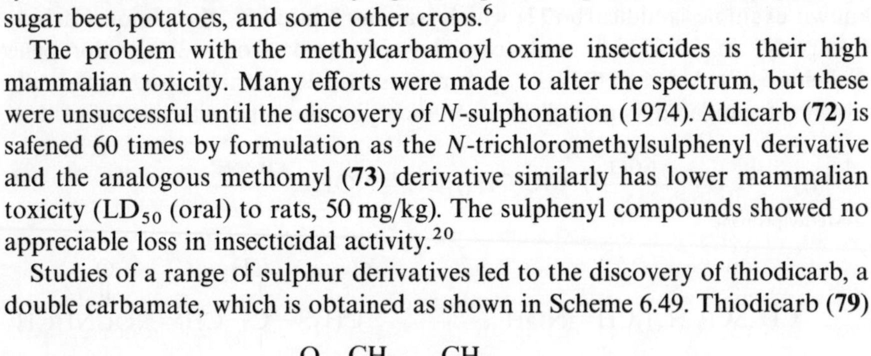

$$2\;\underset{CH_3S}{\overset{CH_3}{>}}C{=}NOH + F{-}\overset{\overset{\displaystyle O}{\|}}{C}{-}\overset{\overset{\displaystyle CH_3}{|}}{N}{-}S{-}\overset{\overset{\displaystyle CH_3}{|}}{N}{-}\underset{\underset{\displaystyle O}{\|}}{C}{-}F \longrightarrow$$

$$\underset{CH_3S}{\overset{CH_3}{>}}C{=}NO\overset{\overset{\displaystyle O}{\|}}{C}{-}\overset{\overset{\displaystyle CH_3}{|}}{N}{-}S{-}\overset{\overset{\displaystyle CH_3}{|}}{N}{-}\underset{\underset{\displaystyle O}{\|}}{C}{-}ON{=}C\underset{SCH_3}{\overset{CH_3}{<}}$$

(79)

Scheme 6.49

(1985) shows high activity against lepidoptera and other pests on cotton, soybeans, corn, and other crops, combined with relatively low mammalian toxicity (LD_{50} (oral) to rats, 431 mg/kg).[6]

Other *N*-sulphenyl-*N*-methylcarbamates, used as commercial insecticides, include carbosulfan and furathiocarb.[3,6] Carbosulfan, a sulphenylated derivative of carbofuran (**68**: X = CH), acts as a contact and systemic insecticide

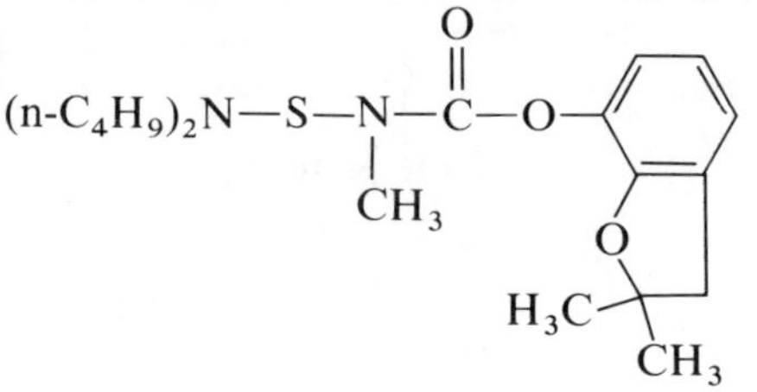

Carbosulfan 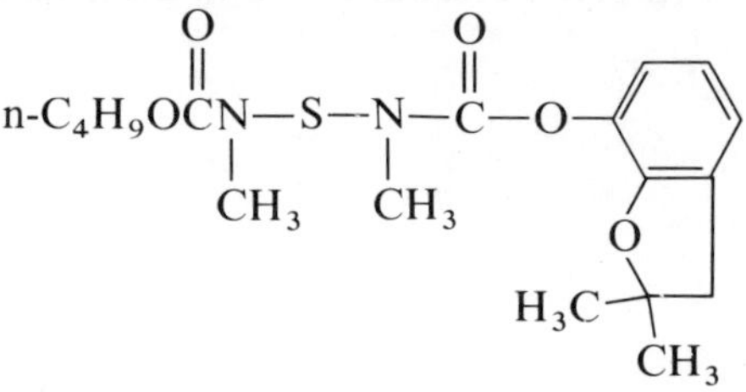   Furathiocarb

which can be applied to the foliage or the soil. It has a significantly lower mammalian toxicity (LD_{50} (oral) to rats, 209 mg/kg) as compared with carbofuran. Furathiocarb (LD_{50} (oral) to rats, 106 mg/kg) is a contact and systemic soil insecticide.

There is considerable potential for the development of new sulphenylated and sulphinylated carbamate insecticides since these compounds generally show lower mammalian toxicity with better residual insecticidal activity and lower phytotoxicity as compared with the parent carbamates. The *N*-sulphenyl-*N*-methylcarbamates are examples of proinsecticides since they appear to be activated *in vivo* by metabolism to the corresponding *N*-methylcarbamates.[3,6] Some sulphenyl derivatives of 3,4-methylenedioxyphenyl-*N*-methylcarbamate were particularly effective against houseflies. The methylenedioxy group appeared to exert a synergistic effect, probably because it inhibits metabolic detoxification by mixed function oxidases (see Chapter 4, page 72).

The metabolism of carbamates in plants and mammals is dominated by hydrolysis to the phenol, oxime, or other hydroxy compound together with

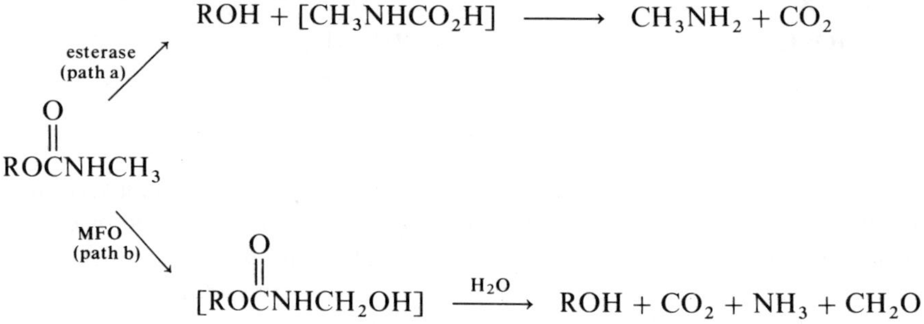

Scheme 6.50

methylcarbamic acid which decomposes to ammonia and carbon dioxide. The phenols and other hydroxy compounds form water-soluble conjugates with sugars (glucuronides) and sulphates which in mammals are excreted in urine.[6]

Carbamates are metabolized by two basic mechanisms, both involving breakage of the carbamate ester linkage, namely by direct esterase attack (path a) or by initial oxidation by mixed function oxidases (MFO) followed by hydrolytic breakdown of an unstable intermediate[16] (path b) (Scheme 6.50).

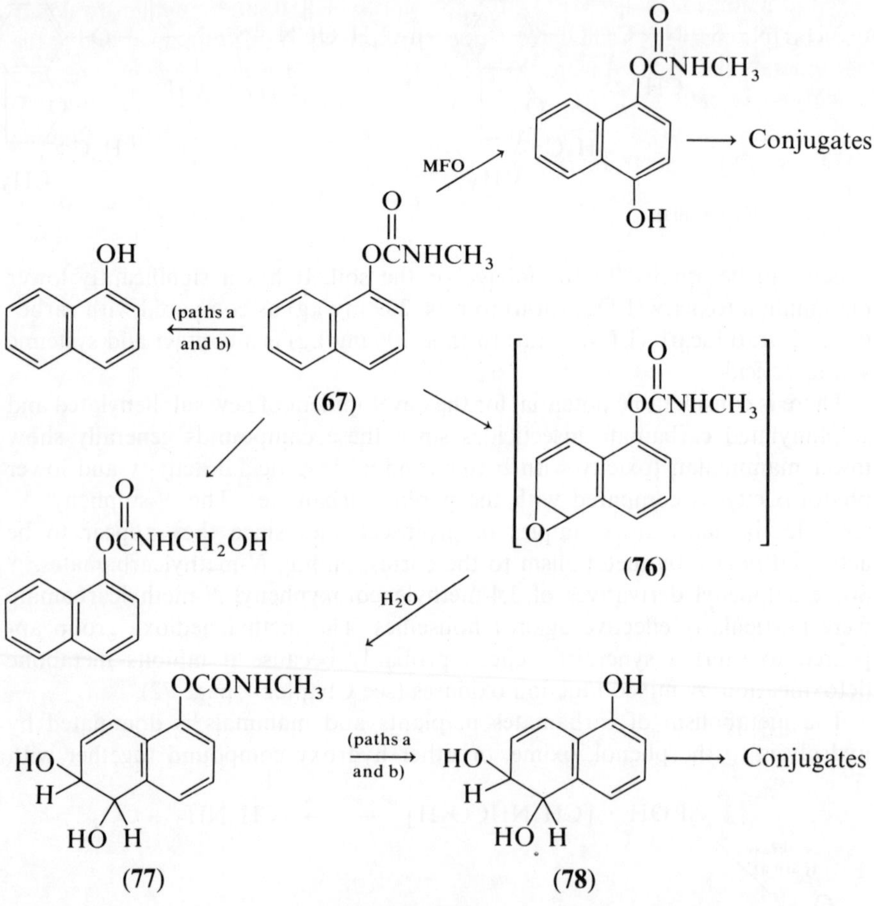

Scheme 6.51

The metabolism of carbaryl (**67**) has been extensively examined in insects and mammals.[6,15] Scheme 6.51 shows the metabolic pathways determined by experiments *in vivo* in rabbits, houseflies, and cockroaches and *in vitro* by rat, mouse, and rabbit liver microsomes. The postulated epoxy compound (**76**) by ring opening gives the diol (**77**), and this is subsequently metabolized to the phenolic diol (**78**).

In rats, cattle, sheep, and man, 70% of an oral dose of carbaryl is eliminated in the urine as the water-soluble glucose conjugates (glucuronides) within 24 hours of treatment.[6,17] The metabolism is complex and thin-layer chromatography of the urine from treated rabbits indicated the presence of some five additional metabolites which may be conjugates.

The oxime carbamate aldicarb (72) is metabolized in cotton plants and houseflies in a similar manner to organophosphorus thioether derivatives such as phorate (see page 133) (Scheme 6.52). Aldicarb (72) is rapidly metabolized by mixed function oxidases (MFOs) to the sulphoxide (80) and much more slowly to the sulphone (81).[16] The latter appears relatively stable in cotton and is the main metabolite isolated from the plants two months after treatment. Enzymic hydrolysis of (80) and (81) yields the corresponding oximes (82) and (83) respectively. The metabolism in mammals is similar except that the sulphone (81) does not appear to be formed.[17]

Carbamates, like organophosphorus compounds, owe their insecticidal properties to inhibition of the enzyme acetylcholinesterase—the resultant accumulation of acetylcholine preventing effective nervous transmission across the synapse (see page 123). The enzyme is poisoned by carbamoylation of the

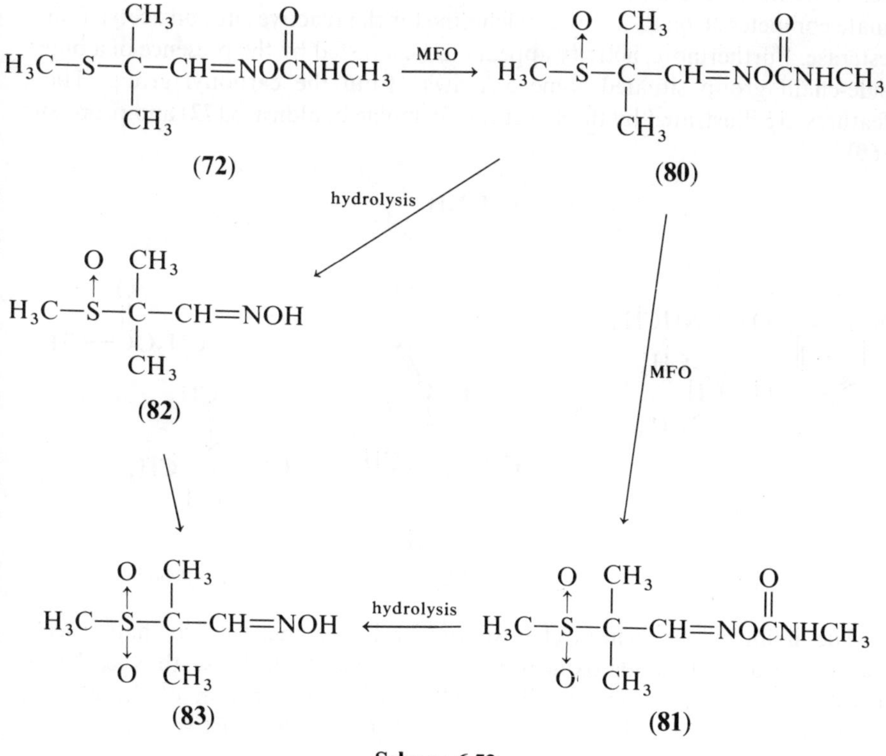

$$\underset{(72)}{\overset{\displaystyle CH_3}{\underset{\displaystyle CH_3}{H_3C-S-\overset{|}{\underset{|}{C}}-CH=NO\overset{\displaystyle O}{\overset{||}{C}}NHCH_3}}} \xrightarrow{MFO} \underset{(80)}{\overset{\displaystyle O\ \ CH_3}{\underset{\displaystyle CH_3}{H_3C-\overset{\uparrow}{S}-\overset{|}{\underset{|}{C}}-CH=NO\overset{\displaystyle O}{\overset{||}{C}}NHCH_3}}}$$

hydrolysis

$$\underset{(82)}{\overset{\displaystyle O\ \ CH_3}{\underset{\displaystyle CH_3}{H_3C-\overset{\uparrow}{S}-\overset{|}{\underset{|}{C}}-CH=NOH}}}$$

MFO

$$\underset{(83)}{\overset{\displaystyle O\ \ CH_3}{\underset{\displaystyle O\ \ CH_3}{H_3C-\overset{\uparrow}{\underset{\downarrow}{S}}-\overset{|}{\underset{|}{C}}-CH=NOH}}} \xleftarrow{hydrolysis} \underset{(81)}{\overset{\displaystyle O\ \ CH_3}{\underset{\displaystyle O'\ \ CH_3}{H_3C-\overset{\uparrow}{\underset{\downarrow}{S}}-\overset{|}{\underset{|}{C}}-CH=NO\overset{\displaystyle O}{\overset{||}{C}}NHCH_3}}}$$

Scheme 6.52

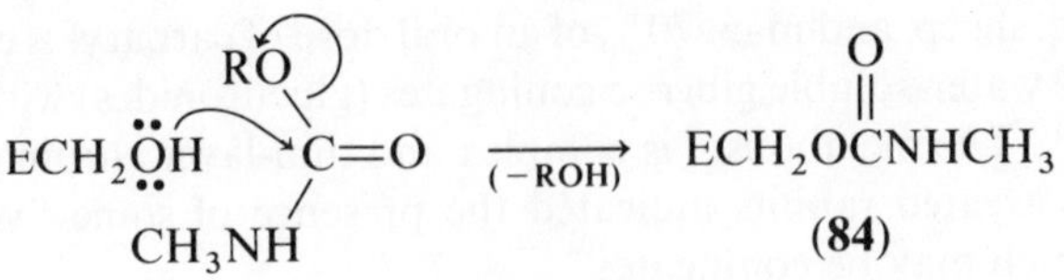

$$\text{ECH}_2\overset{..}{\underset{..}{\text{O}}}\text{H} \quad \overset{\displaystyle R\overset{\displaystyle O}{\frown}}{\underset{\displaystyle CH_3NH}{\diagdown}} C=O \quad \xrightarrow[(-ROH)]{} \quad \text{ECH}_2\text{O}\overset{\displaystyle O}{\overset{\|}{\text{C}}}\text{NHCH}_3$$
$$(84)$$

Scheme 6.53

primary hydroxyl group of a serine residue of the enzyme (Scheme 6.53). The carbamoylated enzyme (84) is only slowly hydrolysed back to the active enzyme. However, unlike the organophosphorus compounds, the structure of the leaving group RO^- is of critical importance in determining the insecticidal activity of carbamates.[11,12] Generally the rate of hydrolytic breakdown of the carbamoylated enzyme is intermediate between that of the acetylated and phosphorylated enzyme so that acetylcholinesterase is inactivated for a significant time. However, as with the organophosphates, the *in vitro* anticholinesterase properties of carbamates often bear little relation to their *in vivo* insecticidal activity because of the importance of such additional factors as ease of penetration and metabolism.

For insecticidal activity, carbamates appear to require a degree of structural resemblance to the natural enzyme substrate acetylcholine, so that the carbamate competes strongly with acetylcholine for the reactive sites on acetylcholinesterase. Furthermore, activity appears to be assisted by the presence of a bulky side-chain group situated some 5 Å away from the carbonyl group. These features are illustrated by the structural formulae of aldicarb (72) and propoxur (69).

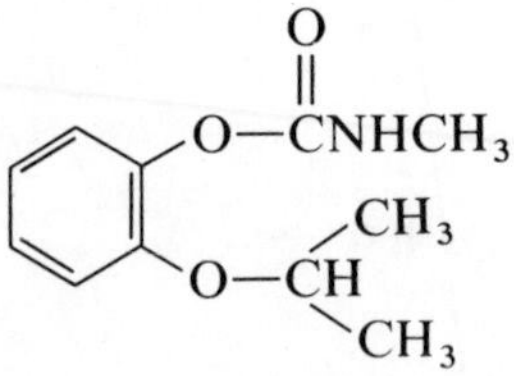

| (69) | (72) | Acetylcholine |

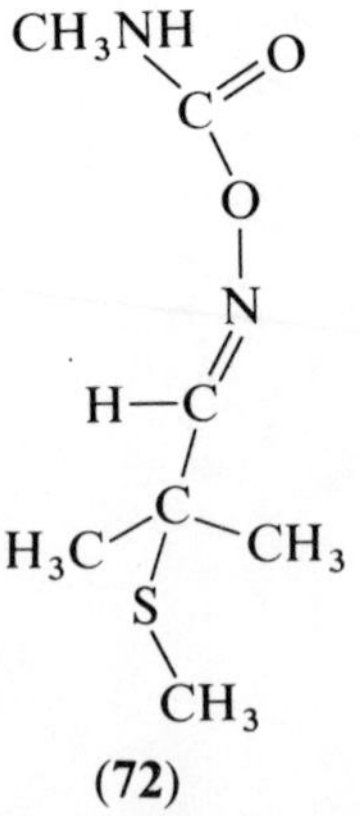

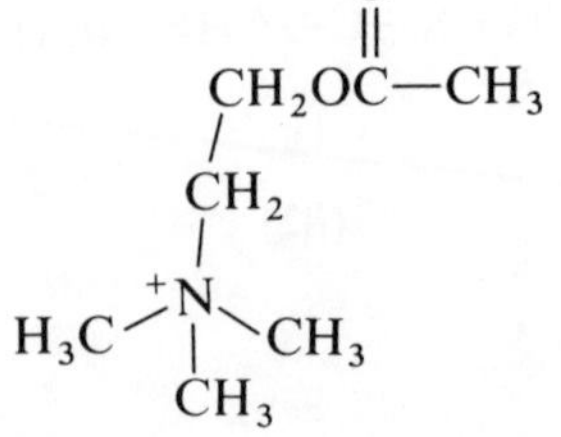

The major biochemical mechanism of insect resistance to carbamate insecticides appears to be detoxification via enzymic hydrolysis; thus carbaryl-resistant houseflies show an abnormally high concentration of the enzyme carbamate esterase which converts carbaryl (67) to the inactive α-naphthol[16] (Scheme 6.54).

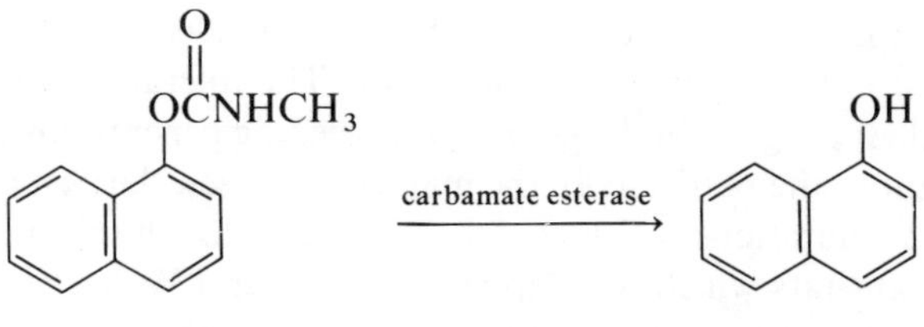

$$\text{OCNHCH}_3 \qquad \xrightarrow{\text{carbamate esterase}} \qquad \text{OH}$$

(67)   α-Naphthol

Scheme 6.54

RESISTANCE OF INSECTS TOWARDS INSECTICIDES

Resistance may be defined as the ability of a given strain of insects to tolerate doses of an insecticide that would kill the majority of a normal population of the same insect species. Some of the best documented cases of insect resistance have been observed with DDT and other persistent organochlorine insecticides (see Chapter 5, page 82), though considerable resistance to organophosphorus and other insecticides has also been noted and has caused serious control problems.[21,22]

By 1946 some strains of DDT-resistant houseflies had been discovered and in 1950 five to eleven species had acquired tolerance to one or more insecticides. In 1969 there were 156 resistant insect species: 55 to DDT, 84 to dieldrin, and 17 to organophosphorus compounds with some insects resistant of all three types of insecticides. By 1980 over 400 species of insects and mites had acquired resistance.[14,23]

Mites, because of their rapid reproduction, are especially liable to quickly become resistant. Over the period 1952 to 1973 some twenty different acaricides were used on American apple orchards. Fifteen of these chemicals are no longer used because they have ceased to be effective as a result of resistance. In Australia cattle ticks have become resistant to the majority of previously effective chemicals. In many areas, wild insect strains now fail to respond to dose rates > 100 times more than that originally adequate for their control.

One of the early examples of an insect acquiring tolerance to an insecticide was recorded in California in the 1920s when scale insects infesting citrus orchards became resistant to hydrogen cyanide. Also, in Queensland cattle ticks failed to be controlled by arsenical dips.

It was, however, not expected that the introduction of the new synthetic insecticides in the late 1940s would induce such rapid insect resistance. The reason was probably that these chemicals had extremely high initial toxicity and so they quickly killed all the susceptible individuals in the pest population leaving the small number of naturally resistant pests available to reproduce explosively with little competition because these non-selective insecticides often eliminated many of the natural predators.

Pesticides do not produce resistance; they merely select resistant individuals already present in the natural pest population. The tolerant individuals confer resistance to their progeny in the genes, so succeeding generations of insects will also be resistant to the pesticide. In the majority of cases, the pesticide probably does not induce mutations which confer resistance, though this may be true for warfarin-resistant rats which have appeared in Central Wales (see Chapter 11, page 291).

In screening a new potential insecticide, it is therefore important to see whether it is effective against strains of the target pest that are already tolerant to established insecticides, and also how quickly a strain resistant to the new chemical develops.[11]

The inheritance of specific resistance is generally comparatively simple and often monofactorial, although the influence of the principal gene may sometimes be modified by secondary genes. Data have been obtained on the genetic characteristics of the resistance of different insect species towards DDT, dieldrin, and organophosphorus compounds.[22] By interbreeding strains of houseflies resistant to organochlorine and phosphorus insecticides, the genes conferring resistance have been studied. In a total of seventeen species the inherited resistance to DDT was monofactorial in thirteen species; to dieldrin in sixteen species; and to organophosphorus compounds in five species. The relatively simple mode of inheritance is probably a further reason for the rapid growth of resistance in field populations of pest insects; often enabling the insect to acquire resistance to several insecticides simultaneously. When a specific detoxification mechanism confers resistance to two compounds, the phenomenon is termed cross-resistance, since this involves the same genes for the two chemicals. Thus when an insect becomes resistant to DDT, it is also generally resistant to the related compounds DDD and methoxychlor, but not to the cyclodiene insecticides (e.g. aldrin) or lindane (HCH) which fall into another cross-resistance group (see Chapter 5). Organophosphorus insecticides can be divided into two major cross-resistance groups, illustrated by parathion and malathion, and the insects are often also resistant to carbamate insecticides (see page 149). A fifth cross-resistance group is afforded by the pyrethroids and DDT.

When different resistance mechanisms exist in a given insect, it is said to show multiple resistance. This can be induced when the insect population has been exposed to different insecticides, and also in some cases when the pressure seems to have come from only one insecticide and the multiple resistance is morphological or behavioural in origin, when it can generally be overcome by quite small increases in dosage of the toxicant.

In the native insect population only a few individuals are preadapted against the insecticide and there was no reason to favour this mutation in the absence of the insecticide. On the other hand, when the insecticide was introduced into the environment, the tolerant strains survived, reproduced, and simultaneously there was a general selection of a genotype better adapted to the prevailing insecticidal environment. The situation was made more serious when a persistent, broad-spectrum insecticide like DDT was used, because this eliminated the natural predators so the resistant strains could multiply very rapidly.

The basis of insect resistance can be behavioural, morphological, or biochemical. Behavioural resistance often involves the insect adapting a pattern of behaviour that brings it into less contact with the toxicant. Thus certain strains of anopheline mosquitoes will not settle on surfaces that have been treated with DDT, while others do not enter buildings that have been sprayed[21] with the chemical. Similarly codling moth larvae have evolved the habit of discarding the initial bite of the apple when they bore into it to avoid the insecticides with which the apples have been treated.

Morphological resistance is associated with biophysical effects and may arise from the insect having an exceptionally thick cuticle, a high rate of excretion of the insecticide, or its storage in adipose tissue.[11] All these factors hinder transport of the toxicant to the site of action; in some instances the sensitivity of the target site may be altered.[14] The resistance of one strain of houseflies towards HCH appears to be related to an increase in the degree of cuticular absorption. Biochemical resistance is almost always due to the resistant strains possessing exceptionally high amounts of enzymes that can detoxify the insecticide. In DDT-resistant insect strains tolerance is often due to an abnormally high concentration of the enzyme DDT-dehydrochlorinase, which converts DDT to the non-insecticidal DDE (see Chapter 5). Mechanisms associated with resistance to organophosphorus compounds are more complex. Many organophosphates, like schradan, have to be metabolized *in vivo* to the active insecticide. They may also be deactivated by enzymes—phosphatases and carboxyesterases. So the toxicity of the given compound depends on the balance of activating and deactivating enzymes within the insect. With malathion, for instance, the low mammalian toxicity is ascribed to the higher carboxyesterase activity in mammals in comparison with the low activity of this enzyme in susceptible insects (see page 129). Insects exhibiting resistance to malathion generally show no cross-resistance to other insecticides, suggesting that tolerance depends on high carboxyesterase activity, which is supported by the discovery that carboxyesterase inhibitors function as malathion synergists and almost eliminate resistance.[2]

The resistance to organophosphates shown by several strains of houseflies and blowflies is associated with exceptionally low levels of aliesterase activity which is controlled by a single gene, whereas normally houseflies have large quantities of an aliesterase.[2] Microsomal mixed function oxidases (MFOs) play a vital role in both the activation and degradation of organophosphates. It has been shown that the remarkable selectivity of chlorfenvinphos (page 138) from one organism to another depends on the differing ability of their liver enzymes to deethylate the chemical. The greater the activity of the liver enzymes, the lower will be the effectiveness of chlorfenvinphos (LD_{50} (oral) to dogs $> 12\,000$; mouse, 170–200; rat, 15 mg/kg[12]). A microsomal oxidation involving NADP and oxygen may be one cause of DDT and organophosphorus resistance in some insects. The level of microsomal oxidases varies considerably from one strain of insects to another. The responsible gene determines the level of NADP-dependent microsomal oxidase activity, which detoxifies certain chlorinated hydrocarbons, pyrethroids, carbamates, and organophosphorus insecticides. The oxidase inhibitor piperonyl butoxide prevented the metabolic degradation of these insecticides.

The resistance shown by houseflies to certain organophosphorus insecticides, such as dichlorvos and bromophos, appears to be related to the decreased penetration of the toxicant to the thoracic ganglionic complex of the resistant flies.

The birth of generations of DDT-resistant insects resulted in the substitution first of lindane or γ-HCH, followed by organophosphates (e.g. malathion) for their control. However, after about six years the pests, in many cases, developed strains showing multiple resistance. Carbamate insecticides like carbaryl were often useful, but unfortunately pests which had acquired tolerance to organophosphates often showed cross-resistance to carbamates. Many important insect pests and disease vectors have multiple resistance and are consequently difficult to control. Red spider mites in many parts of the world are an outstanding example of this problem.

The growth in the population of resistant anopheline mosquitoes has resulted in outbreaks of malaria in areas in which the disease was previously thought to have been practically eradicated. Malaria control campaigns have been also hindered by the emergence of strains of bed bugs that are resistant to DDT.

When a particular insecticide is no longer applied, the resistant strain of insect often reverts to the natural susceptible strain. However, when the original insecticide is reintroduced resistance very quickly reappears. There is some advantage to be gained by changing from one insecticide to another every five or six generations.

The addition of synergists is also often helpful in overcoming resistance; for instance, DDT-dehydrochlorinase inhibitors such as WARF antiresistant (**85**)

$$\text{Cl}-\!\!\left\langle\!\!\bigcirc\!\!\right\rangle\!\!-\!\!\overset{\displaystyle O}{\underset{\displaystyle O}{\overset{\|}{\underset{\|}{S}}}}\!\!-\!\!\text{N(C}_4\text{H}_9)_2$$

(**85**)

have restored the toxicity of DDT to populations of resistant houseflies. Piperonyl butoxide inhibits microsomal enzymes and has been useful against insects that have developed tolerance to some organophosphorus and carbamate insecticides while with malathion-resistant insects, the most effective synergists were triphenyl phosphate, tributyl phosphorotrithioate (DEF), and several dibutylcarbamates.[2] The major biochemical detoxification mechanisms operating in dimethoate-resistant insects can be inhibited by methylene dioxy-

phenyl synergists. However, evidence indicates that insects can become i
to the synergists themselves, which can substantially reduce the effectiv
synergist-insecticide mixtures.

To minimize the spread of insect resistance control should not rely on one insecticide. If possible, treatment should rotate with other insecticides having different modes of action. The chemical should, if possible, only be applied when monitoring shows that the insect pest has attained economically significant levels. The application of insecticides which are selectively toxic to the target pest and are not too persistent in their action will also help to reduce resistance. Many insect pests have now become resistant to pyrethroid insecticides, although severe effects have been restricted to relatively few cases.[6] To prolong the effectiveness of modern synthetic pyrethroids, the nine major companies developing and marketing these chemicals have set up the Pyrethroid Efficacy Group (PEG).[23] The PEG will provide joint monitoring of incidents of resistance to pyrethroids and will recommend strategies which hopefully will delay the onset of resistance and hence prolong the useful life of these insecticides. Measures so far taken include a ban on the use of pyrethroids in animal-rearing houses and against cotton bollworm in some districts of Australia. In other areas, the application of pyrethroids to cotton has been restricted to a maximum of three sprays per season.

Generally the use of mixtures of pyrethroids with other insecticides and rotation with different chemicals may well assist in reducing the problems of resistant insect strains.

REFERENCES

1. Fest, C., and Schmidt, K. J., *The Chemistry of Organophosphorus Pesticides*, Springer-Verlag, Berlin, 1973.
2. Eto, M., *Organophosphorus Pesticides: Organic and Biological Chemistry*, CRC Press, Cleveland, Ohio, 1974.
3. Matolcsy, Gy, Nádasy, M., and Andriska, V., *Pesticide Chemistry*, Elsevier, Amsterdam, 1988.
4. Emsley, J., and Hall, D., *The Chemistry of Phosphorus*, Harper and Row, London, 1976, p. 494.
5. Worthing, C. R., and Hance, R. J. (Eds.), *Pesticide Manual*, 8th edn, British Crop Protection Council, Thornton Heath, 1991.
6. Hutson, D. H., and Roberts, T. R. (Eds.), 'Insecticides', in *Progress in Pesticide Biochemistry and Toxicology*, Vol. 5, Wiley, Chichester, 1987.
7. Ware, G. W., *Pesticides*, W. H. Freeman, San Francisco, Calif., 1983.
8. Cremlyn, R. J., *International Pest Control*, **16** (6), 5 (1974).
9. Martin, H., and Woodcock, D., *The Scientific Principles of Crop Protection*, 7th edn, Arnold, London, 1983.
10. Kay, I. T., Snell, B. K., and Tomlin, C. D. S., 'Chemicals for agriculture', in *Basic Organic Chemistry* (Eds. Tedder, J. M., Nechvatal, A., and Jubb, A. H.), Wiley, London, 1975.
11. Green, M. B., Hartley, G. S., and West, T. F., *Chemicals for Crop Improvement and Pest Management*, 3rd edn, Pergamon, Oxford, 1987.

12. O'Brian, R. D., 'Acetylcholinesterase and its inhibition', in *Insect Biochemistry and Physiology* (Ed. Wilkinson, C. F.), Plenum Press, New York, 1973, p. 271.

13. Hart, R. J., 'Mode of action of agents used against arthropod parasites', in *Chemotherapy of Parasitic Diseases*, Plenum, New York, 1986.

14. Corbett, J. R., Wright, K., and Baillie, A. C., *The Biochemical Mode of Action of Pesticides*, Academic Press, London and New York, 1984.

15. Matsumura, F., and Murti, C. R. M. (Eds.), *Biodegradation of Pesticides*, Plenum Press, London and New York, 1982.

16. Fukuto, T. R., and Sims, J. J., 'Metabolism of insecticides and fungicides', in *Pesticides in the Environment* (Ed. White-Stevens, R.), Dekker, New York, 1971, p. 145.

17. Miyamoto, J., *et al.*, *Pesticide Metabolism; Extrapolation from Animals to Man*, Blackwell, Oxford, 1988.

18. Walker, C. H., 'Variations in the intake and elimination of pollutants', in *Organic Chlorine Insecticides: Persistent Organic Pollutants* (Ed. Moriarty, F.), Academic Press, London, 1965, p. 73.

19. Hassall, K. A., *The Chemistry of Pesticides*, Macmillan, London, 1982.

20. D'Silva, T. J. D., 'Structure–activity relationships in methylcarbamoyl oximes and their sulphenated derivatives', in *Recent Advances in the Chemistry of Insect Control*, London, 1985, p. 205.

21. Woods, A., *Pest Control*, McGraw-Hill, London, 1974, p. 124.

22. Metcalf, R. L., *Pesticide Science*, **26**(4), 333 (1989).

23. Ruscoe, C. N. E., 'Pesticide resistance', in *Rational Pesticide Use* (Eds. Brent, K. J., and Atkin, R. K.), Cambridge University Press, Cambridge, 1987, p. 197.

Chapter 7
Fungicides

Fungi can be classified according to their life style as follows. Group 1 fungi infect plants via the aqueous phase and their growth is facilitated by moist conditions. This group includes the *Oomycetes* taxonomic group of fungi of which the downy mildews and potato late blight (*Phytophthora infestans*) are commercially important examples. Some fungi from other taxonomic groups may also be classified in this group, e.g. *Septoria* responsible for leaf spot and glume blotch of wheat.

Group 2 fungal spores are air borne and dry conditions favour infection; the often hydrophobic mycelium grows on the leaf surface. The group includes some *Ascomycetes*, e.g. powdery mildews.

Group 3 fungi are also air borne, but infection and growth occurs when the plants are damp; the mycelium grows in the surface layers of the plant. This group comprises several different taxonomic groups, some of which are of great economic importance: *Fungi Imperfecti*, e.g. *Botrytis* grey mould and several rots, and scabs, e.g. *Venturia* (apple scab) and *Sclerotinia* (brown rot of fruit).

Other taxonomic groups of fungi infect plants via the seeds or the soil; the former include the *Basidiomycetes*, e.g. the smuts and bunts of cereals, and some *Ascomycetes*. The other fungi of the *Basidiomycetes* group are soil borne, e.g. onion smut, as are some *Fungi Imperfecti*, e.g. *fusarium* and *verticillium*.

Many commercial fungicides used at present belong to the class known as protectant or surface fungicides. They are usually applied to plant foliage as dusts or sprays.[1–4] Such materials do not appreciably penetrate the plant cuticle and are not translocated within the plant, whereas the more recent systemic fungicides or plant chemotherapeutants, as they are sometimes called, are absorbed by the plant via the roots, leaves, or seeds and are translocated within the plant.

Most pathogenic fungi penetrate the cuticle and ramify through the plant tissues, so if the fungicide is to combat the fungal infection a protectant fungicide must be applied *before* the fungal spores reach the plant. However, a few fungi such as powdery mildews are restricted to the surface of the leaf and in these cases surface fungicides may also possess eradicant action.

If a given candidate chemical is to be an effective protectant fungicide, the following conditions must be satisfied:[5]

(a) It must have very low phytotoxicity, otherwise too much damage will be caused to the host plant during application.

(b) It must be fungitoxic *per se* or be capable of conversion into an active fungitoxicant within the fungal spore and must act quickly before the fungal infection penetrates the plant cuticle.

(c) Generally the fungicide must be able to penetrate the fungal spore and reach the ultimate site of action in the fungus.

(d) Most agricultural protectant fungicides are applied as foliar sprays, and they must be capable of forming tenacious deposits that are resistant to the effects of weathering over long periods.

Considering these criteria in rather more detail, the most difficult problem is the attainment of the desired selective toxicity to the fungus because of the close relationship between plants and fungi. In fact none of the protectant fungicides currently on the market is completely non-phytotoxic. A possible route to a non-phytotoxic fungicide would be the discovery of compounds that interfere with the biosynthesis of chitin which is found in the cell walls of most parasitic fungi but is absent in higher plants.[6] The cell membrane is semi-permeable, often consisting of two monolayers of lipids surrounded on either side by a layer of protein, so a degree of lipid solubility is an important factor aiding penetration. For instance, in a series of 2-alkylimidazolines the fungitoxicity reached a maximum with increasing length of the alkyl chain up to C_{17} and then decreased,[6] and there are many similar cases in which addition of inert lipophilic substituents increases the fungicidal activity.[5] Once the compound has gained entry into the fungal cell and reached the critical reactor site, it must then exert its toxic action on the fungus by either a chemical or physical mechanism. With chemical toxicants there will be a chemical reaction with perhaps a vital enzyme that ultimately kills the fungus. On the other hand, with physical toxicants the deleterious effect on the fungus is caused by the compound possessing the correct hydrophilic–lipophilic balance so that it dilutes the biophase and thus physically inhibits vital cellular processes. In both mechanisms, the success of the material does require the optimum oil–water solubility balance, enabling the fungicide to reach the critical site in the fungal cell.

The majority of protectant fungicides are directly toxic to fungi and so will show up as active against spore germination *in vitro* tests. Fungicides may be applied to fruits, foliage, or seeds as dusts or sprays: in dusting a uniform coverage is most important and this requires a small particle size (approximately 5 μm).[4,5] Spraying is a much more widely used method of application; the spray may be a solution or a fine suspension of the material, and in the latter case reduction of particle size leads to increased effectiveness in disease control which has been illustrated in the control of tomato early blight (*Alternaria solani*) by dichlone. A more uniform coverage is also aided by small spray droplets which reduces 'run-off', though too fine a spray results in substantial loss through evaporation. Most fungicides are formulated with wetting agents or 'spreaders' which are particularly beneficial when the chemical is being applied to a waxy leaf surface (see Chapter 2, page 20). The spreading property of the spray on foliage is an important factor in determining the tenacity of the dried deposit on which the persistency of protectant fungicides largely depends. The fungicide, as a dried deposit on the leaf surface, must generally be stable towards photochemical oxidation, hydrolysis, and carbonation, and instability in sunlight probably accounts for the ineffectiveness of chloranil as a foliage fungicide as compared with dichlone which is more stable.[5] Of course, in some cases, the products of decomposition may be more active than the original compound; thus the fungistatic nabam is activated by oxidation on the leaf surface to the active fungicide (see page 166).

The earliest fungicides were inorganic materials like sulphur, lime-sulphur, copper, and mercury compounds. Elemental sulphur has been recognized as fungicidal for at least 180 years; in 1803 the Royal Gardener Forsyth recommended the use of a sulphur spray against mildew on fruit trees. In the nineteenth century sulphur was increasingly employed against powdery mildew on fruit and later for control of powdery mildew on grapes.[1] It may be applied as dusts or sprays, and the more finely divided the sulphur, the more effective is the formulation against the disease; hence colloidal sulphur is formulated by grinding up flowers of sulphur with a mineral diluent (e.g. kaoline) so that there is a sulphur content of 40%, with most particles of less than 6 μm in diameter. A much more widely used liquid sulphur product is 'lime-sulphur' obtained by boiling sulphur with an aqueous suspension of slaked lime. This is a clear orange-coloured liquid consisting mainly of calcium polysulphides which break down on exposure to air to release elemental sulphur. The polysulphide content of a lime-sulphur mixture is the best indication of its effectiveness and the official specification contains at least 24% w/v of the polysulphide.[2,4]

In 1958 the weight of sulphur used against fungi was four times that of all other fungicides combined, although since that time the amount has fallen off due to the development of more modern organic fungicides. However, sulphur and lime-sulphur are still extensively applied against powdery mildews and apple and pear scab.[7]

There has been much speculation regarding the mode of fungitoxic action of sulphur. At first it was assumed that sulphur could not be the toxic agent, although Sempio (1932) took the contrary view, and the fungicidal action was ascribed to the production of various sulphur derivatives. The earliest idea was that the active fungitoxicant was hydrogen sulphide and experiments showed that fungal spores can readily reduce sulphur to hydrogen sulphide which was shown to be toxic to the spores.[5] However, in 1953 this theory was disproved when it was shown conclusively that colloidal sulphur was appreciably more fungitoxic than the equivalent quantity of hydrogen sulphide. Another hypothesis ascribed the fungitoxicity of sulphur to various oxidation products such as sulphur dioxide, sulphuric acid, thiosulphuric acid, or pentathionic acid.[5] The importance of these products has, however, been largely discounted because their toxicity can be accounted for on the basis of their hydrogen ion concentrations. Since no sulphur derivative appears responsible for the activity, we have now come back full circle to the idea of Sempio that sulphur itself as the S_8 molecule may be toxic to fungi.[4,6] Certainly work on plants and fungi indicates that sulphur is not biologically inert and probably becomes involved in biological redox systems. There is considerable evidence that sulphur can penetrate fungal spores and the improved activity observed in the presence of urea, hydrocarbons, soft soap, or lime is probably due to enhanced penetration.[5] Fungal spores can take up large amounts of sulphur, practically all of which is subsequently evolved as hydrogen sulphide, and the toxicity of sulphur is probably due to it functioning as a hydrogen acceptor, thereby interfering with vital hydrogenation and dehydrogenation reactions within the fungal cell. It remains difficult to understand why sulphur has a specific toxicity to fungi; unlike copper and other heavy metal fungicides, sulphur is practically non-toxic to mammals. This may possibly be due to differential availability at the site of action within the fungus, and lipid-rich fungi such as powdery mildews may retain more sulphur than other types of organisms.[4]

Among the heavy metals, only the compounds of copper and mercury have been used widely as fungicides, although silver is the most toxic metal cation to fungi. The relative toxicity of the various metal cations to fungi is as follows: Ag > Hg > Cu > Cd > Cr > Ni > Pb > Co > Zn > Fe > Ca. The toxicity of the metals has been related to their position in the Periodic Table—toxicity within a given group generally increases with the atomic mass. It has also been related to the relative chelating powers of the metals, the stability of the metal sulphide, and the electronegativity of the cations. Studies[5,6] indicate that the latter criterion provides the best correlation because the degree of electronegativity is a measure of the stability of metal bonds with cellular constituents which will in turn affect the stability of metal chelates and sulphides. The degree of fungitoxicity is probably determined by the strength of covalent or coordinate bonding in un-ionized complexes at the cell surface.

Copper sulphate has been used since the eighteenth century as a seed treatment against cereal bunt, although it has later been replaced by the much more effective organomercurial compounds. Copper ions in solution are toxic to all plant life; selective fungicidal action can therefore only be achieved by application of an insoluble copper compound to the foliage. Examples of compounds used include copper oxychloride, copper carbonate, cuprous oxide, and Bordeaux mixture. The latter is the most important of the copper fungicides and was discovered by Millardet in 1882 (Chapter 1, page 4). Bordeaux mixture was effective against vine downy mildew (*Plasmopara viticola*), a disease that had been introduced into France from the United States of America on vine rootstocks imported because they were resistant to *phylloxera*. The mildew threatened the French vineyards and so the situation was extremely serious. The botanist Millardet, who recommended the use of American vines, discovered that he had merely exchanged one disease for another, so his accidental discovery of Bordeaux mixture came at just the right time to control the mildew. Bordeaux mixture, named from the locality of its origin, consists of $CuSO_4$ (4.5 kg) and $Ca(OH)_2$ (5.5 kg) in 454 litres of water. Once the mixture has been prepared, it should be sprayed on to the crop as soon as possible since the fungitoxicity of the mixture decreases on standing. It is rather difficult to apply because the precipitate tends to block the spray nozzles.

The chemistry of Bordeaux mixture and the precise mode of fungicidal action are complex. The proportions of the ingredients used and the method of preparation have considerable influence on the fungicidal effectiveness of the product. In particular, the fineness and the composition control the physical properties of the dried deposit which in turn greatly affects the tenacity of the spray deposit on the leaves.

The active ingredient is probably not cupric hydroxide, but rather a basic copper sulphate approximating to the formula $[CuSO_4·3 Cu(OH)_2]$. Bordeaux mixture is almost insoluble in water, so how is the copper mobilized to kill the fungus? Wain considered[8] that exudates both from the surface of the leaf and from the fungal spores can dissolve appreciable quantities of copper from the dried Bordeaux deposits due to the presence of certain compounds like amino and hydroxy acids which can form chelates with the copper. It has been demonstrated[5] that exudates from the spores of *Neurospora sitophila* react with Bordeaux mixture giving a soluble copper complex which dissociates in solution yielding toxic cupric ions. Similarly,[8] when copper fungicides are employed as seed dressings the exudates from the seed dissolved copper from the dressing, which accounts for the fungicidal protection provided. Soluble copper compounds are too phytotoxic to be useful as foliar sprays and different insoluble copper compounds ('fixed coppers') vary considerably in their fungitoxic properties. These have the advantage of easier application as compared with Bordeaux mixture and are now widely used. Two well-known examples are copper oxychloride, approximately $[CuCl_2·3 Cu(OH)_2]$, which is marketed both as a colloid and as a dispersable powder, the former being used with a wetting agent, and yellow cuprous oxide in finely divided form prepared by reduction of copper salts in alkaline solution. Such 'fixed coppers' are used for control of potato and tomato blight, vine and hop downy mildew, and many other common leaf and fruit diseases of horticultural crops.[7] In Europe, their major use is in the protection of potatoes from late blight (*Phytophthora infestans*), about 15 % of the potato crop being sprayed annually. The use of cuprosan, the first organic cupric combination to show synergism, against vine downy mildew has been reviewed[9] and the mode of action of copper and sulphur fungicides discussed.[10]

The powerful bactericidal properties of mercuric chloride led to its examination as a fungicidal dressing for cereal seeds when it was shown to be effective against *Fusarium* disease of rye. For medicinal purposes mercuric chloride was quickly replaced by less poisonous organomercury derivatives and these were subsequently examined as cereal seed dressings.[3]

Mercury forms stable organic derivatives containing carbon–mercury bonds; the majority of organomercurial seed dressings have the general formula RHgX (where R = alkyl/aryl radical and X = an anion). The fungicidal activity against wheat bunt (*Tilletia caries*) is almost entirely dependent on the nature of the hydrocarbon radical R and the active species is probably RHg^+. Those derivatives containing phenyl or other aryl groups were generally the most effective as seed dressings. Two well-known examples are phenylmercury chloride and the corresponding acetate. Such compounds are extensively applied as seed dressings against wheat bunt and smut on oats and barley.[7] The cereal seeds are either steeped in a suspension of the active ingredient or may be dusted with a powder (see Chapter 2, page 22). Seed dressing is a very efficient method of applying the chemical; with organomercurial seed dressings only a few grams of the active ingredient are needed per hectare. A disadvantage is that the treated seed presents hazards, especially to seed-eating birds (see Chapter 15, page 347). In Britain, where phenylmercury compounds are much more extensively used than alkyl derivatives there have been comparatively few cases of birds poisoned by organomercurials.[3] On the other hand, in Sweden, where alkylmercury compounds are largely used as liquid seed dressings widespread mercury poisoning of birds and other mammals has occurred.

Phenylmercury compounds are also used as sprays against apple and pear scab and canker, as industrial fungicides, and to protect cotton and rice from fungal attack.[7]

Arylmercury compounds may be obtained by direct electrophilic substitution (mercuration) of the aromatic substrate. Phenylmercury acetate, one of the most widely used of all organomercurials, is prepared by heating benzene and mercuric acetate in glacial acetic acid (Scheme 7.1).

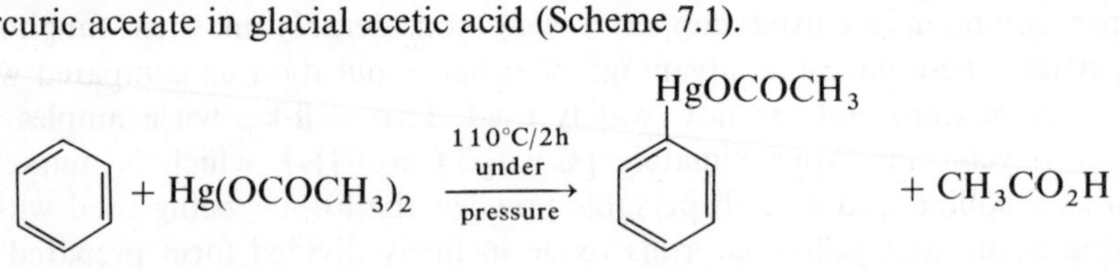

$$\text{Benzene} + Hg(OCOCH_3)_2 \xrightarrow[\text{pressure}]{\substack{110°C/2h \\ \text{under}}} \text{Phenylmercury acetate} + CH_3CO_2H$$

Benzene   Mercuric acetate   Phenylmercury acetate

Scheme 7.1

The organomercurials are valuable seed dressings for cereals. Their main disadvantage is that they are very poisonous and there are fears that their continued large-scale use may result in serious effects on the ecosystem. Efforts are therefore being made to find less toxic seed dressings and several of the triazole fungicides have proved effective and much safer alternatives.

Mercury, like copper and several other heavy metals, is known to affect respiration by poisoning essential sulphydryl respiratory enzymes in the fungal cell; the toxicity can be reversed by the addition of sulphydryl compounds.[5]

However, it is not certain whether organomercurials are toxic *per se* or whether they just function as carriers of toxic mercury ions through lipid barriers to a reactive site.

Some alkylmercury, but not arylmercury, compounds produce severe brain lesions in mammals and they may destroy nervous tissue. Certain mercury compounds also undergo hydrolysis to inorganic mercury salts which are powerful long-term poisons tending to accumulate in the kidneys.[4] Consequently these compounds have been largely phased out as pesticides.

Inorganic tin compounds, like stannic chloride, are generally not fungicidal and in alkyltin derivatives of types $RSnX_3$, R_2SnX_2, and R_3SnX, the fungitoxicity increases as the number of alkyl groups goes up from one to three.[4] As with the organomercurials, the nature of the anion (e.g. Cl, OH, $OCOCH_3$) was not decisive and maximum activity always occurred with alkyl groups containing three or four carbon atoms, but the trialkyltin compounds were too phytotoxic for use as foliage fungicides. Tributyltin oxide, $(C_4H_9)_3Sn$—O—$Sn(C_4H_9)_3$, is, however, used extensively for stabilization of transparent plastics against photochemical change, for rot proofing of fabrics, in antifouling paints to inhibit the growth of barnacles and algae on the bottom of ships, and for preservation of marine timbers.

Triphenyltin compounds were much less phytotoxic and still very strongly fungicidal;[2-5] thus triphenyltin acetate (LD_{50} (oral) to rats, 135 mg/kg) is the active ingredient of the foliage fungicide fentin, in which it is formulated with maneb. Triphenyltin acetate has a wide spectrum of activity and is used specially for control of potato blight and leaf spot diseases of sugar beet and celery;[7] it also deters insects from feeding on the treated foliage. Although fentin has higher phytotoxicity than 'fixed' coppers, it is appreciably more fungicidal and can be successfully used at only about one-tenth of the dose required for the copper fungicides; consequently the fungus is controlled with less damage to the host plant.

These organotin compounds can be prepared by a double decomposition reaction from phenylmagnesium bromide and stannic chloride:

$$3C_6H_5MgBr + SnCl_4 \xrightarrow{(C_2H_5)_2O}$$

$$(C_6H_5)_3SnCl + 3MgBrCl \xrightarrow{CH_3CO_2Na} (C_6H_5)_3SnOCOCH_3 + NaCl$$

Fentin acetate

The aryltin derivatives apparently do not build up in the environment since they undergo biodegradation to triphenyltin(IV) hydroxide. Consequently their application does not appear to damage the environment. Fentin acetate interferes with fungal growth by inhibition of oxidative phosphorylation.

The development of purely organic fungicides really began with the discovery of the fungicidal activity of the dithiocarbamates which had been originally developed as vulcanization agents for the rubber industry. The dithiocarbamates

and their derivatives are one of the most important groups of organic fungicides for controlling plant diseases. They were introduced by Tisdale and
Williams in 1934: the most valuable members were derived from dimethyldithiocarbamic acid since other alkyl groups reduced the fungitoxicity.[5,11]
Thiram or tetramethylthiuram disulphide (1) was the first compound to be
applied as a fungicide and is still used, especially against grey mould on lettuce
and strawberry and as a seed dressing against soil fungi causing damping-off
diseases.[7] Thiram (1) is prepared by the interaction of carbon disulphide and
dimethylamine in the presence of sodium hydroxide solution to give sodium
dimethyldithiocarbamate which is subsequently oxidized by air, hydrogen
peroxide, chlorine, or iodine to thiram:

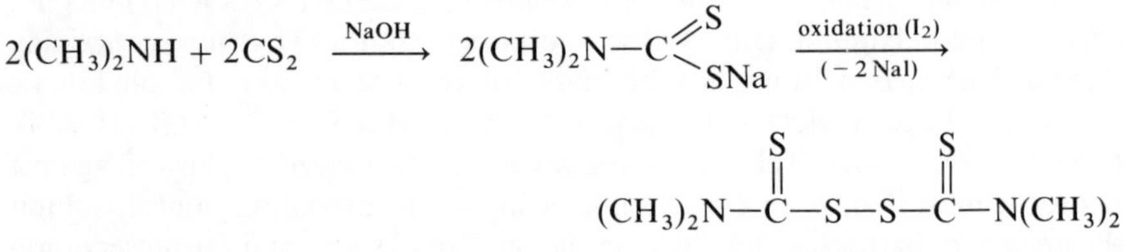

$$2(CH_3)_2NH + 2CS_2 \xrightarrow{\text{NaOH}} 2(CH_3)_2N-C{\overset{\displaystyle S}{\underset{\displaystyle SNa}{\diagdown}}} \xrightarrow[(-2\,\text{NaI})]{\text{oxidation (I}_2)}$$

$$(CH_3)_2N-\overset{\displaystyle S}{\overset{\displaystyle \|}{C}}-S-S-\overset{\displaystyle S}{\overset{\displaystyle \|}{C}}-N(CH_3)_2$$

$$(1)$$

Later work[11] resulted in the discovery of the fungicidal activity of the zinc and
ferric salts of dimethyldithiocarbamic acid, known as ziram or ferbam (**2**;
M = Zn, x = 2 or M = Fe, x = 3).

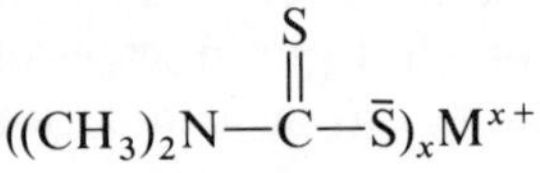

$$((CH_3)_2N-\overset{\displaystyle S}{\overset{\displaystyle \|}{C}}-\bar{S})_x M^{x+}$$

$$(2)$$

Disodium ethylenebisdithiocarbamate or nabam (**3**) is also fungicidal and is
used to control some root rots, but as a foliage fungicide it tends to be
phytotoxic and has little persistence in rain. The ethylenebisdithiocarbamates
(**3**), the most important dithiocarbamates as protectant fungicides,[3] are obtained by reaction of ethylene diamine with carbon disulphide in the presence of
sodium hydroxide:[12]

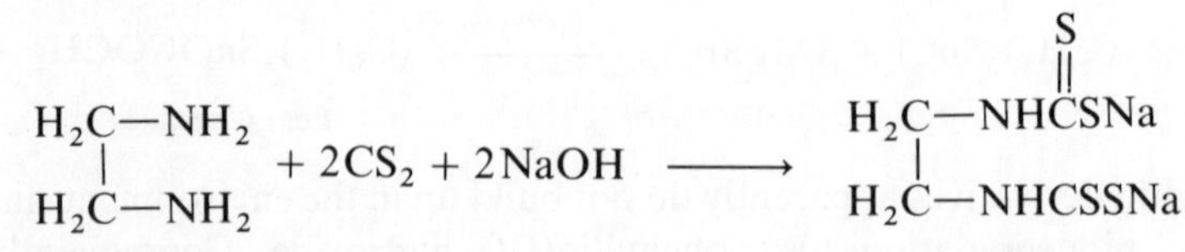

$$\begin{array}{l} H_2C-NH_2 \\ | \qquad\qquad + 2CS_2 + 2NaOH \longrightarrow \\ H_2C-NH_2 \end{array} \qquad \begin{array}{l} \overset{\displaystyle S}{\overset{\displaystyle \|}{}} \\ H_2C-NHCSNa \\ | \\ H_2C-NHCSSNa \end{array}$$

Ethylene diamine   (3)

Nabam, which is water soluble, has been largely replaced by the insoluble zinc
and manganese salts known as zineb (**4**; M = Zn) and maneb (**4**; M = Mn),
which are produced by reaction with an aqueous solution of zinc or manganous
sulphate. They are some of the most widely used organic protectant fungicides

and are applied for the control of a wide range of phytopathogenic fungi such as downy mildews and potato and tomato blight.[7] They have low mammalian toxicities with LD_{50} values to rats of approximately 7000 mg/kg and they now rank in tonnage use only below sulphur and fixed coppers and have partly replaced copper fungicides for control of potato blight.

A more recent compound is sodium *N*-methyldithiocarbamate or metham-sodium (5), valuable as a soil sterilant for control of damping-off diseases, potato cyst eelworm, and weed seedlings.[7]

$$
\begin{array}{c}
\text{S} \\
\parallel \\
\text{H}_2\text{C}-\text{NH}-\text{C}-\bar{\text{S}} \\
\mid \qquad\qquad\qquad \text{M}^{2+} \\
\text{H}_2\text{C}-\text{NH}-\text{C}-\bar{\text{S}} \\
\parallel \\
\text{S}
\end{array}
\qquad
\begin{array}{c}
\text{S} \\
\parallel \\
\text{CH}_3\text{NH}-\text{CSNa}
\end{array}
$$

(4)   (5)

Mancozeb, a coordinated complex of the zinc and manganous salts, was introduced in 1962 (Plate 10). Maneb and mancozeb are now often formulated with systemic fungicides, e.g. carbendazim and metalaxyl, to reduce the development of resistant fungi (see page 210). 1,1-Complexes of zineb with diamines, e.g. 1,2-diaminoethane, were more fungicidal and persistent than zineb against several pathogens, e.g. rice blast and wheat rust fungi.[9]

The uses of dithiocarbamates and their derivatives as fungicides have been reviewed and their mode of fungicidal action has been the subject of considerable study.[4,11] There are some differences in their fungicidal properties which suggest that the *N*,*N*-dimethyldithiocarbamates may have a different mode of action from that of the ethylenebisdithiocarbamates; thus they possess a distinctive spectrum of activity against various species of fungi and histidine has been shown to antagonize the antifungal action of thiram (1) but not that of nabam (3).

The dithiocarbamates, such as thiram and ziram, probably owe their fungitoxicity to their ability to chelate with certain metal ions, especially copper. Studies with the fungus *Aspergillus niger* have demonstrated that in the presence of cupric ions an increase in the concentration of sodium dimethyldithiocarbamate inhibits the growth of the fungus at two levels. The first inhibition level occurs when the copper–dithiocarbamate ratio is reasonably high, approximately 20:1 and is associated with formation of the 1:1 copper–dithiocarbamate complex (6). When the dithiocarbamate concentration is further increased, the fungitoxicity decreases due to formation of the 1:2 complex (7).

(6)   (7)

The second inhibition level requires high concentrations of dimethyldithiocarbamate ($\simeq$ 50 p.p.m.) and is due to the presence of free dimethyldithiocarbamate ions since no cupric ions are needed for this effect. On the other hand, the presence of copper is definitely required for high fungicidal activity with dimethyldithiocarbamates, and the 1:1 chelate (6) can penetrate lipid barriers in the fungal cell, whereas the 1:2 chelate (7) is apparently not fungicidal because it is too insoluble in water to permit penetration and its formation accounts for the bimodal curve observed for the graph of fungitoxicity against sodium dimethyldithiocarbamate concentration (Figure 7.1).

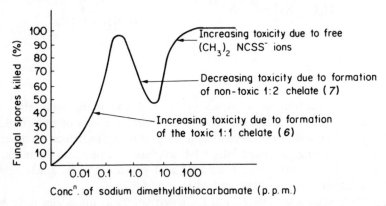

Figure 7.1 Relation between sodium dimethyldithiocarbamate concentration and fungitoxicity

When the 1:1 complex (6) has entered the fungal cell it is probably converted to the free dimethyldithiocarbamate anion which can complex with certain vital trace metals. Alternatively, the complex may itself be fungicidal, and has been shown to interfere with the uptake of oxygen by yeast cells and pyruvate accumulates in *Aspergillus niger* after treatment with sodium dimethyldithiocarbamate. The fungitoxicity of the complex (6) may thus arise from interference with the respiration of the fungus by inactivation of the pyruvate dehydrogenase system (see Chapter 3, page 40).

The ethylenebisdithiocarbamates (e.g. nabam, maneb, and zineb) almost certainly act by a different mechanism involving oxidative decomposition of the chemical on the leaf surface to such products as thiuram disulphide, carbon disulphide, and possibly ethylene diisothiocyanate:[6]

$$\begin{array}{ccc}
\text{H}_2\text{C}-\text{NHCSS}^- & & \text{H}_2\text{C}-\text{NH}-\overset{\displaystyle S}{\overset{\|}{\text{C}}} \\
| & \xrightarrow{\text{oxidation}} & | \qquad\qquad \searrow \\
\text{H}_2\text{C}-\text{NHCSS}^- & & \text{H}_2\text{C}-\text{NH}-\underset{\displaystyle S}{\underset{\|}{\text{C}}} \nearrow
\end{array} \;\; \xrightarrow{(-\text{H}_2\text{S})} \;\; \begin{array}{c}
\text{H}_2\text{C}-\text{N}{=}\text{C}{=}\text{S} \\
| \\
\text{H}_2\text{C}-\text{N}{=}\text{C}{=}\text{S}
\end{array}$$

Ethylenebisdithiocarbamate   Ethylene diisothiocyanate

It has been shown[4] that pure ethylenebisdithiocarbamates are inactive *before* they have been exposed to air and also that isothiocyanates are fungicidal by virtue of their ability to react with vital thiol compounds within the fungal cell. The inhibitory action of nabam on the germination of fungal spores is strongly antagonized by the addition of thiols such as thiolglycollic acid and cysteine.

In 1951 Kittleson working for the Standard Oil Company of America discovered that certain compounds containing the *N*-trichloromethylthio group are powerful surface fungicides.[13] The best-known example is captan or *N*-(trichloromethylthio)-4-cyclohexene-1,2-dicarboximide (**8**) which is easily synthesized from the readily available starting materials butadiene and maleic anhydride (Scheme 7.2).

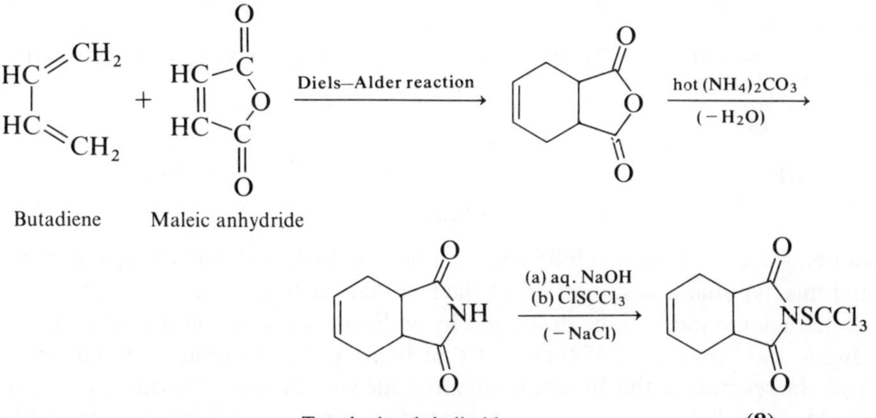

Butadiene   Maleic anhydride

Tetrahydrophthalimide   (**8**)

Scheme 7.2

Trichloromethylsulphenyl chloride is obtained from carbon disulphide:

$$CS_2 + 3Cl_2 \xrightarrow{I_2} Cl_3CSCl + SCl_2$$

Captan (**8**) is a very effective and persistent foliage fungicide, especially for control of apple and pear scab, black spot on roses, and as seed dressings against many soil and seed-borne diseases. It reduces *Gloeosporium* rot of apples, stem rot of tomatoes, and *Botrytis* mould on soft fruit,[1,7] but has little action against powdery mildews. Analogues that have been subsequently developed as foliar fungicides include folpet (**9**) and difolatan (**10**), which are more effective against potato blight. These are some of the safest of all fungicides, with LD_{50} (oral) values to rats of about 10 000 mg/kg.

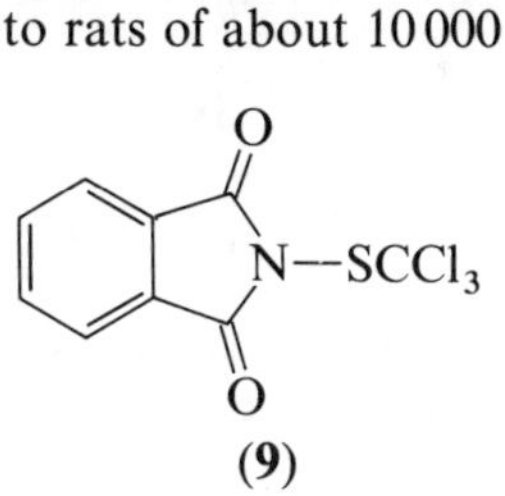

(**9**) 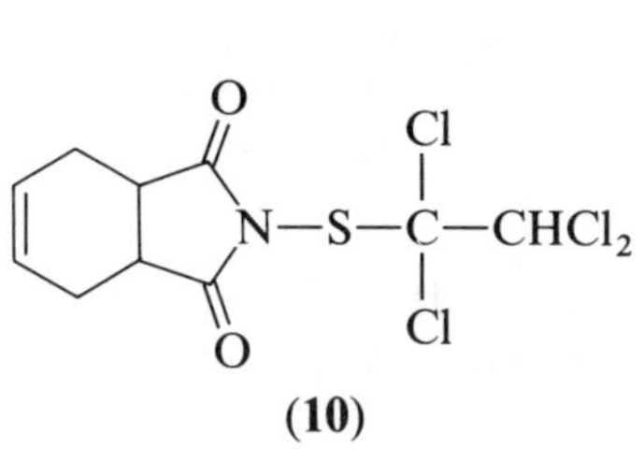   (**10**)

All the compounds originally listed by Kittleson contain the N—S—CCl$_3$ group and he ascribed their fungitoxicity to the presence of this group. In contrast Rich[14] considered that the toxophore was probably the —CO—N(R)—CO— grouping and the —S—CCl$_3$ group was merely a convenient shaped charge to facilitate penetration of the compound into the fungal cell. The latter idea was, however, disproved by the discovery of a large number of fungicidal trichloromethylthio compounds which do not contain the imide moiety.

Lukens and Sisler (1958) showed that captan (**8**) interacts with cellular thiols to produce thiophosgene which they regard as the ultimate toxicant[6] (Scheme 7.3). The evolved thiophosgene would finally poison the fungus by

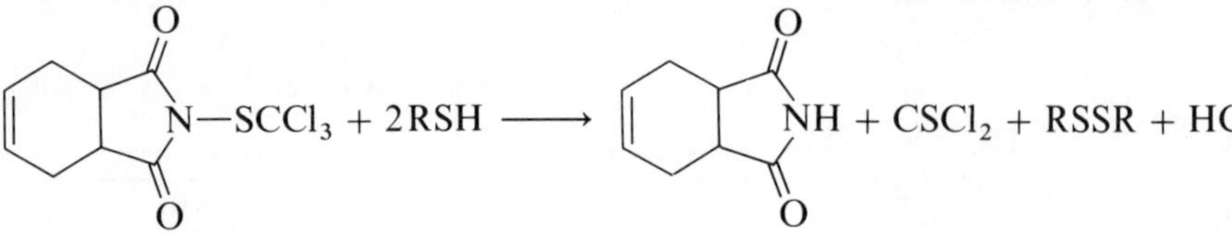

(**8**)   Thiophosgene

Scheme 7.3

combination with vital sulphydryl-, amino-, or hydroxyl-containing enzymes, and this hypothesis is supported by the fact that the fungitoxicity of captan and related compounds can be destroyed by addition of thiols. On the other hand, Owens and Novotny (1959) argued that the toxicity of captan to fungi arises from the reaction of the intact captan molecule with unprotected thiol groups in the fungal cell because they concluded that captan would react faster with cellular thiols than thiophosgene. The fungicidal activity of compounds of type R—S—CCl$_3$ probably depends on rupture of the R—S bond by reaction with cellular thiols which oxidizes thiol groups and simultaneously releases toxic products (e.g. thiophosgene) into the fungal cell. For maximum fungitoxicity, the toxic products must be released at the critical site of action in the fungal cell which in turn will be governed by such factors as the strength of the R—S bond and the oil/water solubility balance, and both of these factors will be controlled by the nature of the R group.[13]

Another theory proposed by Rich[14] suggested that captan owes its fungitoxicity to the formation of the transitory *N*-chlorotetrahydrophthalimide which subsequently reacts with vital enzymes in the fungal cell:

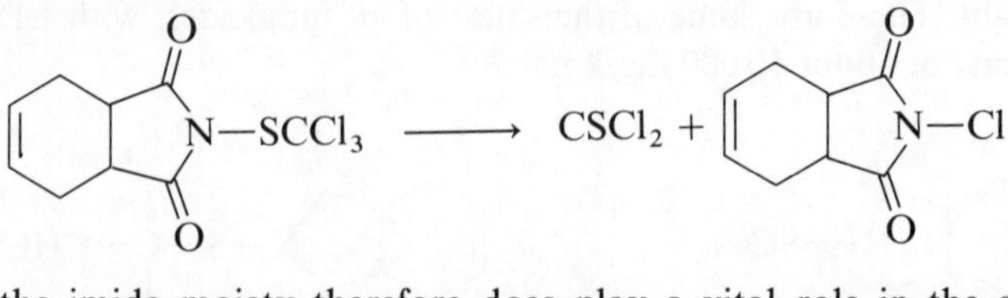

Perhaps the imide moiety therefore does play a vital role in the activity of captan; certainly a large number of fungicidal compounds are known[15] that contain the imino group. One important mechanism of the fungitoxicity is probably the reaction with vital cellular thiols, but these trichloromethylthio compounds do not appear to act by a single biochemical mechanism. The idea of multisite action is supported by the fact that although *N*-trichloromethylthio fungicides have been extensively used for some 30 years there has been little evidence of the emergence of resistant fungal strains (Chapter 7, page 210).

Captan probably acts non-specifically because *Neurospora crassa* conidia accumulated the fungicide when they were exposed to a dose that killed 50% of the spores. The precise nature of the interactions between the toxicant and the fungal cell has not been definitely established, but thiols are the most likely reaction sites, since the intact captan molecule is unlikely to react with amino or hydroxy groups at cellular pH conditions but reacts easily with thiols.

The reactivity of captan (**8**) and the aromatic analogue folpet (**9**) with various biochemical systems has been clearly demonstrated. Thus interaction of 20 μM solution of folpet with isolated yeast glyceraldehyde-3-phosphate dehydrogenase resulted in 82% reduction of enzymic activity due to combination with thiol groups in the enzyme. On the other hand, folpet also caused substantial inhibition of the activity of isolated α-chymotrypsin which does not contain thiol groups, suggesting that other reactions may be significant in the fungicidal action of these compounds. Experiments with rat liver mitochondria showed that both captan and folpet inhibited a number of mitochondrial reactions, including oxidative phosphorylation and the oxidation of the reduced form of nicotinamide adenine dinucleotide (NADH$_2$).

Dichlofluanid, introduced by Bayer AG (1965), is structurally related to captan, but does not contain the trichloromethylthio group. It can be synthesized from dimethylamine (Scheme 7.4). Dichlofluanid is a broad-spectrum protective fungicide which is much less sensitive to light than captan. It is effective

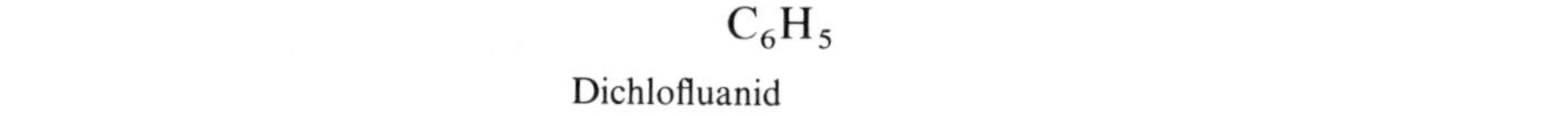

$$(CH_3)_2NH + SO_2Cl_2 \xrightarrow[(-HCl)]{NaOH} (CH_3)_2NSO_2Cl \xrightarrow[NaOH]{C_6H_5NH_2,} (CH_3)_2NSO_2NHC_6H_5$$

$$(CH_3)_2NSO_2{-}N{\overset{\displaystyle/SCCl_2F_2}{\diagdown}} \quad \xleftarrow[NaOH]{ClSCCl_2F_2,}$$

$$\underset{\displaystyle C_6H_5}{\big|}$$

Dichlofluanid

Scheme 7.4

against all phytopathogenic fungi and is specially valuable to control *Botrytis cinerea* and *venturia* spp. and downy mildews. The *p*-tolyl analogue, tolyfluanid, has a similar fungicidal spectrum, but is chiefly used to control scab and has some acaricidal activity.[12] The fungitoxicity is considered to be associated with non-specific reaction with cellular thiols.

PHENOLS

The majority of phenols, especially those containing chlorine, are toxic to microorganisms; their bactericidal action has been known for a long time and many phenols are also fungicidal. However, the majority of phenols are too phytotoxic to permit their use as agricultural fungicides. They are widely used as industrial fungicides: cresols contribute to the fungicidal action of creosote oil as a timber preservative (see Chapter 1, page 5); chlorinated phenols such as pentachlorophenol and its esters are widely used as industrial biocides for the protection of such materials as wood and textiles from fungal attack.[16] One of the earliest organic fungicides was salicylanilide or Shirlan (11) (1931) used to inhibit the growth of moulds on cotton and against a number of leaf diseases, such as tomato leaf mould.[2,5,8] This has been superseded by other compounds, but is structurally closely related to one group of modern systemic fungicides (see page 190).

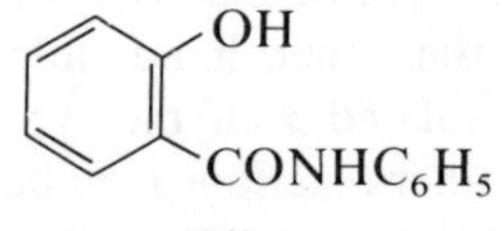

$$\text{(11)} \quad \text{OH, CONHC}_6\text{H}_5$$

(11)

Dinitrophenols are very versatile pesticides, thus 2,4-dinitro-*o*-cresol (DNOC) was first used as an insecticide in 1892 and in 1933 as a selective herbicide (see Chapter 1, page 5). This is of course much too phytotoxic to be used as a foliage fungicide, but dinocap, a mixture of 2,4-dinitro-6-(octyl)phenyl-crotonate (12) (30–35%) together with 2,6-dinitro-4(octyl)phenylcrotonate (70–65%), introduced in 1946, is a non-systemic aphicide and contact fungicide which is effective for control of powdery mildew on many horticultural

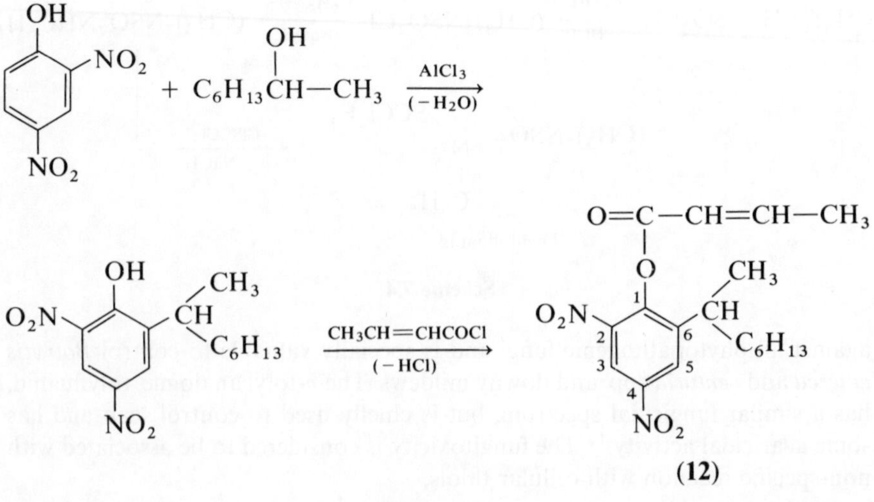

$$\text{OH, NO}_2, \text{NO}_2 \; + \; \text{C}_6\text{H}_{13}\text{CH}-\text{CH}_3 \; (\text{OH}) \; \xrightarrow[(-\text{H}_2\text{O})]{\text{AlCl}_3}$$

$$\text{O}_2\text{N}, \text{OH}, \text{CH}-\text{CH}_3, \text{C}_6\text{H}_{13}, \text{NO}_2 \; \xrightarrow[(-\text{HCl})]{\text{CH}_3\text{CH}=\text{CHCOCl}}$$

$$\text{O}{=}\text{C}-\text{CH}{=}\text{CH}-\text{CH}_3$$

$$\text{O}_2\text{N}, \text{O}, \text{CH}-\text{CH}_3, \text{C}_6\text{H}_{13}, \text{NO}_2$$

(12)

Scheme 7.5

crops.[7,17] Unlike DNOC, which has a high mammalian toxicity (LD_{50} (oral) to rats, 30 mg/kg) and has done considerable damage to the environment (see Chapter 15, page 344), dinocap (**12**) has a comparatively low mammalian toxicity (LD_{50} (oral) to rats, 980 mg/kg). It is made by condensation of 2,4-dinitrophenol and isooctyl alcohol followed by esterification with crotonyl chloride (Scheme 7.5). Binapacryl (**13**) is closely related to dinocap and is used for control of red spider mites and powdery mildew on apples.[4,7] *o*-Phenyl-phenol or 2-hydroxybiphenyl is used on citrus fruit wrappings to inhibit rot, for the disinfection of seed boxes, and for control of apple canker.

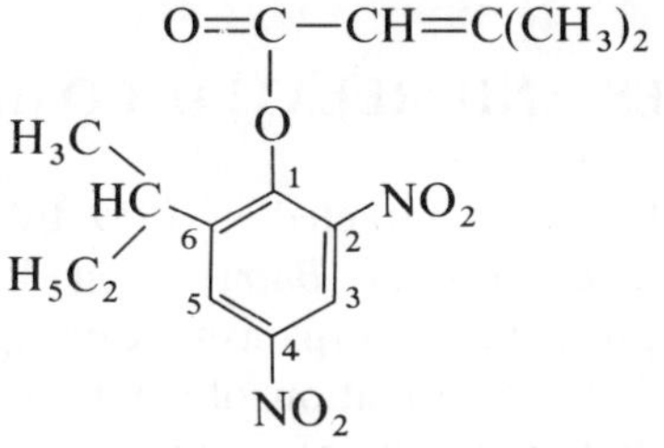

$$O{=}C{-}CH{=}C(CH_3)_2$$

(**13**)

The fungicidal action of the various phenols depends on their ability to uncouple oxidative phosphorylation and thus prevent the incorporation of inorganic phosphate into ATP without affecting electron transport. This action probably occurs at the mitochondrial cell wall and causes the cells to continue to respire, but they are soon deprived of the ATP necessary for growth. The activity of phenolic esters like (**12**) and (**13**) presumably arises from *in vivo* hydrolysis to the dinitrophenol within the fungal cell, the rest of the molecule merely functioning as a convenient shaped charge conferring the right degree of oil/water solubility to aid penetration of the fungal spore.[4]

Oxine or 8-hydroxyquinoline is a protectant fungicide which, when suitably formulated, appears to possess limited systemic action. The sulphate has been used against *Rhizoctonia* and *Fusarium* on horticultural crops and the benzoate against the Dutch elm disease. Oxine has a striking capacity to form chelates with metals and the copper chelate (**14**) is a potent fungicide which is effective

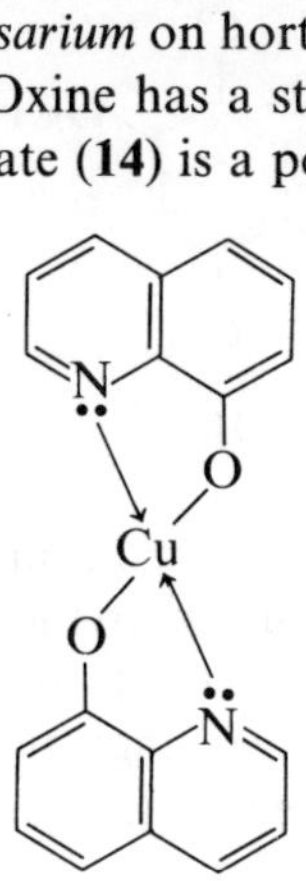

(**14**)

against a range of phytopathogenic fungi. The 1:2 chelate (14) has greater lipophilicity than oxine and the fungitoxicity of oxine is probably due to interaction with trace amounts of copper forming the chelate (14) which penetrates the fungal cell and then equilibrates with the 1:1 cupric ion–oxine complex that may be the actual toxicant. The proposed mode of action resembles that of the dithiocarbamates and oxine also exhibits a similar bimodal dosage-response graph (see page 166).[2,4,17]

CHLOROBENZENES AND RELATED COMPOUNDS

2,6-Dichloro-4-nitroaniline was marketed in 1959 by Boots Limited and is especially valuable for control of *Botrytis* on lettuce, tomatoes, and strawberries,[7,16] and against fungal organisms causing postharvest decay of fruits.[4] It is made by direct chlorination of *p*-nitroaniline. Other fungicidal chloronitrobenzenes include pentachloronitrobenzene (PCNB) called quintozene (15), which was introduced in the 1930s but has only recently become widely used as a soil fungicide against many pathogenic fungi such as damping-off diseases;[7] it is prepared by iodine-catalysed chlorination of nitrobenzene at 60°C. 1,2,4,5-Tetrachloro-3-nitrobenzene (TCNB) or tecnazene[16] is specially useful for control of *Fusarium* rot of potatoes and inhibits sprouting during storage. Chloroneb (16) is used to control soil-borne fungi, e.g. *pythium* and *rhizoctonia*, as seed dressings or by soil application. These aromatic fungicides[18] are used primarily as soil fungicides and they exhibit cross-resistance to the modern dicarboximide fungicides (see page 174). Chlorothalonil (16a) is a broad-spectrum foliar or soil-applied fungicide used in many crops (Plate 10).

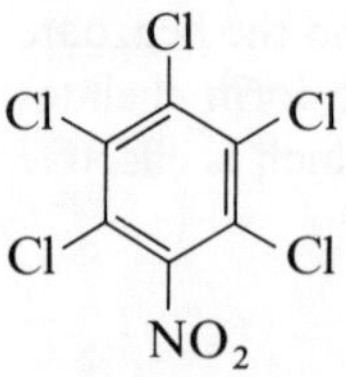

(15) 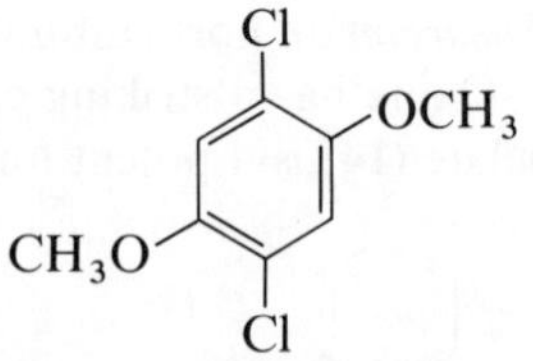 (16) 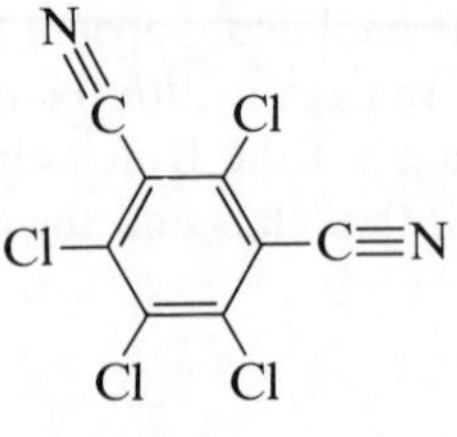 (16a)

The aromatic fungicides have a common mode of action because fungi that are resistant to one member are resistant to the other members of the group. The fungicidal action is generally considered to be due to increased lipid peroxidation of the membranes of sensitive fungi; they interact with flavin enzymes blocking electron transport and induce the generation of free radicals resulting in membrane peroxidation.[18]

QUINONES

A number of quinones occur in plants and are also products of fungal metabolism.[2] Certain members of this group have found use as agricultural fungicides such as tetrachlorobenzoquinone or chloranil, useful as a seed dressing, although as a foliage fungicide it is too rapidly decomposed by sunlight. The related derivative of naphthaquinone known as dichlone (17) is more stable in light and has been used as a seed dressing and a foliage spray, especially against downy mildew and apple scab, and as an algicide. However, these compounds tend to be rather too phytotoxic for extensive use as foliar fungicides. Dichlone (17) is manufactured by bubbling chlorine gas through a solution of 1,4-aminonaphthalenesulphonic acid in aqueous sulphuric acid (Scheme 7.6).

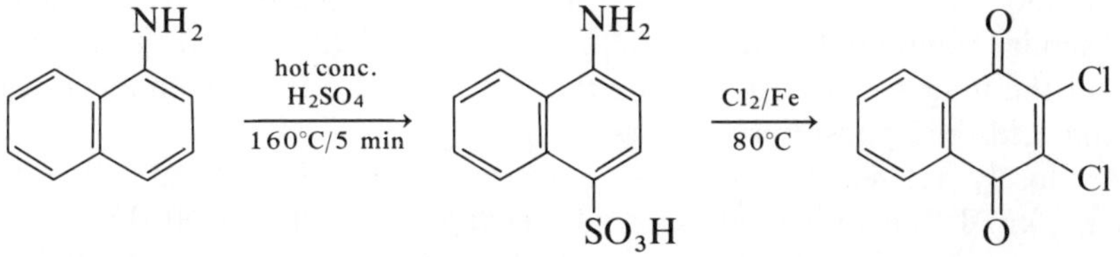

α-Naphthylamine   (17)

Scheme 7.6

Quinones are α, β-unsaturated ketones and probably owe their fungicidal properties to an addition reaction with vital sulphydryl-containing respiratory enzymes in the fungal cell, since their fungitoxicity can be antagonized by the addition of sulphydryl compounds like cysteine or glutathione.[6] The overall toxic mechanism therefore probably consists in binding the enzyme to the quinone nucleus by substitution or addition at the double bond[5,11,12] to give (18) (Scheme 7.7).

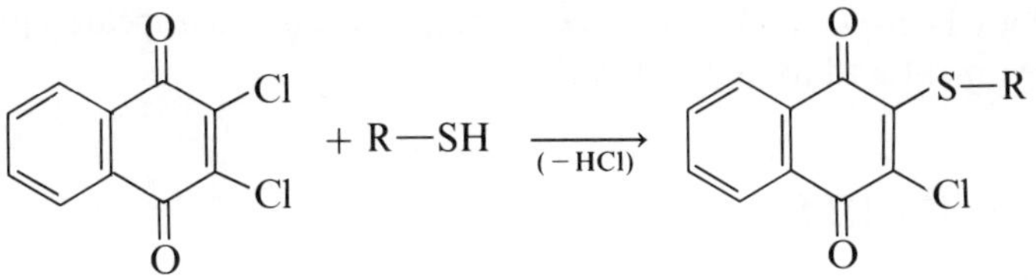

(17)   (18)

Scheme 7.7

Dodine (n-dodecylguanidine acetate) (19) has been known as a bactericide since 1941[5] and more recently has been shown to have fungicidal activity, especially against apple and pear scab, cherry and blackcurrant leaf spot, and black spot on roses.[7] Dodine (19) is a cationic surfactant and is generally formulated as a wettable powder containing the acetate. Dodine is prepared as shown in Scheme 7.8.[2,17] It is a protectant foliage fungicide of low mammalian

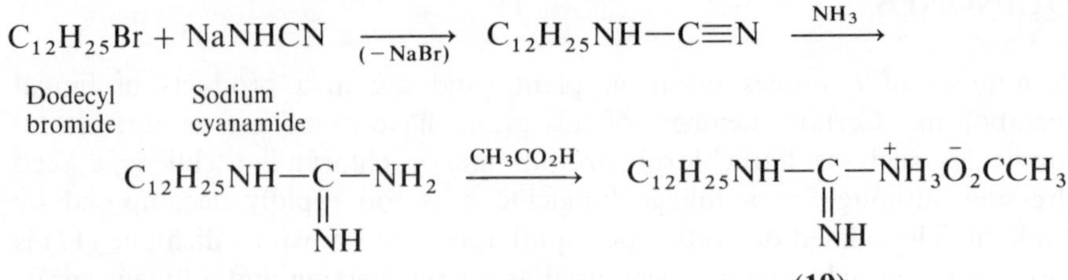

$$C_{12}H_{25}Br + NaNHCN \xrightarrow[(-NaBr)]{} C_{12}H_{25}NH-C{\equiv}N \xrightarrow{NH_3}$$

Dodecyl   Sodium
bromide   cyanamide

$$C_{12}H_{25}NH-\underset{\underset{NH}{\|}}{C}-NH_2 \xrightarrow{CH_3CO_2H} C_{12}H_{25}NH-\underset{\underset{NH}{\|}}{C}-\overset{+}{N}H_3\bar{O}_2CCH_3$$

(19)

Scheme 7.8

toxicity (LD_{50} (oral) to rats is about 1500 mg/kg). The fungitoxicity of a series of alkylguanidine acetates against yeast and *Monilinia fructicola* has been studied:[4–6] maximum activity occurred in the $C_{12}-C_{13}$ homologues whereas the C_{10} homologue had the greatest phytotoxicity. The fungicidal activity of dodine and similar compounds probably depends on their ability to alter the permeability of the fungus cell wall, causing the loss of vital cellular components, e.g. amino acids and phosphorus compounds.

A closely related fungicide is guazatine (**20**) (LD_{50} (oral) to rats $\simeq$ 500 mg/kg). It is mainly used as a seed dressing for cereals at 0.60–0.80 g a.i./kg of seed and against postharvest diseases. A similar mode of action to dodine is

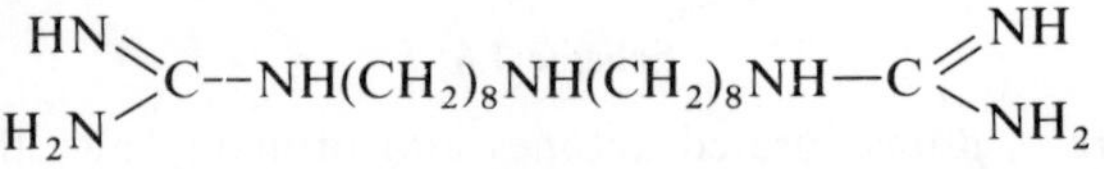

$$\underset{H_2N}{\overset{HN}{>}}C-NH(CH_2)_8NH(CH_2)_8NH-C\underset{NH_2}{\overset{NH}{<}}$$

(20)

indicated. As the triacetate salt, guazatine (**20**) controls apple canker, fusarium, and sclerotinia diseases of wheat and citrus; the compound appears to inhibit lipid biosynthesis.

Alkenals are reported[19] to be a new class of fungicides; these are formed in plants as a result of peroxidation of polyunsaturated fatty acids by lipoxidase.

2-Ethylhex-2-en-1-al and derivatives were fungicidal and appeared promising for soil treatment and food preservation.

DICARBOXIMIDES

The structures (**21**) to (**26**) of some important members of this group of fungicides are shown in Table 7.1. All members contain the 3,5-dichlorophenyl moiety and fungicidal activity depends on the presence of the two chlorine atoms in the 3,5-positions. Procymidone (**21**), the hydantoin iprodione (**23**), and vinclozolin (**24**) have been extensively used as wettable powders for control of *Botrytis* and *Sclerotinia* spp. in cereals, fruit, and vegetables, but their use has declined due to the development of resistance in *Botrytis cinerea*.

More recently metomeclan (**22**), chlozolinate (**25**), and myclozoline (**26**) have been introduced. To extend the antifungal spectrum several commercial mix-

Table 7.1

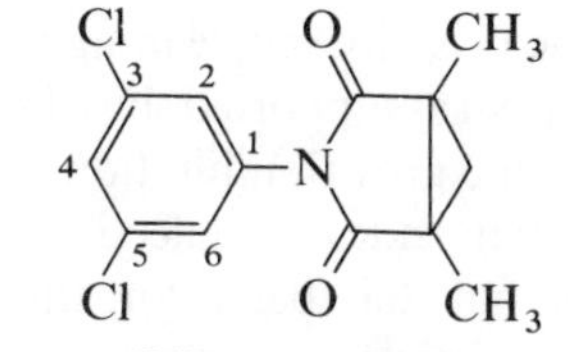

(21) (1976)

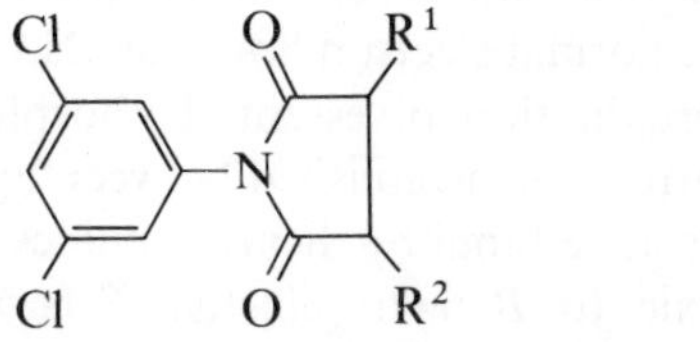

(22) $R^1 = H$, $R^2 = CH_2OCH_3$ (1984)

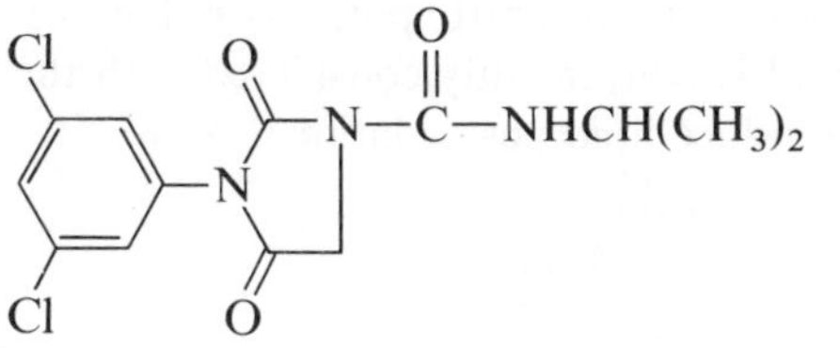

(23)

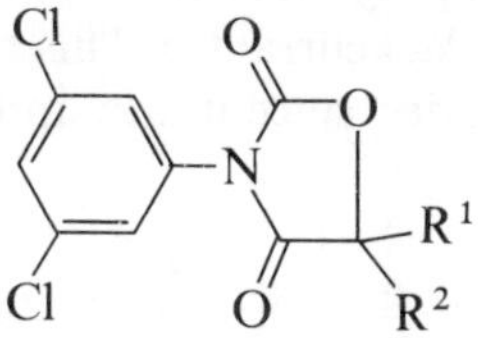

(24) $R^1 = CH{=}CH_2$, $R^2 = CH_3$ (1975)
(25) $R^1 = CO_2C_2H_5$, $R^2 = CH_3$ (1980)
(26) $R^1 = CH_2OCH_3$, $R^2 = CH_3$

tures are available, for instance with maneb, thiram, or thiophanate-methyl. The oxazolidins (e.g. (24) to (26)) may be obtained by reaction of 3,5-dichlorophenyl isocyanate with the appropriately substituted methyl glycollate[12] (Scheme 7.9). The dicarboximides (21) to (26) have low mammalian toxicities (LD_{50} (oral) to rat >3500 mg/kg) and show essentially protective activity against representatives of the following types of fungi: *Botrytis, Sclerotinia, Monilinia, Alternaria, Phoma*, and *Rhizoctonia*.

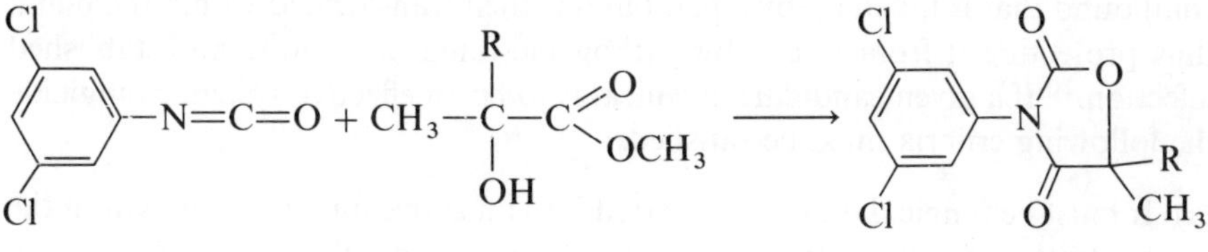

(24)–(26)

Scheme 7.9

In greenhouse crops and those under plastic sheets, loss of activity due to resistance has been widely reported. On the other hand, in the field resistance does not develop as rapidly and in spite of resistant strains of *Botrytis* in vineyards, these fungicides remain reasonably effective.

There is little knowledge regarding the precise mode of action of the dicarboximide fungicides;[18] they inhibit spore germination, mycelial growth, and DNA synthesis, but these do not appear to represent the primary mode of action.[12] The cross-resistance of these fungicides with the chlorobenzene fungicides suggested that the primary action could be lipid peroxidation. Dicarboximides, like the chlorobenzenes, may react with flavin enzymes so that the normal electron flow from $NADPH_2$ to cytochrome C is blocked, leading to peroxidation of essential phospholipids. This theory was supported by the correlation established between lipid peroxidation and the growth inhibition of sensitive fungi by dicarboximides. The dicarboximide fungicides are selectively toxic to *Botrytis cinerea, Sclerotinia,* and *Monilia,* but the basis for this selectivity is not clear from the proposed mechanism of action.

The isoxazole, drazoxolon (**27**) (ICI, 1960), is valuable as a seed dressing against damping-off diseases and as a foliar spray to control powdery mildew on roses and blackcurrants.[3] This compound is conveniently considered with the dicarboximides, since it also appears to induce lipid peroxidation.

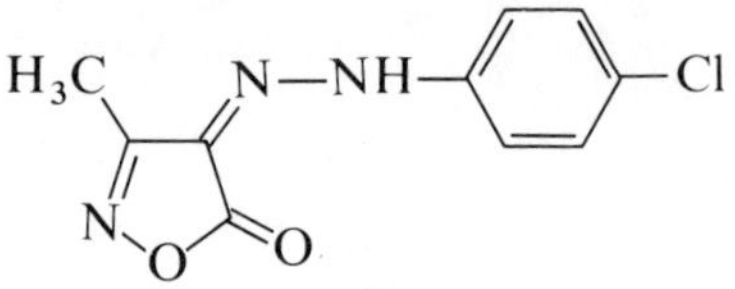

(**27**)

SYSTEMIC FUNGICIDES OR PLANT CHEMOTHERAPEUTANTS

The idea of controlling plant diseases by the internal treatment of plants is not new (see Chapter 1, page 7) but it is only since the 1960s that commercially viable systemic fungicides have come onto the market. A systemic fungicide is a compound that is taken up by a plant and is then translocated within the plant, thus protecting it from attack by pathogenic fungi or limiting an established infection.[20] If a given candidate chemical is to be an effective systemic fungicide the following criteria must be satisfied:

(a) It must be fungicidal or be converted into an active fungitoxicant within the host plant. Some compounds appear to act by modifying the resistance of the host to fungal attack.
(b) It must possess very low phytotoxicity. This requirement is especially important with a systemic fungicide since the chemical is brought into intimate contact with the host plant.
(c) It must be capable of being absorbed by the roots, seeds, or leaves of the plant and then translocated, at least locally, within the plant. For most commercial purposes the compound will be applied as a foliar spray or a seed dressing. Unfortunately uptake from the leaves is generally much more difficult than translocation by root treatment. Consequently many materials that give promising fungicidal results by root treatment fail to exhibit

comparable activity in spray tests. Conditions b and c are the major obstacles to be overcome by a candidate systemic fungicide.

The earlier protectant or surface fungicides, applied as foliar sprays, formed dried deposits on the leaves of the host plant, protecting it from fungal attack. However, the deposits are of course gradually removed by the effects of weathering and cannot protect new plant growth formed after spraying, or any part of the plant not covered by the spray. These disadvantages can be overcome by the use of systemic fungicides which since they penetrate the plant cuticle they also offer the possibility of controlling an established fungal infection. Therefore systemic fungicides should exhibit both protectant and eradicant activity.

Systemic antifungal action has been demonstrated in many compounds,[20-22] for instance sulphonamides, antibiotics, phenoxyalkanecarboxylic acids, 6-azauracil, and phenylthiourea, although their discovery made little impact on the large-scale control of fungal diseases because they were either too expensive, not sufficiently active under field conditions, or caused phytotoxic damage.

The development of systemic fungicides has largely arisen from the tremendous advances in the systemic chemotherapy of human diseases based on the discoveries of the antibacterial action of a *Penicillium* mould by Fleming (1929) and of Prontosil by Domagk (1935) which led to the production of antibiotics and sulphonamide drugs respectively. Plant pathologists considered that as bacteria and fungi are closely related, these materials might also show systemic activity against plant pathogenic diseases, and accordingly a number of synthetic bactericides and antibiotics were examined as potential systemic fungicides.

Sulphonamides

These are the most important class of synthetic bactericides from the viewpoint of systemic antifungal properties. Sulphonamides were first studied by Hassebrauk (1938); he showed that wheat rust could be controlled by root treatment with *p*-aminobenzenesulphonamide (sulphanilamide), and the systemic activity of sulphonamides has been confirmed by several workers.[20,21] Crowdy and his collaborators[21] investigated root uptake, translocation, and detoxification of sulphonamides in various plant species and showed that the behaviour of a given sulphonamide may vary from one plant to another. When a series of N^4-acylsulphonamides (28) was applied via the roots of wheat plants for control of wheat stem rust (*Puccinia triticina*), maximum disease control occurred with the n-butyl or pentyl derivatives (29; $R = C_4H_9$ or C_5H_{11}). The higher

$$RCONH-\!\!\left\langle\!\!\bigcirc\!\!\right\rangle\!\!-SO_2NH_2$$

(28)

homologues were ineffective due to low aqueous solubility, or because they were so rapidly hydrolysed in the plant that severe phytotoxic symptoms developed.

Sulphonamides have been used mainly against rust diseases on cereals, but comparatively large doses are needed and there is a danger of phytotoxic damage to the host plant. A major disadvantage is the fact that sulphonamides are fungistatic rather than fungicidal and as soon as treatment stops the fungus starts to develop.[4] The antibacterial action of sulphonamides can be suppressed by addition of *p*-aminobenzoic acid (PABA) and PABA will also reverse the antifungal action of sulphanilamide against *Trichotyton purpureum*. PABA may therefore be an essential metabolite for fungi (as well as bacteria) and the antifungal action of sulphonamides may be due to these compounds acting as antimetabolites, interfering with some stage in the synthesis of folic acid.[4,20] The effectiveness of different sulphonamides varies: sulphanilamide enters plants better than sulphaguanidine, and although both sulphanilamide and sulpha-thiazole are rapidly absorbed by plants only sulphanilamide is appreciably translocated from the roots.[21,22]

Antibiotics

Antibiotics are chemicals produced by living organisms that are selectively toxic to other organisms. The study of the biological activity of the metabolic products of microorganisms was stimulated by the successful development of penicillin for medicinal use. The properties of some 300 antibiotics have now been listed;[23] more than 100 of these are produced by fungi. These substances play an important role in the biological control of soil pathogens, but the majority have very complex structures and are often too unstable for practical application as pesticides.[23] The first antibiotics examined against pathogenic fungi were those employed in human chemotherapy.

The first successful antibiotic against human diseases was penicillin (**29**) discovered by Fleming (1929) and first used in medicine by Chain and Florey (1940), but it has never achieved commercial significance as a systemic fungicide.[9] Gliotoxin (**30**), an antifungal antibiotic produced by the soil fungus *Trichoderma viride*, inhibited the growth of *Botrytis* and *Fusarium* spores at concentrations of 2–4 p.p.m., but the compound was too unstable for use as a soil fungicide.[11]

Streptomycin (**31**), isolated from the culture filtrates of certain strains of *Streptomyces griseus*, is used for control of bacterial pathogens of plants and is especially effective against bacterial diseases of stone fruits.[21] Streptomycin (**31**) is also employed as a spray against downy mildew on hops, but an interval of at least eight weeks must elapse between application and harvest.[7] Streptomycin is active too against peach blast fungus, and the copper chelate is very effective as a foliage spray against *Phytophthora infestans* on tomato.[20]

Streptomycin does not appear to undergo metabolic activation in plants and this observation combined with the very low *in vitro* activity suggested that it may act indirectly by inducing a change in the tissue of the host plant.[4] The activity of plant polyphenolases may be increased, which would probably help the plant combat fungal infections. Streptomycin is easily taken up by plant roots, but unfortunately is rather phytotoxic due to its inhibition of chlorophyll synthesis, and tends to produce chlorosis. Application must therefore be at exactly the correct dosage to prevent the pathogen without causing damage to the host plant. The high cost of streptomycin is another major obstacle to its commercial development as a pesticide.

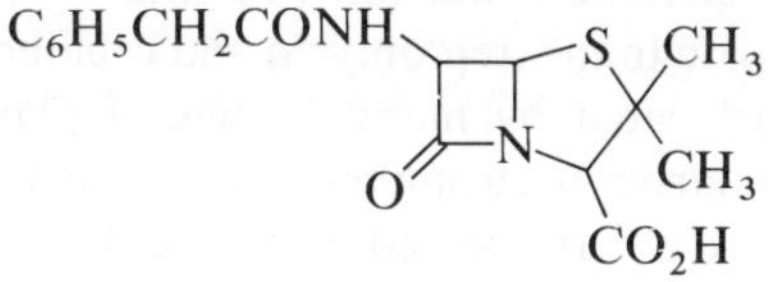

(29)

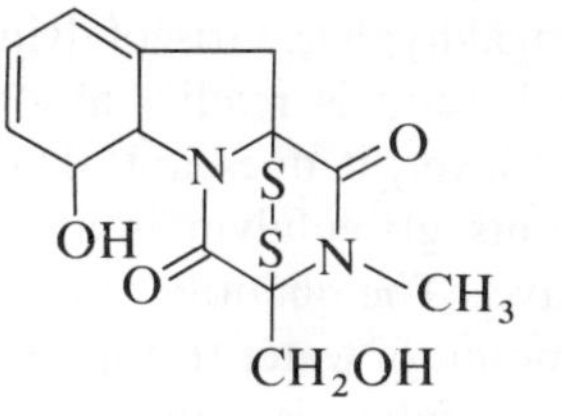

(30)

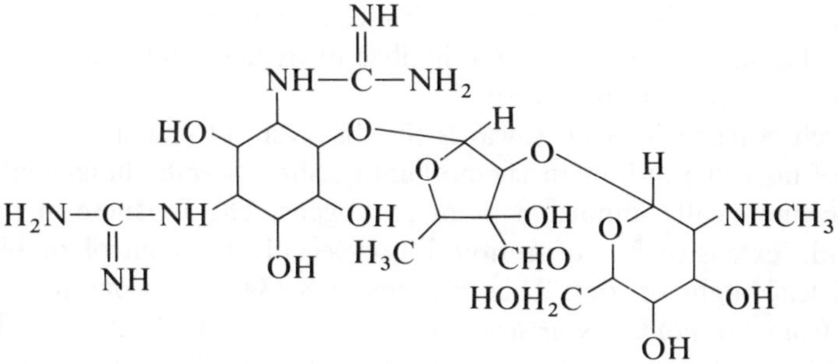

(31)

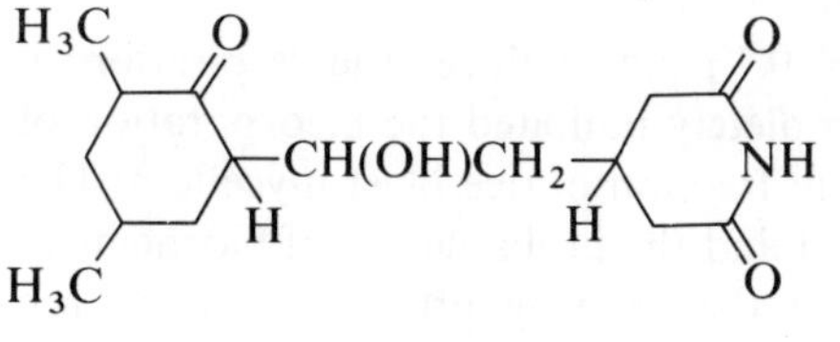

(32)

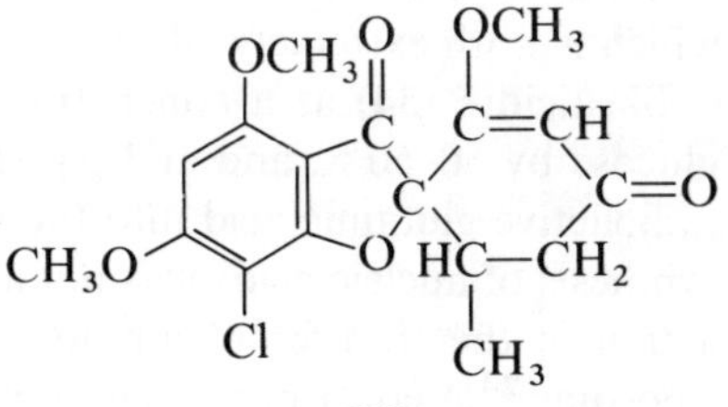

(33)

Cycloheximide (32), an antifungal antibiotic obtained from the culture filtrates of *Streptomyces griseus*,[20,21] is active against plant pathogenic fungi at concentrations of 1–5 p.p.m., and will eradicate cherry leaf spot and wheat stem rust.[17] Cycloheximide is absorbed by the roots of tomato plants and translocated to the leaves in sufficient quantities to be antifungal. The control of wheat stem rust and cherry leaf spot was promising but, like streptomycin, practical application of cycloheximide is limited by the risk of phytotoxic damage to the host plant. The fungitoxicity is probably due to the ability of the antibiotic to inhibit the transfer of amino acids from aminoacyl transfer RNA to the ribosomal protein (see Chapter 3, page 47). Cycloheximide is also toxic to animals and plants but not to bacteria, possibly due to differences in ribosomal structure.

Griseofulvin (33), isolated in 1939[11] from the mycelium of *Penicillium griseofulvum*, is an important antifungal antibiotic showing a wide spectrum of activity, especially against *Botrytis cinerea* on lettuce and *Alternaria solani* on tomato. The fungicidal effect is generally greatest with fungi having chitinous cell walls and is associated with a characteristic distortion ('curling') of the fungal hyphae. Griseofulvin is less phytotoxic than streptomycin and cycloheximide, and is readily absorbed and translocated by many species of plants. However, it does not always behave similarly; in cucumber and broad bean plants griseofulvin is absorbed and quickly translocated unchanged to the leaves. The quantity of the antibiotic moving is apparently proportional to the amount of water transpired, whereas in tomatoes translocation is much slower. Griseofulvin is also promising in the systemic treatment of superficial fungal infections in animals, including man.[20] The primary fungicidal action of the antibiotic is probably interference with cell division and not with chitin synthesis, because the curling of the hyphae of treated fungi was not apparent until some 6 hours after treatment.[24]

Research is being directed towards the discovery of new antibiotics which, though of no value in human chemotherapy, show specific fungicidal activity against economically important plant pathogens. The Japanese in particular have made extensive use of antibiotics, especially for control of blast and bacterial leaf blight on rice. Thus blasticidin S (34), a pyrimidine derivative isolated from *Streptomyces griseochromogenes*, gives excellent control of rice blast[11] (*Piricularia oryzae*) and also inhibits certain bacteria, but the compound is rather phytotoxic and has been replaced to some extent by kasugamycin (35), which is both extremely effective and safer to use.

Blasticidin (34) at a concentration of 0.1 p.p.m. reduces the respiration of glucose by 50–60%, and at 1 p.p.m. completely inhibited the incorporation of radioactive glutamic acid into the protein fraction of rice blast mycelium. The synthesis of nucleic acids was not affected and the probable site of biochemical action is the transfer of amino acids to the growing protein chain on the ribosome.[24] A contributory factor is the inhibition of respiration since this limits the supply of energy to the fungal cell. Kasugamycin (35) shows selective toxicity against rice blast at low pH and also interferes with protein biosynthesis *in vivo* as well as in cell-free systems. The process inhibited is probably the binding of the aminoacyl transfer RNA complex with the messenger RNA and ribosome to a complex.[24]

The antibiotic polyoxin D (36) is another pyrimidine derivative which is toxic towards several fungi, including rice blast. Polyoxin D does not affect respiration, or nucleic acid, or protein biosynthesis, but strongly inhibits the incorporation of radioactive glucosamine into the fungal cell walls. Studies with

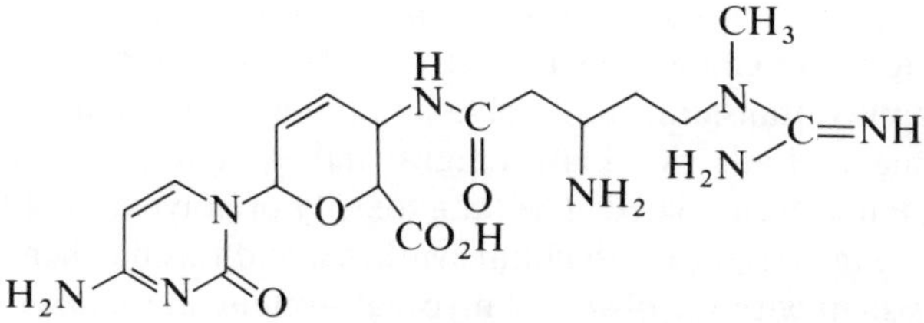

(34)

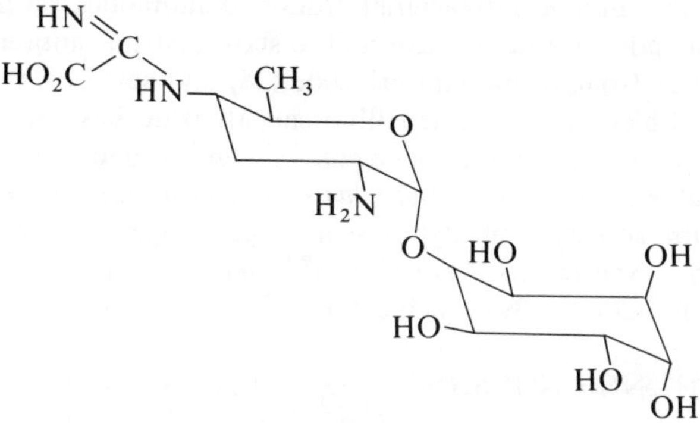

(35)

Neurospora crassa showed[24] that a concentration of 100 p.p.m. of the antibiotic depressed both fungal growth and the incorporation of ^{14}C-glucosamine into chitin, an important structural material for many fungal cell walls and insect cuticles. Chitin is essentially a polymer of *N*-acetylglucosamine and is synthesized from uridine diphosphate *N*-acetylglucosamine (37) by a reaction in which

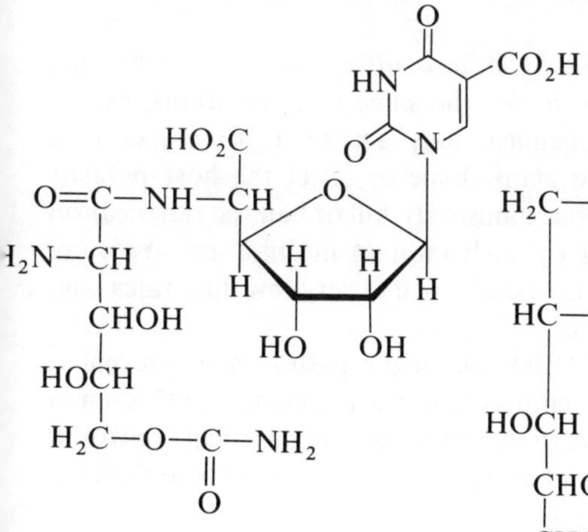

(36)

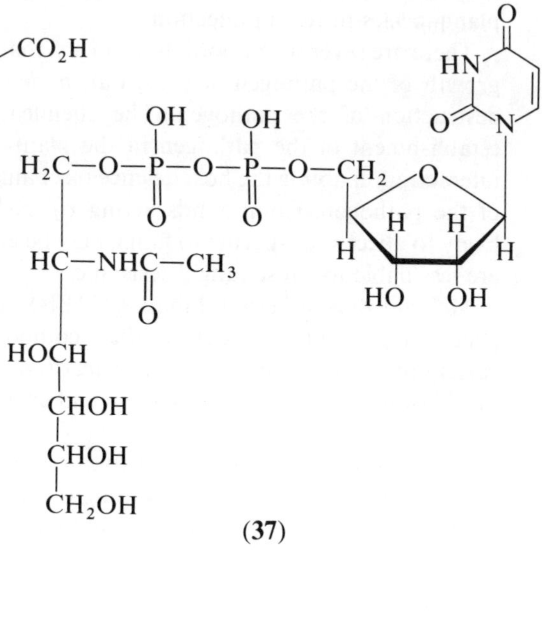

(37)

N-acetylglucosamine units are transferred from UDP-*N*-acetylglucosamine (**37**) to the growing chitin chain. This is controlled by the enzyme chitin-UDP *N*-acetylglucosaminyl transferase and polyoxin D competitively inhibits the action of this enzyme in both rice blast fungus and *N. crassa* so that UDP-*N*-acetylglucosamine accumulated. The fungitoxicity of polyoxin D (**36**) is almost certainly due to interference with chitin synthesis, and this biochemical mode of action is consistent with the observed mycelial swelling in treated fungi and the lack of toxicity shown against mammals and plants.

In addition to synthetic antibacterial drugs and antibiotics, several synthetic organic compounds have been discovered to show systemic antifungal properties, for instance 1-phenylthiosemicarbazide (**38**) and some of its derivatives, such as phenylthiourea.[25–27] Phenylthiosemicarbazide has low mammalian toxicity and good systemic properties against cucumber mildew, apple scab, late blight on potatoes, and *Botrytis* and *Cladosporium* on tomatoes. On the other hand, the *in vitro* activity of phenylthiosemicarbazide against fungal spores was much less and experiments have shown[27] that it is activated *in vivo* to phenylazothioformamide (**39**) which is the active fungitoxicant:

$$C_6H_5NH\!-\!NHCSNH_2 \quad \xrightarrow[(-2H)]{in\ vivo} \quad C_6H_5N\!=\!NCSNH_2$$

$$(\mathbf{38}) \qquad\qquad\qquad\qquad\qquad (\mathbf{39})$$

In contrast, phenylthiourea appears to act indirectly so as to increase the resistance of the host plant to fungal infection; probably by inhibition of a pectolytic enzyme secreted by the fungus to dissolve the plant cells it is attacking,[25] in rice blast fungi it acts to prevent the pathogen penetrating the host plant. Wain[8] has suggested that such inactivation of plant pectolytic enzymes may be a factor contributing to the natural immunity shown by certain plant species to fungal infection.

There are several compounds known[18] that have little fungicidal effect on the growth of the pathogen *in vitro*, but *in vivo* the same concentrations lead to destruction of the pathogen. The chemical may act so as to prevent the establishment of the pathogen in the plant tissue or affect the host–parasite interaction, enabling the host defence mechanisms to kill or halt the ramification of the pathogen. Compounds acting by such indirect mechanisms are more likely to affect sites specific to fungi, may be effective at very low dose rates, and are less liable to cause fungal resistance.

Melanin biosynthesis inhibitors (MBIs) act on the pathogen to prevent it penetrating the plant epidermis; these compounds block melanin synthesis in a variety of *Ascomycetes* and *Fungi imperfecti*. They provide practical control of rice blast fungus (*Piricularia oryzae*) and experimental control of some *Colletrichum* spp.[28]

Some MBIs are shown in Table 7.2. If the mechanism of appressorial penetration adopted by rice blast fungus was more widely used by phytopathogenic fungi, this group of MBIs would have a major impact on the control of

Table 7.2 Compounds inhibiting melanin biosynthesis

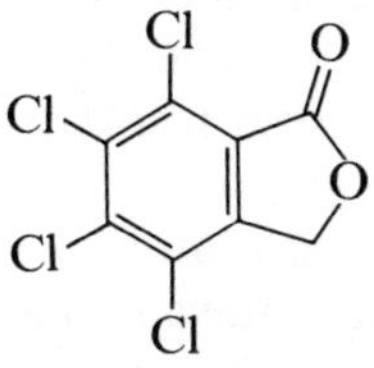

(40) 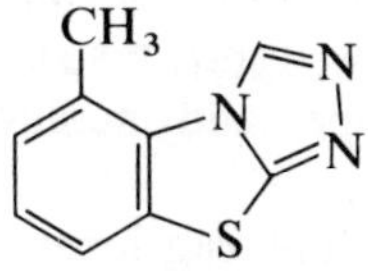 (41) (42) (43) 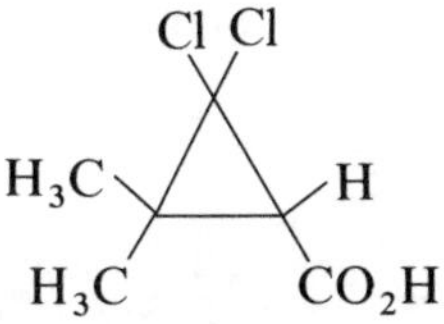 (44)

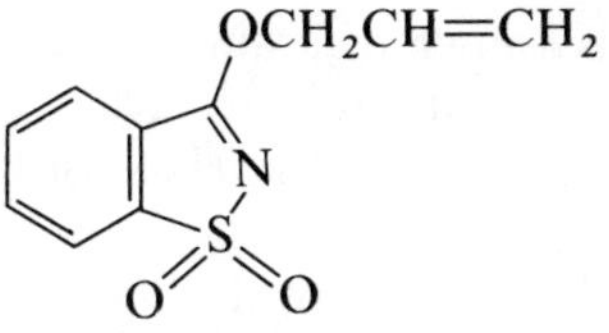

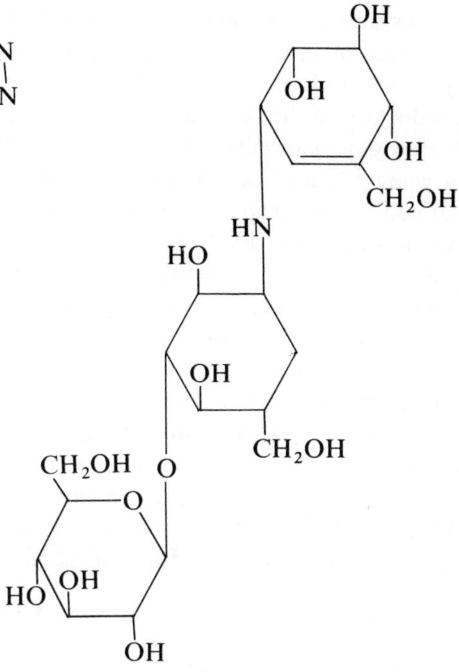

fungicidal diseases. Pentachlorobenzyl alcohol and the tetrachloro compound (40) do not impair spore germination, but prevent penetration of epidermal barriers by appressoria of *P. oryzae*.

Tricyclazole (41) prevents penetration by blocking the polyketide pathway leading to melanin biosynthesis, which causes the appressorial wall of *P. oryzae* not to have sufficient rigidity to effect penetration.

Validamycin (42) is an aminoglucoside antibiotic, active mainly against *Rhizoctonia* diseases, and has been widely applied to control rice sheath blight (*R. Solani*).[18] The activity again possibly arises from the compound reducing the penetration of the rice plant by the fungi, thus giving time for the host's defence system to operate.

Probenazole (43), related to saccharin, is effective by root application against rice blast and bacterial leaf disease (*Xanthomonas oryzae*). It is not fungitoxic *in vitro* and probably acts indirectly by enhancing the resistance response of the host plant.

2,2-Dichloro-3,3-dimethylcyclopropanecarboxylic acid (44) experimentally controls rice blast disease, not by preventing epidermal penetration by the fungus but by promoting a marked increase in host resistance response once penetration has occurred. It has been discovered that solutions of certain amino acids, e.g. D-alanine, D- and DL-leucine and D- and DL-phenylalanine enhanced the resistance of apple leaves to scab, but all the L isomers were inactive. Later work demonstrated that a similar effect was shown strongly by α-aminoisobutyric acid (AIB); the application of copper and mercury salts also increases production of the phytoalexin pisatin in peas.

Phenoxyalkanecarboxylic acids are rapidly translocated from the roots to the growing points of bean plants. Most of these compounds showed little antifungal activity either *in vitro* or *in vivo*, but a few exhibited higher *in vivo* activity which is probably due to alteration of the metabolism of the host plant, making it less favourable to the fungus, possibly by alteration in the level of reducing sugars present in the sap. Comparatively little is known regarding the effect of alteration in the balance and concentration of naturally occurring plant constituents, such as amino acids, phenols, and sugars, on the level of resistance shown by the host plant to the invading pathogen.

Phytoalexins are natural antifungal compounds produced by the host plant in response to fungal infection and they may play an important role in the host's defence mechanism. A number of phytoalexins have been isolated: for example, the phytoalexin 6-methoxy-2(3)-benzoxalinone (**45a**) from maize and wheat seedlings[20] and pisatin (**45b**) from peas.

(**45a**)   (**45b**)

The resistance of some varieties of crop plants towards attack by certain species of fungi may be due to their ability to accumulate phytoalexins in the host–pathogen complex following infection. However, more detailed knowledge of the biochemical differences between resistant and susceptible varieties of the same plant species is needed before their precise role can be established. The isolation and structural elucidation of phytoalexins provides a possible route to the discovery of novel systemic fungicides and their subsequent molecular modification may lead to simpler and more potent compounds.

Other methods of controlling fungi include the genetic breeding of new resistant varieties of crop plants. The resistance is generally associated with the presence of phytoalexins; it may be feasible to control plant diseases by treatment with systemic chemicals that stimulate the production of phytoalexins by the host plant.

One of the most interesting sources of natural fungitoxicity is the soil itself; the soil microbial population is exceedingly complex and soil microorganisms such as fungi and actinomycetes can synthesize appreciable amounts of antibiotics which probably diffuse into the soil solution regulating the growth of other soil organisms.[22] In sterile soil, which has been inoculated with suitable microorganisms, the formation of fungistatic antibiotics can be demonstrated easily because now there are no other organisms breaking them down. The production of fungistatic compounds has also been shown in natural soil, and the concept of a widespread soil fungitoxicity has been developed.[20] In addition, a large number of compounds, such as amino acids, can diffuse out from the roots of higher plants into the surrounding soil (the rhizosphere). These root exudates will affect the neighbouring soil microflora and this may increase or decrease the resistance of the plant roots to the attack of pathogenic soil organisms, or alter the pathogenicity of the fungus.

Bison and Novelty are two different varieties of flax; the former is resistant, while the latter is susceptible, to soil-borne *Fusarium* wilt disease. Bison exudates stimulated the growth of the soil fungus *Trichoderma viride*, which is known to produce fungistatic antibiotics (see page 178). Therefore the *Fusarium* wilt fungus was checked by this variety of flax; possibly in the future plant diseases might be controlled by cultivation of suitable antibiotic-producing organisms in the soil. The ideal systemic fungicide might be one which, after application to the foliage, travelled down to the roots and exuded into the rhizosphere inhibiting root parasites without interfering with naturally beneficial microorganisms, such as actinomycetes. Chemical soil sterilants, like formaldehyde and carbon disulphide, probably not only destroy the parasitic fungi but also do not kill certain soil saprophytes, so that after fumigation these multiply rapidly producing a diffusible toxin that finally destroys the fungal parasites.

These ideas open up fascinating possibilities for future control of fungi, but it must be admitted that to date the discovery of commercial systemic fungicides has largely arisen from exploitation of the results from more or less random screening of synthetic organic compounds.

There are now a number of systemic fungicides on the market[18,20,21,27] which may be divided into benzimidazoles, thiophanates, carboxamides, pyrimidines, pyridines, morpholines, azoles, and organophosphorus compounds.

Benzimidazoles

These represented a new era in fungicide use when they were introduced in the late 1960s. The most important members of this group are methyl-l-(butylcarbamoyl)benzimidazole-2-carbamate or benomyl (**46**) and 2-(4′-thiazolyl)benzimidazole or thiabendazole (**47**; X = S).

Benomyl (**46**), introduced in 1967, was synthesized from cyanamide and methyl chloroformate (Scheme 7.10). Benomyl and thiabendazole are both wide-spectrum systemic fungicides, active against many pathogenic fungi, including powdery mildews and soil-borne pathogens, *Verticillium alboatrum* on cotton and black spot on roses,[7] but they are not effective against the

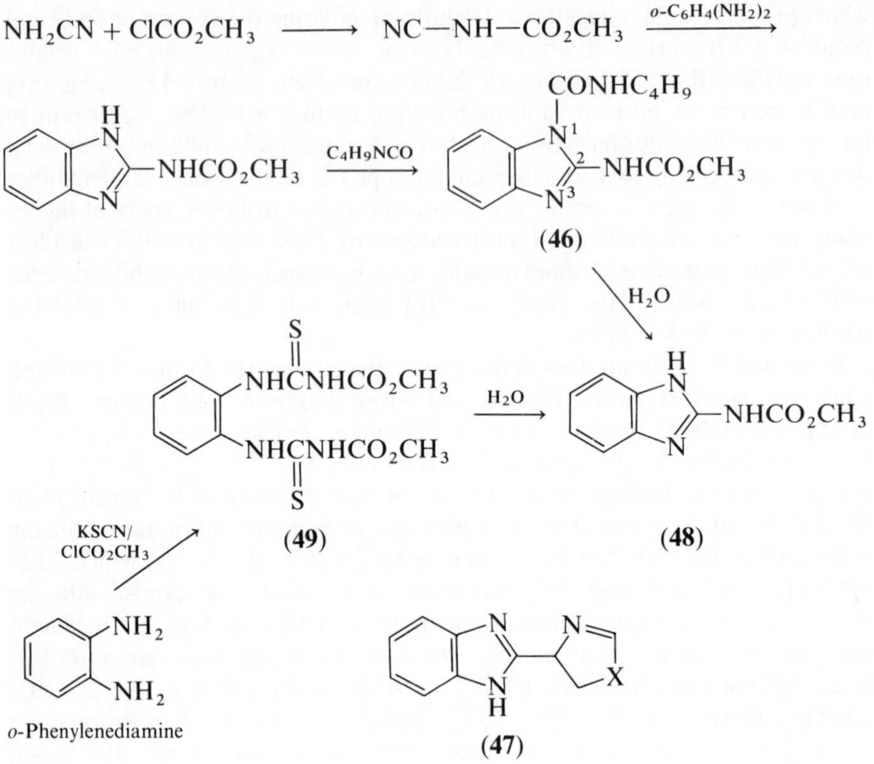

$$NH_2CN + ClCO_2CH_3 \longrightarrow NC-NH-CO_2CH_3 \xrightarrow{o\text{-}C_6H_4(NH_2)_2}$$

(structure) —NHCO₂CH₃ $\xrightarrow{C_4H_9NCO}$ (structure, **46**) with CONHC₄H₉ on N¹, NHCO₂CH₃ at position 2, N³

(**46**) $\searrow$ H₂O

(structure, **49**): NHCNHCO₂CH₃ (C=S) and NHCNHCO₂CH₃ (C=S)

$\xrightarrow{H_2O}$ (structure, **48**): benzimidazole —NHCO₂CH₃

$\xrightarrow[ClCO_2CH_3]{KSCN/}$ (**49**)

(structure) o-Phenylenediamine: NH₂, NH₂

(structure, **47**): benzimidazole—CH with ring containing N=, CH₂—X

Scheme 7.10

phycomycetes group of fungi, such as potato blight, vine downy mildew, or damping-off diseases. Benomyl is the more active compound and is widely applied as a foliar spray, seed dressing, or to the soil for control of grey mould (*Botrytis cinerea*), apple scab (*Venturia inaequalis*), canker and powdery mildew (*Podosphaera leucotricha*), *Gloeosporium* storage rots, leaf spot (*Cercospora beticola*) on sugar beet, the major fungal diseases of soft fruits, and some pathogens of tomato and cucumber.[20,27]

Thiabendazole (**47**; X = S) is good against postharvest diseases of apples and pears, and as a seed dressing against common bunt of wheat (*Tilletia caries*), when the fungicide may persist in the plants for several months. Thiabendazole has also been used against stem rot of bananas, blue and green mould on citrus fruits, and is reported[20] to be more effective than benomyl for eradication of Dutch elm disease, although neither chemical was sufficiently active to be a practical means of large-scale control.[21] Thiabendazole is translocated from either the roots or leaves of growing plants and also moves from the leaves to the roots.

Fuberidazole (**47**; X = O) is a valuable seed dressing against *Fusarium* diseases (e.g. of rye and peas), wheat rust (*Puccinia triticina*), and barley powdery mildew (*Erysiphe graminis*).[17]

The activity of selected benzimidazoles has been reviewed[20,21] and the fungitoxicity is clearly associated with the benzimidazole nucleus, which probably accounts for the almost identical *in vitro* fungicidal spectrum shown by benomyl (46) and thiabendazole (47; X = S). In aqueous solution benomyl (46) is rapidly hydrolysed to the strongly fungicidal methylbenzimidazole-2-carbamate (48), and this is probably the active fungitoxicant, since the fungicidal properties of benomyl are increased after standing in aqueous media. In this connection, it is interesting that methylbenzimidazole-2-carbamate (carbendazim) (48) is used as a wide-spectrum systemic fungicide and may be formulated as a 50% wettable powder for control of *Botrytis*, *Gloeosporium* rots, powdery mildews, and apple scab. Carbendazim is absorbed by the roots and foliage of plants and is quicker acting than benomyl.

With thiabendazole (47; X = S) and fuberidazole (47; X = O) there is no evidence of *in vivo* metabolism. When benomyl (46) is used for control of *Verticillium* wilt of cotton and potatoes by soil treatment, the activity is enhanced by addition of surfactants, probably because these improve the mobility of the metabolite (48) through the soil. Both benomyl and thiabendazole have been used, by injection into the tree, for control of Dutch elm disease; the latter is also used as an anthelmintic.

Thiophanates

These are an important group of systemic fungicides based on thiourea.[27] Thiophanate is 1,2-bis(3-ethoxycarbonyl-2-thioureido)-benzene and thiophanate-methyl (49) is the methyl analogue which is obtained by condensation of potassium thiocyanate, methyl chloroformate, and *o*-phenylenediamine. Both are effective against such pathogenic fungi as apple powdery mildew, apple and pear scab, sheath blight of rice, *Cercospora* leaf spot of sugar beet, and *Botrytis* and *Sclerotinia* on various crops.[7,20] They also showed a high level of persistent systemic activity by root uptake against barley and cucumber mildew. The overall fungicidal spectrum of the thiophanates resembles that of the benzimidazoles; in particular benomyl (46) and thiophanate-methyl (49) are closely similar in their antifungal properties. The latter compound, like benomyl, on treatment with water gave methylbenzimidazole-2-carbamate (48). Similarly thiophanate afforded ethylbenzimidazole-2-carbamate. The carbamate (48) has recently been isolated from plants treated with thiophanate-methyl (49) and it is concluded that the carbamates (e.g. compound (48)) are the active fungitoxicants in both benomyl and the thiophanates. This idea is supported by the inactivity of 1,3- and 1,4-bis(3-ethoxycarbonyl-2-thioureido)benzenes which are incapable of cyclization to the benzimidazolecarbamate (48).

The thiophanates are not themselves fungicidal but are converted to the active benzimidazole derivatives (e.g. compound (48)). The fungitoxicity of both benomyl and thiophanate-methyl appears to be related to the retention in the roots of either benomyl or, in the case of the thiophanates, of an intermediate between (48) and (49). These compounds then function as reservoirs of the active toxicant (48) which is subsequently released gradually to other parts of the plant. A similar effect is believed to operate when benomyl is applied as a foliar spray due to the effects of weathering of the deposit on the leaf surface.

The benzimidazoles are effective at relatively low doses and all have similar patterns of selective fungitoxicity: most *Ascomycetes*, some of the *Basidiomycetes*, and *Deuteromycetes* are sensitive but none of the *Phycomycetes*.[18] They are safe to animals (LD_{50} (oral) to rats >7500 mg/kg), bees, and plants but show a high toxicity to certain species of fungi, insects, mites, and worms. Serious resistance problems appeared with benzimidazole fungicides in the 1970s due to the fact that they were applied intensively and are single-site inhibitors. Benzimidazoles can be rendered ineffective for several seasons after the emergence of resistant fungal strains. The problem of resistance stimulated the studies on the mode of action of the benzimidazole fungicides and it was demonstrated that benzimidazoles interfere with DNA synthesis in certain fungi.

Thus the application of a 5 μM solution of methyl benzimidazole-2-carbamate (48) to *Neurospora crassa conidia* resulted in 85% inhibition of DNA synthesis in 8 hours.[24] However, later work[18] indicated that this was a secondary effect which was primarily due to blockage of nuclear division. In preventing the nuclear division of sensitive fungi, benzimidazoles show a remarkable similarity to the effects of the secondary plant metabolite colchicine, which acts by binding to the cellular protein tubulin.

In vivo binding experiments using ^{14}C-labelled carbendazim and mycelial extracts of *Aspergillus nidulans* demonstrated the binding of the fungicide to β-tubulin and resultant interference with the microtubule assembly.

The activity of the benzimidazole fungicides is concluded to arise from inhibition of nuclear division due to their action on the microtubule assembly. In agreement with this theory, only sensitive fungi showed evidence of binding to β-tubulin and the affinity of the compound to the target site on the tubulin determines whether a given benzimidazole is or is not fungitoxic. The development of resistance in fungi is considered to result from the mutant strains possessing an altered microtubule assembly[18] with reduced binding affinity for the fungicide.

The parent benzimidazoles, or some of their metabolites, may be mutagenic, and benomyl is known to be mutagenic to *Aspergillus nidulans*; such mutagenicity may account for the emergence of strains of *Botrytis cinerea* on cyclamen that are resistant to benomyl[27] (see page 211).

Carboxin and Related Compounds

Oxathiins are another group of heterocyclic compounds with interesting systemic fungicidal properties; examples are carboxin (5,6-dihydro-2-methyl-

1,4-oxathiin-3-carboxanilide) (**50**) and the sulphone analogue known as oxycarboxin (**51**).[20] Carboxin is prepared by reaction of α-chloroacetoacetanilide and 2-thioethanol followed by cyclization (Scheme 7.11). Oxycarboxin (**51**) is obtained by subsequent oxidation of carboxin (**52**) with hydrogen peroxide.

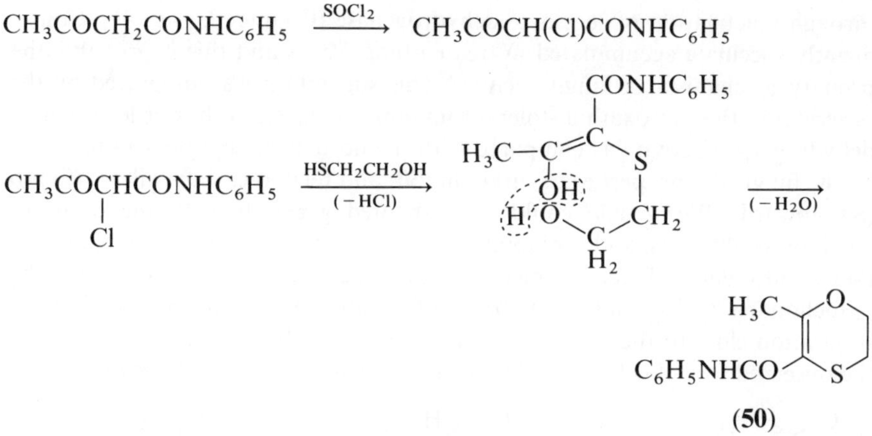
CH3COCH2CONHC6H5
SOCl2
CH3COCH(Cl)CONHC6H5
CH3COCHCONHC6H5
Cl
HSCH2CH2OH
(−HCl)
CONHC6H5
H3C
OH
H
O
S
CH2
H2
(−H2O)
H3C
O
C6H5NHCO
S

(50)

Scheme 7.11

Both of these compounds are primarily active against the *Basidiomycetes* class of fungi which includes such economically important pathogens as the rusts, smuts, and bunts of cereals, and the soil fungus *Rhizoctonia solani.*[27] Carboxin (**50**) was introduced in 1966 and is an extensively used and highly effective seed dressing (1–2 g a.i./kg of seed) for the eradication of loose smut of barley and wheat.[7] These diseases cannot be satisfactorily controlled by organomercurial seed dressings because the mycelium of the fungus is deeply within the seed. Carboxin also shows activity against oat smuts, seedling blight of wheat, leaf stripe of barley, brown foot rot of oats, and malsecco disease of lemon trees.[20,25] Carboxin can be formulated with other fungicides, e.g. thiram and copper oxine, to broaden the fungicidal spectrum. Oil-dispersable formulations are better than wettable powders as they more effectively penetrate the host and pathogen. Oxycarboxin (**51**) has systemic activity against rust diseases of cereals and vegetables, and seed treatment or soil application has delayed the onset of wheat leaf and stem rust by some two months; it is better than carboxin on rusts but less effective against smuts.

Carboxin is absorbed and translocated by plant roots. In water, soil, and plants (barley, wheat, and cotton), the compound was oxidized to the corresponding sulphoxide but further oxidation to the sulphone (**51**) was not observed; in the roots of Pinto beans hydrolysis of the carboxanilide group also occurred.

The sulphoxide is much less fungicidal so the oxidation causes loss of activity. Eventually the sulphoxide disappears, probably by binding to lignin by complex formation, leaving no residue in cereals.

The fungitoxicity of carboxin and eight analogues was examined in relation to their effects on metabolic pathways. All the active compounds strongly inhibited glucose and acetate oxidative metabolism and RNA and DNA synthesis, although the latter may arise from lack of cellular energy due to inhibition of respiration. Carboxin also interfered with succinate oxidation in sensitive fungi through inactivation of succinate dehydrogenase (Chapter 3, page 41). Consequently succinate accumulated in treated fungal cells and this is probably the primary mechanism of fungitoxicity.[24] This suggestion was supported by the observation that an oxathiin-tolerant mutant fungus strain has little succinate dehydrogenase activity as compared with the normal susceptible strain.

The fungicidal properties of carboxin (**50**) and related carboxanilides (**51**) to (**54**) towards *Rhizoctonia* could be correlated generally with the oil/water partition coefficients, and the uptake of carboxin by fungi is closely related to their lipid content. Electron microscopy revealed that carboxin damaged the mitochondria and vacuolar membrane of sensitive fungi, causing inhibition of respiration close to the site of succinate oxidation. This probably accounts for the observed loss of cellular materials from treated hyphae of *R. solani*.[21,27]

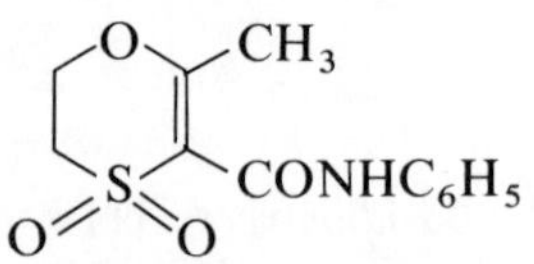

(**51**) 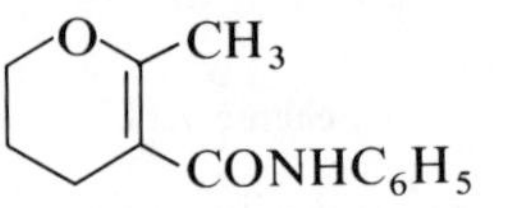   (**52**) 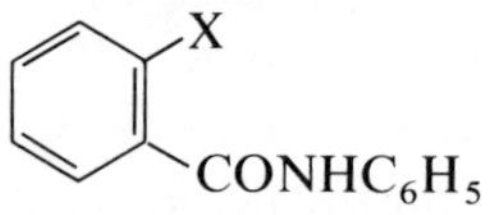   (**53**)

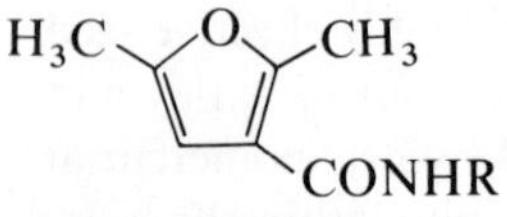

(**54**)

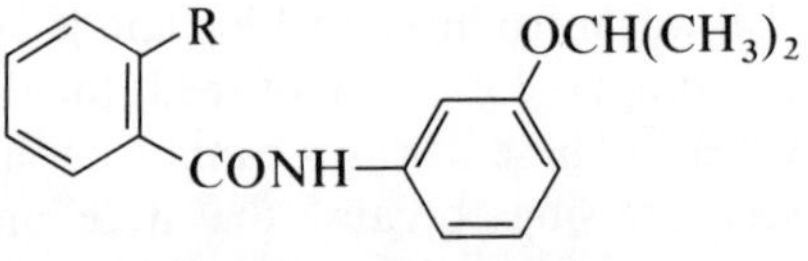

(**55**) R = CH$_3$ (1981)
(**56**) R = CF$_3$ (1982)

Several other substituted carboxanilides show specific fungicidal activity towards *Basidiomycetes*, although the 2-methyl group appears to be essential for activity; one or both of the hetero atoms can be dispensed with. Thus 2-methyl-5,6-dihydropyran-3-carboxanilide (**52**) is rather more effective than carboxin (**50**) as a seed dressing against smut diseases of barley and oats and is also active against rusts and *Rhizoctonia* spp. Oxycarboxin (**51**) uniquely controls some species of *Deuteromycetes* and *Phycomycetes* as well as *Basidiomycetes*.

2-Methyl and 2-iodobenzanilide (**53**; X = CH$_3$ or I) are valuable systemic fungicides for control of a range of *Basidiomycetes*, especially rust diseases of cereals, coffee, tobacco, vegetables, and ornamentals, and for dressing seed potatoes against *Rhizoctonia* spp. These compounds are structurally closely related to Shirlan or salicylanilide (**53**; X = OH), one of the first organic surface fungicides (see Chapter 1, page 5), still used on a small scale against *Cladosporium fulvum* on tomatoes.

2,5-Dimethylfuran-3-carboxanilide (**54**; R = C_6H_5) and the cyclohexylamide (**54**; R = C_6H_{11}) have been shown to have outstanding systemic fungicidal properties as seed dressings against loose smut of barley and wheat.[17,18] Mepronil (**55**) and flutolanil (**56**) are useful for control of sheath blight in rice.

The oxathiin ring is not essential for systemic fungicidal activity and can be replaced by other rings, e.g. benzene, furan, pyran, thiazole, or thiophen, without loss of activity. The essential feature of the carboxamide fungicides is the *N*-phenyl-2-butenamide group (**57**). The double bond in compound (**57**) may be

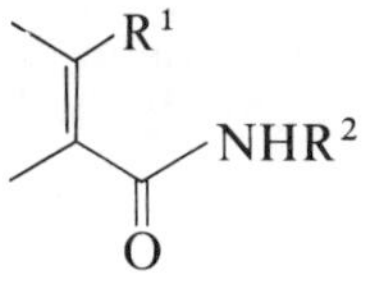

$(R^1 = CH_3, CF_3, Cl, I;$
$R^2 = phenyl, cyclohexyl)$

(**57**)

part of a planar or non-planar ring or an acyclic system, but the R^1 group must be *cis* with respect to the carboxamido moiety. The intact molecules are probably responsible for the fungitoxicity.[12]

There has been no widespread resistance of fungal pathogens in the field to carboxin or oxycarboxin; these compounds also stimulate the growth of plants contributing to increased crop yields. The primary mode of action of carboxin and related compounds probably involves the blocking of succinate oxidation in the mitochondria of sensitive fungi. The active compounds interact with receptors in the succinate–ubiquinone oxidoreductase system of the mitochondrial electron transport chain.

Genetic studies demonstrated that carboxin-resistant mutants of *Aspergillus nidulans* and *Ustilago maydis* showed decreased sensitivity of the succinate oxidase system towards carboxin and related compounds. The steric shape of the molecules appears to be crucial in obtaining binding of the fungicide molecule to the reactive sites of the enzyme.[18]

Aminopyrimidines

Many valuable crops are attacked by powdery mildews; hence there are large potential markets for compounds that can control these diseases. Some 30 years ago screening at Jealott's Hill Research Station in England indicated that a series of 2-amino-4-hydroxypyrimidines had specific systemic activity against powdery mildews. These structure–activity studies led to the development of dimethirimol and ethirimol.

Dimethirimol is 5-butyl-2-dimethylamino-4-hydroxy-6-methylpyrimidine (**58**) and is prepared by condensation of ethyl α-butylacetoacetate with *N*,*N*-dimethylguanidine (from dimethylamine and cyanamide) (Scheme 7.12). Dimethirimol, discovered in 1965, showed outstanding systemic activity by root application against certain powdery mildews, such as those of cucumber, melon,

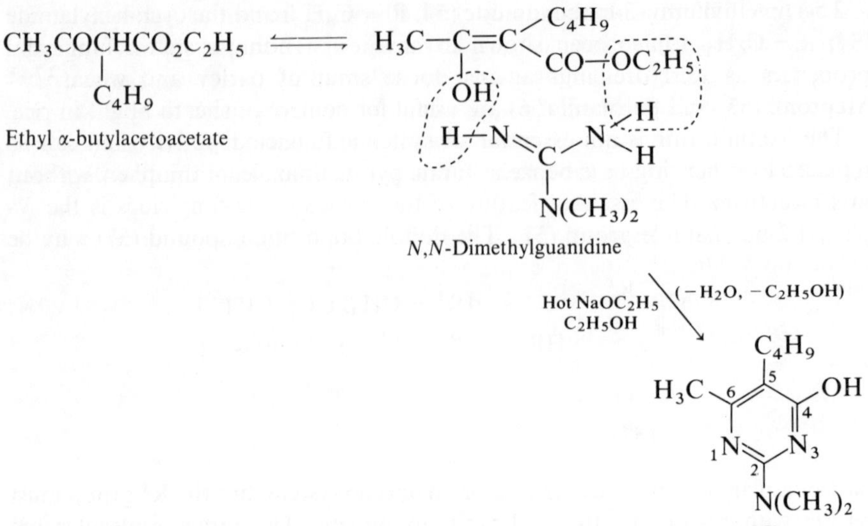

$$CH_3COCHCO_2C_2H_5 \rightleftharpoons$$
$$\underset{C_4H_9}{|}$$

Ethyl α-butylacetoacetate

$$H_3C-C=C\underset{\overset{|}{OH}}{\overset{C_4H_9}{\diagdown}}CO-OC_2H_5$$

$$H-N=\underset{\overset{|}{N(CH_3)_2}}{C}N-H$$

N,N-Dimethylguanidine

Hot NaOC$_2$H$_5$
C$_2$H$_5$OH

$(-H_2O, -C_2H_5OH)$

$$H_3C\underset{1}{\overset{6}{\diagdown}}\underset{2}{\overset{5}{\diagup}}\underset{N}{\overset{C_4H_9}{\diagup}}OH$$
$$\underset{N(CH_3)_2}{N_3}$$

(58)

Scheme 7.12

and some ornamentals, but only slight activity against powdery mildew of roses and vines, and was not toxic to most other pathogenic fungi.[21] A single application by soil treatment with an aqueous solution of dimethirimol hydrochloride against cucumber powdery mildew (*Sphaerotheca fuliginea*) controlled the disease for up to eight weeks.

The closely related pyrimidine, ethirimol, containing the ethylamino group attached to the 2-carbon atom, is effective as a seed dressing against barley powdery mildew, a major disease of barley in Europe.[17,18] Large-scale trials in the United Kingdom on spring and winter barley showed that seed treatment with ethirimol substantially reduced the incidence of mildew and resulted in increased crop yields.

Both of these pyrimidines are remarkably specific for powdery mildews and are absorbed by plant roots and translocated via the transpiration stream to all parts of the plant, eradicating established infection and giving protection for long periods against fungal attack. They are not rapidly degraded in the soil which can thus function as a reservoir providing a slow release of the toxicant, but in plant tissue both compounds are quickly metabolized by progressive *N*-dealkylation. Dimethirimol (**58**) is first converted into the fungicidal monomethyl derivative, and then to the almost inactive 2-amino compound, followed by conjugation as glycosides and phosphates. Oxidation of the 5-butyl group to hydroxybutyl derivatives also occurs and these transformations afford a complex series of metabolites, some of which are fungitoxic.[20,21] Little structural variation appears possible without loss of activity. Thus, the presence of a 4-oxygen function (carbonyl or hydroxy) and the 2-alkylamino group are essential, the size of the alkyl groups being critical—those in dimethirimol and ethirimol represent optimum sizes.

The *N,N*-dimethylsulphamate ester of ethirimol, known as bupirimate, has been developed for control of powdery mildews. It shows specific systemic, protectant, and eradicant fungicidal activity by spray treatment against powdery mildews on a number of crops, being particularly effective against apple powdery mildew. Bupirimate moves readily from spray deposits on the stems into the young leaves. In leaves and in aqueous solution bupirimate is degraded first to ethirimol.

The biochemical mode of action of bupirimate has not been reported but is presumably similar to that of ethirimol.

The use of dimethirimol to control cucumber powdery mildew in glasshouses soon resulted in the development of resistant strains of the fungus. Control of powdery mildews in the field has not suffered the same problems; however, a gradual increase in the tolerance of barley powdery mildew towards ethirimol has been observed.[18,29] The use of this chemical on autumn-sown wheat was discontinued in 1973 to reduce the carryover of resistant strains. The broad-spectrum triazole fungicides (see page 196) introduced in 1978 have now largely replaced ethirimol. A mixture of ethirimol with the triazole flutriafol and thiabendazole is now available (1984); such fungicide mixtures should maintain the usefulness of the aminopyrimidines for many years. The fungicidal action of the 2-aminopyrimidines was antagonized by addition of the purine adenine or the corresponding ribonucleoside adenosine. The chemicals therefore appeared to interfere with purine metabolism. It was shown that germinating *Erysiphe graminis* conidia incorporated adenine and adenosine to produce nucleic acids, but in the presence of ethirimol the formation of nucleic acids was prevented. Several enzymes are involved in this process and these were examined in cell-free extracts of *E. graminis* conidia, but only the enzyme adenosinedeaminase (ADAase) was significantly inhibited by the fungicide.

This enzyme is present in many fungi, but only that from powdery mildews was sensitive to ethirimol, probably accounting for the extreme specificity of these fungicides. It was observed that those 2-aminopyrimidines that were poor inhibitors of ADAase were also poor fungicides. Structure–activity studies provided further indications that the enzyme ADAase was the primary site of action of these fungicides. The inhibition of the enzyme was pH dependent and non-competitive; indeed, ethirimol and dimethirimol show little spatial resemblance to adenosine and the fungicides do not bind tightly to the enzyme. The 5-butyl group appears essential for fungitoxicity and provides the optimum lipophilicity for membrane permeability and enzyme binding.[18]

Piperazine, Pyridine, Pyrimidine Imidazole, and Triazole Fungicides

These fungicides are considered together because they all share a common biochemical target, namely the biosynthesis of ergosterol; they are demethylation inhibitors (DMIs) to distinguish them from the morpholine fungicides which interfere with a later stage in the fungal sterol biosynthesis.[18] The history, chemistry, and development of azole fungicides have been reviewed.[29,30] Triforine (**59**) is the only piperazine derivative that is a commercial fungicide; it has systemic activity as a foliar spray against powdery mildews on cereals, apples and cucumber, apple scab, and *Monilinia* spp.[17] Triforine is easily absorbed and translocated from leaves and roots and there is some downward movement from leaves to the roots. It may also be formulated as a seed dressing to control barley powdery mildew. Apparently oxidation of the aldehyde groups does not occur *in vivo*; triforine (**59**) is manufactured from piperazine;

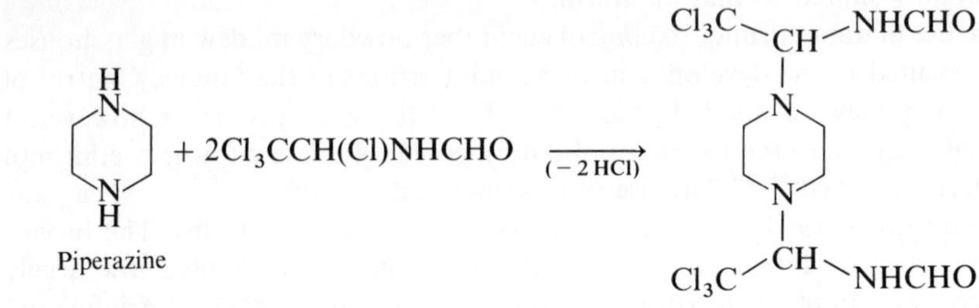

$$\text{Piperazine} + 2Cl_3CCH(Cl)NHCHO \xrightarrow[(-2HCl)]{} \quad (59)$$

Buthiobate (**60**) and pyrifenox (**61**) are pyridine fungicides; buthiobate (**60**) is used mainly in Japan against powdery mildews, while pyrifenox (**61**) controls a wide range of leaf spot pathogens of fruits and vegetables.

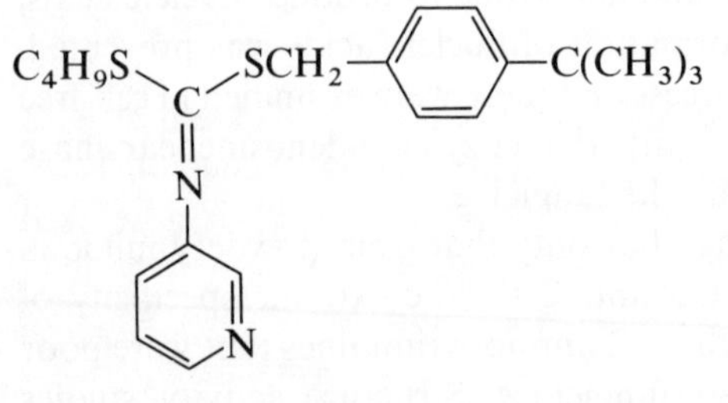

(**60**) 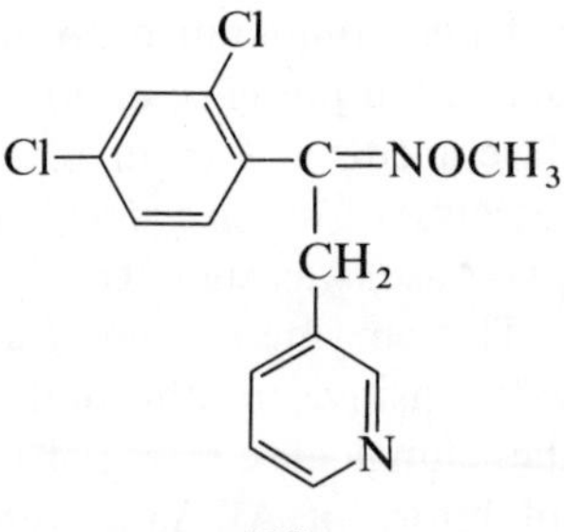   (**61**)

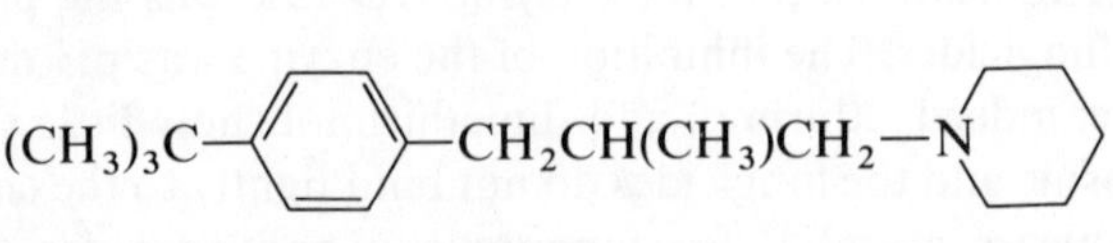

$$(CH_3)_3C-\!\!\!\langle\;\rangle\!\!\!-CH_2CH(CH_3)CH_2-N\langle\;\rangle$$

(**62**)

Fenpropidin (**62**), a piperidine derivative, was introduced in 1986 and is a new systemic fungicide active as a foliar spray against powdery mildews and rusts on cereals at 0.5 kg a.i./ha, and is also highly promising for control of Dutch elm disease. It is quickly absorbed from the roots and leaves of cereals. Fungicidal action is due to fenpropidin (**62**) blocking two essential steps in sterol biosynthesis, namely reduction of the Δ^{14} double bond and the $\Delta^8 \to \Delta^7$ isomerization. The mode of action is different from that of other azole fungicides (e.g. the triazoles), which act by inhibition of the C-14 demethylation step. The development of resistance to fenpropidin should therefore be more difficult and consequently the fungicide is predicted to have a longer useful lifespan.[31]

The series of 3-substituted pyrimidine-5yl methanol fungicides, triarimol (63), fenarimol (64), and nuarimol (65), were introduced by Eli Lilly in the late 1960s. Triarimol has been withdrawn due to undesirable toxicological properties. Fenarimol (64), a systemic and protective fungicide, is applied as a foliar spray to control a broad spectrum of powdery mildews, scabs, rusts, and leaf spot pathogens on fruits. Nuarimol (65) is used against powdery mildews on cereals, sugar beet, and swedes. It is also formulated as a seed dressing in mixtures with other fungicides, e.g. imazalil, to control seed and soil-borne pathogens.[17]

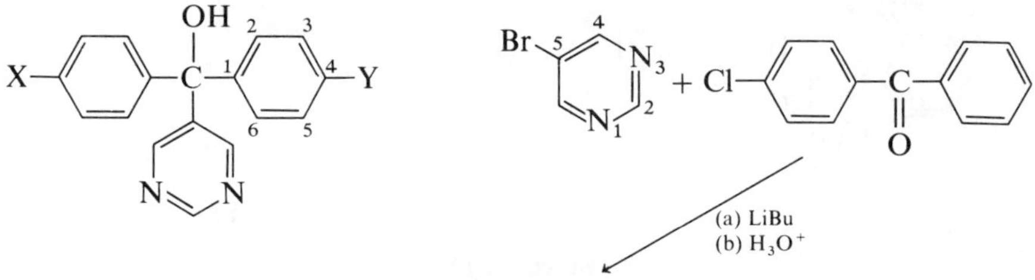

(63) X = H, Y = 2,4-Cl$_2$
(64) X = Cl, Y = 2-Cl
(65) X = F, Y = 2-Cl

Scheme 7.13

These compounds are prepared by condensation of 5-bromopyrimidine with the appropriate benzophenone as indicated for fenarimol (64) (Scheme 7.13).

Imazalil (66), the first imidazole agricultural fungicide (1969), is now used mainly as a seed dressing for cereals, but it can be applied as a spray to control powdery mildews in cucumbers, marrows, and ornamentals and also fusarium in

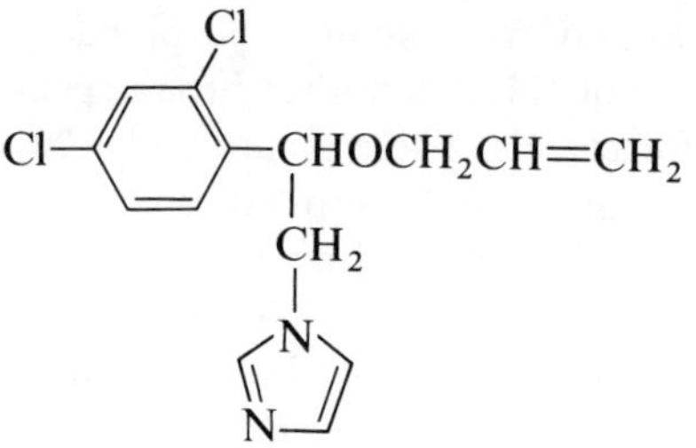

(66)

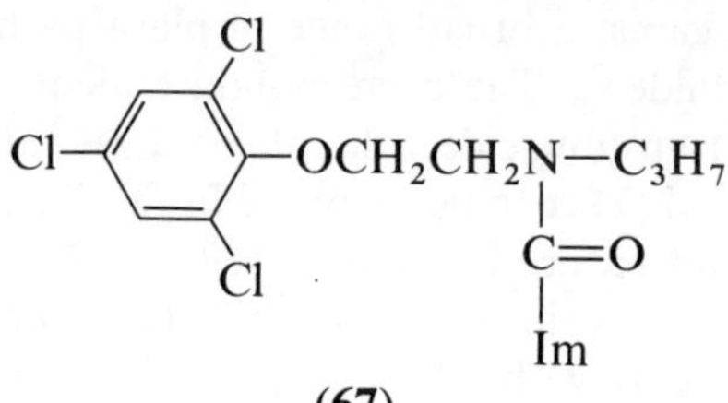

(67)

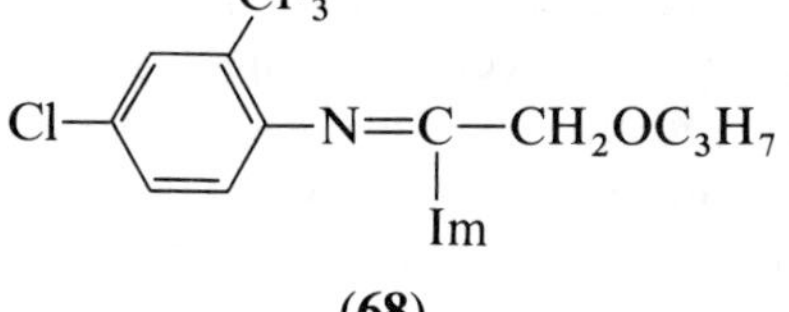

(68)

(Im = imidazole residue)

seed potatoes. Prochloraz (Boots Company, 1977) (**67**) is a broad-spectrum fungicide with good activity against *Ascomycetes* and *Fungi Imperfecti* but rather less activity against *Basidiomycetes*. It is also valuable for disease control in winter wheat, barley, and oilseed rape and shows no cross-resistance to benzimidazole fungicides which permits its use against fungi that have become resistant to carbendazim. Triflumizole (1978) (**68**) controls a wide range of pathogens, e.g. powdery mildews, scab, and *Monilinia* spp.[30] It is, like prochloraz (**67**), a contact fungicide with some translaminar activity and is particularly effective against apple scab. Triflumizole (**68**) can be prepared as shown in Scheme 7.14.

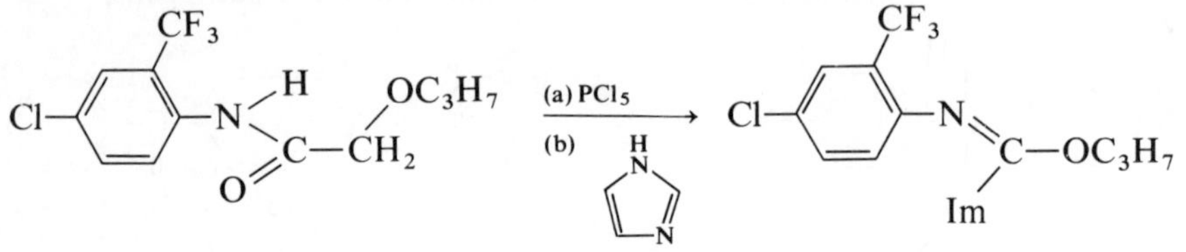

Scheme 7.14

The azole fungicides were discovered in the late 1960s[29] when *N*-tritylmorpholine was marketed as a molluscicide for control of water snails (see Chapter 13, page 311). *N*-Acetyl and *N*-tritylimidazoles and triazoles were found to be active against human and plant pathogenic fungi. In the triazoles, only the 1-substituted 1,2,4-triazoles had good fungitoxicity, but the trityl residue was not essential. The triazoles were discovered to be superior to the corresponding imidazoles for control of phytopathogenic fungi. Fluotrimazole (**69**) was introduced as a protectant fungicide against powdery mildews on cereals and fruits; the compound can be prepared from *m*-xylene (Scheme 7.15).

The related triazole triadimefon (**70**) has eradicant and systemic activity against a broad range of plant pathogens, notably rusts, smuts, and powdery mildews. The corresponding alcohol triadimenol (**71**) is another broad-spectrum fungicide valuable as a seed dressing for cereals. These compounds (**70**) and (**71**) can be prepared from pinacolone (Scheme 7.16). Both plants and fungi can reduce compound (**70**) to (**71**) and this can be regarded as an activation process since triadimenol (**71**) is more fungicidal; in wheat leaves 60% reduction occurs within two days. The four enantiomers of triadimenol (**71**) (two chiral carbon atoms) differ considerably in their fungitoxicity; the 1*S*, 2*R*, enantiomer is the most active. Triadimefon (**70**) (1973) was the first broad-spectrum triazole fungicide, effective at rates of 0.1–2 p.p.m. against a number of *Ascomycetes*, *Basidiomycetes*, and *Fungi Imperfecti*.[18]

This discovery led to the introduction of several 1,2,4-triazole fungicides that are structurally related to triadimenol (**71**) (Table 7.3). Bitertanol (**72**) is used

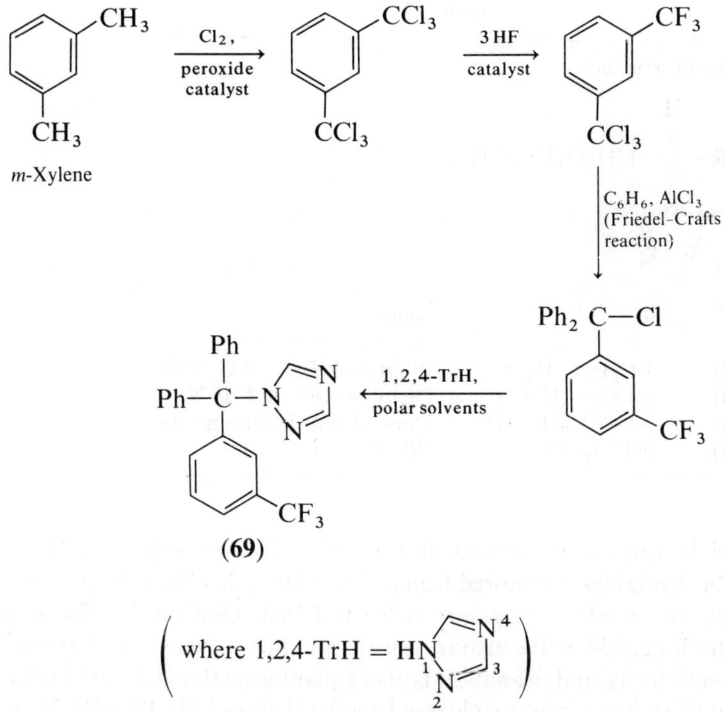

m-Xylene

Scheme 7.15

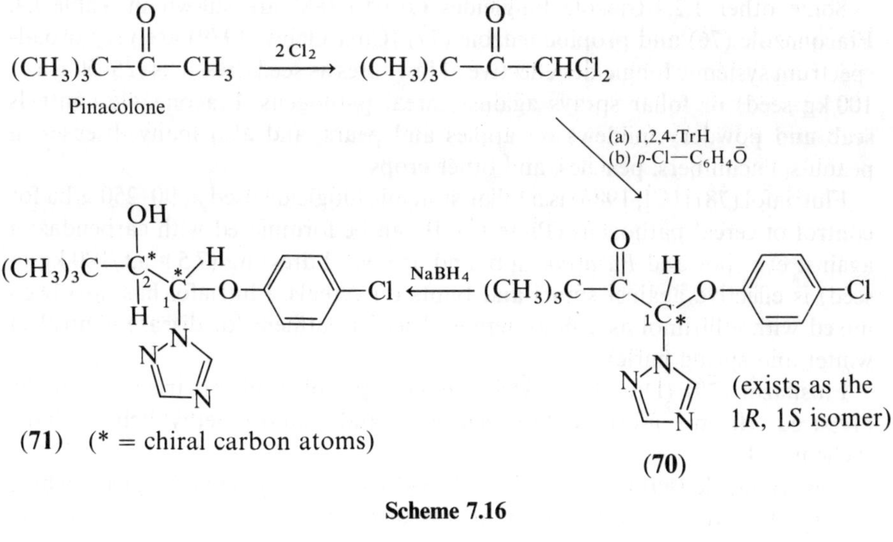

$$\left(\text{where 1,2,4-TrH} = \text{HN} \underset{2}{\overset{1}{\underset{}{}}} \overset{N^4}{\underset{3}{}} \right)$$

Pinacolone

(71) (* = chiral carbon atoms)

(exists as the 1R, 1S isomer)

(70)

Scheme 7.16

Table 7.3

General structure

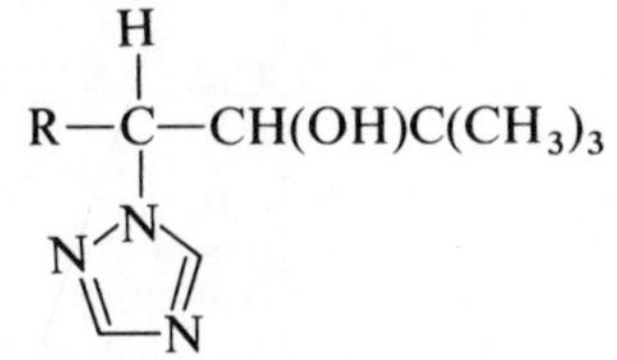

$$R-\underset{\underset{\overset{\displaystyle N}{\|}}{\overset{\displaystyle H}{|}}}{C}-CH(OH)C(CH_3)_3$$

	R	Name
(72)	$4\text{-}C_6H_5C_6H_4O$	Bitertanol (Bayer AG, 1978)
(73)	$2,4\text{-}Cl_2C_6H_3CH_2$	Diclobutrazol (ICI, 1979)
(74)	$2,4\text{-}Cl_2C_6H_3CH{=}$	Diniconazol (Sumitomo, 1983)
(75)	$(CH_3)_3CCOCH_2$	PP-969 (ICI, 1983)

mainly against *Venturia* scab on fruits, leaf diseases (*Puccinia* spp.) and *Monilinia* diseases in vegetables and stored fruits. Diclobutrazole **(73)** is applied as a foliar fungicide on cereals and against coffee leaf rust. Diniconazol **(74)** is a broad-spectrum fungicide active against powdery mildews, rusts, and leaf spot diseases in cereals, fruits, and peanuts; it is also a plant growth regulator (PGR), like the related triazole paclobutrazole (see Chapter 9, page 278). PP-969 **(75)** is a water-soluble compound, which is very mobile in woody plants. A single application as a soil drench controlled coffee, banana, or apple leaf diseases for up to 30 weeks.

Some other 1,2,4-triazole fungicides **(76)** to **(88)** are shown in Table 7.4. Etaconazole **(76)** and propioconazole **(77)** (Ciba-Geigy, 1979) are very broad-spectrum systemic fungicides effective at low rates as seed dressings (25–50 g a.i./ 100 kg seed) or foliar sprays against cereal pathogens. Etaconazole controls scab and powdery mildews on apples and pears, and also many diseases in peanuts, cucumbers, peaches, and other crops.

Flutriafol **(78)** (ICI, 1983) is a foliar systemic fungicide used at 90–250 g/ha for control of cereal pathogens (Plate 11). It can be formulated with carbendazim against eye spot and *Fusarium* spp. and as a seed dressing (7.5 g a.i./100 kg of seed) is effective against smuts and bunts on cereals. Flutriafol has also been mixed with ethirimol as a non-mercurial seed treatment for disease control in winter and spring barley.

Flusilazol **(79)** (Du Pont, 1984), an example of a silicon triazole, can be prepared from *p*-bromofluorobenzene and chloromethyldichlorosilane (Scheme 7.17).

Hexaconazole **(80)** (ICI, 1986) is a broad-spectrum systemic fungicide which is very effective as a foliar spray at low dose rates (10–20 p.p.m.) to control cucumber, apple and vine powdery mildew, apple scab, coffee rust, peanut leaf

Table 7.4 Some triazole fungicides

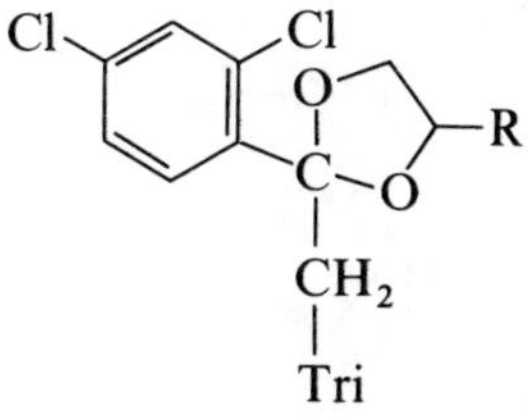

(**76**) R = C_2H_5
(**77**) R = $n\text{-}C_3H_7$

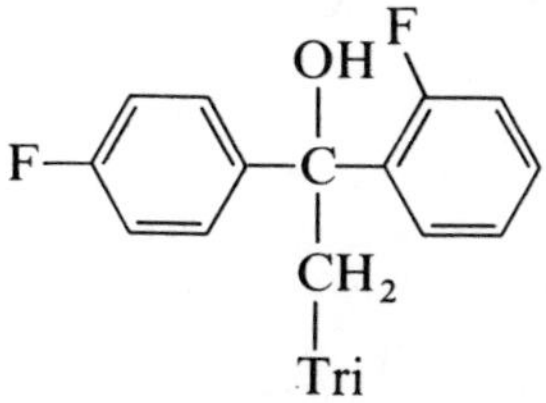

(**78**)

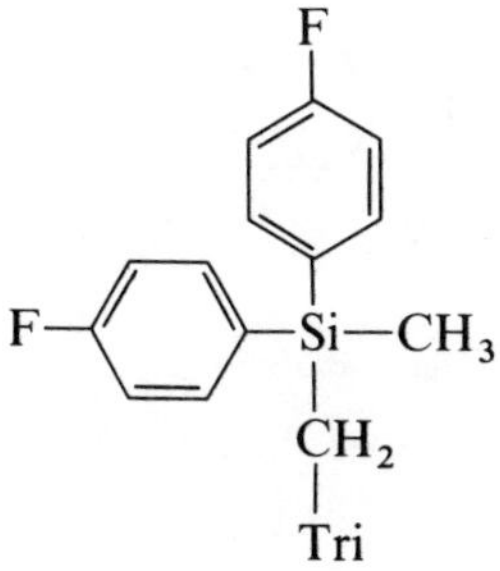

(**79**)

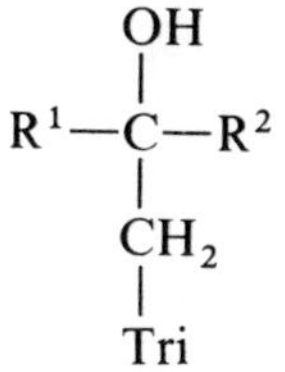

$$R^1-\underset{\underset{\displaystyle Tri}{|}}{\overset{\overset{\displaystyle OH}{|}}{\underset{\displaystyle CH_2}{|}}}R^2$$

(**80**) $R^1 = 2,4\text{-}Cl_2C_6H_3$, $R^2 = n\text{-}C_4H_9$
(**81**) $R^1 = 4\text{-}ClC_6H_4$, $R^2 = CH\!\!-\!\!\triangle$, CH_3
(**82**) $R^1 = 4\text{-}ClC_6H_4(CH_2)_2$, $R^2 = C(CH_3)_3$

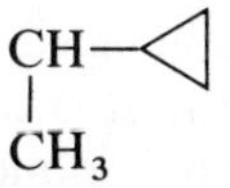

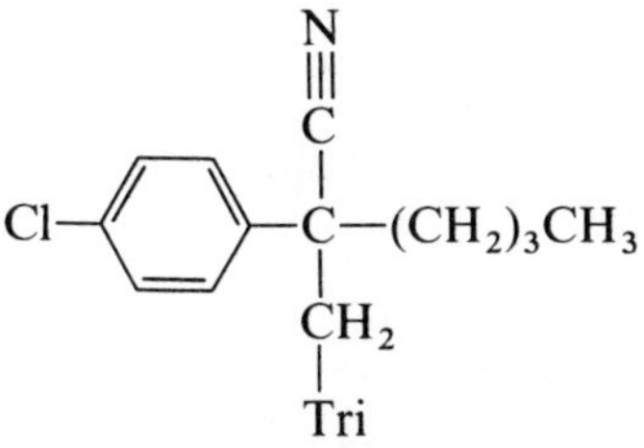

(**83**)

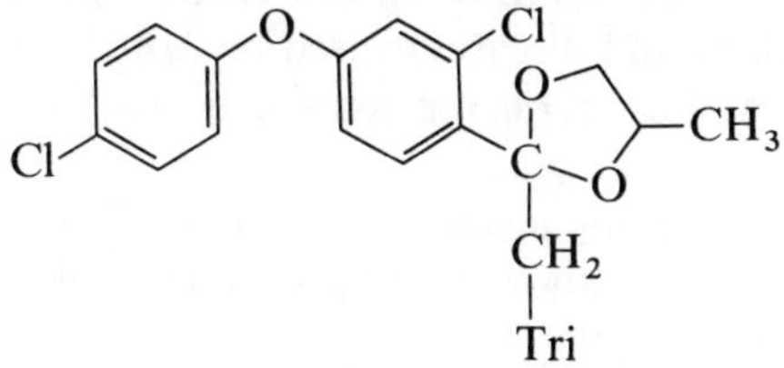

(**84**)

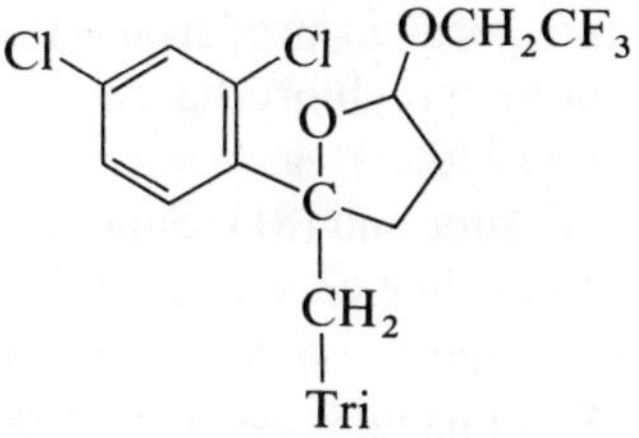

(**85**)

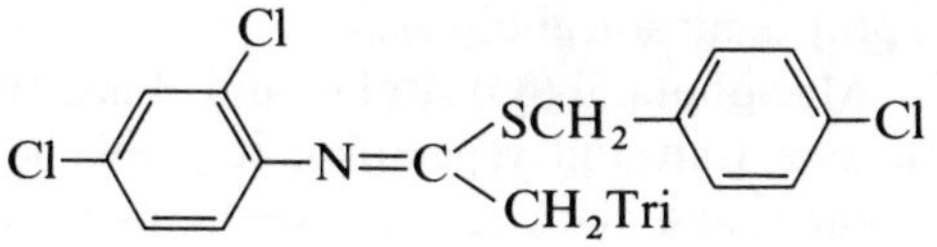

(**86**)

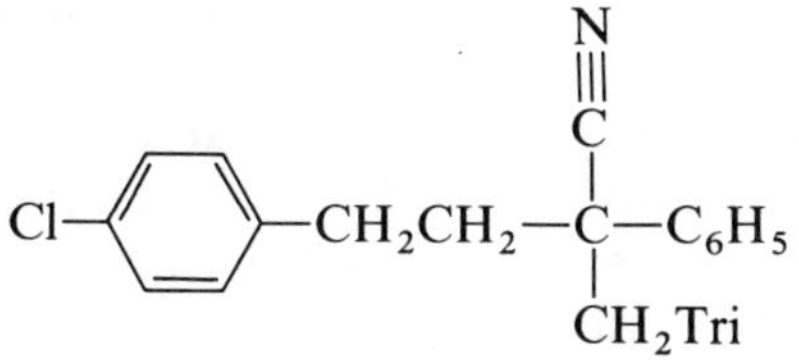

(**87**)

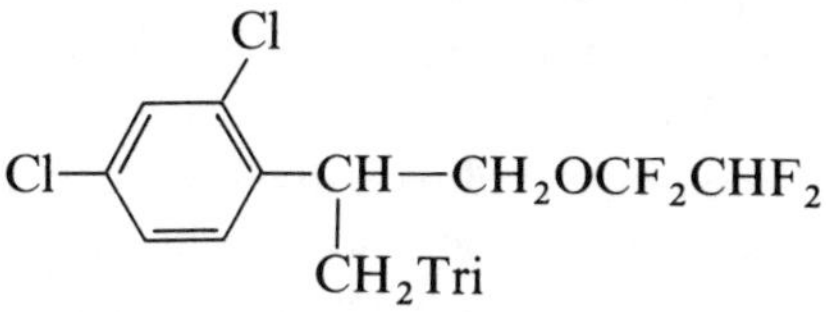

(**88**)
(where Tri = 1,2,4-triazolyl radical)

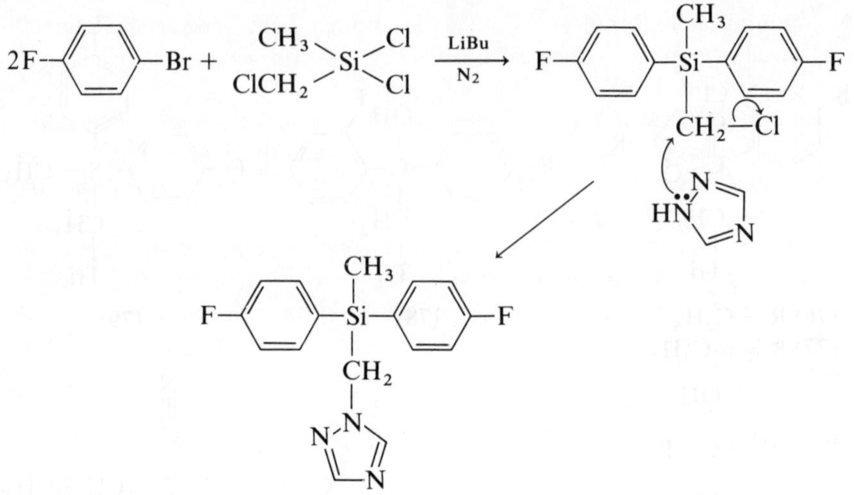

(79)

Scheme 7.17

spot, brown rot (*Monilinia fructicola*), and diseases of vegetables.[32] Hexaconazole has outstanding protective activity combined with systemic, curative, and antisporulent properties (Plates 12 and 13). It is compatible with dithiocarbamates and chlorothalonil and hence can be formulated as mixtures with these fungicides.

Compound (**81**) (Sandoz, 1986) is a systemic fungicide with excellent activity at 40–100 g a.i./ha against powdery mildews and diseases caused by *Monilinia*, *Cercospora*, *Venturia*, rusts, *Rhizoctonia*, and *Sclerotium* in cereals, grapes, fruit trees, peanuts, rice, and vegetables.

Compound (**82**) is a highly potent seed dressing against cereal smuts, bunts, and other seed-borne diseases, e.g. *Septoria*; it can also be applied as a foliar spray against leaf diseases.

Myclobutanil (**83**) (Rohm and Haas, 1986) is a broad-spectrum fungicide for use on fruit and vegetables. It is excellent against apple scab, vine powdery mildew, and black rot; as a seed dressing (10–20 g a.i./100 kg of seed) myclobutanil controls cereal seed-borne pathogens.[33]

Compound (**84**) (Ciba-Geigy, 1988) is an exceptionally broad-spectrum fungicide showing activity against all the diseases of wheat, sugar beet, peanuts, potatoes, pomefruit, grapes, and vegetables at comparatively low dose rates (30–250 g a.i./ha).[34]

Compound (**85**) (Rhône-Poulenc, 1988) is active against a wide range of fungi belonging to the *Ascomycetes*, *Basidiomycetes*, and *Fungi Imperfecti* classes, especially powdery mildews, rusts, scabs, and leaf spots at 10–100 g a.i./ha.

Compound (**86**) (1988) has a similar fungicidal spectrum with very high activity against all powdery mildews.

Compound (87) (Rohm Haas, 1988) is another broad-spectrum fungicide giving excellent control of *Septoria* and *Puccinia* spp. on cereals at 75 g a.i./ha or as a seed dressing (10–20 g a.i./100 kg of seed).

Compound (88) (Montedison, 1988) controls powdery mildews of wheat and barley by a single application of 75–125 g a.i./ha; activity is retained for six weeks or more after spraying. The compound afforded better control than the standard propiconazole treatment.

The piperazine, pyridine, pyrimidine, imidazole, and triazole fungicides owe their fungitoxicity to their ability to inhibit ergosterol biosynthesis. Ergosterol is a major sterol in many fungi where it plays a vital role in membrane structure and function. These fungicides specifically act as demethylation inhibitors (DMIs) to distinguish them from the morpholine fungicides (page 202) which also interfere with sterol biosynthesis but at a later stage.[33–35]

Studies with fungi treated with sublethal doses of DMIs showed that ergosterol biosynthesis was profoundly affected, but there was little interference with other metabolic processes. Analysis of these experiments showed no sterol intermediates beyond the C-14 demethylation step, which indicated that the primary mode of action of this group of fungicides was to prevent removal of the 14α-methyl group (step 1). This is a stage in the conversion of 24-methylenedihydrolanosterol into ergosterol and is achieved by three $NADPH_2$-dependent oxygenase reactions (Scheme 7.18).

$$14\alpha\text{-CH}_3 \xrightarrow{\text{step 1}} 14\alpha\text{-CH}_2\text{OH} \xrightarrow{\text{step 2}}$$

$$14\text{-CHO} \xrightarrow{\text{step 3}} [14\text{-CO}_2\text{H}] \xrightarrow[(-CO_2)]{} 14\text{-H}$$

Scheme 7.18

The first step is catalysed by a specific form of microsomal cytochrome P 450 enzyme and it is concluded that the triazole fungicides and related compounds poison this enzyme. All the DMIs, apart from triforine (59), contain an unsubstituted, and relatively unhindered, nitrogen atom in the 4-position of the heterocyclic ring and there is evidence that this nitrogen atom binds to the haem iron atoms of cytochrome P 450.[37] Studies of the interaction of diclobutrazol (73) and its isomers with cytochrome P 450 showed good correlation between binding and fungicidal activity; thus the more fungicidal (2*R*, 3*R*) isomer showed a higher bonding to the enzyme than the less active (2*S*, 3*S*) isomer.[18] In *Aspergillus nidulans*, treatment with 0.01–0.03 μM amounts of imazalil (66) temporarily inhibited the sterol demethylation system. In this way, the treated fungi are deprived of ergosterol and their growth is probably impaired by the accumulation of 14-methylated sterols.

The combination of these effects causes disruption of the fungal cell wall, inhibition of hyphal growth, and eventual death of the fungus.

The triazoles are currently a rapidly growing class of fungicides with some 20 commercial examples and many more at the experimental stage. Several triazoles can be used instead of the environmentally hazardous organomercurial seed dressings (see Chapter 15, page 347).

All the DMIs can penetrate the plant cuticle or the seed coat and are often further translocated upwards in the xylem. The risk of the development of resistant fungi is low and the main problems to date have been confined to powdery mildews, but the compounds remained fully effective against other fungal pathogens. Resistance is observed to develop relatively slowly in comparison with the benzimidazoles or acylalanine fungicides. The resistant fungal strains generally exhibited reduced fitness as compared with the wild (susceptible) strains. Genetic studies with *Aspergillus nidulans* detected ten different loci coding for resistance; hence it is assumed that more than one gene is responsible. The resistant strains were generally cross-resistant to other DMI fungicides, consistent with the concept of a common mode of action.

Morpholine Fungicides

The first two members of this group were dodemorph (**89**) and tridemorph (**90**); both have systemic fungicidal activity as a foliar spray (0.75 l/ha) against powdery mildews. Dodemorph (**89**) is mainly used to control the disease on roses and tridemorph (**90**) for cereals. The commercial morpholine fungicides (**89**) to (**93**) are shown in Table 7.5.

Fenpropiomorph (**91**) is effective as a foliar spray against powdery mildew and rust on cereals showing contact and systemic activity. It can be formulated with iprodione for control of *Botrytis* which has become resistant to benzimidazole fungicides. Both dodemorph (**89**) and tridemorph (**90**) are absorbed and translocated by the roots and leaves of oat and barley plants providing protection from powdery mildew (*Erysiphe graminis*) for 3–4 weeks.[7,18]

The morpholine fungicides, like the triazoles, are inhibitors of ergosterol biosynthesis (EBIs), but this may not be the only basis for their fungitoxicity.[18] The mode of action probably involves a combination of blockage of sterol

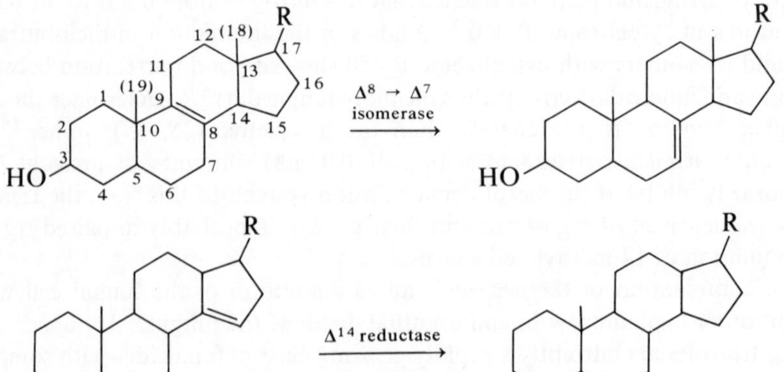

Scheme 7.19

Table 7.5

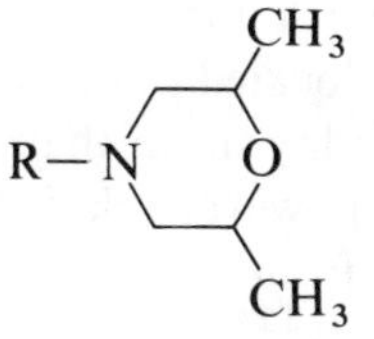

$$R-N \underset{CH_3}{\overset{CH_3}{\diagdown O}}$$

General structure

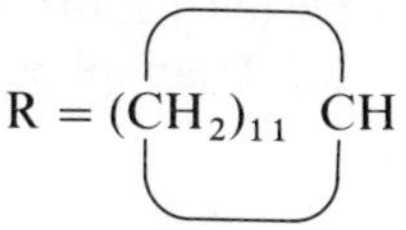

R = (CH$_2$)$_{11}$ CH   R = CH$_3$(CH$_2$)$_{12}$

(89) Dodemorph (BASF, 1965)   **(90)** Tridemorph (BASF, 1966)

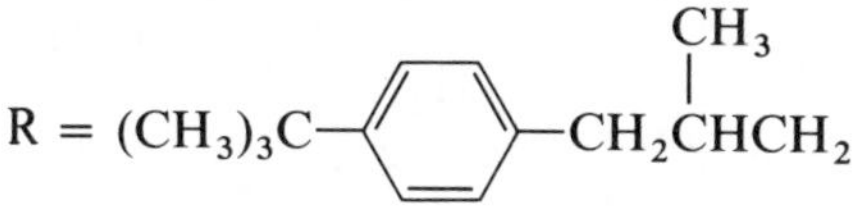

R = (CH$_3$)$_3$C—⟨ ⟩—CH$_2$CHCH$_2$
                              |
                             CH$_3$

(91) Fenpropimorph (BASF, 1980)

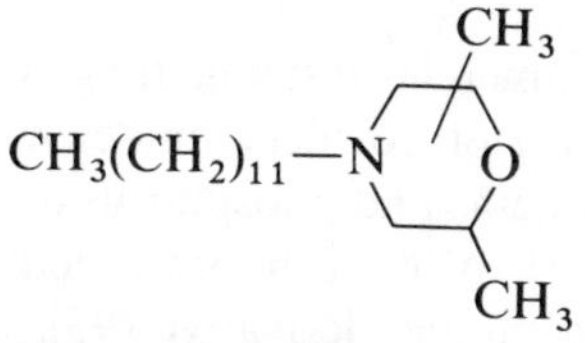

$$CH_3(CH_2)_{11}-N \underset{CH_3}{\overset{CH_3}{\diagdown O}}$$

(92) Aldimorph (1980)

$$OHCNH-\underset{}{\overset{CCl_3}{\underset{|}{CH}}}-N \diagup O$$

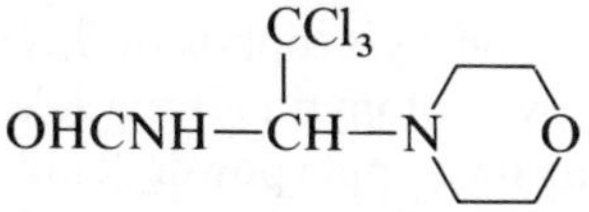

(93) Trimorphamide (1981)

synthesis and inhibitory action on NADH$_2$-oxidase and succinate reductase enzymes. The latter effects almost certainly account for the antibacterial properties of tridemorph, since bacteria contain only small amounts of sterols.[18,27]

The blockage of ergosterol biosynthesis by the morpholine fungicides probably occurs by inhibition of $\Delta^8 \to \Delta^7$ isomerase enzyme. This is the most frequently mentioned site of inhibition in the biosynthetic pathway and occurs later than the 14-demethylation stage blocked by the triazoles and related compounds[24] (page 201) (Scheme 7.19). However, in some fungi evidence indicates that the morpholine fungicide prevents the NADPH$_2$-dependent reduction of the Δ^{14} double bond. For instance, in rice blast fungus (*Piricularia oryzae*) tridemorph **(90)** acts by inhibition of $\Delta^8 \to \Delta^7$ isomerase while fenpropimorph **(91)** blocks Δ^{14}-double-bond reduction.[36]

Organophosphorus Compounds

This group has produced many valuable systemic insecticides (Chapter 6, page 139) which are readily translocated in plants from either the roots or from spray deposits on the leaves. Today more than 100 organophosphorus compounds show fungicidal action; however relatively few are of practical use as fungicides. Many of them suffer from strong phytotoxicity, and are also often very specific against fungus species.[12]

One of the first organophosphorus fungicides was triamiphos (94) which is obtained by condensation of the sodium salt of 3-amino-5-phenyl-1,2,4-triazole with bis(dimethylamino)phosphoryl chloride[17] (Scheme 7.20). Triamiphos (94)

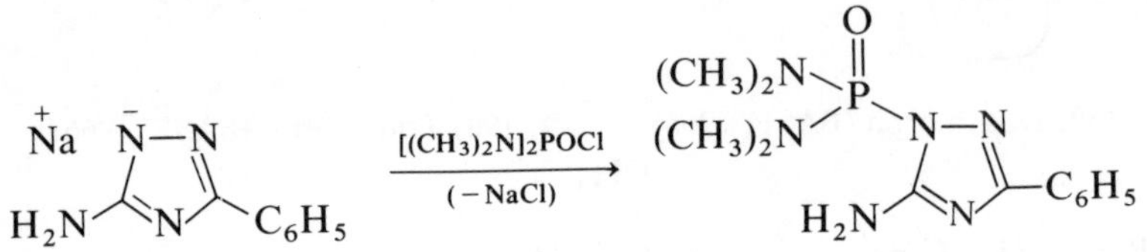

Scheme 7.20

has been claimed (1960) to be the first commercial systemic fungicide[21,26] and has limited systemic action against a range of powdery mildews. Since then many combinations of heterocyclic ring systems and phosphorus moieties have been synthesized and evaluated in an effort to obtain more effective systemic fungicides.[21,27] Two examples are the pyrimidine derivative (95), which gave almost complete control of cucumber powdery mildew, wheat rust, and vine downy mildew and was also insecticidal,[21] and pyrazophos or 2-(O,O-diethyl-thionophosphoryl)-5-methyl-6-carbethoxypyrazolopyrimidine (96), which has given promising results by foliar spraying against apple powdery mildew in field trials.

Tolclofos-methyl (97) is effective against *Rhizoctonia* and other soil-borne diseases as a drench in vegetables and against black scurf and canker in seed potatoes. The mode of action is unknown but it shows cross-resistance to the chlorobenzene fungicides (page 172).

S-Benzyl-O,O'-diisopropylphosphorothiolate or kitazin P (98) is prepared by condensation of sodium benzylthiolate and O,O'-diisopropylphosphoro-

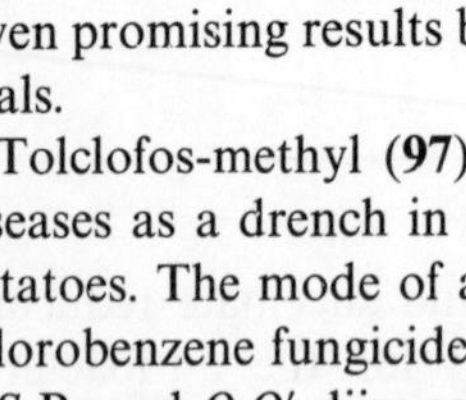

(95) 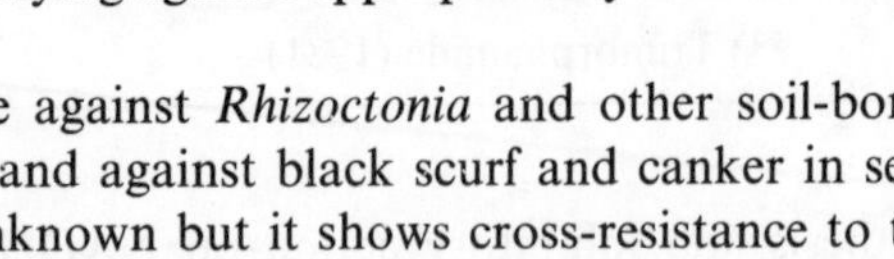   (96)

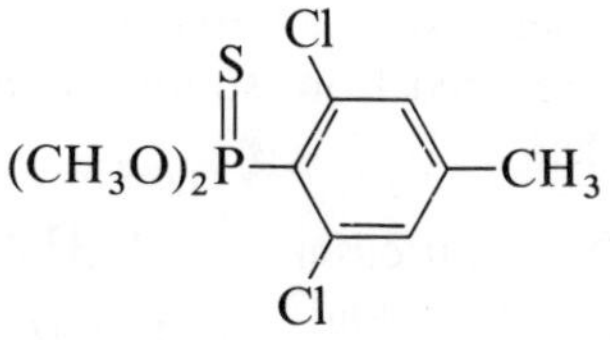

$$(CH_3O)_2\overset{\overset{S}{\|}}{P}-$$

(97)

chloridate[17,38] (Scheme 7.21). Kitazin P (98), introduced in 1968, is a systemic rice fungicide applied as granules to paddy water to control rice blast (*Piricularia oryzae*), and it inhibits mycelial growth in tissue.[12] Kitazin P (98) is

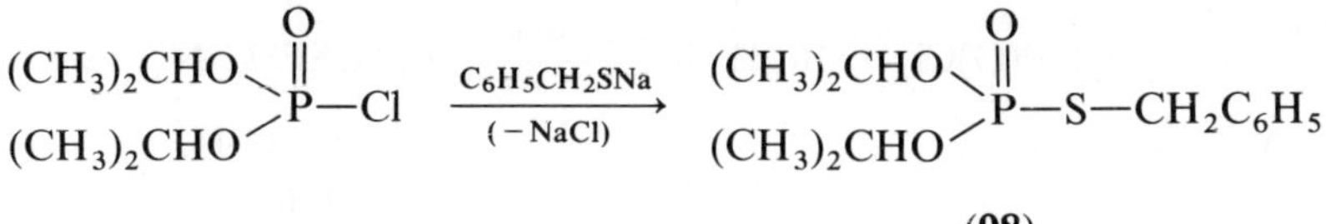

$$\begin{array}{c}(CH_3)_2CHO \\ (CH_3)_2CHO\end{array}\!\!\!\searrow\!\!\overset{\overset{O}{\|}}{P}\!-\!Cl \xrightarrow[(-NaCl)]{C_6H_5CH_2SNa} \begin{array}{c}(CH_3)_2CHO \\ (CH_3)_2CHO\end{array}\!\!\!\searrow\!\!\overset{\overset{O}{\|}}{P}\!-\!S\!-\!CH_2C_6H_5$$

(98)

Scheme 7.21

metabolized in rats, cockroaches, rice plants, soils, and *P. oryzae* (Scheme 7.22), chiefly by S—C bond cleavage to give the diisopropyl hydrogen phosphorothioate (route a). In rice plants, this is converted to the phosphate. In rice blast fungus, the most interesting metabolite was the *m*-hydroxy derivative (route b).

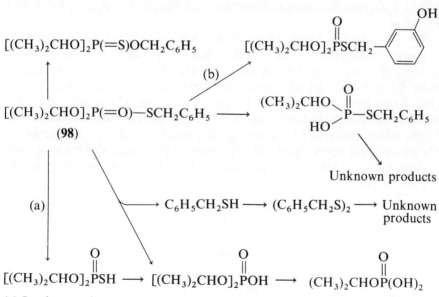

(a) In plants only
(b) In fungi only

Scheme 7.22

Another fungicide used against rice diseases is *O*-butyl-*S*-ethyl-*S*-benzyl-phosphorodithioate **(99)**, prepared from sodium *O*-butyl-*S*-benzylphosphoro-dithioate and ethyl bromide:[38]

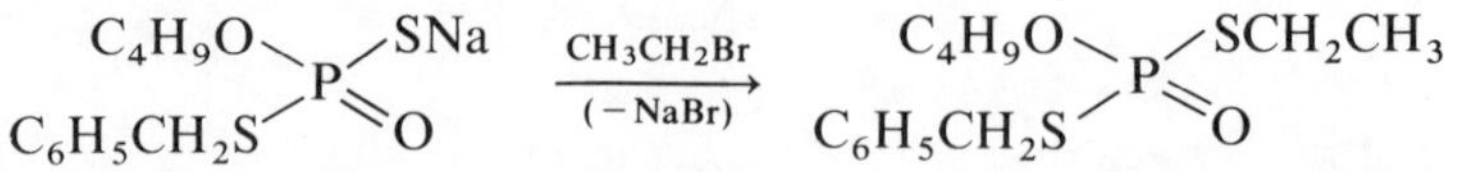

$$\underset{C_6H_5CH_2S}{\overset{C_4H_9O}{>}}P\overset{SNa}{\underset{O}{\lessgtr}} \quad \xrightarrow[(-\,NaBr)]{CH_3CH_2Br} \quad \underset{C_6H_5CH_2S}{\overset{C_4H_9O}{>}}P\overset{SCH_2CH_3}{\underset{O}{\lessgtr}}$$

(99)

Edifenphos **(100)**, which is also very effective against rice blast, is prepared from ethylphosphorodichloridate and phenylthiol:

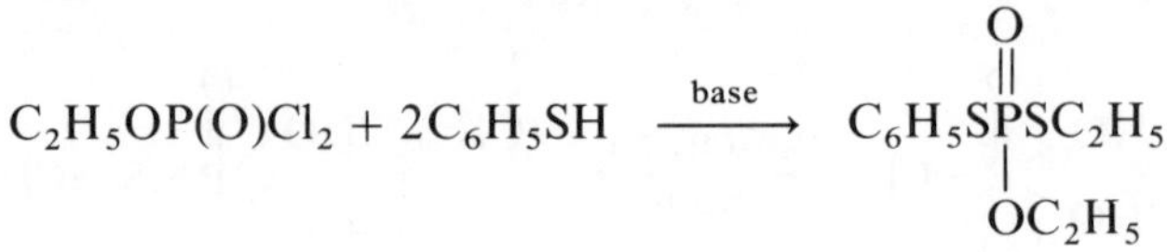

$$C_2H_5OP(O)Cl_2 + 2C_6H_5SH \quad \xrightarrow{base} \quad \underset{\underset{(100)}{OC_2H_5}}{\overset{\overset{O}{\parallel}}{C_6H_5SPSC_2H_5}}$$

Kitazin and edifenphos have been shown to block the synthesis of phospholi-pids by inhibiting the *S*-adenosylmethionine-dependent methylation of phos-phatidylethanolamines to the corresponding phosphatidylcholine, an important constituent (ca. 50%) of the phospholipids in *P. orzae*.[18,29] The reduction in phospholipids alters membrane structure in the fungus, increasing permeability with consequent loss of vital cellular components and eventually killing the fungus. It appears that this is an important mode of action for organophosphorus fungicides.

The phosphonate fosetyl **(101)** (Rhône-Poulenc, 1977) is a water-soluble compound (LD_{50} (oral) to rats, 2600 mg/kg) giving systemic control of downy mildews of vines, tropical crops, and vegetables. Fosetyl is also active against some *Oomycetes* rotting diseases of pineapples, citrus, and avocados. Fosetyl **(101)** is translocated upwards and downwards in plants so foliar application can prevent infections in the lower parts of the crop. Fosetyl is often formulated with folpet, which appears to be a synergistic mixture. Fosetyl, or its metabolite phosphonic acid, probably acts directly on the target fungus, slowing its growth and hence allowing the crop plant's natural defence mechanisms time to finally kill the fungus.

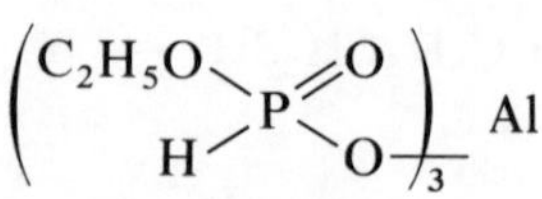

$$\left(\underset{H}{\overset{C_2H_5O}{>}}P\overset{O}{\underset{O}{\lessgtr}}\right)_{\!3}\ Al$$

(101)

Recent studies[18] demonstrated that phosphonic acid is highly fungitoxic and probably exerts a primary effect on the pathogen.

Alkyl phosphonates are a novel group of fungicides which were discovered in 1983. The methyl 3-isononyloxypropyl ammonium phosphonate **(101a)** is useful

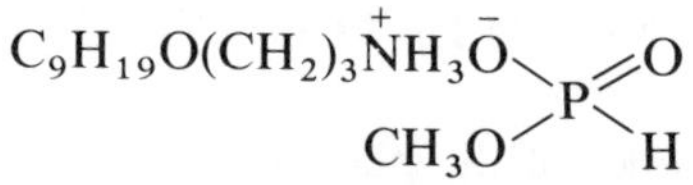

$$C_9H_{19}O(CH_2)_3\overset{+}{N}H_3\overset{-}{O}\diagdown_P\diagup^O$$
$$CH_3O\diagup^{}\diagdown H$$

(101a)

as a seed dressing to control seed-borne pathogens in winter wheat, maize, and rye. It can also be applied as a spray against *Plasmopara tobacina* in tobacco.

The fungicidal properties of compounds containing imino groups, e.g. captan, are well known. An organophosphorus fungicide containing this group is ditalimfos **(102)**, obtained by condensation of potassium phthalimide with *0,0*-diethylphosphorochloridothionate[27] (Scheme 7.23). Ditalimfos **(102)** is

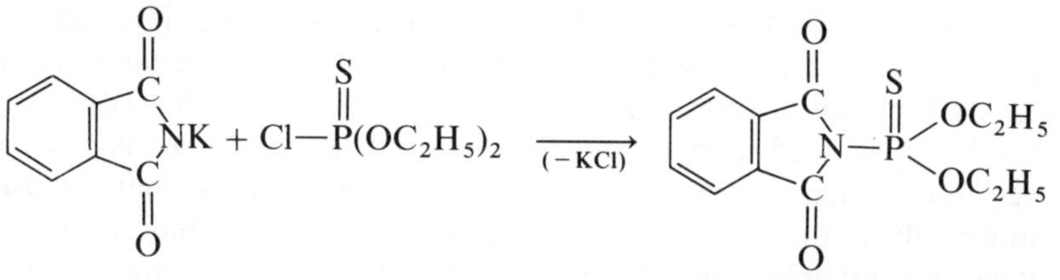

(102)

Scheme 7.23

extremely effective against a wide range of powdery mildews[21] and has a very low mammalian toxicity (LD_{50} (oral) to rats, 4930 mg/kg).

The fungitoxicity is lost by replacement of P = S by P = O, and also if there is a bridging atom or group, e.g. —S—CH$_2$— or —O—, between the nitrogen and phosphorus atoms. Certain non-phosphorus *N*-substituted imides like folpet (page 167) are fungicidal and Tolkmith therefore argued that the toxophore in phthalimidophosphorothioates was not the phosphorus-containing moiety but the unsaturated carboximide ring, and the activity is probably due to acylation of a vital enzyme system (H—Ez) with opening of the carboximide ring (Scheme 7.24).

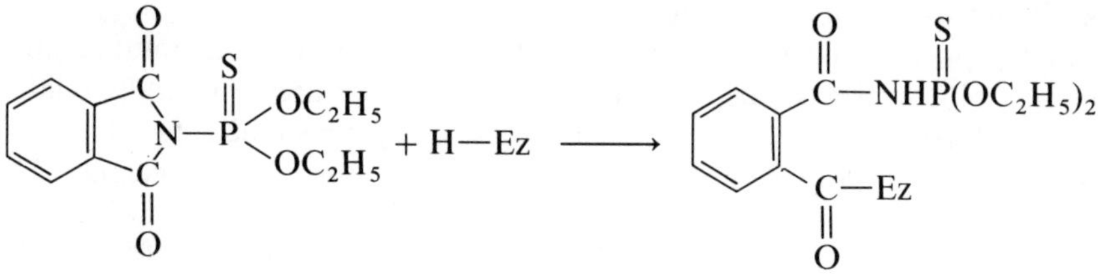

(102)

Scheme 7.24

PHENYLAMIDES AND RELATED COMPOUNDS

Table 7.6(a) shows examples of the phenylamide group of fungicides (**103**) to (**109**); the group consists of the acylalanines, butyrolactones, and oxazolidinones. These compounds show protective and systemic activity against *Oomycetes*; this group of fungi comprise many of the plant parasitic species causing foliar, root, and crown diseases in a wide range of crops, e.g. downy mildews and late blight, which can result in rapid epidemics with massive crop losses.[18,36]

The first members of this group, metalaxyl (**103**) and furalaxyl (**104**), were introduced by Ciba-Geigy in 1977. Metalaxyl can be applied against late blight (*Phytophthora infestans*) on potatoes and tomatoes. A wettable powder formulation with mancozeb is widely used as a foliar spray against blight on potatoes (Plate 14). Metalaxyl (**103**) has the broadest spectrum of fungicidal activity of this group of fungicides, it is good against downy mildew on vines, lettuce, maize, and pythium diseases, and can also be formulated as a seed dressing. Furalaxyl (**104**) is chiefly applied as a drench against *Phytophthora* and *Pythium* diseases on ornamentals. Benalaxyl (**105**) shows a similar fungicidal spectrum.

Ofurace (**106**) is formulated with manganese zinc ethylene dithiocarbamate as a contact and systemic fungicide against potato blight and downy mildew on oilseed rape. Oxadixyl (**107**), mixed with mancozeb, is used to control potato blight. In 1988 Ciba-Geigy[39] introduced (**109**) to combat soil-borne *Phytophthora* and *Pythium* spp. causing root and stem disorders of such crops as tobacco, citrus, avocado, soft fruit, and ornamentals. Season-long control is often achieved by a single soil treatment at 0.25–1 kg a.i./ha prior to planting the crop. The compound is rapidly absorbed by plant roots and is appreciably more stable in soil as compared with metalaxyl because it is less subject to microbial degradation.

The phenylamide fungicides are generally formulated as wettable powders mixed with surface fungicides to combat resistance problems which appeared when the single products were used continuously against potato blight (*Phytophthora infestans*) or vine downy mildew (*Plasmopara viticola*). The formulated mixtures of phenylamides and dithiocarbamates have the advantage that they are effective at substantially lower rates (0.25 kg/ha) as compared with 2 kg/ha for the dithiocarbamates alone.

All the phenylamides show cross-resistance to one another, pointing to a common biochemical mode of action. The compounds are structurally related to the chloroacetamide group of herbicides, e.g. propachlor (see Chapter 8, page 233), which provided initial compounds for fungicidal screening. The phenylamide fungicides therefore derived from a programme in which the antifungal properties were optimized with concurrent elimination of herbicidal activity; the latter appeared to be associated with the chloroacetyl group.[18] High fungitoxicity required the alanine methyl ester or an equivalent structure. The *R* isomers had higher fungicidal activity than the *S* enantiomers, whereas the opposite was true for the herbicidal properties.

Table 7.6 Phenylamides and related compounds

(a) Acylalanines

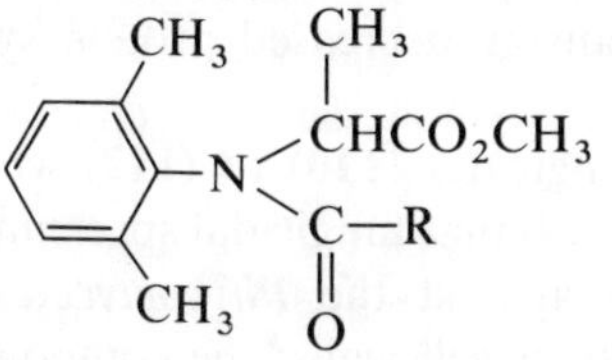

(103) R = —CH$_2$OCH$_3$

(104) R = [furan ring]

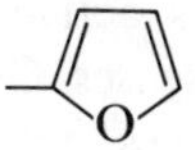

(105) R = —CH$_2$C$_6$H$_5$ (Montedison, 1982)

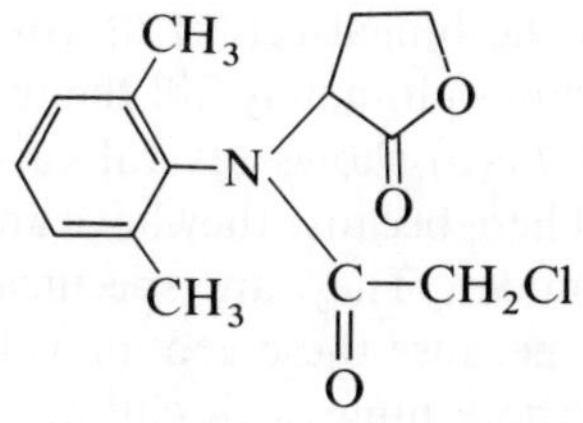

(106) (Chevron, 1978)

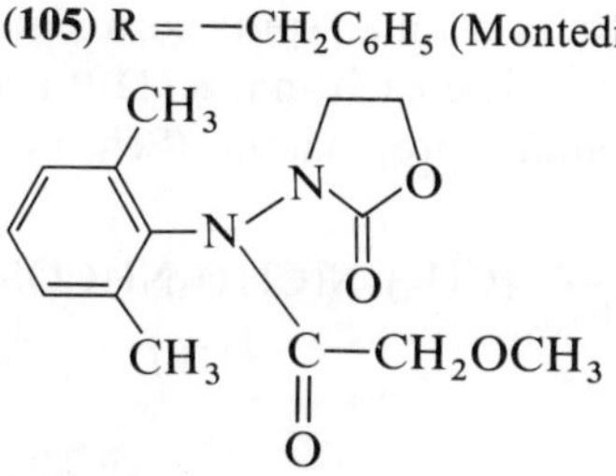

(107) (Sandoz, 1983)

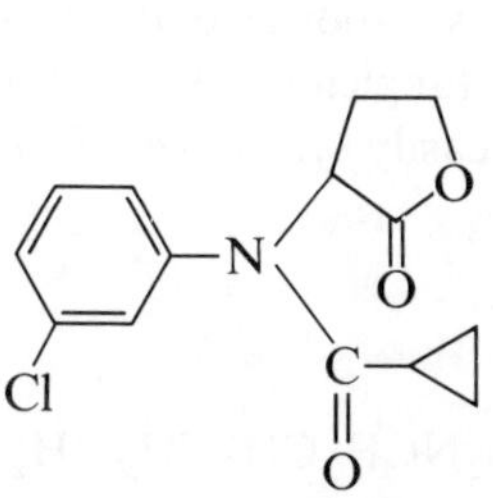

(108) (Schering, 1982)

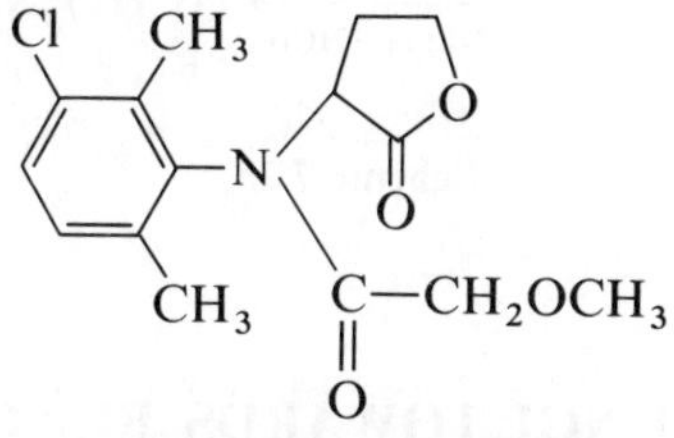

(109) (Ciba-Geigy, 1988)

(b) Carbamates

(CH$_3$)$_2$N(CH$_2$)$_3$NHCOSCH$_2$CH$_3$.HCl
(110) (Schering, 1976)

(CH$_3$)$_2$N(CH$_2$)$_3$NHCOOC$_3$H$_7$HCl
(111) (Schering, 1981)

CH$_3$CH$_2$NHCONHCOC(=NOCH$_3$)(CN)

(112) (Dupont, 1978)

Metalaxyl has been shown[18] to cause single-site inhibition of the synthesis of ribosomal (r)-RNA in sensitive fungi; it is therefore concluded that this is the primary mode action of metalaxyl and related phenylamides. Loss of r-RNA deprives the fungal cell of its ribosomes, causing decreased protein synthesis which would ultimately kill the fungus.

Table 7.6(b) shows several carbamate fungicides (110) to (112) which are included here because they have an essentially similar fungicidal spectrum to the phenylamides. They are specifically active against the *Phycomycetes* fungi, possibly because these contain cellulose in their cell walls; the compounds do not kill those fungi with chitinous cell walls.[12] Prothiocarb (110) is specially used on ornamentals, while propamocarb (111) is also used on fruit and vegetables; they are soil-applied systemic fungicides active by root uptake against downy mildew, *Phytophthora* and *Pythium* diseases. Cymoxanil (112) is valuable to control fungal diseases of potatoes and vines, but it has no activity against soil-borne pathogens. It is applied as a foliar spray formulated with surface fungicides like mancozeb or folpet.[12] The carbamates (110) and (111) can be easily synthesized from 3-(dimethylamino)propylamine (Scheme 7.25).[12]

$$\text{(CH}_3)_2\text{NCH}_2\text{CH}_2\text{CH}_2\text{NH}_2 \xrightarrow[\substack{\text{ClCOSC}_2\text{H}_5, \\ \text{NEt}_3(-\text{HCl})}]{} (\text{CH}_3)_2\text{N(CH}_2)_3\text{NHCOSC}_2\text{H}_5$$

(110)

3-(Dimethylamino)propylamine

$$\xrightarrow[\substack{\text{ClCOOC}_3\text{H}_7, \\ \text{NEt}_3\,(-\text{HCl})}]{} (\text{CH}_3)_2\text{N(CH}_2)_3\text{NHCOOC}_3\text{H}_7$$

(111)

Scheme 7.25

RESISTANCE OF FUNGI TOWARDS FUNGICIDES

Organisms possess the capacity to adapt to changing environmental conditions; microorganisms such as fungi and bacteria reproduce extremely rapidly, so they are able to change more quickly to different conditions than higher organisms.[21,40]

The development of strains of bacteria resistant to antibiotics was observed shortly after their introduction in human chemotherapy, and now many pathogenic protozoa can no longer be controlled by drugs that were formerly successful. Insects and mites can also adapt towards certain synthetic insecticides such as the organochlorine compounds (Chapter 5, page 85) and this has caused severe control problems.

On the other hand, there have been comparatively few examples of fungi acquiring resistance to the surface fungicides; for instance, although organomercurials were introduced as fungicides in 1913 not many species of fungi have become resistant to them.

Fungi, like several other organisms, may become resistant to toxicants and less sensitive cells can emerge by mutation or otherwise. The frequency of mutation is generally 10^{-4} to 10^{-10}, but it can increase considerably when the fungus is exposed to mutagenic agents. Administration of a fungicide will favour the resistant cells by elimination of competition from sensitive cells. Under the selection pressure by the fungicide, build-up of the resistant pathogen occurs which may result in failure of disease control.This has occurred many times since the introduction of systemic fungicides because these often have a specific single-site action.[21,25,40]

Thus dimethirimol was introduced into Holland to control cucumber powdery mildew in greenhouses in 1968, but by 1970 the mildew had acquired resistance to the fungicide. The fungus *Botrytis cinerea* causes severe losses of cyclamen in glasshouses; treatment with surface fungicides proved ineffective but spraying with benomyl initially gave excellent disease control. However, by 1971, the fungi had become resistant so that even a dose of 1000 p.p.m. of benomyl did not completely kill the fungus, whereas the susceptible strain was eliminated by 0.5 p.p.m. of the fungicide. The resistant fungi exhibited cross-resistance to other benzimidazole fungicides, such as thiabendazole, fuberidazole, and thiophanate methyl. Benzimidazoles were used widely for crop protection. They are single-site inhibitors and in many fungi highly resistant strains can be obtained by mutation of a single gene, e.g. ultraviolet irradiation of nine strains of *Aspergillus nidulans* induced resistance to benomyl in five strains.[18,25]

Failure of disease control due to resistance under field conditions has been reported for >100 plant pathogens,[40] e.g. with acylalanines, benzimidazoles, carboxanilides, hydroxypyrimidines, and antibiotics, like kasugamycin used against rice blast. In other cases, moderate loss in sensitivity has been observed without rapid loss of control, e.g. with EBI systemic fungicides like triazoles and dicarboximides. Many systemic fungicides act by specific interference at a single site with biosynthetic processes in fungi. In such cases, resistance is often due to a single gene mutation, resulting in a modified target site with greatly reduced affinity for the fungicide. In contrast, the majority of conventional surface fungicides, e.g. heavy metal compounds and dithiocarbamates, were general toxicants (multisite inhibitors) and the development of resistance to such compounds would require many mutations by the fungi.[18]

Other mechanisms of resistance include alterations in the fungal cell so that the toxicant cannot reach the site of action within the cell. The cell may be modified such that it has an increased capacity to detoxify the fungicide or the fungicide may become bound to other cellular components before reaching the site of action. The rate of build-up of the resistant pathogen population can vary considerably; it may occur rapidly, e.g. with benomyl and metalaxyl, or may take several years, e.g. in rice blast fungi towards kitazin P or in apple scab

(*Venturia inaequalis*) towards dodine. Factors determining the speed of resistance are the relative fitness of resistant as compared with the sensitive strains, the nature of the pathogen, and the selection pressure by the fungicide. With benzimidazoles and acylalanines, the resistant fungal strains had similar fitness to the wild sensitive strains which favours the shift to the resistant pathogen population.[41]

Resistance builds up faster in an abundantly sporulating pathogen on aerial plant parts, as compared with a pathogen that shows little sporulation and spreads slowly like certain soil-borne root or foot diseases. Thus resistance to metalaxyl rapidly developed in *Phytophthora infestans* on potatoes so that control by the single compound had to be abandoned. The level of resistance has important practical implications in the disease control—if it is low or moderate, continued applications of the fungicide may achieve satisfactory disease control, even when the majority of the pathogen population has become less sensitive to the fungicide. Resistant strains of *Botrytis cinerea* have often been found in areas frequently treated with dicarboximide fungicides, but this has not become a severe problem in the field due to a loss of vigour in the resistant strains of the pathogen.

Generally it appears that phytopathogenic fungi do not possess the genes or polygenic systems to develop resistance to many of the protectant fungicides (e.g. heavy metal derivatives, dithiocarbamates, quinones, phthalimides, and chlorothalonil). Sudden and complete loss of fungicidal potency occurs in cases of major gene resistance (e.g. benzimidazoles, acylalanines); however, major genes for resistance do not always lead to complete loss of control because the resistant fungal strains may have substantially lower fitness. In order to predict the useful life of a chemical fungicide, it is important to recognize whether genetic variation for sensitivity is available to the target organisms and, if so, what is the precise genetic control of such variation.

To avoid the spread of resistant pathogens, a mixture of fungicides, often conventional as well as systemic, should be applied which act at different sites in the fungal cell and against which many resistance mechanisms are required to overcome the fungitoxicity.

BIODEGRADATION OF FUNGICIDES

The majority of fungicides are not very stable and do not approach the stability of the organochlorine insecticides. The stability in soil will depend on the chemical structure, the nature of the soil, and the general climatic conditions. Some examples of the half-lives in days (shown in brackets) of some fungicides are: pentachloronitrobenzene (quintozene) (117–1059), benomyl (90–365), triphenyltin acetate (140), chloroneb (30–90), thiram (50), and captan (3–4).[42]

The dimethyldithiocarbamates form one of the most important worldwide group of fungicides (e.g. maneb, zineb, and mancozeb), readily decompose chemically in acidic soils to give dimethylamine and carbon disulphide, and these products are also formed by the action of soil microorganisms.

The carboximides (e.g. captan, folpet, and captafol) are readily hydrolysed under neutral or alkaline conditions; thus captan is degraded to tetrahydrophthalimide, hydrogen sulphide, carbon dioxide, and chloride anion.

The soil fungicide chloroneb (16) degrades as indicated in Scheme 7.26. Some fungi can carry out the first step in either direction and the microbial resynthesis of the a.i. from the degradation product may account for the long-term effectiveness of chloroneb in soil.

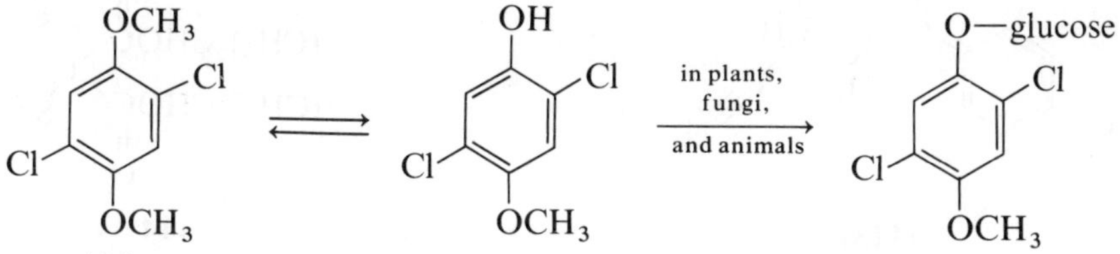

Scheme 7.26

The benzimidazole systemic fungicides (benomyl, thiabendazole, and thiophanate-methyl) are all converted to carbendazim (48) which is the active ingredient. Carbendazim itself can be used as a fungicide; the field performance may be different from that when benomyl or thiophanate-methyl is used due to factors such as retention, tissue penetration, and redistribution. Benomyl and thiophanate-methyl are rapidly converted to carbendazim in various soil types; the a.i. is, however, resistant to degradation and remains for several months as the major soil metabolite with some 2-aminobenzimidazole. The rate of degradation of thiophanate-methyl to carbendazim is reduced by steam treatment of the soil, which illustrates the role played by soil microorganisms in the degradative process. In plant tissues, the pathway appears to be as shown in Scheme 7.27.

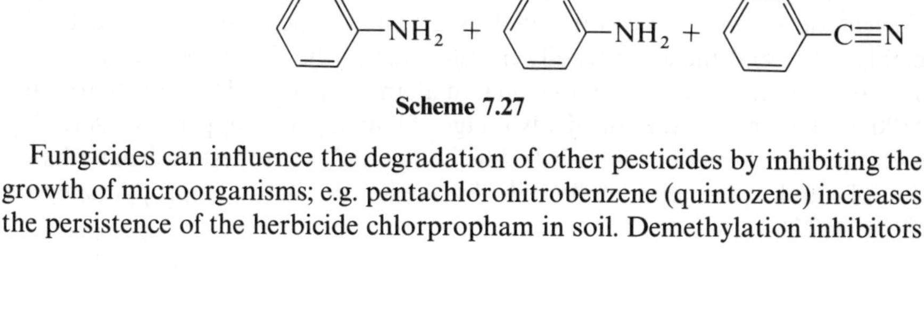

Scheme 7.27

Fungicides can influence the degradation of other pesticides by inhibiting the growth of microorganisms; e.g. pentachloronitrobenzene (quintozene) increases the persistence of the herbicide chlorpropham in soil. Demethylation inhibitors

(DMIs), like the fungicides imazalil, tridimefon, and the triazoles, act by inhibition of the C-14 demethylation step in ergosterol biosynthesis. Such fungicides probably bind to the enzyme cytochrome P 450 and they may consequently interfere with the oxidative metabolism of other pesticides.

MISCELLANEOUS COMPOUNDS

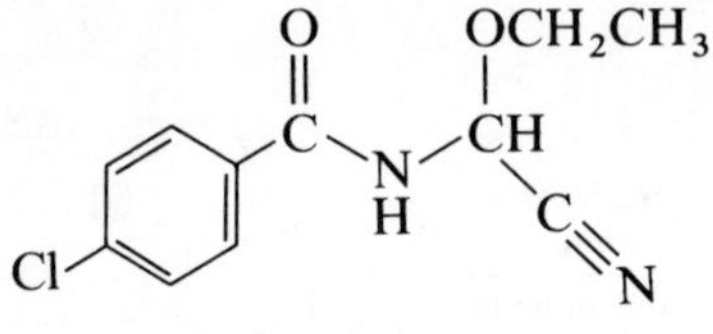

(113)

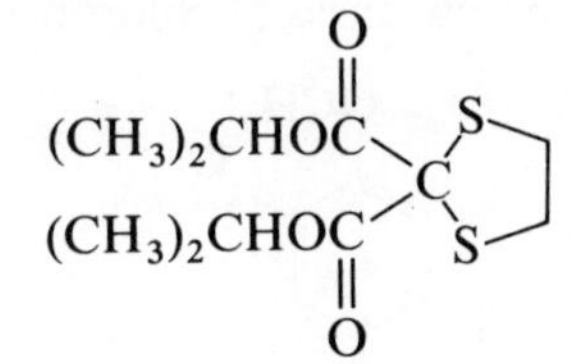

(114)

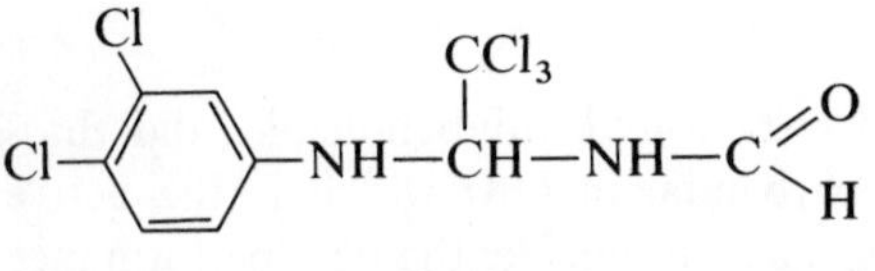

(115)

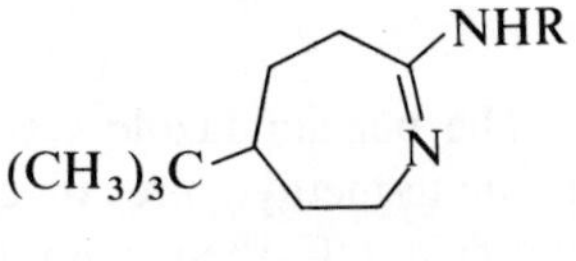

(116)

(R = a long chain alkyl group, C_{10} to C_{12})

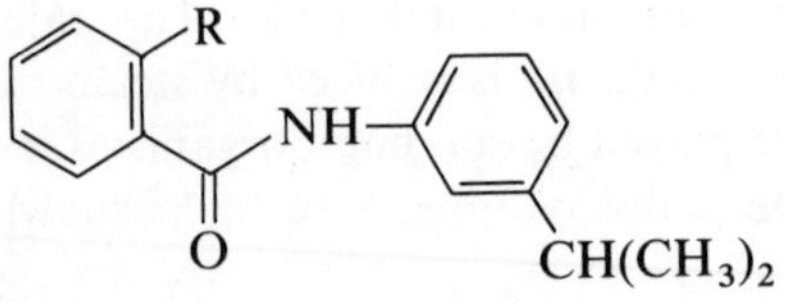

(117)

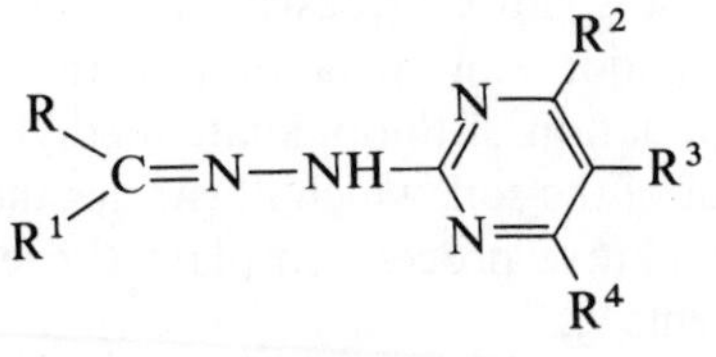

(118)

(*RS*)-4-Chloro-*N*-[cyano(ethoxy) methyl] benzamide **(113)** is a novel highly active fungicide, effective against a broad range of *Oomycetes* pathogens, e.g. vine downy mildew and leaf spot. It is formulated as a wettable powder and penetrates leaf and stem tissue with a new biochemical mode of action.[43]

Isoprothiolane **(114)** is a selective systemic fungicide against rice blast fungus (*Piricularia oryzae*); it is formulated as granules which can be added to the paddy water and the a.i. is readily translocated to the rice leaves via the roots. The dithiolane **(114)** has a low mammalian toxicity (LD_{50} (oral) to rats, 1190 mg/kg) and has the added advantage of killing plant hoppers by interfering with the insect's metamorphosis.[12] The compound is synthesized from ethane-dithiol and phosgene, followed by reaction with isopropylmalonate (Scheme 7.28).

Plate 1 Aerial spraying of flowering cotton (South Africa). Aerial spraying is 7–8 times more energy efficient than ground spraying and it enables larger areas to be sprayed quickly, economically and without risk of damage to crops from machinery. (Photograph: Shell Photographic Service.)

Plate 2 Application of pesticides by use of a LK sprayer mounted on a Mercedes Benz MB tractor. (Photograph: Ciba-Geigy Agrochemicals.)

Plate 3 Spraying wheat with a fungicide, e.g. flutriafol or propiconazole. (Photograph: ICI Plant Protection Division.)

Plate 4 Spraying a Venezuelan banana plantation with Shell pesticides. (Photograph: Shell Photographic Service.)

Plate 5 Insect pests *(Bucculatrix* and *Heliothis)* on cotton controlled by 'Karate' (cyhalothrin) or 'Cymbush' (cypermethrin). Left, untreated; right, treated. (Photograph: ICI Plant Protection Division.)

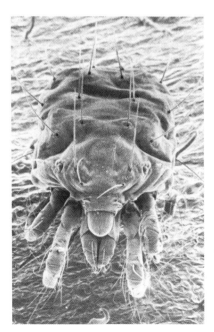

Plate 6 Scanning electron micrograph of a red spider mite. (Photograph: Agrochemicals Division of the Boots Company Ltd.)

Plate 7 Safe spraying of vegetables with 'Gardona' (tetrachlorvinphos—an insecticide from Shell; it can be applied safely without protective clothing and is effective against caterpillars and flea beetles. (Photograph: Shell Ltd.)

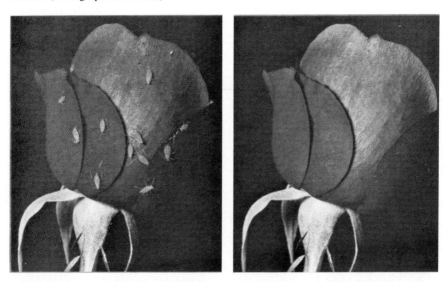

Plate 8 Aphids on rose. Left, untreated; right, treated with the carbamate insecticide pirimicarb. (Photograph: ICI Plant Protection Division.)

Plate 9 Spraying 'Ficam W', (propoxur), a methyl carbamate insecticide, in a food larder for control of cockroaches and other domestic pests. The non-odorous and non-staining properties of 'Ficam W' ensure that there is no danger of food contamination, as long as food preparation surfaces are covered and food containers closed. (Photograph: Fisons Agrochemical Division.)

Plate 10 Left, untreated peppers infected with Phytophthora in Spain; right, peppers protected by a fungicide, e.g. mancozeb or chlorothalonil. (Photograph: ICI Plant Protection Division.)

Plate 11 *Ustilago* smut on wheat and barley: Left, controlled by the fungicide 'Impact' (flutriafol); right, untreated. (Photograph: ICI Plant Protection Division.)

Plate 12 Grapes infected by *Uncinula necator* fungi in Toulouse, France. (Photograph: ICI Plant Protection Division.)

Plate 13 Healthy grapes at Toulouse, France, which have been treated with the fungicide 'Anvil' (hexaconazole) (Compare with Plate 12). (Photograph: ICI Plant Protection Division.)

Plate 14 Control of potato blight by 'Fubol' (a mixture of mancozeb and metalaxyl). Background; sprayed potatoes; foreground, untreated potatoes destroyed by blight. (Photograph: Ciba-Geigy Agrochemicals.)

Plate 15 Control of the weed (paspalum) in pineapples (South Africa) by 'Fusilade' (fluazifop-butyl). Left, untreated; right, treated. (Photograph: ICI Plant Protection Division.)

Plate 16   Weed control in cotton by 'Fusilade' (fluazifop-butyl). Left, untreated; right, treated. (Photograph: ICI Plant Protection Division.)

Plate 17   Control of chickweed in brassica with 'Butisan S' (metazachlor). Left, untreated; right, treated. (Photograph BASF Ltd.)

Plate 18   Untreated strips of wild oats in barley in a field near Seville, Spain, which had been partly treated with 'Metaven' (flamprop-methyl). Left hand side, treated; right hand side, untreated. (Photograph: Shell Photographic Service.)

Plate 19 Control of mayweed in wheat with 'Dicurane' (chlortoluron). Left, treated; right, untreated. (Photograph: Ciba-Geigy Agrochemicals Ltd.)

Plate 20 Selective weed control in maize by 'Gesaprim' (atrazine). Front, untreated; back, treated. (Photograph: Ciba-Geigy Agrochemicals Ltd.)

Plate 21 Weed control in sugarbeet with pyramin (chloridazon). Left, untreated; right, treated. (Photograph: BASF Ltd.)

Plate 22   Rice field trials (Japan) using 'Argold' (cinmethylin) for weed control in transplanted rice seedlings. Applications of 60g ai/ha were made from 7 days after transplanting. (Photograph: Shell Photographic Service.)

Plate 23   Weed control in potatoes by treatment with either 'Gramoxone' (paraquat) or 'Fusilade' (fluazifop-butyl). (Photograph: ICI Plant Protection Division.)

Plate 24 Weed control in soya (late) by 'Flex' (fomesafen). Left, untreated; right, treated. (Photograph: ICI Plant Protection Division.)

Plate 25 Control of wild oat *(Avena fatua)* in winter wheat by spraying with 'Grasp' (traloxydim). Top, untreated; bottom, treated. (Photograph: ICI Plant Protection Division.)

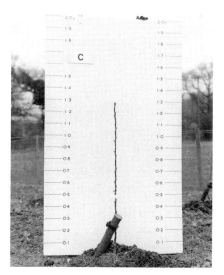

Plate 26 Conference pear—unsprayed nursery stock. (Photograph: Rhône-Poulenc Agriculture Ltd.)

Plate 27 Conference pear after spraying with the PGR 'MB 25105' (propyl-3-t-butylphenoxyacetate). Comparison with plate 26 clearly shows the increased lateral growth caused by the chemical. (Photograph: Rhône-Poulenc Agriculture Ltd.)

Plates 28 and 29 Bramley apples on MM111 rootstock at the beginning of their third year in the orchard. One tree has been pruned by hand (Plate 28), the other has been treated in the nursery and in the second year in the orchard with 900 p.p.m. of the PGR 'MB 25105' (propyl-3-t-butylphenoxyacetate (Plate 29). (Photographs: Rhône-Poulenc Agriculture Ltd.)

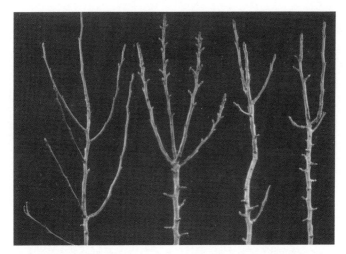

Plate 30   Pear: control (left) compared with the pear after treatment with the PGRs 'Cyclocel' (chlormequat chloride) (second left) and 'Cultar' (paclobutrazol) at rates of 1000 and 2000 p.p.m. (third and fourth left). (Photograph: ICI Plant Protection Division.)

Plate 31   Bramley apple: control (top); treated with 'Cultar' (paclobutrazol) (bottom). (Photograph: ICI Plant Protection Division.)

Plate 32 Geranium: control (left) and right after treatment with the PGR 'Bonzi' (paclobutrazol). (Photograph: ICI Plant Protection Division.)

Plate 33 Fuschia 'Beacon'. Control (left) and after treatment with 'Bonzi' (paclobutrazol) at 2 p.p.m. (centre) and 40 p.p.m. (right). (Photograph: ICI Plant Protection Division.)

Plate 34 A pheromone trap. (Photograph: ICI Plant Protection Division.)

Plate 35 The effect of the insect growth regulator (IGR) azadirachin on *Pieris brassica* pupa; left, untreated; right, treated with the IGR. (Photograph: ICI Plant Protection Division.)

Plate 36 The effect of the IGR azadirachin on *Pieris brassica* pupa. (Photograph: ICI Plant Protection Division.)

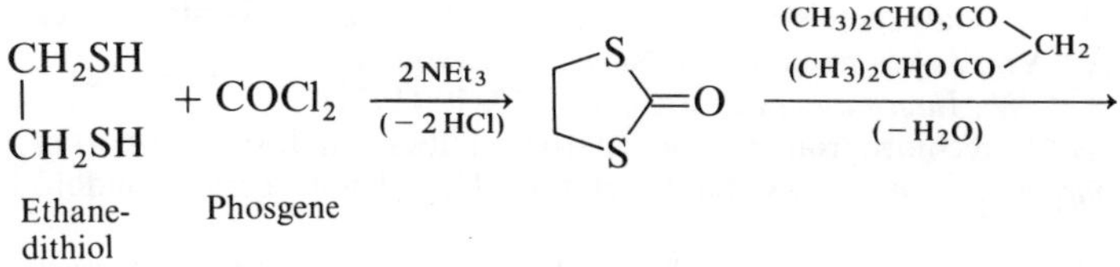

$$\underset{\substack{\text{Ethane-}\\\text{dithiol}}}{\underset{\text{CH}_2\text{SH}}{\overset{\text{CH}_2\text{SH}}{|}}} + \underset{\text{Phosgene}}{\text{COCl}_2} \xrightarrow[(-2\text{HCl})]{2\,\text{NEt}_3} \underset{\text{S}}{\overset{\text{S}}{\left[\!\!\!\!\right.}}\!\!=\!\!O \xrightarrow[(-\text{H}_2\text{O})]{\overset{(\text{CH}_3)_2\text{CHO, CO}}{(\text{CH}_3)_2\text{CHO CO}}\!\!\!\diagup\!\!\text{CH}_2} \quad (114)$$

Scheme 7.28

Formaldehyde, as the 40% aqueous solution (formalin), has long been used as a seed dressing and soil sterilant (Chapter 10, page 284). The fungitoxicity is due to reaction with the amino groups of fungal spore proteins. Other aldehydes show some protectant antifungal action; however, chloraniformethan (115) has valuable systemic fungicidal activity, particularly against cereal powdery mildews, and has a very low mammalian toxicity (LD_{50} (oral) to rats $\simeq$ 2500 mg/kg).[12]

Some substituted azepines (116) have systemic fungicidal activity against leaf spot, powdery mildew, and rust diseases.

Isopropylbenzanilides (117) gave excellent control of *Rhizoctonia solani* at 20 p.p.m. Aliphatic aldehyde and ketone pyrimidinyl hydrazones (118) containing a wide range of organic radicals (R^1-R^4) were useful antifungal agents. The optimum activity against rice blast fungus (*Piricularia oryzae*) was shown by the diisobutylketone-4,6-dimethyl derivative (118); $R = R^1 = CH_3CH(CH_3)CH_2$—; $R^2 = R^4 = CH_3$; $R^3 = H$).

REFERENCES

1. Woods, A., *Pest Control: A Survey*, McGraw-Hill, London, 1974, p. 93.
2. Green, M. B., Hartley, G. S., and West, T. F., *Chemicals for Crop Improvement and Pest Management*, 3rd edn, Pergamon Press, Oxford, 1987, p. 152.
3. Martin, H., and Woodcock, D., *The Scientific Principles of Crop Protection*, 7th edn, Arnold, London, 1983, p. 122.
4. Hassall, K. A., *The Biochemistry and Uses of Pesticides*, Macmillan, London, 1990, p. 286.
5. Cremlyn, R. J. W., *International Pest Control*, **5**, 10 (1963).
6. Lukens, R. J., *Chemistry of Fungicidal Action*, Chapman and Hall, London, 1971.
7. *The UK Pesticide Guide*, British Crop Protection Council, Bracknell, England, 1989.
8. Wain, R. L., *Some Chemical Aspects of Plant Disease Control*, Royal Institute of Chemistry, Monograph No. 3, 1959.
9. Thiolliére, J., *Monograph British Crop Protection Council*, 1985; *Chemical Abstracts*, **104**, 47020n (1986).
10. Foltzer, G., Deruche, C., and Barnavon, M., *Monograph British Crop Protection Council*, **1**, 31 (1985); *Chemical Abstracts*, **104**, 2056n (1986).
11. Metcalf, R. L., 'Chemistry and biology of pesticides', in *Pesticides in the Environment* (Ed. White-Stevens, R.), Dekker, New York, 1971, p. 1.
12. Matolcsy, Gy., Nádasy, M., and Andriska, V., *Pesticide Chemistry*, Elsevier, Amsterdam, 1988, p. 273.
13. Cremlyn, R. J. W., *Pest Articles and News Summaries* (B), **13**, 255 (1967).

14. Rich, S., 'Fungicidal chemistry', in *Plant Pathology* (Eds. Horsfall, J. G., and Dimond, A.), Vol. 2, Academic Press, New York, 1960, p. 588.
15. Cremlyn, R. J. W., *International Pest Control*, **13**, 12 (1971).
16. Green, M. B., 'Polychloroaromatic and heteroaromatics of industrial importance', in *Polychloroaromatic Compounds* (Ed. Suschitzky, H.), Plenum Press, London, 1974, p. 433.
17. Worthing, C. R., and Hance, R. J. (Eds.), *Pesticide Manual*, 9th edn, British Crop Protection Council, Croydon, 1991.
18. Lye, H. (Ed.), *Modern Selective Fungicides*, Longman, Harlow, 1987.
19. Lye, H., Banasiak, L., and Casperson, G., *Biol. Rundsch.*, **25**(4), 239 (1987); *Chemical Abstracts*, **108**, 17649q (1988).
20. Cremlyn, R. J. W., *J. Sci. Food Agric.*, **12**, 805 (1961).
21. Marsh, R. W. (Ed.), *Systemic Fungicides*, 2nd edn, Longman, London, 1977.
22. Cremlyn, R. J. W., *International Pest Control*, **15**(2), 8 (1973).
23. Goodman, R. N., *Advances in Pest Control Research*, **5**, 1 (1962).
24. Corbett, J. R., Wright, K., and Baillie, A. C., *The Biochemical Mode of Action of Pesticides*, Academic Press, London and New York, 1984.
25. van der Kerk, G. J. M., *World Review of Pest Control*, **2**(3), 29 (1963).
26. Woodcock, D., *Chemistry in Britain*, **4**(7), 294 (1968).
27. Woodcock, D., *Chemistry in Britain*, **7**(10), 415 (1971).
28. Trinci, A. P. J., and Riley, J. F. (Eds.), *The Mode of Action of Antifungal Agents*, Cambridge University Press, 1984.
29. Green, M. B., and Spilker, D. A. (Eds.), *Fungicidal Chemistry: Advances and Practical Applications*, American Chemical Society, Washington, 1986.
30. Worthington, P. A., *Proceedings of British Crop Protection Conference, Brighton*, **3**, 955 (1984).
31. Bohnen, K., *et al.*, *Proceedings of British Crop Protection Conference, Brighton*, **1**, 27 (1986).
32. Shephard, M. C., *et al.*, *Proceedings of British Crop Protection Conference, Brighton*, **1**, 19 (1986).
33. Efthimiadis, P., *Proceedings of British Crop Protection Conference, Brighton*, **3**, 887 (1988).
34. Ruess, W., *et al.*, *Proceedings of British Crop Protection Conference, Brighton*, **2**, 543 (1988).
35. Carlie, W. R., *Control of Crop Diseases*, Arnold, London, 1988.
36. Berg, D., 'The biochemical mode of action of fungicides', in *Human Welfare and the Environment*, Vol. 1, Pergamon Press, Oxford, 1983, p. 55.
37. Kato, T., 'Sterol biosynthesis in fungi: a target for broad-spectrum fungicides', in *Chemistry of Plant Protection* (Ed. Haug, G.), Springer-Verlag, Berlin, p. 1 (1986).
38. Fest, C., and Schmidt, K. J., *The Chemistry of Organophosphorus Pesticides*, Springer-Verlag, Berlin, 1973.
39. Margot, P., Eckhardt, W., and Dahman, H., *Proceedings of British Crop Protection Conference, Brighton*, **2**, 527 (1988).
40. Geoghiou, G. P., and Saito, T. (Eds.), *Pest Resistance to Pesticides*, Plenum Press, London, 1983.
41. Brent, K. J., and Atkin, R. K., 'The resistance of fungi and weeds', in *Rational Pesticide Use*, Cambridge University Press, 1987.
42. Sisler, H. D., 'The Biodegradation of agricultural fungicides', in *Biodegradation of Pesticides* (Eds. Matsumura, F., and Murti, C. R. K.), Plenum Press, London, 1982, p. 133.
43. Heaney, S. P., Shephard, M. C., Crowley, P. J., and Shearing, S. J., *Proceedings of British Crop Protection Conference, Brighton*, **2**, 551 (1988).

Chapter 8
Herbicides

Weeds may be defined as plants growing where man does not wish them to be. As soon as man began organized agriculture, which in parts of the world goes back some 10 000 years, though extensive agriculture is only about 5000 years old, weeds started to compete with the crop plants for moisture, nutrients, and light. Much of the energy expended in arable farming goes in such mechanical operations as ploughing, harrowing, and general soil cultivation, all of which both remove weeds and provide suitable conditions for the efficient growth of the crop plant. It is, therefore, somewhat ironic that the majority of our common weeds today were rare plants before man became an agriculturalist and provided conditions on which they too thrived.[1] Traditional mechanical means of weed control reached the peak of efficiency in Britain by the end of the nineteenth century when plenty of cheap agricultural labour was still available and simple horse-drawn machines kept the fields free from weeds. With the rapid growth of industrialization, the resultant drift of labour from the country-side to the factories meant a shortage of manpower on the farms and consequently agricultural wages increased and crops became weedier and weedier and different crop rotations were introduced to try and cut down weeds, not always with success. This situation provided the stimulus for the development both of more efficient mechanical means of weed control and the introduction of chemical weed killers (herbicides).

The idea of controlling weeds with chemicals is not new; for more than a century chemicals have been employed for total weed control—the removal of all plants from such places as railway tracks, timber yards, and unmetalled roads. Crude chemicals such as rock salt, crushed arsenical ores, creosote, oil wastes, sulphuric acid, and copper salts were used in massive doses.[2,3] Under these conditions, all plants were killed. What was much more needed were chemicals that would selectively kill the weeds, but not harm the crop plants. In the early 1900s some selective control of broad-leaved weeds in cereals was achieved by spraying with a sufficient concentration of such general plant toxicants as soluble copper salts and sulphuric acid. The selectivity was based on

217

physical factors—the larger, rougher surfaces of weed leaves were more effectively wetted by the spray as compared with the narrow, smooth cereal leaves in which there was much greater run-off of the toxicant. Copper salts are no longer used, though sulphuric acid is still applied on a limited scale for the destruction of potato haulms at the end of the season. The weed-killing properties of other inorganic compounds such as sodium chlorate, borates, and arsenic compounds (e.g. sodium arsenite) have been known for a long time.[3] These compounds all function as total herbicides and treated areas remain toxic to plants for months or, in some cases, years. Sodium arsenite was formerly applied extensively as a potato haulm destroyer.

Creosote and other crude tar oils have been employed as total herbicides for a long time. Later petroleum oils were purified by shaking them with concentrated sulphuric acid to remove unsaturated compounds, followed by fractional distillation. Such purified petroleum oils of suitable boiling points were used as contact herbicides for postemergence control of many annual broad-leaved weeds and grasses in carrot, parsley, and parsnip, and for pre- or postemergence weed control in forest nurseries.

With vegetable crops, spraying must be carefully timed, because if carried out too early there is a danger of phytotoxic damage to the crops and if too late tainting of the crop may occur. Such hydrocarbon oils are physical toxicants. They affect photosynthesis, disrupt mitosis, and change the permeability of cell membranes; prolonged application at high concentrations causes irreversible cell damage leading to death. Consequently they are also used as insecticides and acaricides.[4]

Most of the inorganic herbicides were too phytotoxic to be used safely for selective postemergence weed control in crops. In a few cases, selectivity is based on *biochemical differences* between weeds and the crop plants, such that the latter have a unique defence mechanism. The absence of β-oxidases from leguminous plants enables 2,4-DB to be applied safely for selective weed control in these crops (see page 222). Maize detoxifies the triazine herbicides (e.g. atrazine and simazine) by enzymic hydrolysis but weeds do not possess this enzyme; hence triazines are extensively used for weed control in maize (see page 243). The application of biotechnology will extend the range of biochemical selectivity (see page 369).

Selectivity may also be based on *morphological differences*: dicotyledonous plants, e.g. broad-leaved weeds, have a larger surface area, with the meristemic tissue exposed to the herbicidal spray. On the other hand, monocotyledonous plants, e.g. grasses, possess narrow upright leaves which form a protective sheath around the meristematic tissue: this partly accounts for the selective toxicity of the phenoxyacetic acid herbicides, e.g. MCPA and 2,4-D, towards broad-leaved weeds.

Physicochemical factors may be important: with soil-applied herbicides like triazines (see page 241), the low aqueous solubility means that the chemical only penetrates some 5 cm downwards in the soil and so germinating weed seeds are killed but more deep-rooted crops are unaffected. The bipyridinium herbicides, e.g. paraquat (see page 253) and glyphosate (page 257), are strongly bound to soil colloids by cation exchange. Consequently they enter the soil solution which means that only those plants directly hit by the spray are damaged and there is no uptake of the toxicant from the soil.

The first important discovery in the field of selective weed control was the introduction of 2-methyl-4,6-dinitrophenol (1; R = CH_3) (DNOC) in France in 1933. This is a contact herbicide and when sprayed onto a cereal crop it killed the majority of annual weeds infesting the crop without causing appreciable damage to the cereals. Unfortunately most perennial weeds, like couch grass and creeping thistle, were not killed because, although their top growth was desiccated, DNOC is not translocated in plants and accordingly their extensive root systems survived and in due course sent up further shoots. The related compound dinoseb (1; R = $CH_3CH_2(CH_3)CH—$) was mainly used for weed control in peas, beans, seedling lucerne, and cotton.[3,4] Both compounds are generally applied as high-volume sprays because even at 890 l/ha they are usually not completely dissolved in the spray tank. These dinitro compounds played a big part in increasing food production during World War II. DNOC also finds use as a general insecticide, especially as a winter wash for fruit trees (see Chapter 5, page 80).

A serious disadvantage is that these substances are very poisonous to mammals. For DNOC (1; R = CH_3), LD_{50} (oral) to rats is 30 mg/kg, and consequently they can only be applied by operators wearing full protective clothing. They are used at fairly high concentrations of about 11 kg/ha and they have caused considerable damage to wild life and some human fatalities.[5] However, on contact with plants or soil, they are fairly rapidly degraded to non-toxic substances and consequently do not accumulate along food chains.[6]

DNOC is now little used although dinoseb (1; R = $CH_3CH_2(CH_3)CH—$) is still employed on a limited scale as a selective herbicide in peas and cotton.[4] These dinitrophenols owe their herbicidal action on plants to their ability to uncouple oxidative phosphorylation and hence inhibit ATP synthesis[7] (see Chapter 3, page 41). This process is common to plants and mammals so it is not surprising that dinitrophenols have high mammalian toxicities.

CARBOXYLIC ACID HERBICIDES

The discovery of the phenoxyacetic acid herbicides derived from the work of Kögl and his collaborators (1934) who showed that indole-3-acetic acid or IAA (2) promotes cell elongation in plants. Auxin or IAA was isolated from plants and this stimulated the search for other compounds of related structure which had growth-regulating activity on plants, but which, unlike IAA, were not rapidly metabolized in plants. 1-Naphthylacetic acid was found to be active, as was 2-naphthoxyacetic acid. This aroused interest in aryloxyacetic acids as potential plant growth regulators.[8] In 1942 Zimmerman and Hitchcock showed that certain chlorinated phenoxyacetic acids such as 2,4-dichlorophenoxyacetic acid or 2,4-D (**3**; X = Y = Cl, Z = H, n = 1) were more active than the natural growth hormone IAA (**2**) and furthermore were not rapidly degraded in the plant. Consequently, 2,4-D could be applied externally to the plant and, since it was not internally regulated like IAA, produced markedly abnormal growth resulting in the death of the plant. This discovery may be regarded as the beginning of the organic herbicide industry, since the previous herbicides had been mainly inorganic compounds. The herbicidal phenoxyacetic acids were found to be much more active against broad-leaved weeds (dicotyledons) than against cereals and grasses (monocotyledons). The selectivity may arise partly from the fact that the sprays adhere much better to the rougher surfaces of broad-leaved weeds than to the narrow waxy leaves of grasses; but despite its vital importance the selectivity of these compounds is not clearly understood.[9-11] It may be associated with differences in plant structure and rates of translocation in monocotyledonous and dicotyledonous plants.[9] The phenoxyacetic acids were the first really effective selective herbicides and were, like the organochlorine and organophosphorus insecticides, the product of war-directed research. They came into use at a time when the maximum home food production with a much reduced agricultural labour force was a vital factor in the war effort[2,7,8] and their use revolutionized agriculture.

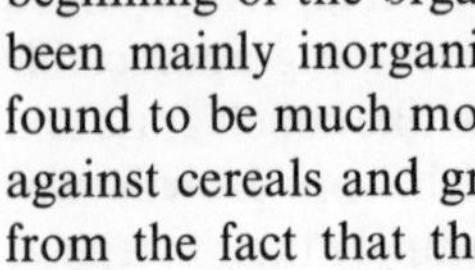

(1) 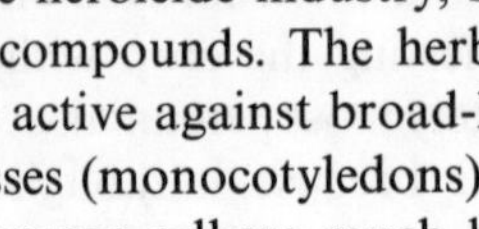   (2) 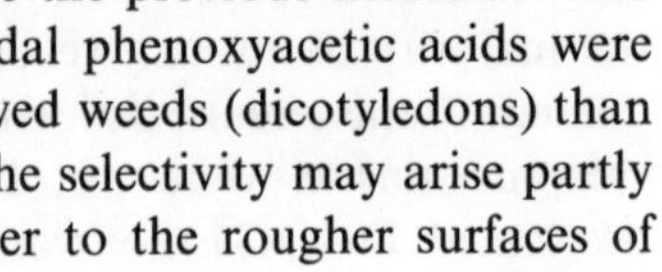   (3)

The phenoxyalkanoic acids are best described as auxin-type growth regulators and are easily and cheaply obtained by the Williamson-type synthesis from the corresponding phenol; for instance, the preparation of 2,4-D (**3**; X = Y = Cl, Z = H, n = 1) is as follows (Scheme 8.1).

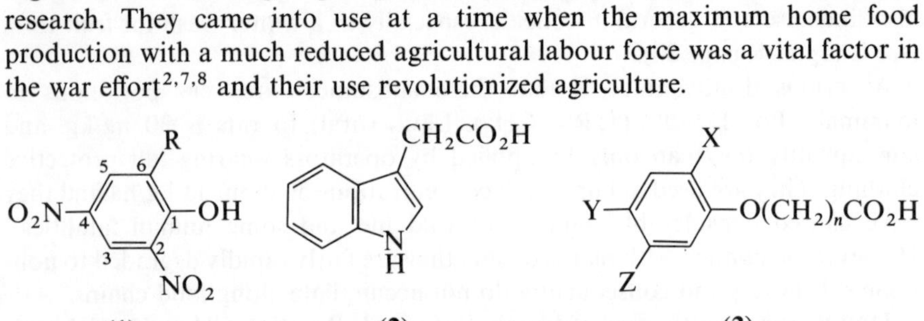

Scheme 8.1

Another early example of great importance was 2-methyl-4-chlorophenoxy-acetic acid $(3; X = CH_3, Y = Cl, Z = H, n = 1)$ (MCPA).[12] This is used for postemergence control of many broad-leaved weeds in cereals, grassland, and asparagus.[4,12] Gardeners often use MCPA on lawns when the annual broad-leaved weeds die rapidly while perennial weeds such as daisies, dandelions, and plantains grow fast and grotesquely and eventually die. The hormone-type chemical induces very rapid RNA production and consequently the plant grows itself to death—in other words, the growth completely outstrips the available nutrients. Plants in dry climates are more susceptible and can be effectively treated at lower dose rates than those in conditions of greater rainfall.

In Britain MCPA is used on a larger scale than 2,4-D, whereas in America the reverse is true. MCPA has the important advantage that the sodium salt is much more soluble in water than that of 2,4-D. Also local overdosage is less liable to produce phytotoxic damage to the crop, but the main reason why MCPA was more extensively used in Britain was the greater availability of the starting material *o*-cresol from coal tar distillation.

Another important member of this series of selective herbicides is 2,4,5-trichlorophenoxyacetic acid (2,4,5-T) $(3; X = Y = Z = Cl, n = 1)$. This translocated herbicide is employed for control of woody plants and can be used for selective weed control in conifers;[4,12] it is more persistent in soil than either MCPA or 2,4-D. The compound is contaminated with the highly toxic, teratogenic dioxin (see Chapter 15, page 345); in the United Kingdom 2,4,5-T can now only be used when the dioxin content is <0.5 p.p.m., and in many countries 2,4,5-T is banned.

The analogous 2-phenoxypropionic acids show similar herbicidal activity but are much more effective against chickweed (*Stellaria media*) and cleavers (*Gallium aparine*) in cereals.[4] Well-known examples are dichlorprop (**4a**), mecoprop (**4b**), and fenoprop (**4c**); like 2,4,5-T, the latter is effective against woody plants and also as a selective herbicide in rice and sugar cane. They are obtained by condensation of 2-chloropropionic acid with the appropriate phenol (Scheme 8.2). Due to the presence of the chiral carbon atom (C*), they

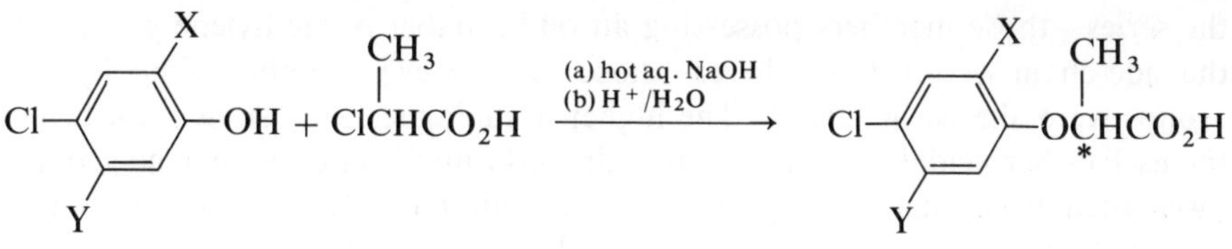

	X	Y
(**4a**)	Cl	H
(**4b**)	CH_3	H
(**4c**)	Cl	Cl

Scheme 8.2

can exist in (+) and (−) optical isomeric forms, but only the former has appreciable herbicidal activity. These compounds have proved useful to kill certain weeds that have acquired tolerance to 2,4-D and MCPA.

The 4-(aryloxy)phenoxypropanoates (5) were introduced by Hoechst AG (1975) for selective preemergence control of annual grass weeds in cereals and a variety of broad-leaved crops, which is a completely different herbicidal spectrum from the usual phenoxyacetic acid herbicides; later a further group of analogues was brought onto the market, e.g. fluazifop-butyl (6) by ICI (1980), which is extensively used for postemergence control of annual and perennial grass weeds in broad-leaved crops—soybean, sugar beet, cotton, oilseed rape, vines, fruits, and vegetables at 0.125-0.5 g/ha[13] (Plates 15 and 16). Such selective grass–weed herbicides are sometimes termed graminicides. These compounds may act as antiauxins and compete for the IAA binding site. More probably, their activity results from blockage of fatty acid biosynthesis due to their selective inhibition of the acetyl CoA carboxylase in grasses (see page 264).

$$X$$
$$CH_3$$
$$Cl-\langle\rangle-O-\langle\rangle-OCHCO_2R$$

(5)

	X	R	
(5a)	Cl	CH_3	Diclofop-methyl
(5b)	H	$CH_2CH(CH_3)_2$	Clofop-isobutyl

$$CF_3-\langle N\rangle-O-\overset{CH_3}{CH}-\langle\rangle-CO_2C_4H_9\text{-}n$$

(6)

Research by Professor Wain of Wye College (University of London) on the herbicidal properties of a series of ω-phenoxyalkanecarboxylic acids of type $(3; X = Y = Cl, Z = H, n > 1)$ showed that herbicidal activity alternated along the series—those members possessing an odd number of methylene groups in the side-chain were active, whereas those with an even number of methylene groups were almost inactive.[14] The higher homologues were shown not to be themselves herbicidal, but those containing odd numbers of methylene groups owed their herbicidal activity to *in vivo* β-oxidation within the plant by the enzyme β-oxidase to give 2,4-D which was the active herbicidal entity. Biochemically the process is similar to the β-oxidative degradation of fatty acids which is a well-established metabolic route in mammals (Scheme 8.3).

Wain found[14] that different plant species vary considerably in their ability to perform this enzymic β-oxidation, which provides an additional mechanism for increasing selectivity. Several species of leguminous plants are resistant to the phenoxybutyric acid herbicides because they largely lack the β-oxidases to

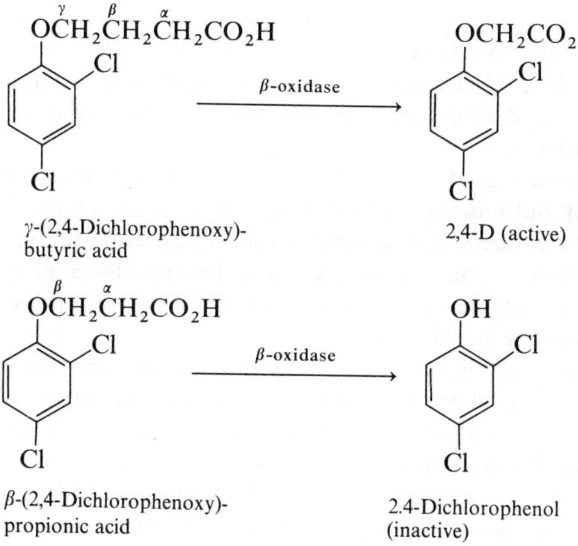

$\overset{\gamma}{O}CH_2\overset{\beta}{C}H_2\overset{\alpha}{C}H_2CO_2H$ OCH_2CO_2H

$\xrightarrow{\beta\text{-oxidase}}$

γ-(2,4-Dichlorophenoxy)-
butyric acid   2,4-D (active)

$\overset{\beta}{O}CH_2\overset{\alpha}{C}H_2CO_2H$ OH

$\xrightarrow{\beta\text{-oxidase}}$

β-(2,4-Dichlorophenoxy)-
propionic acid   2.4-Dichlorophenol
(inactive)

Scheme 8.3

convert them *in vivo* to the active phenoxyacetic acid derivatives, so γ-(2,4-dichlorophenoxy)butyric acid (2,4-DB) and the related 2-methyl-4-chlorophenoxybutyric acid (MCPB) are widely used for selective postemergence weed control in leguminous crops (e.g. peas, clover, and lucerne) and in undersown cereals where the analogous phenoxyacetic acid derivatives would damage the legumes.[4,12] The phenoxybutyric acids are synthesized by reaction of the appropriate substituted phenol with butyrolactone.[13] Other important features of the chemical structure of herbicidal phenoxyacetic acids are as follows:

(a) *The carboxylic acid group.* For high activity the molecule must generally possess either the $-CO_2H$ group or a group that is easily converted to it within the plant tissues.

(b) *The unsaturated ring system.* For activity the ring appears to require at least one unsaturated bond adjacent to the side chain; thus cyclohexanoxyacetic acid is inactive, but the unsaturated derivative is active:

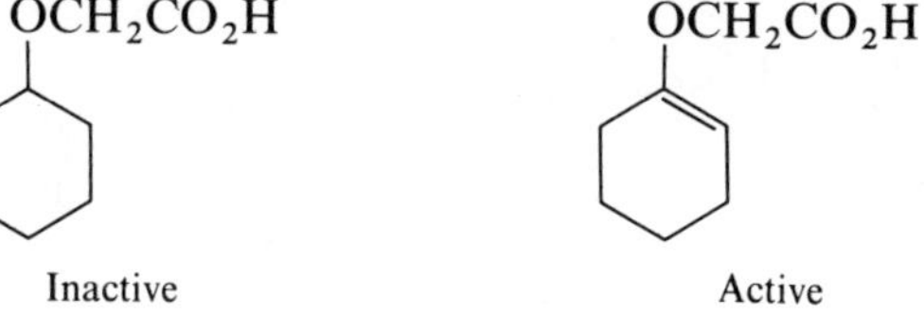

$\overset{\alpha}{O}CH_2CO_2H$ $\overset{\alpha}{O}CH_2CO_2H$

Inactive   Active

(c) *The side-chain.* For a phenoxyalkanecarboxylic acid to show herbicidal activity depending on cell elongation, the presence of a hydrogen atom attached to the α-carbon is necessary, although this criterion does not appear to hold for the benzoic acid herbicides (see page 225).

(d) *Substitution in the phenyl ring.* In the phenoxyacetic acids, the introduction of chlorine atoms into the 2-, 4-, or 2,4-positions greatly enhances activity, although the 2,4,6-trichloro derivative is almost completely inactive. Poor activity is found for compounds containing the chlorine atoms in the 2,6- or 3,5-positions.[13]

It has been argued[14] that one free *o*-position is an essential requirement for activity, but some very active compounds are known, like 2,4-dichloro-6-fluorophenoxyacetic acid in which all the *o*-positions are substituted. There is therefore some uncertainty regarding the importance of specific nuclear positions on growth-regulating properties; however, *at least one nuclear position* must be unsubstituted.

(e) *Optical isomers.* When an alkyl group is introduced into the side-chain of a phenoxyacetic acid, the resultant chiral carbon atom makes resolution into (+) and (−) optical isomers possible. For instance, with mecoprop (**4b**), the (+) isomer was much more active than the (−) isomer. Generally the *dextro* isomers are more herbicidal and Wain has suggested[14] that this implies that the active stereoisomer is capable of a specific interaction with some chiral component in the plant cell concerned with the production of growth response.

Thimann argued that the activity of auxins like IAA and the benzoic acid derivatives depends on the presence of a fractional positive charge situated 0.55 nm from the negative charge of the carboxyl group. Figure 8.1 illustrates how these criteria are satisfied by IAA, 2,4-D, 2,3,6-trichlorobenzoic acid, and picloram.[15–17] This hypothesis is supported by 2-chloroindole acetic acid which is more active than IAA since the chlorine atom would be expected to increase the positive charge on the nitrogen atom. In 2,4-D the two chlorine atoms increase the positive charge on the six carbon atom which explains why the 2,4,6-trichloro derivative is inactive since now that the 6-position is blocked no combination with the presumed site of action in the plant is possible. In the benzoic acid derivatives, auxin activity is only possible when the fractional positive charge is located at the 4-position to fulfil the spacing requirement. Presumably the correct spacing of the polar centres allows the auxins to bind to active receptor sites in the plant cell.

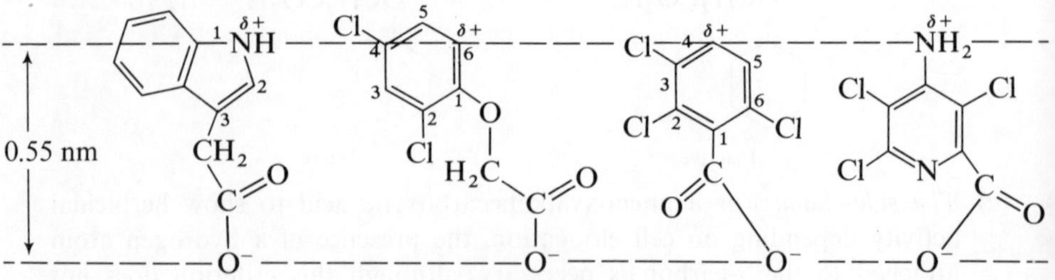

Figure 8.1

Wain and Fawcett (1969) suggested that a three-point contact between auxin and the receptor was required; the concept of a chiral three-dimensional receptor is more realistic than the original view and would account for the inactivity of ($-$) 2,4-dichlorophenoxypropionic acid (**4a**).

Katekar (1979) considered that both the delocalized electrons of the aromatic nucleus and the substituent chlorine atoms may interact directly with a relatively positive area of the auxin binding site.[7]

The phenoxyalkanoic acids and the benzoic acid herbicides act as persistent IAA mimics preventing normal plant growth which depends on the presence of optimum quantities of auxins. Phenoxyacetic acids (e.g. 2,4-D) stimulate RNA and protein synthesis at low concentrations (10^{-7} M) while at higher concentrations (10^{-3} M) oxidative phosphorylation is inhibited. Diclofop-methyl (**5a**) acts by leaf contact when the chlorophyll is reduced; it is a powerful auxin antagonist and can damage plant cell membranes, apparently more so in wild oats than in wheat.[7,9,10]

In 1984, the estimated market size for the phenoxyalkanoic acid herbicides was \$375 million with 2,4-D as the major individual herbicide in this group. Since 1972, the demand for these herbicides has been decreasing, due partly to increasing numbers of resistant weeds and to the growing importance of wild species not controlled by these chemicals.[15]

The major cause of the herbicidal or auxin-like activity of phenoxyacetic acids and the various chlorinated benzoic acid derivatives is their effect on cell division. 2,3,6-Trichlorobenzoic acid or 2,3,6-TBA was originally reported (1954) as a total weed killer in America, but work by Fisons Limited in Britain showed that this compound can function as a translocated selective herbicide against many broad-leaved annual and perennial weeds in cereals.[4,12] The crop is not damaged provided the dosage of the chemical is sufficiently low.[3] Commercially 2,3,6-TBA is obtained from toluene[8] (Scheme 8.4). *o*-Chlorotoluene is separated from the *p* isomer by fractional distillation; further chlorination gives a mixture of isomers containing some 60% of the required 2,3,6-trichlorotoluene which is finally oxidized to the required compound. It was found that 4-chlorobenzoic acids were almost completely inactive, but the 2,6-dichloro- and 2,3,5,6-tetrachlorobenzoic acids had a similar, but less powerful, herbicidal activity to that of 2,3,6-TBA.

2-Methoxy-3,6-dichlorobenzoic acid (dicamba)[8] finds use as a translocated herbicide against many broad-leaved weeds, especially bindweed, chickweed,

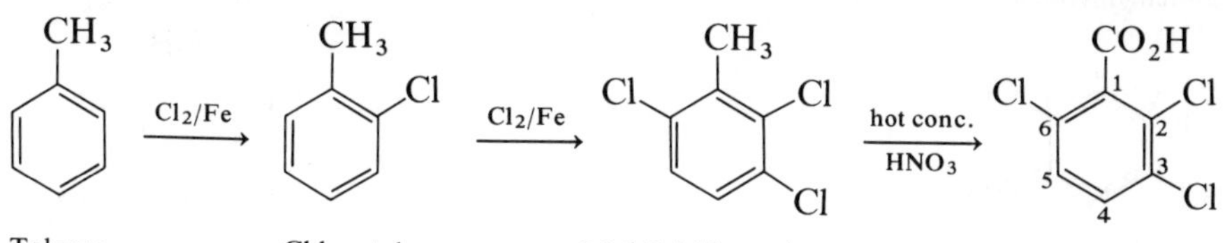

Toluene   *o*-Chlorotoluene   2,3,6-Trichlorotoluene   2,3,6-TBA

Scheme 8.4

and mayweed in cereals,[12] and controls bracken for several years after one application. It is manufactured from 1,2,4-trichlorobenzene (Scheme 8.5). Dicamba has given good postemergence control of triazine-resistant weed species, especially bindweed.

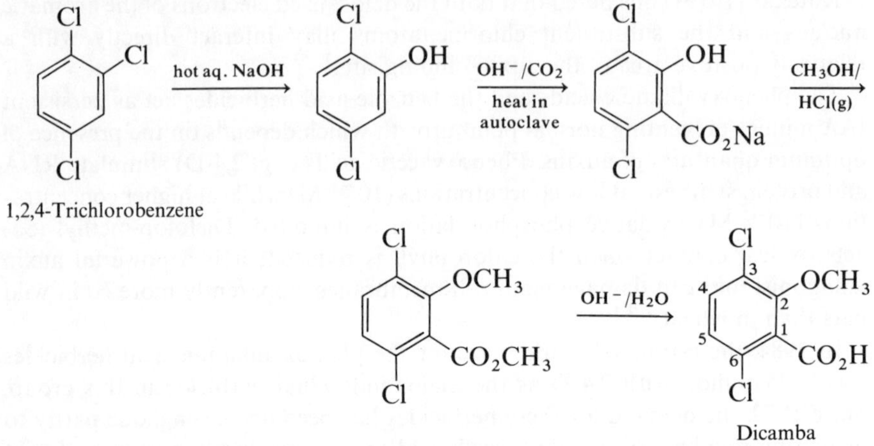

1,2,4-Trichlorobenzene

Dicamba

Scheme 8.5

Picloram (4-amino-2-carboxy-3,5,6-trichloropyridine) is a remarkably stable compound, in spite of the free amino group, and is one of the most persistent herbicides available. Picloram is a translocated herbicide used for control of perennial broad-leaved weeds such as docks and ragwort, and may be effective at concentrations as low as 140 g/ha.[4] It is also the best compound for killing trees by application to the bark at the base of the tree. Picloram is manufactured from 2-methylpyridine (Scheme 8.6).

Clopyralid (2-carboxy-3,6-dichloropyridine) was introduced by the Dow Chemical Company (1975) as a selective postemergence herbicide for weed control in graminaceous and some broad-leaved crops, e.g. brassicas, beet,

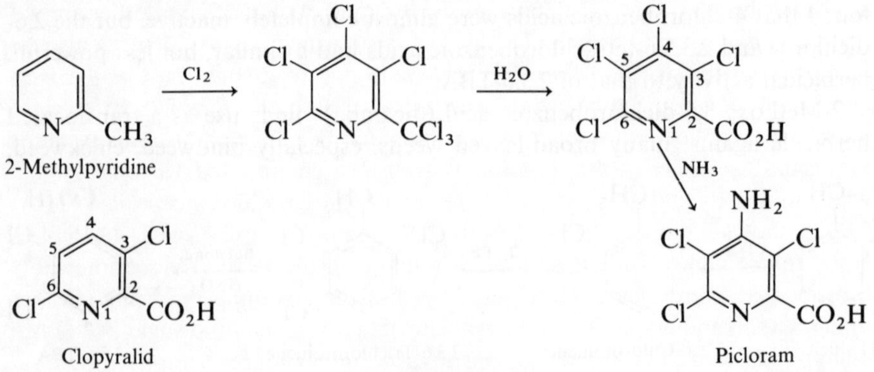

2-Methylpyridine

Clopyralid

Picloram

Scheme 8.6

strawberry, and onion. This is a foliar, translocated herbicide, generally formulated as the monoethanolamine salt, which is readily soluble in water and is sprayed at rates of 50–400 g a.i./ha.[4,13]

In the aromatic carboxylic acids, the optimum positions of chlorine substitution for herbicidal activity depend on the nature of the compounds; in the phenoxyacetic acids the preferred positions of substitution are 2,4- or 2,4,5-, whereas the 2,6-derivatives are not herbicidal. In the benzoic acid derivatives, on the other hand, 2,6-disubstitution appears essential, while 2,3,6-trisubstitution leads to maximum activity, but 4-substitution destroys almost all herbicidal action. These observations are in general agreement with Thimann's theory (see page 224).

Both phenoxyacetic acid and the aromatic carboxylic acid herbicides are hormone-type herbicides and are often active at comparatively low concentrations.[9,10] Phenoxyacetic acid derivatives generally cause contortions of leaf stalks and stems and the roots become stumpy. The benzoic acid herbicides, on the other hand, result in very narrow leaves and buds with extremely brittle stems but with no abnormal root growth. Both types of compound cause swollen stems.[15,16]

CHLOROALIPHATIC ACIDS

A number of chlorinated aliphatic acids have been recognized for some time to be herbicidal. The two most active members of this group are dalapon (sodium 2,2-dichloropropionate) (7) and trichloroacetic acid (TCA). α-Chlorination is an essential requirement for activity, as chlorination in other positions does not lead to active compounds. Replacement of chlorine by other halogens and increasing chain length both reduce activity.

Dalapon (sodium 2,2-dichloropropionate) (7), introduced in 1953, is obtained by direct chlorination of propionic acid by passing chlorine gas into the hot acid until the calculated increase in weight for the replacement of two α-hydrogen atoms has been attained. Dalapon is used for control of couch and other annual and perennial grasses before planting or sowing most crops.[4] It may also be employed for the selective control of grasses in potatoes, sugar beet, and carrot, and to kill sedges and other water weeds. It ceases to affect growth six to eight weeks after application and acts as a contact herbicide which also has systemic properties; in couch grass dalapon is translocated into the rhizome system where it inhibits bud growth. The related compound, trichloroacetic acid (TCA) (1947), shows a similar selective herbicidal action towards grasses but is less active than dalapon, since it is only a soil-acting herbicide and is not appreciably translocated from foliage. However, TCA has the advantage of being cheaper than dalapon and is specially used against wild oats in peas and sugar beet. Both compounds need to be applied in considerably higher concentrations (11–33 kg/ha) as compared with the hormone-type herbicides at 0.36–1.8 kg/ha. Dalapon is also absorbed by foliage, and both compounds are not readily attacked or metabolized and they persist in plants for many weeks.[5]

These chloroalkanoic acids are known to affect many biological processes in plants, including carbohydrate, lipid, and nitrogen metabolism. Both dalapon (7) and TCA cause the precipitation of proteins and modification of protein structure is considered to be the primary mode of action. Dalapon has also been shown to interfere with RNA synthesis.[7,15]

The reactive chlorine atoms may also condense with the sulphydryl groups of enzymes; the presence of the reactive chlorine atoms is certainly essential for herbicidal activity. Thus when the hygroscopic dalapon absorbs water, the resultant hydrolysis causes loss of herbicidal activity:

$$\underset{\underset{Cl}{|}}{\overset{\overset{Cl}{|}}{H_3C-C-CO_2Na}} + H_2O \longrightarrow \underset{\text{Pyruvic acid}}{H_3C-\overset{\overset{O}{||}}{C}-CO_2H} + HCl + NaCl$$

(7)

Hydrolysis at room temperature is very slow but at 50°C it is much faster, so it is essential to store dalapon in moisture-proof containers and to apply dalapon solutions immediately, especially in hot climates.

AROMATIC CARBAMATES AND THIOCARBAMATES

The arylcarbamate herbicides were introduced by Imperial Chemical Industries in Britain (1946) to combat monocotyledonous weeds, such as grasses, in peas and beet which could not be controlled by dinitrophenols or phenoxyacetic acid herbicides. The first compound was *O*-isopropyl-*N*-phenylcarbamate or propham (8; X = H, R = (CH$_3$)$_2$CH—) but, although effective against seedling grasses, it was not much good against established plants and the selectivity towards grasses was not as clear-cut as that shown by TCA or dalapon.[3,7,10] However, later studies of the 3-chloro derivative known as chlorpropham (8; X = Cl, R = (CH$_3$)$_2$CH—) revealed that several crop plants are not harmed by this compound when it is applied to the soil, so it is useful as a soil-acting preemergence herbicide for the control of many seedling weeds or established chickweed in bulbs, fruit trees, rhubarb, peas, and sugar beet,[4,12] and to inhibit sprouting in stored potatoes. Generally chlorpropham is of little value against established weeds, but the acetylenic carbamate (8; X = Cl, R = ClCH$_2$C≡C—CH$_2$—), known as barban, is effective against wild oats, with postemergence control in wheat and barley.[4,12] It was the first compound to clearly discriminate between wild oats and the cereals, but it must be applied at the right stage of crop growth. Later compounds do not demand such precision in timing for their application.

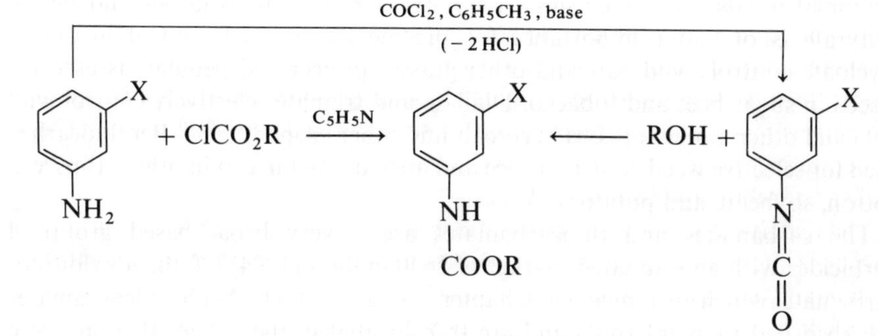

$$COCl_2, C_6H_5CH_3, base$$
$$(-2 HCl)$$

(8)

Scheme 8.7

The N-phenylcarbamates can be prepared by reaction of the aromatic amine and the alkyl chloroformate, or by reaction of the appropriate phenyl isocyanate with the alcohol. The phenyl isocyanate is obtained by reaction of carbonyl chloride (phosgene) with the amine (Scheme 8.7). 3,4-Dichlorophenylcarbamate is used for selective weed control in rice and several other carbamate herbicides have been introduced (1968–78) by Schering AG for the selective postemergence control of broad-leaved weeds (cf. the O-aryl carbamates which are insecticides; Chapter 6, page 142).

Table 8.1

Name	R^1	R^2	R^3
EPTC	n-C_3H_7	n-C_3H_7	C_2H_5
Butylate	$CH_2CH(CH_3)_2$	$CH_2CH(CH_3)_2$	C_2H_5
Cycloate	C_2H_5	Cyclohexyl	C_2H_5
Pebulate	C_2H_5	n-C_3H_7	n-C_3H_7
Vernolate	n-C_3H_7	n-C_3H_7	n-C_3H_7
Diallate	$CH(CH_3)_2$	$CH(CH_3)_2$	CH_2—CCl=$CHCl$
Triallate	$CH(CH_3)_2$	$CH(CH_3)_2$	CH_2—CCl=CCl_2
Benthiocarb	C_2H_5	C_2H_5	$PhCH_2$

Several thiocarbamates **(9)** are herbicidal (Table 8.1); they are synthesized by reaction of the appropriate amine with phosgene:[13]

$$R^1R^2NH + COCl_2 \xrightarrow{\text{base}} R^1R^2NCOCl \xrightarrow[(-NaCl)]{R^3SNa} R^1R^2N-COSR^3$$

(9)

Thiocarbamates have low aqueous solubility combined with good volatility and are valuable as soil-applied herbicides to control germinating broad-leaved and grass weeds; they are the most effective herbicides against nutsedge (*Cyperus rotundus*), a serious weed problem. EPTC (1954) controls annual and perennial weeds, e.g. couch grass (*Agropyron repens*) in potatoes and beans. Butyrate is of major importance for preemergence weed control in maize. Cycloate controls wild oats and other grasses in beet and pebulate is used for weeds in sugar beet and tobacco. Diallate and triallate selectively control wild oats and other annual grasses in cereals and other crops.[3,4,10,13] Benthiocarb is used for selective weed control in rice and appears useful also in other crops, e.g. cotton, soybean, and potatoes.[13]

The carbamates and thiocarbamates are a very broad-based group of herbicides with an estimated market of $640 million (1984) (cf. the alkyldithio-carbamates which are fungicides; Chapter 7, page 163). The *N*-phenylcarbamates are absorbed by plant roots and are translocated in the xylem; they interfere with photosynthesis and affect meristematic activity causing deformation of roots, shoots, and buds. They inhibit cell division or mitosis and the primary mode of action is probably the disorganization of the microtubular assembly.[7,15]

There is much evidence demonstrating that thiocarbamates reduce the production of cuticular wax by plants and interference with lipid biosynthesis probably represents the primary mode of action of this group of herbicides.

The sulphonyl carbamate asulam, introduced by May and Baker Limited in 1965 for the control of docks in grassland and against bracken,[4] is believed,[7] like the *N*-phenylcarbamates, to act by inhibition of cell division since the shoots of treated plants do not behave normally. Asulam is prepared from aniline according to Scheme 8.8.

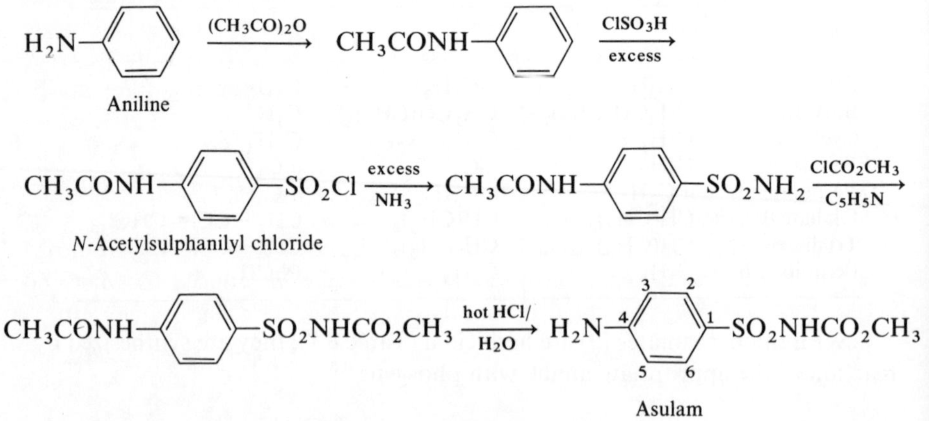

$$H_2N-\langle\text{benzene}\rangle \xrightarrow{(CH_3CO)_2O} CH_3CONH-\langle\text{benzene}\rangle \xrightarrow[\text{excess}]{ClSO_3H}$$

Aniline

$$CH_3CONH-\langle\text{benzene}\rangle-SO_2Cl \xrightarrow[NH_3]{\text{excess}} CH_3CONH-\langle\text{benzene}\rangle-SO_2NH_2 \xrightarrow[C_5H_5N]{ClCO_2CH_3}$$

N-Acetylsulphanilyl chloride

$$CH_3CONH-\langle\text{benzene}\rangle-SO_2NHCO_2CH_3 \xrightarrow[H_2O]{\text{hot HCl/}} H_2N-\langle\text{benzene}\rangle-SO_2NHCO_2CH_3$$

Asulam

Scheme 8.8

Asulam is used at 1–7 kg/ha for selective and total weed control. Sugar beet is comparatively tolerant to the herbicide. Derivatives containing CH_3, CH_3NH, CH_3CONH, or NO_2 groups in the 4-position are also herbicidal. The 4-nitro analogue, nisulam, has a similar herbicidal spectrum to asulam, but is safer for use on dicotyledonous crops (e.g. alfalfa, peas, and potatoes).[13]

NITRILES

Dichlobenil (2,6-dichlorobenzonitrile) was introduced as a soil-acting herbicide in 1960.[8,13] It can be prepared from 2,6-dichlorotoluene (Scheme 8.9).

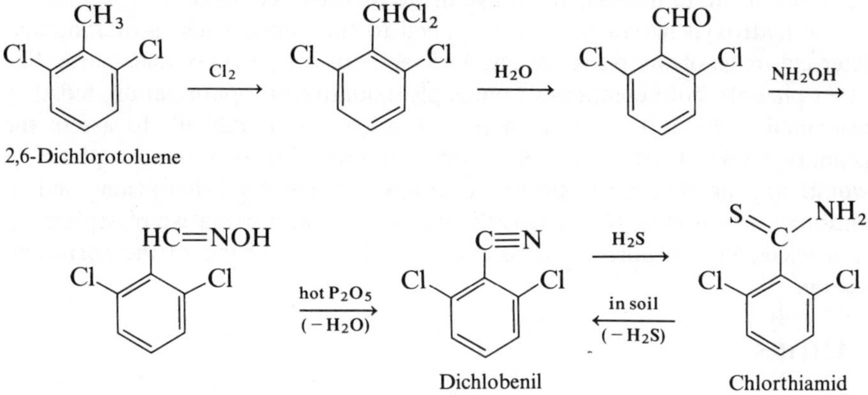

2,6-Dichlorotoluene

Dichlobenil   Chlorthiamid

Scheme 8.9

Dichlobenil is generally applied to the soil as granules and prevents the germination of annual weeds, the shoot growth of perennials, e.g. bracken, and controls aquatic weeds.[15] It is absorbed by the plant roots and is transported in the xylem. Chlorthiamid is obtained from dichlobenil by reaction with hydrogen sulphide and is converted into dichlobenil in soil. It was introduced by Shell (1963) as a total herbicide and for selective weed control in soft fruit; it has more residual action than dichlobenil. Dichlobenil is not apparently hydrolysed to the corresponding carboxylic acid *in vivo* and does not function as an auxin-type herbicide like 2,3,6-TBA. It inhibits actively dividing plant tissue and evidence indicates that the herbicidal action is due to the inhibition of cellulose biosynthesis because the compound prevents the incorporation of glucose into the cell wall glucans.[7]

Several hydroxynitriles are herbicidal; two well-known examples are bromoxynil (**10**; X = Br) and ioxynil (**10**; X = I). These were introduced by May and Baker (1963) and are prepared from *p*-hydroxybenzaldehyde (Scheme 8.10).

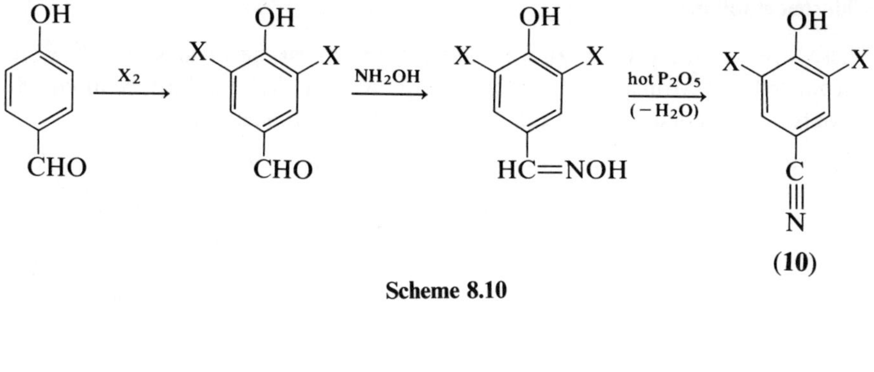

(**10**)

Scheme 8.10

Both compounds are foliage-applied herbicides which selectively control dicotyledonous weeds in cereals; as the octanoate esters they are valuable against such problem weeds as chickweed, bindweed, and *Veronica* spp.[4] They are often formulated as mixtures with other herbicides, e.g. MCPA, dichlorprop, or isoproturon, to increase the range of weed species controlled.

The hydroxybenzonitriles when applied to the foliage cause scorching and later chlorosis of the plant tissues. The mode of action is complex since, like other phenols, both compounds uncouple oxidative phosphorylation, but they also inhibit the Hill reaction in photosynthesis. It is difficult to assess the primary mode of action, but from observations of their effects on plants, it would appear that both effects contribute to the total herbicidal action, although with ioxynil (**10**; X = I) the interference with oxidative phosphorylation (electron-transport chain) is probably the major basis of the herbicidal action.[7,13]

AMIDES

Acylanilides

Herbicidal activity was observed in chloroanilides; the most successful examples were 3,4-dichloropropionanilide (propanil) (**11**; R^1 = Cl, R^2 = C_2H_5) and pentanochlor (**11**; R^1 = CH_3, R^2 = $CH(CH_3)(CH_2)_2CH_3$). These are contact herbicides used for preemergence control of annual grass and broad-leaved weeds in rice and a wide variety of horticultural crops.[4] They are prepared by condensation of the appropriate acid chloride with the amine:

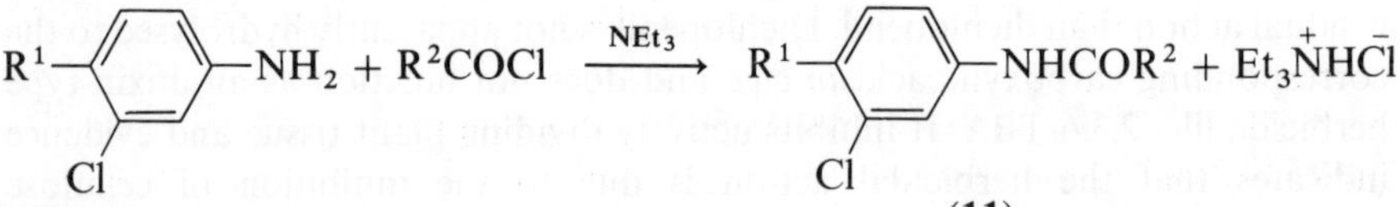

$$R^1-\!\!\!\left\langle\!\!\bigcirc\!\!\right\rangle\!\!-NH_2 + R^2COCl \xrightarrow{\text{NEt}_3} R^1-\!\!\!\left\langle\!\!\bigcirc\!\!\right\rangle\!\!-NHCOR^2 + Et_3\overset{+}{N}H\overset{-}{Cl}$$

$$\text{Cl} \qquad\qquad\qquad \text{Cl}$$
$$(\textbf{11})$$

The acylanilides owe their activity to inhibition of the Hill reaction in photosynthetic electron transport.[7,17] Selectivity depends on much more rapid degradation in the crop plants as compared with the grass weeds.

Chloroacetanilides

These are an interesting and important group of herbicides which were introduced by Monsanto. Allidochlor or *N,N'*-diallyl-2-chloroacetamide is prepared by condensation of diallylamine with chloroacetyl chloride:

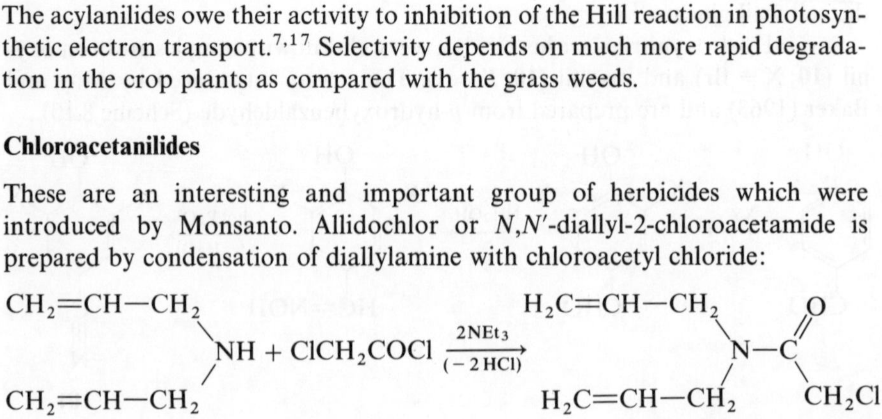

$$\begin{array}{cc}
CH_2\!=\!CH\!-\!CH_2 & H_2C\!=\!CH\!-\!CH_2 \qquad O \\
\qquad\qquad\diagdown & \qquad\qquad\qquad\diagdown \quad \diagup\!\!\diagup \\
\qquad\qquad NH + ClCH_2COCl \xrightarrow[(-2\,HCl)]{2NEt_3} & \qquad\qquad N\!-\!C \\
\qquad\qquad\diagup & \qquad\qquad\qquad\diagup \qquad \diagdown \\
CH_2\!=\!CH\!-\!CH_2 & H_2C\!=\!CH\!-\!CH_2 \qquad CH_2Cl
\end{array}$$

Allidochlor

Allidochlor is a selective preemergence herbicide applied at 4–5 kg a.i./ha to control annual grass weeds and some broad-leaved weeds. It is selective in several crops, e.g. maize, millet, soybeans, sugar cane, and vegetables.[13] It acts on grasses by inhibition of certain enzymes, protein synthesis, and respiration. Allidochlor is readily absorbed by the roots of plants, translocated and rapidly degraded as shown in Scheme 8.11.[16]

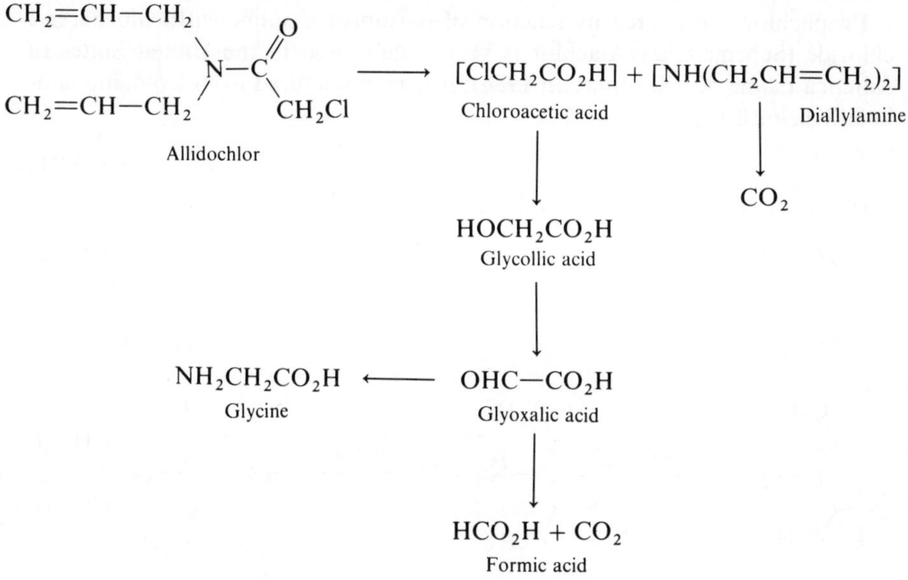

$$CH_2{=}CH{-}CH_2 \diagdown \atop CH_2{=}CH{-}CH_2 \diagup N{-}C \underset{CH_2Cl}{\overset{O}{}} \longrightarrow [ClCH_2CO_2H] + [NH(CH_2CH{=}CH_2)_2]$$

Allidochlor

Chloroacetic acid   Diallylamine

$$HOCH_2CO_2H$$
Glycollic acid

$$CO_2$$

$$NH_2CH_2CO_2H \longleftarrow OHC{-}CO_2H$$
Glycine   Glyoxalic acid

$$HCO_2H + CO_2$$
Formic acid

Scheme 8.11

Some other important herbicidal chloroacetamides are shown in Table 8.2; these have the following general structure:

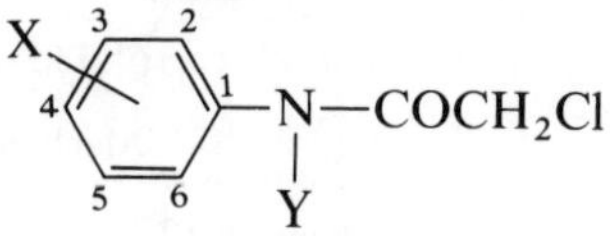

$$X \diagdown \underset{5 \quad 6}{\overset{3 \quad 2}{\bigcirc}} {-}N{-}COCH_2Cl \atop Y$$

Table 8.2

Name	X	Y
Propachlor	H	$CH(CH_3)_2$
Alachlor	$2,6\text{-}(C_2H_5)_2$	CH_2OCH_3
Butachlor	$2,6\text{-}(C_2H_5)_2$	$CH_2OC_4H_9$
Pretilachlor	$2,6\text{-}(C_2H_5)_2$	$(CH_2)_2OC_3H_7$
Diethatyl ether	$2,6\text{-}(C_2H_5)_2$	$CH_2CO_2C_2H_5$
Metolachlor	$2\text{-}CH_3, 6\text{-}C_2H_5$	$CH(CH_3)CH_2OCH_3$
Metazachlor	H	$CH(CH_3)C{\equiv}CH$

The group is of considerable commercial utility: propachlor, butachlor, and metolachlor are used as preemergence herbicides against annual broad-leaved weeds and grasses in many crops, e.g. maize, groundnuts, soybeans, sugar cane, and cotton. Alachlor kills grass weeds in the same crops, but requires more moisture than propachlor or metolachlor to exert its full effect. Pretilachlor has been developed for weed control in transplanted rice. Metazachlor is used as a selective herbicide in brassicas, rape, onion, and soybean (Plate 17).

Propachlor is prepared by reaction of *N*-isopropylaniline with chloroacetyl chloride (Scheme 8.12). Alachlor is very widely used in the United States of America (some 50 000 tonnes in 1986); it is manufactured from 2,6-diethylaniline (Scheme 8.12).

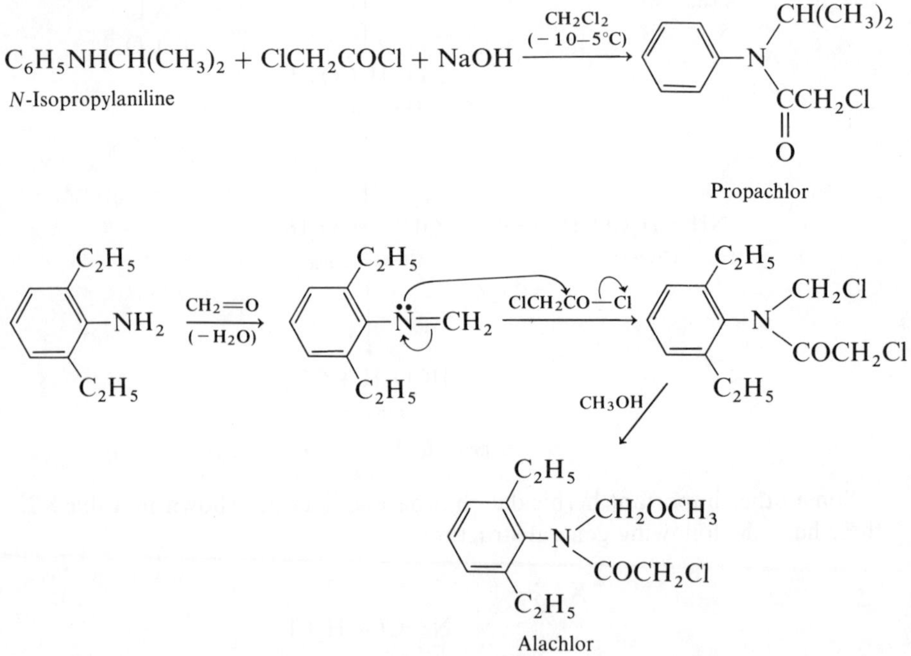

$$C_6H_5NHCH(CH_3)_2 + ClCH_2COCl + NaOH \xrightarrow[(-10-5^\circ C)]{CH_2Cl_2}$$

N-Isopropylaniline

Propachlor

Alachlor

Scheme 8.12

The herbicidal action of the chloroacetanilides probably involves the reaction of the chlorine atom of the α-chloromethyl group with vital cellular components such as sulphydryl and amino groups.[7] Such reactions result in inhibition of cell division and protein synthesis, causing wilting and death of susceptible plants.[13]

Benzamides

In the period 1969–74, Shell Research introduced three herbicides, benzoylprop-ethyl (**12a**), flamprop-isopropyl (**12b**), and methyl (**12c**) for the postemergence control of wild oats in cereals and some other crops (e.g. sugar beet, beans, mustard)[4,15] (Plate 18). These benzamides provide more reliable control of wild oats than barban. Wild oats are a serious weed problem in cereals, because the increasing extent of cereal monoculture encourages wild oat infestation, as did the widespread application of the phenoxyacetic acid herbicides which removed most of the competing broad-leaved weeds.

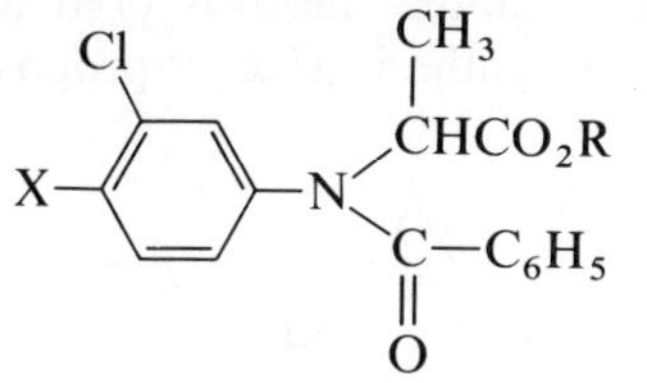

(**12a**) X = Cl, R = C_2H_5
(**12b**) X = F, R = $CH(CH_3)_2$
(**12c**) X = F, R = CH_3

The compounds owe their activity to conversion to the parent acids; their selectivity depends on the rate of hydrolysis relative to the rate of subsequent detoxification. The hydrolysis in wheat and barley is slower than in wild oats so the herbicidally active free acids can be removed as inactive conjugates before a sufficient concentration to damage the plant has accumulated. However, the primary mode of action of these benzamides is unknown.[7]

Other amides

These are soil-applied herbicides, especially active against germinating grasses. Naptalam (*N*-1-naphthylphthalamic acid) (1950) and diphenamid (*N*,*N*-dimethyldiphenylacetamide) are used for preemergence weed control in a range of crops.[13,15] Propyzamide (**13**) and butam (**14**) were introduced in 1965 for preemergence control of grass weeds; the latter is prepared from the methylacetic acid (Scheme 8.13).

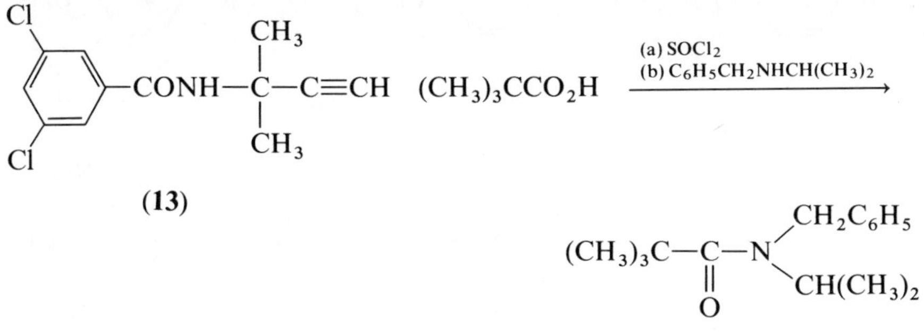
Cl
CH_3
CONH—C—C≡CH
CH_3
Cl
$(CH_3)_3CCO_2H$
(a) $SOCl_2$
(b) $C_6H_5CH_2NHCH(CH_3)_2$
$(CH_3)_3C$—C—N
O
$CH_2C_6H_5$
$CH(CH_3)_2$

(**13**)

(**14**)

Scheme 8.13

More recently May and Baker (1985) produced diflufenican (**15**) as a persistent herbicide to control difficult broad-leaved and grass weeds in cereals at doses of 50–250 g/ha.[15,18] Various modes of action have been identified for different members of this group of herbicides. Naptalam interferes with cell growth and development by affecting auxin transport, while the mode of action of butam and diphenamid is unknown. Propylzamide interferes with cell division[7] and diflufenican causes chlorosis of treated plants resulting from inhibition of carotenoid biosynthesis which is proposed as the primary mode of action.[18]

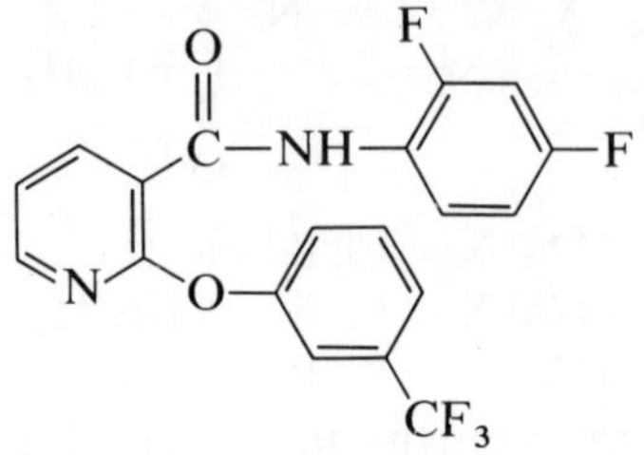

(**15**)

DINITROANILINES

The herbicidal properties of the dinitrophenols led to examination of dinitroanilines for herbicidal activity. In this series, the positions of the nitro groups were critical; activity was only associated with the 2,6-dinitro derivatives. The active dinitroanilines are used as soil-incorporated herbicides which are particularly effective against germinating grasses and may damage the buds of rhizomatous weeds.[15]

Several of these compounds (**16**) can be synthesized from toluene (Scheme 8.14). Some illustrative examples (**16a–e**) are shown in Table 8.3.

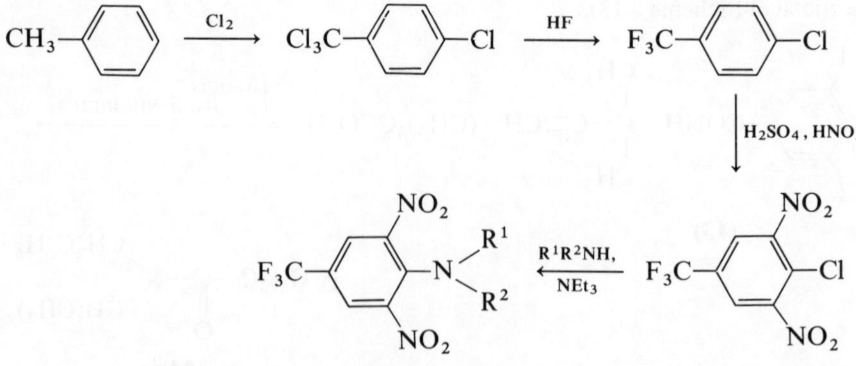

(**16**)

Scheme 8.14

Table 8.3

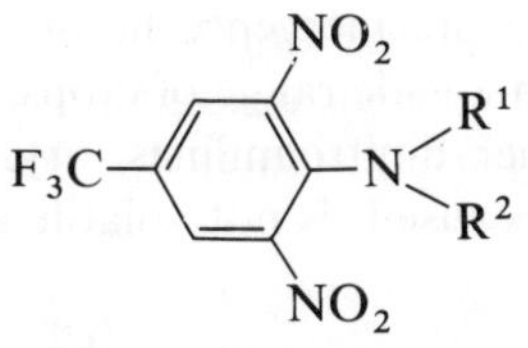

General structure (16)

		R^1	R^2
(16a)	Trifluralin	C_3H_7	C_3H_7
(16b)	Benfluralin	C_2H_5	C_4H_9
(16c)	Profluralin	C_3H_7	$CH_2CH(CH_3)_2$
(16d)	Ethalfluralin	C_2H_5	$CH_2C(CH_3){=}CH_2$
(16e)	Fluchloralin	C_3H_7	CH_2CH_2Cl

The first member of the trifluoromethyl series (Table 8.3) was trifluralin (**16a**) which was introduced by Eli Lilly in 1960; this very important herbicide is used for the control of annual grass and broad-leaved weeds in a wide range of crops, e.g. cotton, groundnuts, soybeans, brassicas, beans, and cereals.[4,15] When formulated with the urea isoproturon, trifluralin (**16a**) is specially valuable for weed control in winter cereals. Some 25 000 tonnes per annum are currently manufactured in the United States of America.

The presence of the trifluoromethyl group is not essential for herbicidal activity and variations in the substituents in the 3,4-positions can cause subtle changes in the selectivity and degrees of activity against specific weed species. Consequently a wide range of compounds has been developed; some illustrative examples (**17a–e**) are shown in Table 8.4.

Table 8.4

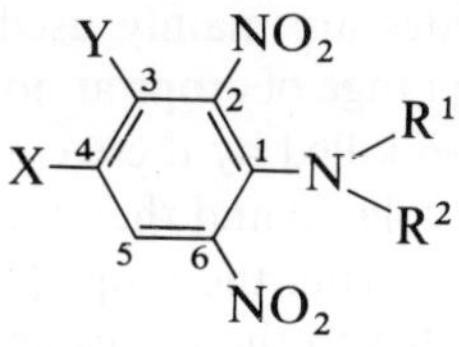

General structure (17)

		X	Y	R^1	R^2
(17a)	Oryzalin	SO_2NH_2	H	C_3H_7	C_3H_7
(17b)	Nitralin	SO_2CH_3	H	C_3H_7	C_3H_7
(17c)	Butralin	$CH(CH_3)_2$	H	H	$CH(CH_3)C_2H_5$
(17d)	Pendimethalin	CH_3	CH_3	H	$CH(C_2H_5)_2$
(17e)	Dipropalin	CH_3	H	C_3H_7	C_3H_7

Oryzalin **(17a)** can be obtained from chlorobenzene as shown in Scheme 8.15.[13] Oryzalin is a preemergence herbicide effective against mono- and dicotyledonous weeds in a wide range of crops, e.g. soybean, rice, tobacco, and cotton. Unlike the other dinitroanilines, oryzalin does not need to be incorporated into the soil because it is not volatile and is stable in sunlight.

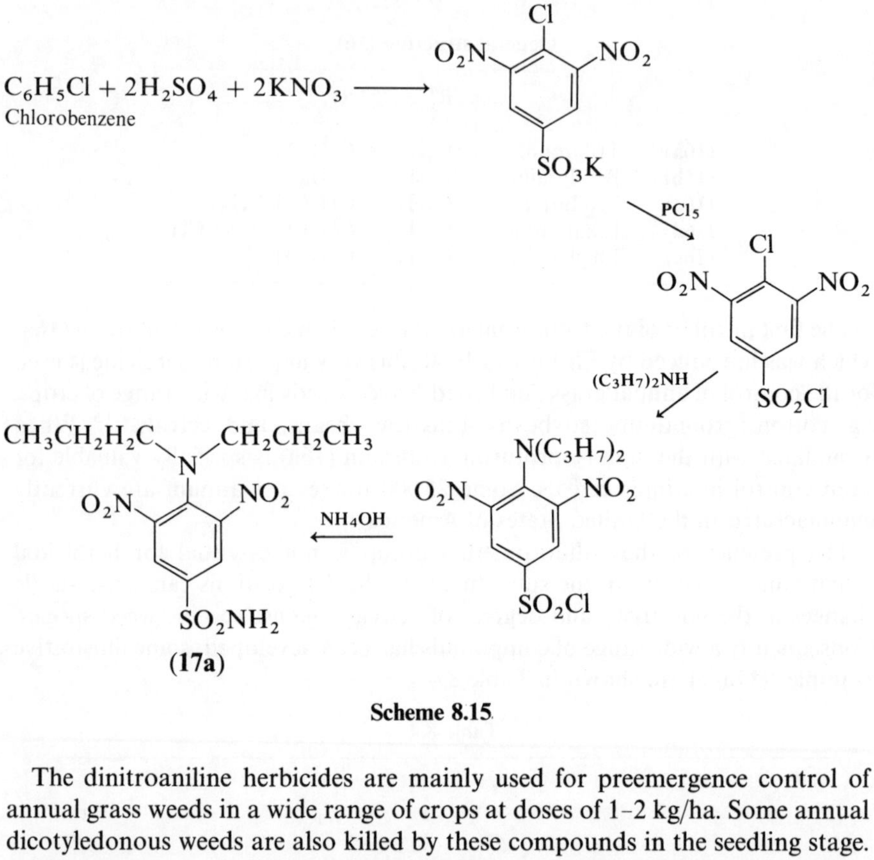

$$C_6H_5Cl + 2H_2SO_4 + 2KNO_3 \longrightarrow$$
Chlorobenzene

Scheme 8.15

The dinitroaniline herbicides are mainly used for preemergence control of annual grass weeds in a wide range of crops at doses of 1–2 kg/ha. Some annual dicotyledonous weeds are also killed by these compounds in the seedling stage.

The compounds are fairly volatile and they are applied by soil incorporation to a depth of 6–8 cm before sowing the crop. The herbicides are absorbed by plant roots and are translocated to the leaves and shoots; however, in dry soil conditions, their action is poor due to low aqueous solubility. Pendimethalin **(17d)** was developed by American Cyanamid (1974) for control of 'difficult' weeds, e.g. *Gallium aparine*, *Veronica*, and *Viola* spp. in many crops.[15]

The dinitroanilines are known[13,15] to interfere with photosynthesis and respiration; however, the primary mode of action is considered to be their effect on plant cell tubulin, preventing its assembly into microtubules.[17] Trifluralin **(16a)** at 1 μM caused 90–100% inhibition of cell division in algae. These compounds therefore act as mitotic poisons.

ARYLUREAS

These are an important group of some 30 soil-acting herbicides[3] (Table 8.5). Du Pont initially discovered the herbicidal properties of monuron in 1952. The aryl ureas are manufactured from the appropriate arylamine (Scheme 8.16).

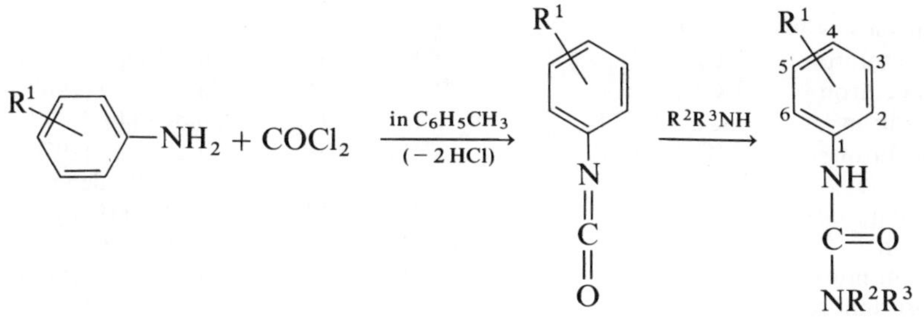

Scheme 8.16

The arylureas (18) (Table 8.5) are persistent herbicides; for instance one application of diuron to the soil may prevent weed germination for up to one year. At high rates of application, they are useful as total weed killers but at low rates many can be used for selective weed control in a wide range of crops. Individual compounds display different activities; linuron has good contact activity and hence will kill emergent weed seedlings.

Monolinuron is used as a preemergence herbicide to control broad-leaved weeds and annual grasses in vegetables, orchards, and vineyards.

Chlorotoluron, isoproturon, and difenoxuron are used at 1–2 kg/ha for selective control of grass and broad-leaved weeds in cereals (Plate 19); the latter is also used in onions. The heterocyclic ureas thiazafluron, terbuthiuron, and ethidimuron are used as total weed killers; they are especially good against woody plants. Fluometuron is valuable for weed control in cotton and S3552 is effective against dicotyledonous weeds in soybeans.[10,13,15]

The arylurea herbicides were estimated to have a market size of $550 million in 1984; slow but steady growth has occurred since and this is expected to continue.[15] The most important commercial ureas are linuron, diuron, and fluometuron.

The herbicidal action of the arylureas is due to inhibition of the Hill reaction in photosynthetic electron transport with consequent inhibition of ATP and $NADPH_2$ formation.[7,15] This results in irreversible damage to photosynthetic processes and a permanent lack of food production. The majority of the active compounds contained small 1-alkyl or alkoxy groups, e.g. 1,1-dimethyl-3-phenylureas; the introduction of a chlorine atom into the *para* position of the phenyl ring often enhanced activity[10] (see Table 8.5).

Ureas are easily absorbed from the soil by the plant roots and are rapidly translocated to the stems and leaves in the transpiration stream. In soil, diuron

Table 8.5 The structures of some arylurea herbicides

Name	R^1	R^2	R^3	Company
Fenuron	H	CH_3	CH_3	Du Pont (1957)
Monuron	4-Cl	CH_3	CH_3	Du Pont (1952)
Diuron	3,4-Cl_2	CH_3	CH_3	Du Pont (1954)
Neburon	3,4-Cl_2	CH_3	C_4H_9	Du Pont (1957)
Chlorotoluron	3-Cl, 4-CH_3	CH_3	CH_3	Ciba-Geigy (1969)
Fluometuron	3-CF_3	CH_3	CH_3	Ciba-Geigy (1960)
Metoxuron	3-Cl, 4-OCH_3	CH_3	CH_3	Sandoz (1968)
Isoproturon	4-$(CH_3)_2CH$	CH_3	CH_3	Ciba-Geigy (1972) Rhône-Poulenc (1972)
Monolinuron	4-Cl	CH_3	OCH_3	Hoechst (1958)
Linuron	3,4-Cl_2	CH_3	OCH_3	Du Pont (1960)
Metobromuron	4-Br	CH_3	OCH_3	Ciba-Geigy (1963)
Chlorbromuron	3-Cl, 4-Br	CH_3	OCH_3	Ciba-Geigy (1961)
Chloroxuron	4-ClC_6H_4O—	CH_3	CH_3	Ciba-Geigy (1961)
Difenoxuron	4-$MeOC_6H_4O$—	CH_3	CH_3	Ciba-Geigy (1970)
S3552	4-$MeC_6H_4(CH_2)_2O$—	CH_3	OCH_3	Sumitomo (1980)

Heterocyclic ureas

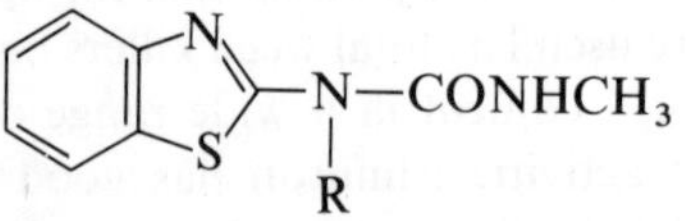

Benthiazuron (R = H) (Bayer AG, 1966)

Methabenthiazuron (R = CH_3) (Bayer AG, 1968)

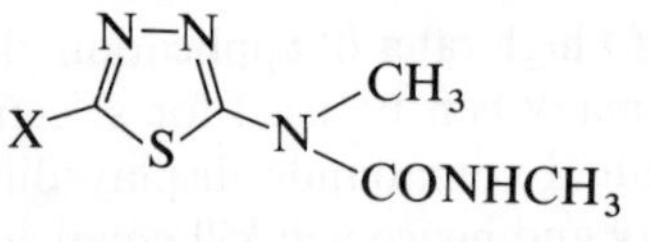

Thiazafluron (X = CF_3) (Ciba-Geigy, 1972)

Tebuthiuron (X = $(CH_3)_3C$) (Eli Lilly, 1974)

Ethidimuron (X = $C_2H_5SO_2$) (Bayer AG, 1975)

Isocarbamid (Bayer AG, 1973)

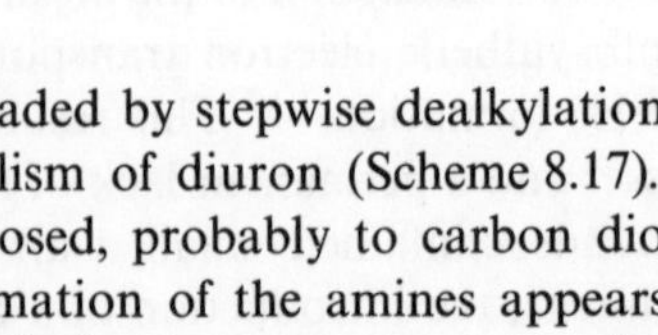

Methazole (Fisons, 1968)

and other ureas are microbiologically degraded by stepwise dealkylation and deamination, as illustrated by the metabolism of diuron (Scheme 8.17). The aromatic chloroamines are finally decomposed, probably to carbon dioxide, ammonia, and halogen. In plants, the formation of the amines appears less common, although the preliminary N-dealkylation steps occur to give the monoalkylurea, and it is concluded that this is either oxidized or becomes conjugated to carbohydrates.[13,16]

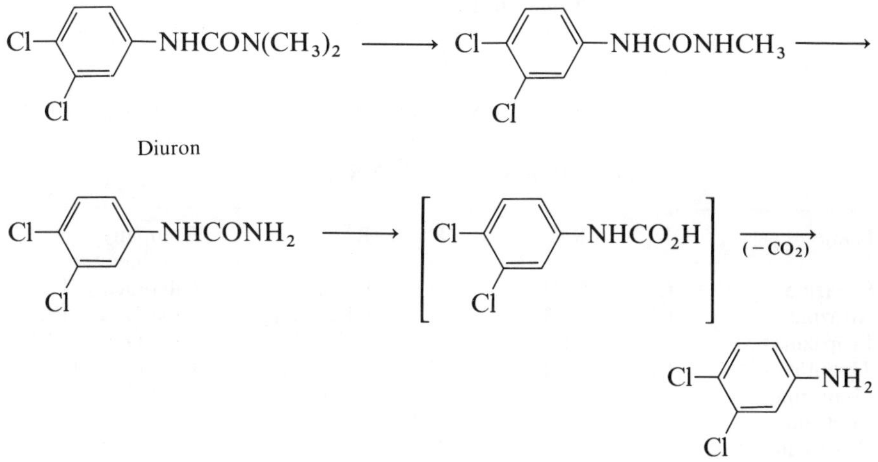

Scheme 8.17

TRIAZINE HERBICIDES

It is convenient to deal with these herbicides directly after the arylureas as they share the same mode of action and rather similar herbicidal properties. The herbicidal substituted s-triazines were discovered by the Swiss firm J. R. Geigy Limited in 1952.[10] Two well-known examples are simazine and atrazine (Table 8.6). Like the ureas, these are persistent soil-acting herbicides which can be applied in large concentrations (5–20 kg/ha) as total weed killers in industrial sites or on paths, etc., but in lower concentrations (1–4 kg/ha) can be used for the selective control of many germinating weeds in a variety of crops especially maize, sugar cane, pineapple, and sorghum, and around fruit bushes[4,8,15] (Plate 20). They are taken up by the roots of the emergent weed seedlings, causing them to turn yellow and die, but owing to their low aqueous solubility they do not appreciably penetrate to lower levels of the soil and consequently they have little effect on deep-rooted crops, such as tree and bush fruits (see Chapter 2, page 31). Because of low aqueous solubility, e.g. simazine with 3.5 mg/l at 20°C, they do not penetrate below about 5 cm in soil. Compounds like atrazine and simazine are widely used for the long-term control of annual grass and broad-leaved weeds in crops such as citrus, coffee, tea, and cocoa. Prometryne is used for selective pre- and postemergence control of annual weeds in vegetables, cotton, and sunflowers. Desmetryn provides postemergence control of annual weeds including fat-hen in brassicas. Methoprotryne, terbutryn, cyanazine, and eglinazine ethyl provide effective control of grass weeds in cereals. Cyanazine also controls a wide spectrum of broad-leaved weeds in many crops, e.g. maize, cereals, onions, peas, potatoes, sugar cane,

Table 8.6 Triazine herbicides

$$\begin{array}{c} X \\ \underset{1}{N} \underset{2}{\overset{}{\diagup}} \underset{3}{N} \\ R^1HN \underset{6}{\diagup} \underset{5}{N} \underset{4}{\diagup} NR^2R^3 \end{array}$$

Name	X	R^1	R^2	R^3	Company
Simazine	Cl	C_2H_5	H	C_2H_5	Ciba-Geigy (1956)
Atrazine	Cl	C_2H_5	H	$CH(CH_3)_2$	Ciba-Geigy (1958)
Propazine	Cl	$CH(CH_3)_2$	H	$CH(CH_3)_2$	Ciba-Geigy (1960)
Terbuthylazine	Cl	C_2H_5	H	$C(CH_3)_3$	Ciba-Geigy (1966)
Cyanazine	Cl	C_2H_5	H	$C(CH_3)_2CN$	Shell (1971)
Trietazine	Cl	C_2H_5	C_2H_5	C_2H_5	Fisons (1972)
Procyazine	Cl	Cyclopropyl	H	$C(CH_3)_2CN$	Ciba-Geigy (1972)
Eglinazine ethyl	Cl	C_2H_5	H	$CH_2CO_2C_2H_5$	Nitrokemia (1972)
Proglinazine ethyl	Cl	$(CH_3)_2CH$	H	$CH_2CO_2C_2H_5$	Nitrokemia (1972)
Terbumeton	OCH_3	C_2H_5	H	$C(CH_3)_3$	Ciba-Geigy (1966)
Secbumeton	OCH_3	C_2H_5	H	$CH(CH_3)C_2H_5$	Ciba-Geigy (1966)
Prometon	OCH_3	$CH(CH_3)_2$	H	$CH(CH_3)_2$	Ciba-Geigy (1959)
Desmetryn	SCH_3	CH_3	H	$CH(CH_3)_2$	Ciba-Geigy (1964)
Ametryne	SCH_3	C_2H_5	H	$CH(CH_3)_2$	Ciba-Geigy (1964)
Simetryne	SCH_3	C_2H_5	H	C_2H_5	Ciba-Geigy (1964)
Terbutryn	SCH_3	C_2H_5	H	$C(CH_3)_3$	Ciba-Geigy (1966)
Prometryne	SCH_3	$CH(CH_3)_2$	H	$CH(CH_3)_2$	Ciba-Geigy (1962)
Methoprotryne	SCH_3	$CH(CH_3)_2$	H	$(CH_2)_3OCH_3$	Ciba-Geigy (1964)
Aziprotryne	N_3	SCH_3		$CH(CH_3)_2$	Ciba-Geigy (1967)

Two unsymmetrical triazines

$$\begin{array}{c} NH_2 \\ \diagdown \\ CH_3S \underset{3}{\overset{4}{-}}N \overset{5}{-} O \\ \underset{2}{N} \underset{1}{-} N \overset{6}{-} C(CH_3)_3 \end{array}$$

Metribuzin
(Bayer AG, Du Pont, 1971)

$$\begin{array}{c} H_2N \\ CH_3 \underset{3}{\overset{4}{-}}N \overset{5}{-} O \\ \underset{2}{N} \underset{1}{-} N \overset{6}{-} C_6H_5 \end{array}$$

Metamitron
(Bayer AG, 1975)

soybeans, and cotton. Aziprotryne is used for weed control in a wide range of vegetables and acts both through the soil and the foliage.[4,10,15]

Two unsymmetrical 1,2,4-triazines metribuzin and metamitron (Table 8.6) have been developed; the former is used for pre- and postemergence weed control in potatoes, tomatoes, lucerne, and raspberry and the latter against weeds in beet.

Altogether some 25 triazines are commercial herbicides, by far the most important being atrazine; at one period this was made on a larger scale than any other herbicide (45 000 tonnes per annum in the United States).

Many varieties of maize and sugar cane are resistant to the herbicidal action of triazines, such as atrazine (**19**; R = C_2H_5, R′ = $(CH_3)_2CH—$) and simazine (**19**; R = R′ = C_2H_5) because these plants contain an enzyme that detoxifies the compounds by hydrolysis in the plant tissues (Scheme 8.18). The products of this enzymic hydrolysis are not herbicidal, so these triazines are very valuable for selective weed control in these crops.[10,15] Owing to their relatively low aqueous solubility, atrazine and simazine are more effective when they are applied to damp soil. Simazine and propazine have the most persistent action; when applied at 10–15 kg/ha the residual life is two to three years.[13]

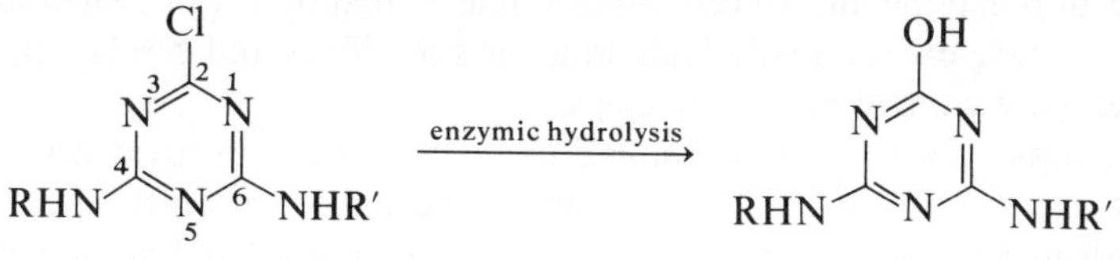

(19)

Scheme 8.18

The triazine herbicides are obtained by reaction of cyanuric chloride with the appropriate nucleophilic reagents. The chlorine atoms may be successively replaced, since, as each is substituted, further replacement of the remaining chlorines becomes progressively more difficult.[3] The general synthesis of triazines from chlorine and sodium cyanide is illustrated in Scheme 8.19.

$$Cl_2 + NaCN \longrightarrow CNCl + NaCl$$

Chlorine Sodium cyanide Cyanogen chloride

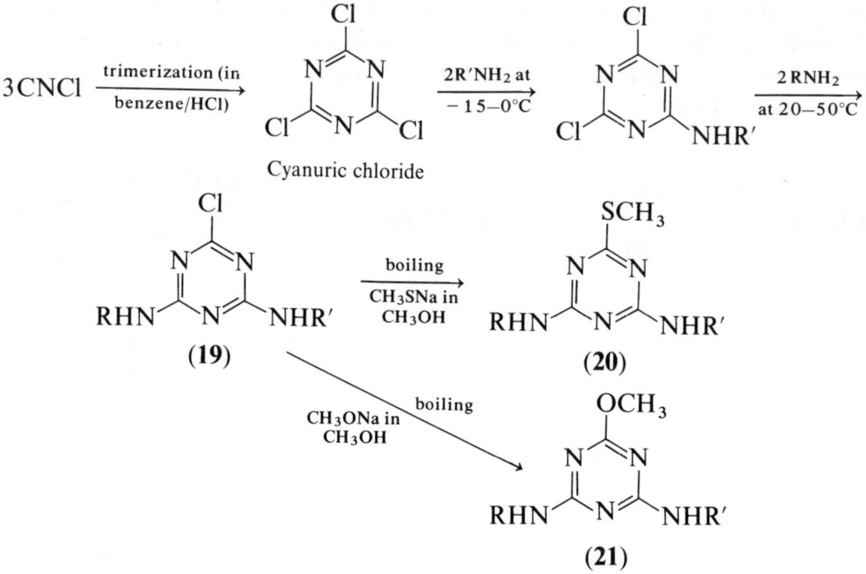

Scheme 8.19

It was also possible to obtain useful herbicidal s-triazines by replacing the 2-chlorine atom of compound (**19**) by the methoxy group; an example is prometon which can be used as a total weed killer. Attempts to substitute the 2-chlorine with groups other than methoxy and methylthio did not generally lead to effective herbicides. However, considerable variations in the *N*-alkylamino side-chains are possible and these extend the range of persistency and biological activity. For instance, terbuthylazine is more persistent than simazine and is used as a preemergence herbicide in peas, potatoes, vineyards, orchards, and for selective weed control in millet.[13] With the 2-methylthio triazines, an alkoxy-amino group may be introduced; an example is methoprotryne. Such methoxy derivatives have exceptionally high aqueous solubilities and can be applied for postemergence control in certain crops.

An example of a triazine containing the azido group is provided by azipro-tryne (Table 8.6); this is an effective selective herbicide to control dicotyledon-ous weeds in cabbage, beans, sunflower, maize, and onions and rapidly degrades in soil.

Metribuzin (**22**) and metamitron are examples of unsymmetrical 1,2,4-triazine (1,2,4-triazinone) herbicides (Table 8.6). The former is prepared by condensa-tion of 2-oxo-3,3-dimethylbutyric acid with thiocarbazide (Scheme 8.20). Both are pre- and postemergence herbicides useful for selective control of annual grasses and broad-leaved weeds at 500–700 g/ha. Metribuzin (**22**) can be used in many crops, e.g. soybean, potatoes, tomatoes, carrots, and cereals, while metamitron is especially useful in beets. Both compounds act by inhibition of photosynthetic electron transport.[13]

Triazines are absorbed by clay minerals in the soil which reduces their concentration in the soil solution (see Chapter 2, page 33). They are readily absorbed by plant roots from the soil solution and generally the 2-chlorotri-azines have lower aqueous solubilities than the corresponding 2-methylthio and 2-methoxytriazines.

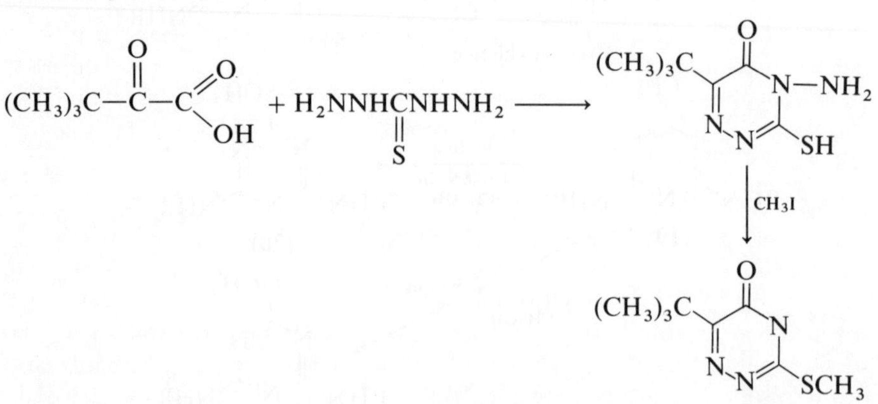

Scheme 8.20

Triazines are metabolized in plants and in the soil by both chemical and microbiological processes. Thus when ^{14}C-side-chain-labelled simazine (**19**; R = R' = C_2H_5) was added to a nutrient solution containing corn, cotton, or soybean plants appreciable amounts of [14]CO_2 were liberated and *N*-dealkylated metabolites were identified. *N*-Dealkylation has been established as a principal degradation mechanism in both resistant and susceptible plants.[16] The monodealkylated triazines retain their phytotoxicity, but become inactive when the 2-chloro group is hydrolysed to the hydroxy. Hydrolysis of the 2-methylthio- and 2-methoxytriazines was also observed in plants. In soils, triazines similarly suffer *N*-dealkylation and hydrolysis of the 2-substituent; then the free amino groups are replaced by hydroxy groups and ultimately there is cleavage of the triazine ring with liberation of carbon dioxide. Certain soil microorganisms can utilize triazines as sources of carbon and nitrogen. Biguanides are possible intermediates in the ring cleavage, being formed after hydrolytic removal of the 2-carbon atom as carbon dioxide, and these would be susceptible to further hydrolysis[13,16] (Scheme 8.21).

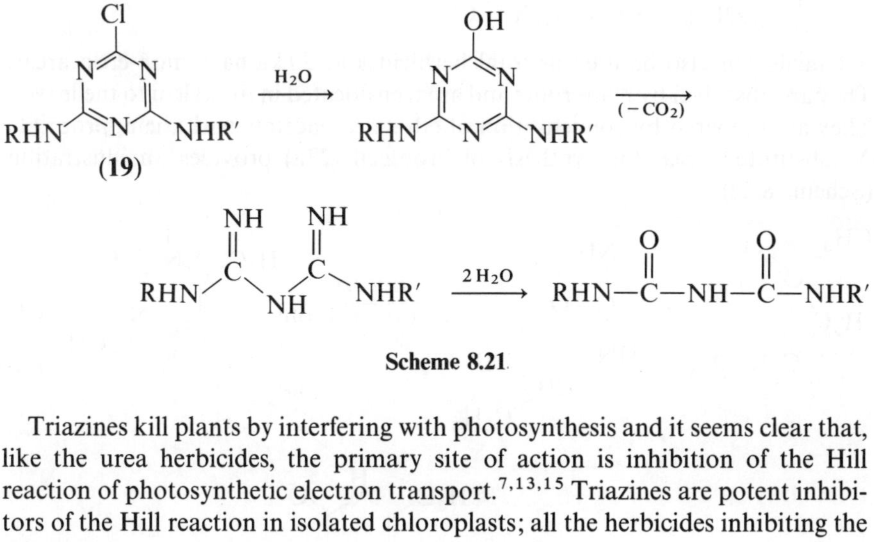

Scheme 8.21

Triazines kill plants by interfering with photosynthesis and it seems clear that, like the urea herbicides, the primary site of action is inhibition of the Hill reaction of photosynthetic electron transport.[7,13,15] Triazines are potent inhibitors of the Hill reaction in isolated chloroplasts; all the herbicides inhibiting the Hill reaction possess the common structural feature:

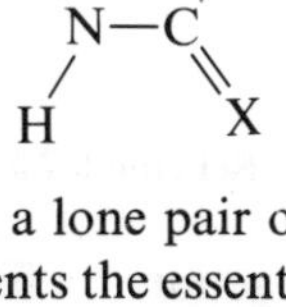

in which X is an atom possessing a lone pair of electrons (either N or O). It is possible that this grouping represents the essential toxophore in these herbicides and is responsible for binding them to a vital enzyme involved in the Hill reaction, thus preventing the photolysis of water and so depriving the plant of its energy supply.

URACILS (PYRIMIDINES)

Unsubstituted uracils have no herbicidal activity but certain derivatives substituted in the 3,5,6-positions are active. The most important examples are bromacil (**23a**), terbacil (**23b**), and lenacil (**23c**). These compounds were introduced by Du Pont (1963) and are applied to the soil; they are used for selective weed control at 2.2–6 kg/ha in cane fruit, asparagus, hops, sugar, apple, and citrus crops. Lenacil is used especially for weed control in beet, spinach, and strawberries.[4,13,15]

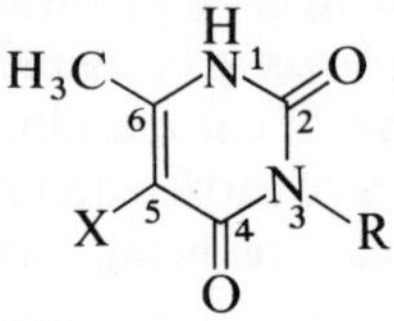

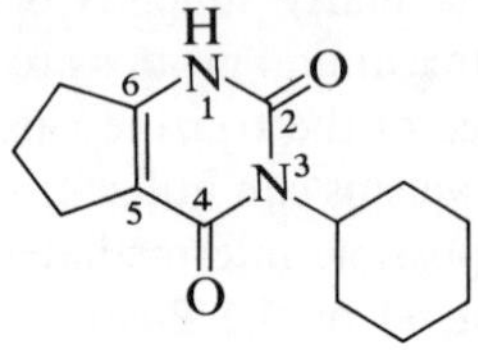

(**23a**) R = CH(CH$_3$)C$_2$H$_5$; X = Br   (**23c**)
(**23b**) R = C(CH$_3$)$_3$, X = Cl

Uracils can also be used as total herbicides at 20 kg/ha in non-crop areas. They are absorbed by plant roots and are translocated in the xylem to the leaves. They are prepared by condensation of ethyl acetoacetate with the appropriate N-substituted urea; the synthesis of bromacil (**23a**) provides an illustration (Scheme 8.22).

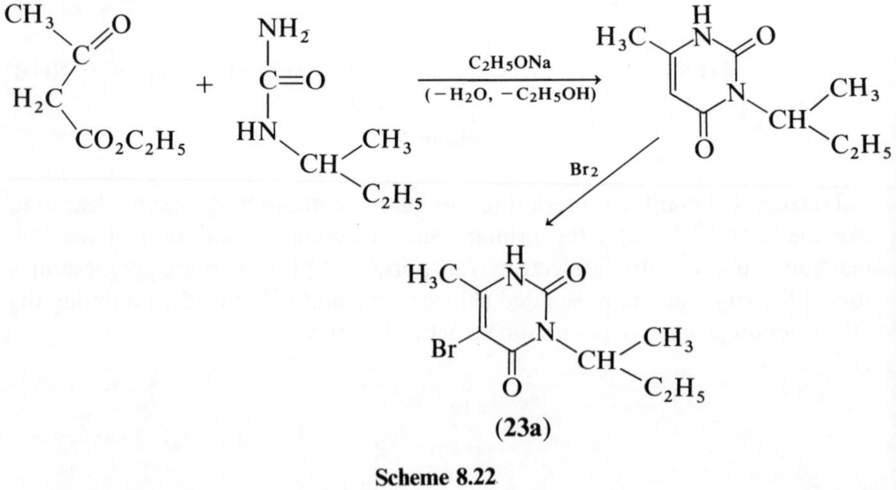

(**23a**)

Scheme 8.22

Uracils owe their herbicidal action to inhibition of the Hill reaction in photosynthetic electron transport.[7,13,15] Several miscellaneous heterocyclic systems are associated with herbicidal activity; the following are some of the more important ones.

TRIAZOLES

The most significant member of this group is amitrole (3-amino-1,2,4-triazole) (**24**) introduced by Amchem Limited (1955) as a foliage-applied non-selective herbicide for use in uncropped land and orchards; it is effective against couch grass and *Rumex* sp., showing good penetration to underground rhizomes. Amitrole (**24**) is a valuable chemical for use in fallow land to kill perennial weeds but the crops, e.g. winter wheat, must not be sown sooner than two weeks after treatment.[4] Amitrole is synergized by addition of ammonium thiocyanate and it is also often formulated with other herbicides, e.g. simazine, diuron, and dicamba.[4] It is prepared by condensation of aminoguanidine and formic acid (Scheme 8.23).

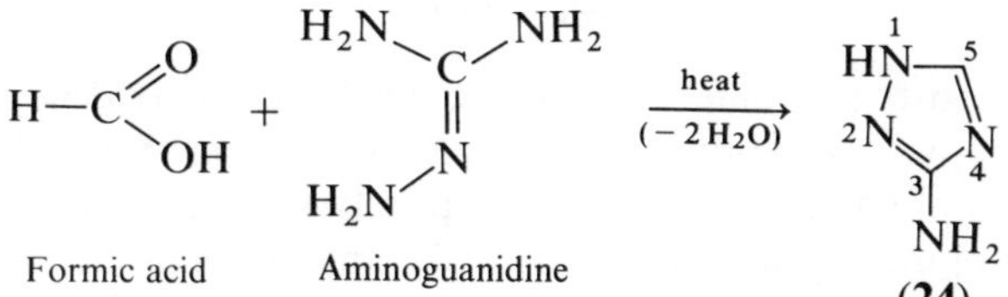

Formic acid   Aminoguanidine

(24)

Scheme 8.23

Low concentrations of amitrole stimulate plant growth, but higher concentrations cause chlorosis and death. It is a bleaching herbicide, like the pyridazinones, and the primary mode of action is probably interference with carotenoid biosynthesis leading to photooxidation of the plant chlorophyll and disruption of chloroplast structure.[7]

Amitrole was suspected of being carcinogenic to rats but this was not substantiated by later studies. However, its use in fodder and food plant crops has been banned worldwide as a precautionary measure.[13]

PYRIDAZINES

Several *N*-phenylpyridazinones are herbicidal; the most important member is chloridazon (**25**), which is prepared by condensation of phenylhydrazine and mucochloric acid (Scheme 8.24), the latter being obtained by oxidative chlorination of furfural. Chloridazon (**25**) was introduced by BASF (1962) as a soil-applied herbicide for pre- or postemergence broad-leaved weed control in beet;[4] it is readily absorbed by the roots and translocated in the xylem, but is not appreciably translocated from the foliage (Plate 21).

The compound is inactivated in beet by metabolism to an aminoglucosyl derivative—hence its value for selective weed control in this crop.[15] In soil,

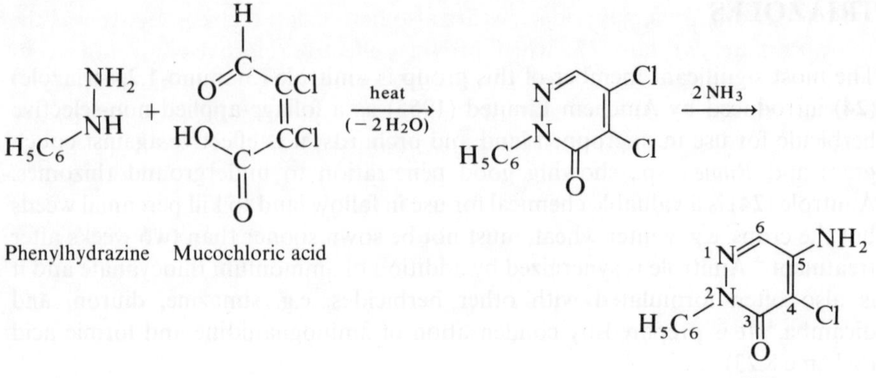

Phenylhydrazine   Mucochloric acid

(25)

Scheme 8.24

chloridazon persists for some five months, but is gradually decomposed to 5-amino-4-chloropyridazin-3-one which is inactive (Scheme 8.25). The biological mode of action of chloridazon is not completely known: experiments showed that < 10 μM of the herbicide resulted in 50% inhibition of the Hill reaction in photosynthesis, and this is believed to be the primary mode of action.[7] However, it also reduces carbon dioxide assimilation, oxygen uptake, and affects carotenoid synthesis.[13]

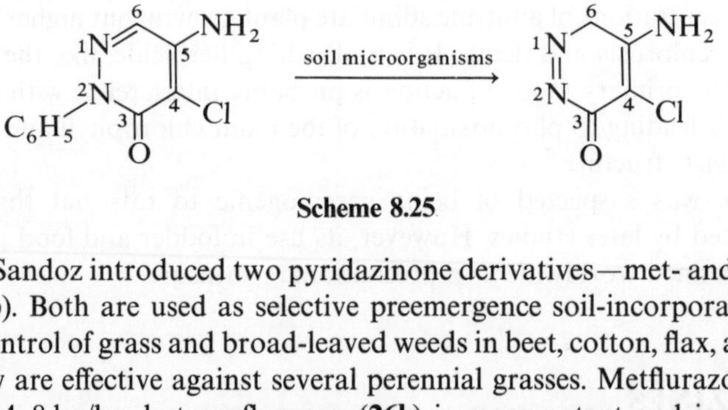

Scheme 8.25

In 1972 Sandoz introduced two pyridazinone derivatives—met- and norflurazon (**26a, b**). Both are used as selective preemergence soil-incorporated herbicides for control of grass and broad-leaved weeds in beet, cotton, flax, and stored fruits. They are effective against several perennial grasses. Metflurazon (**26a**) is applied at 4–8 kg/ha, but norflurazon (**26b**) is more potent and is applied at lower rates (1–4 kg/ha). Both these pyridazinones interfere with photosynthesis, but it is concluded that blocking carotenoid biosynthesis is the primary mode of

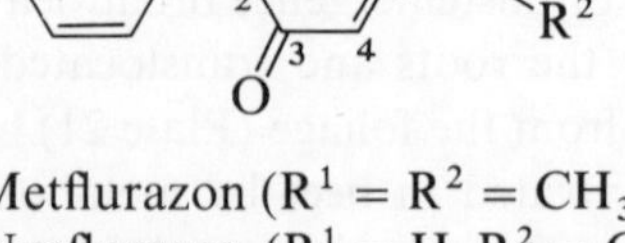

(**26a**) Metflurazon ($R^1 = R^2 = CH_3$)
(**26b**) Norflurazon ($R^1 = H, R^2 = CH_3$)

action. Certain 3-phenylpyridazines are also herbicidal; the most active was the thiocarbonate, pyridate (**27**), which provides effective selective control of weeds in cereals, e.g. maize, rice, and some other crops at 1–1.5 kg/ha. It is active against triazine-resistant weeds; the mode of action is believed to involve inhibition of the Hill reaction in photosynthesis.[13]

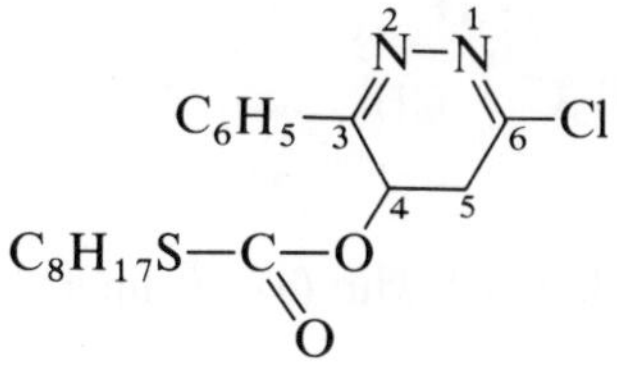

(**27**)

Certain 3-phenoxypyridazines, e.g. the 3-tolyloxy derivative (credazine), have a useful activity; they are used as selective preemergence herbicides for weed control in rice, potato, and strawberry at 2–3 kg/ha.

OTHER HETEROCYCLIC HERBICIDES

Endothal

Endothal (**28**) is an example of a herbicidal oxygen heterocycle which is prepared from furan and maleic anhydride by the Diels–Alder reaction (Scheme 8.26). Endothal (**28**) is mainly the *exo-cis* isomer, the most active of the three stereoisomers. It is formulated as the disodium salt which is reasonably soluble in water, and the aqueous solution is readily absorbed by plant roots. Endothal was developed by the Sharples Chemical Corporation (1954) for pre- and postemergence weed control in beet crops. The mode of action is not definitely known, but in certain plants endothal caused inhibition of both lipid and protein biosynthesis.[7]

Cinmethylin (**29**) (Figure 8.2) is a member of a novel herbicide group called cineoles. It is an effective systemic herbicide for the selective control of grass

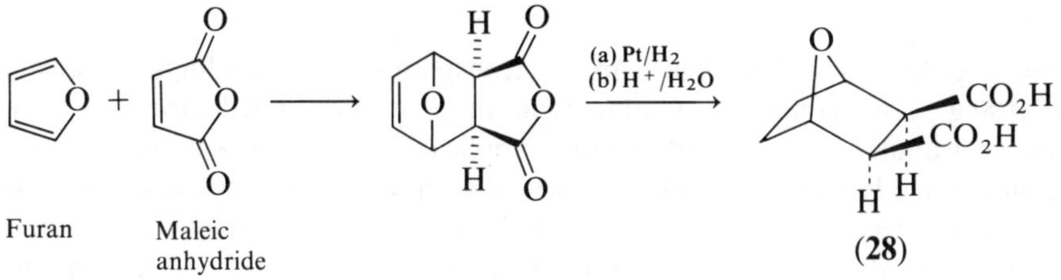

Scheme 8.26

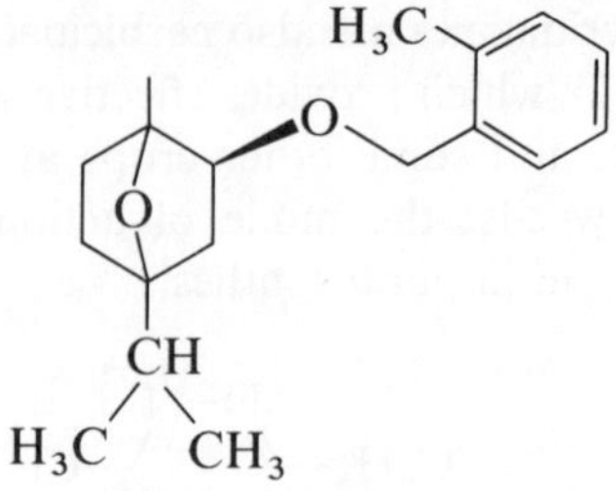

(29)

Cinmethylin (*exo* isomer)

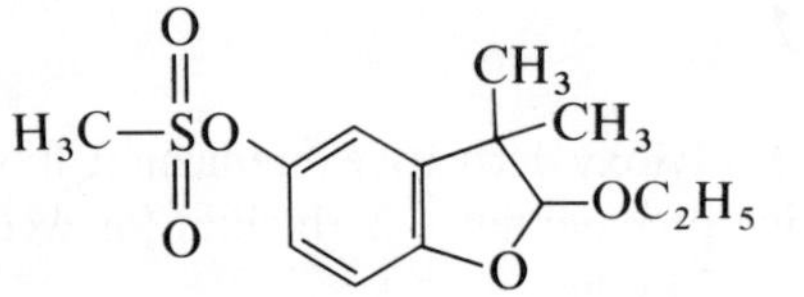

(30)

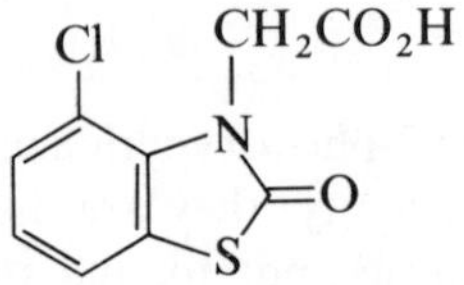

(31)

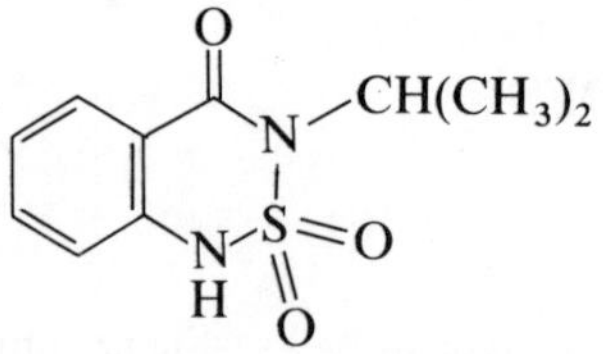

(32)

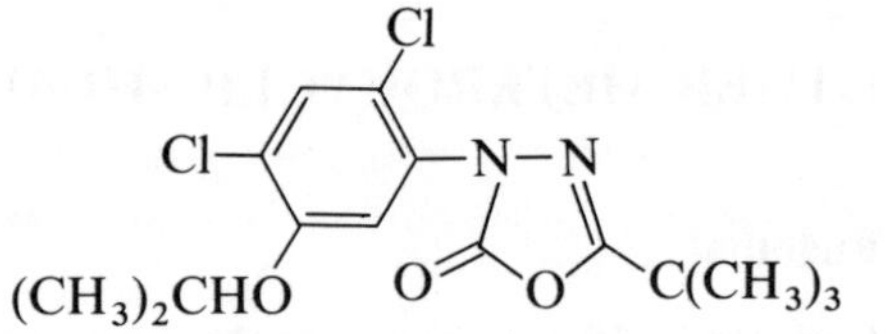

(33)

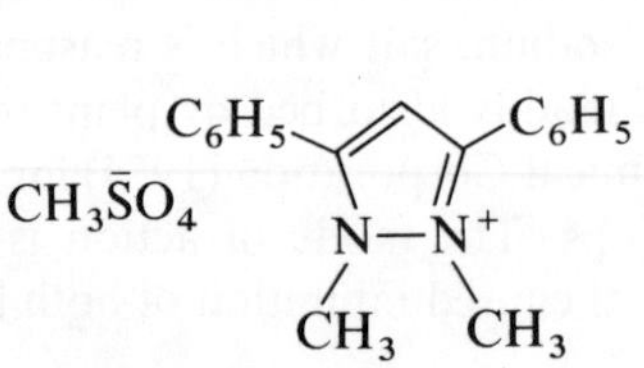

(34)

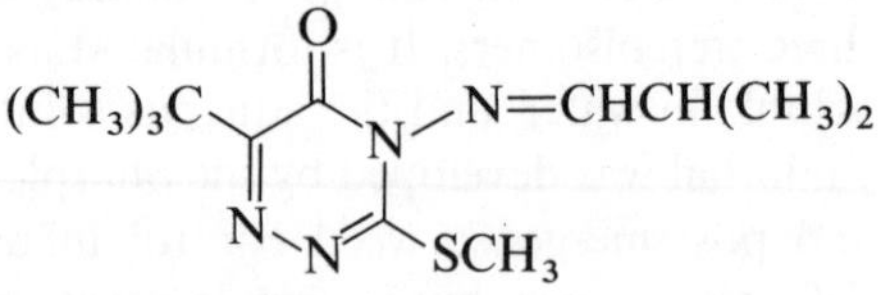

(35)

Figure 8.2

weeds, especially *Echinochloa* spp., in transplanted rice seedlings at rates of 25–100 g a.i./ha (Plate 22). Cinmethylin may be formulated either as slow-release granules or as a liquid for spray application. The compound is absorbed from the paddy water through the roots of germinating or emerging weeds and is translocated in the xylem. Susceptible weeds are killed by disruption of meristematic development in the growing points of shoots and roots. Cinmethylin has a low mammalian toxicity (LD_{50} (oral) to rats, 3960 mg/kg); it is metabolized by loss of carbon dioxide and does not persist beyond the season of use.

Ethofumesate

The benzofuran derivative, ethofumesate (**30**) (Schering, 1980) (Figure 8.2), selectively controls annual broad-leaved and grass weeds in pasture and as mixtures with other herbicides, e.g. bromoxynil or ioxynil; it is useful against weeds in beet. Ethofumesate (**30**) appears to alter the composition of plant cuticular wax and this may be the mode of action.[7]

Benzolin (**31**) (Boots, 1965) and bentazone (**33**) (BASF, 1968) are used at 0.5–3 kg/ha for postemergence control of dicotyledonous weeds in cereals, linseed, beans, and other crops.[4,13] They show good activity against several problem weeds, e.g. mayweeds, cleavers, chickweed. Many graminaceous crops, e.g. rice, barley, maize, and soybean, are resistant to bentazone so it is particularly valuable for selective weed control in these crops. Benzolin (**31**), when formulated with clopyralid (see page 226), has proved valuable for postemergence weed control in oilseed rape. Benzolin (**31**) acts by interference with normal cell growth and development, while bentazone (**32**) affects photosynthesis by inhibition of carbon dioxide fixation and electron transport.[7,13]

Oxadiazon (**33**) (Rhône-Poulenc, 1969) is used at 1–4 kg/ha as a pre- and postemergence herbicide against broad-leaved and grass weeds in a wide range of crops, e.g. rice, cotton, sugar cane, apples, pears, cane fruit, sunflower.[4,13,15] The mode of action appears to be inhibition of photosynthetic electron transport.[7] Oxadiazon (**33**) can be prepared from pivaloyl chloride as indicated in Scheme 8.27.

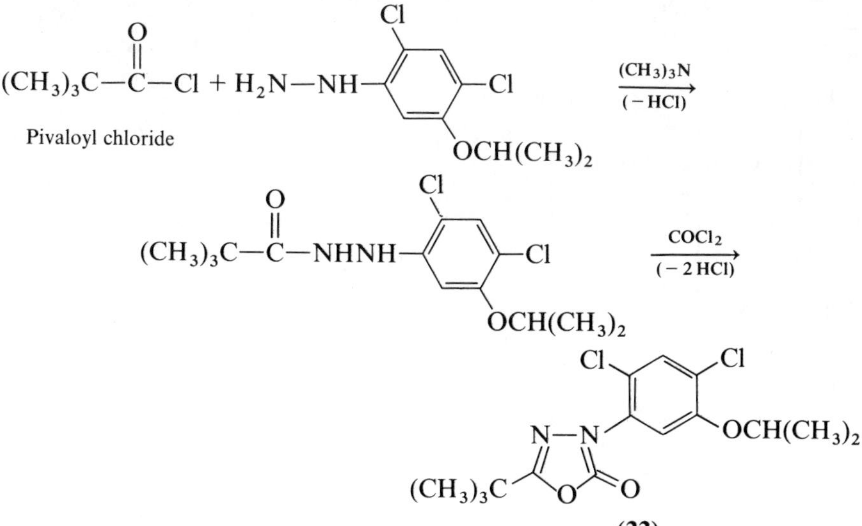

Scheme 8.27

Difenzoquat (**34**) (Cyanamid, 1973) (Figure 8.2) is used as an aqueous spray for postemergence control of wild oats in cereals. Difenzoquat is specifically active against wild oats (*Avena* spp.). To kill different grasses and broad-leaved weeds, it must be formulated with other herbicides, e.g. 2,4-D, MCPA, or dichlorprop. The efficiency of difenzoquat is directly related to the type and concentration of the surfactant used in the aqueous spray; e.g. the optimum control of wild oats was achieved with a non-ionic alkyl polyglycol ester emulsifier (see Chapter 2, page 19).

The primary mode of action of difenzoquat (**34**) is unknown; however, it does inhibit several biochemical processes, e.g. ATP production, DNA, RNA synthesis and photosynthesis.[7] Isomethiozin (**35**) (Bayer, 1975) also controls grass weeds in cereals; the herbicidal action is associated with interference with photosynthesis.

BIPYRIDINIUM HERBICIDES

These compounds were developed from the observation that quaternary ammonium germicides, like cetyltrimethylammonium bromide, will desiccate young plants.[19] The two most important examples, diquat (**36**) and paraquat (**37**), were introduced by the Plant Protection Division of Imperial Chemical Industries Limited in 1958. These herbicides are synthesized from pyridine according to Scheme 8.28. The formation of paraquat provides an interesting example of a commercial synthesis based on a free-radical reaction.

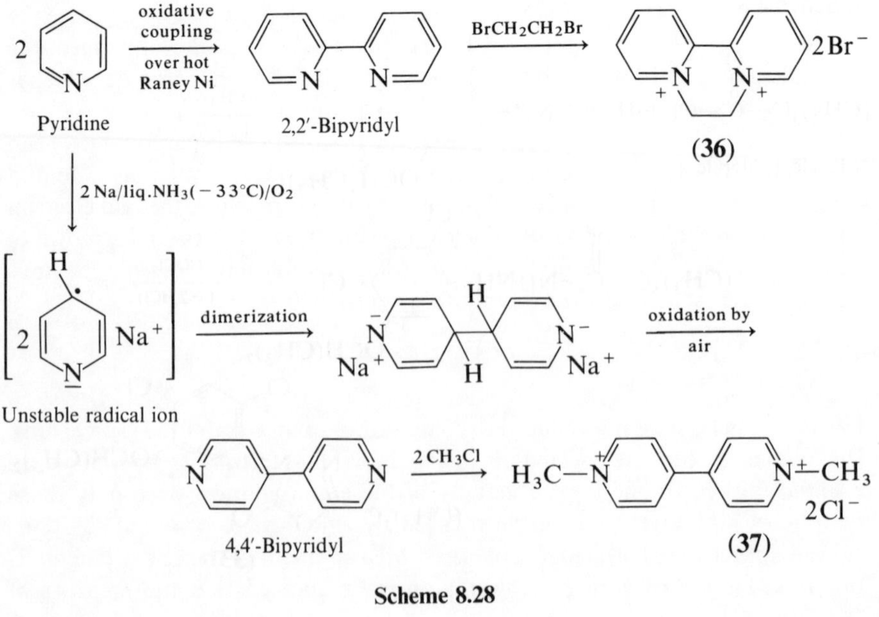

Scheme 8.28

The bipyridinium herbicides are rapidly translocated from the foliage, but not from plant roots, because they are immediately deactivated on contact with the soil due to cation exchange with clay constituents which bind the compounds firmly to the soil particles; in this way they act as contact herbicides, rapidly killing all green plant growth on which they fall[19] (Plate 23).

Both paraquat and diquat are widely employed as plant desiccants; at 0.5–1.5 kg/ha paraquat is rather more effective against grasses than diquat[4,15] and can be used to kill weeds before sowing crops. The rapid deactivation by the soil enables paraquat to be used in 'chemical ploughing' in which the seed is sown after spraying with paraquat without the need for ploughing. This technique is of special importance in areas where soil erosion is a serious problem and is becoming increasingly attractive because it reduces energy costs.[3] Paraquat is useful in renovating degenerate pastures, while diquat is widely applied for potato haulm destruction and against aquatic weeds. Both compounds are easily soluble in water since they are salts and paraquat is very useful to the gardener because of its safety in application, only killing those plants directly hit by the spray. Paraquat shows only moderate mammalian toxicity—LD_{50} (oral) to rats is 150 mg/kg (diquat is considerably less toxic) —but when large amounts are ingested they cause proliferation of human lung cells leading to respiratory failure and ultimately death. At present there appears to be no adequate antidote. The bipyridinium herbicides are almost inactive in the dark, but are extremely active in sunlight, when photochemical decomposition products can be detected in treated leaves.[5,13]

Extensive studies[10,15,19] have demonstrated that in solution these compounds are almost completely dissociated into ions, and in chloroplasts during the process of photosynthesis the positive ion of paraquat is reduced to relatively stable, water-soluble, radical cations. In the presence of oxygen, the free radicals are reorganized to the original ion and hydrogen peroxide, which is probably the ultimate toxicant destroying the plant tissue.

Support for this suggestion comes from the discovery that paraquat (**37**) can be chemically reduced by sodium dithionite or zinc dust to intensely coloured solutions containing resonance-stabilized radical ions in which the odd electron can be delocalized over all twelve nuclear carbon atoms with partial removal of the positive charges on the nitrogen atoms (only possible if the heterocyclic rings are coplanar) (Scheme 8.29) from photosystem 1 or

$$PQ^{2+} + e \longrightarrow PQ^{+\bullet} \xrightarrow{\ O_2\ } PQ^{2+} + O_2^{\bullet -}$$
(**37**)

Diquat can be similarly reduced to coloured resonance-stabilized radical ions. The reduction can be effected in isolated plant chloroplasts and plays an essential part in the herbicidal activity of these compounds, since only those quaternary bipyridyls that can be reduced by zinc dust to coloured solutions showed appreciable herbicidal properties. Measurement of the redox potentials for the reduction of various bipyridinium salts showed that the majority of

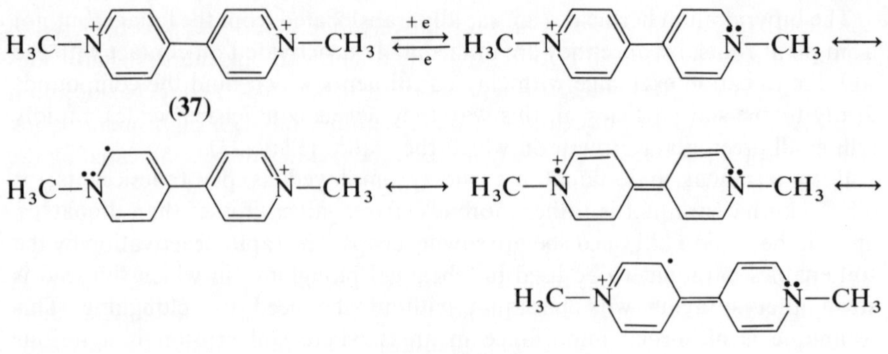

(37)

Scheme 8.29

herbicidal derivatives had redox potential values within the range -0.350 to -0.450 eV; such compounds can be reduced by the reduction potential of light reaction I of photosynthesis (see Chapter 3, page 42). In other words, only those bipyridinium compounds in which the two heterocyclic rings are essentially coplanar are herbicidal. Any twisting causes loss of activity. Consider the derivatives of diquat and paraquat shown below:

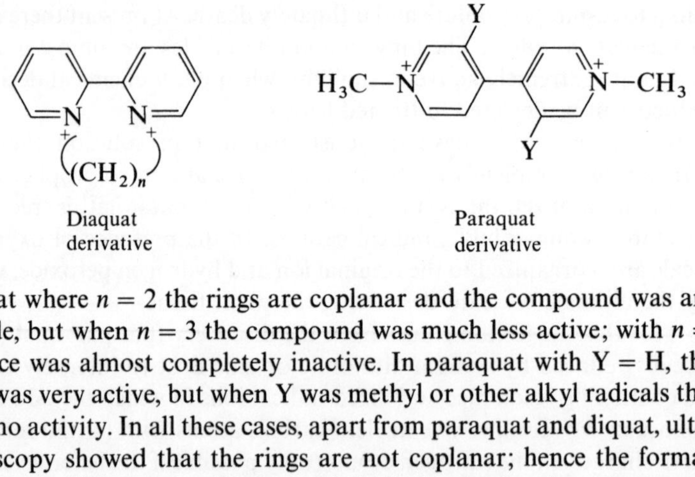

Diaquat
derivative

Paraquat
derivative

In diquat where $n = 2$ the rings are coplanar and the compound was an active herbicide, but when $n = 3$ the compound was much less active; with $n = 4$, the substance was almost completely inactive. In paraquat with $Y = H$, the compound was very active, but when Y was methyl or other alkyl radicals there was almost no activity. In all these cases, apart from paraquat and diquat, ultraviolet spectroscopy showed that the rings are not coplanar; hence the formation of resonance-stabilized free radicals on biological reduction is no longer possible.

Photosynthesis is basically an oxidation–reduction process coupled to ATP formation, and so photosynthesis could provide sufficient reduction potential to reduce diquat and paraquat to the stabilized free radicals; this reduction by isolated plant chloroplasts in the absence of oxygen has been demonstrated. The bipyridinium herbicides are only active in the presence of both light and oxygen, because in the absence of oxygen the free radicals are extremely stable and are unlikely to participate in radical reactions leading to phytotoxic damage. On the other hand, in the presence of oxygen diquat or paraquat present in chloroplasts is quickly reduced to the radical ions which are subsequently reoxidized to form the superoxide anion; however, this is relatively inactive and is not considered to be the ultimate toxicant. It is found that treatment with bipyridinium herbicides causes increased levels of hydrogen peroxide, which, unlike the superoxide anion ($O_2^{\cdot-}$), is very nucleophilic and can readily diffuse through plant membranes. The hydrogen peroxide is formed from the superoxide anion as indicated by the following equation:

$$2H^+ + 2O_2^{\cdot-} \rightleftharpoons H_2O_2 + O_2$$

The majority of evidence indicates that hydrogen peroxide is the essential toxicant and kills the plant by lipid peroxidation[7,15,19] involving a chain reaction which destroys the cellular membranes. Paraquat and diquat are total herbicides and later research has investigated the possibility of introducing some degree of selectivity into this class of herbicides.[19]

DIPHENYL ETHERS

A number of substituted diphenyl ethers (**38**) have been developed as herbicides.[13,15] They are generally prepared by the Williamson synthesis from *p*-chloronitrobenzene and the appropriate substituted sodium phenolate (**39**) (Scheme 8.30).

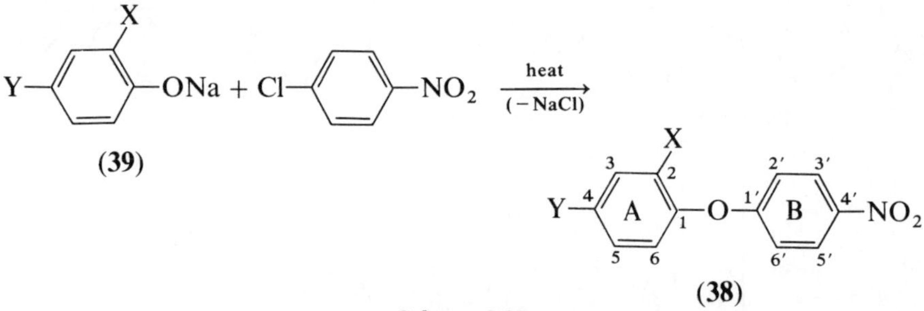

Scheme 8.30

The best-known example is nitrofen (**38**; X = Y = Cl), introduced by the Rohm and Haas Company (1960). Nitrofen is a valuable preemergence herbicide against several annual broad-leaved and grass weeds in brassicas (e.g. kale, rape, cabbage, brussels sprouts) and winter-sown wheat.[4,13] The pesticide is applied as a thin layer on the surface of the soil at a rate of (3.4–4.5 kg/ha); herbicidal activity is, however, quickly lost when the chemical is mixed with the soil.

Another example is fluorodifen (**38**; X = NO_2, Y = CF_3), developed by Ciba-Geigy Limited (1968). This is applied as a contact pre- or postemergence herbicide, and is promising for preemergent weed control in soybeans or pre- or postemergence application in rice (3.3–4.4 kg/ha). On the majority of crops, other than rice, activity remains for 8-12 weeks, especially in dry soil.[13]

Bifenox (38; X = Y = Cl, $3'$-CO$_2$CH$_3$) was introduced by Mobil (1970) for weed control in maize, rice, and soybeans at 1.7–2.2 kg/ha.[13,15] Oxyfluorfen (38; X = Cl, Y = CF$_3$, $3'$-OC$_2$H$_5$), developed by Rohm and Haas Company (1980), is applied at 0.2–0.4 kg/ha for preemergence weed control in cotton, maize, wheat, soybeans, vineyards, and orchards.[10,13]

Fomesafen (38; X = Cl, Y = CF$_3$, $3'$-CONHSO$_2$CH$_3$) was introduced by ICI (Jealott's Hill Research Station, 1982) (Plate 24). It is used as a selective preemergence herbicide for control of broad-leaved weeds in soybean at 0.15–0.5 kg/ha. It provides weed control for the whole season and the activity is enhanced by addition of a non-ionic surfactant to the spray solution.[13] The compound has low mammalian toxicity (LD$_{50}$ (oral) to rats, 1500 mg/kg) and does not harm earthworms.

Diphenyl ether herbicides are formulated as granules or wettable powders; they are rapidly absorbed from the soil by plant roots and are degraded in plant tissue. The major metabolites of fluorodifen are shown in Scheme 8.31; the unknown compounds I and II may be amino acid conjugates of the phenols.[16]

The demand for diphenyl ether herbicides has grown steadily since 1972 by about 20% per annum and this is expected to continue. In 1984, the estimated market share was $280 million, mainly in the United States and the Far East.[15]

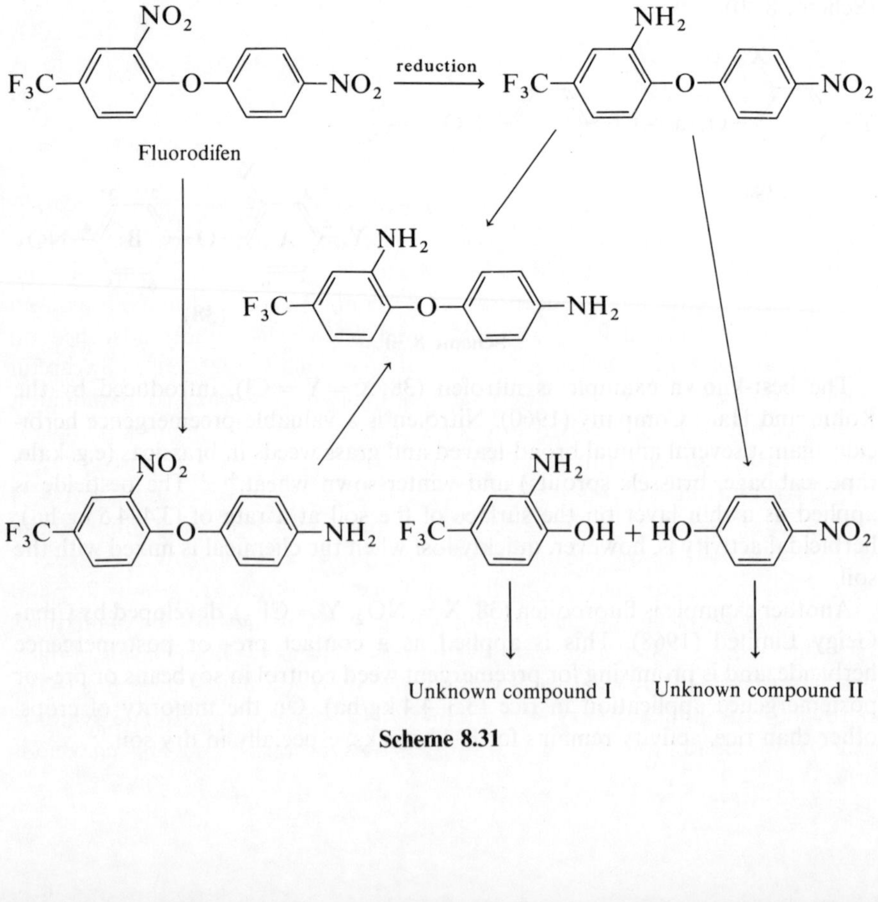

Scheme 8.31

The nitrodiphenyl ether herbicides are only active in the presence of light (like the bipyridinium herbicides) and cause chlorosis of leaf tissue. They inhibit the Hill reaction in photosynthesis and photophosphorylation; however, the primary mode of action probably involves their photosynthetic reduction to radical species which initiate destructive reactions in lipid membranes leading to cell leakage.[7]

ORGANOPHOSPHORUS COMPOUNDS

The value of organophosphorus compounds as systemic insecticides is well established and several members show remarkable selectivity to insects (Chapter 6). More recently organophosphorus compounds have shown some herbicidal action, by far the most important member commercially being glyphosate (**40**).[20] It was introduced by Monsanto in 1971 and is obtained from phosphorus trichloride and formaldehyde (Scheme 8.32).

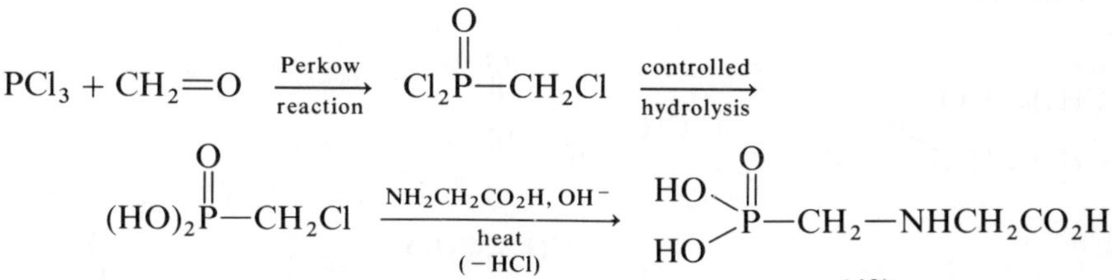

$$PCl_3 + CH_2{=}O \xrightarrow[\text{reaction}]{\text{Perkow}} Cl_2\overset{\displaystyle O}{\overset{\|}{P}}{-}CH_2Cl \xrightarrow[\text{hydrolysis}]{\text{controlled}}$$

$$(HO)_2\overset{\displaystyle O}{\overset{\|}{P}}{-}CH_2Cl \xrightarrow[\substack{\text{heat} \\ (-HCl)}]{NH_2CH_2CO_2H,\ OH^-} \overset{HO}{\underset{HO}{>}}\overset{\displaystyle O}{\overset{\|}{P}}{-}CH_2{-}NHCH_2CO_2H$$

(**40**)

Scheme 8.32

Glyphosate (**40**) is a very potent, non-selective postemergence herbicide which kills mono- and dicotyledonous annual and perennial weeds.[13,20] It is rapidly translocated in the phloem from the foliage to the roots or rhizomes and is therefore effective against deep-rooted perennial weeds, e.g. couch grass and brambles, at a rate of 0.7–5.6 kg/ha.[4,15] Glyphosate is rapidly adsorbed on contact with soil and inactivated, similarly to paraquat (page 253), so by careful application selective control of weeds in crops may be possible. Like paraquat, it can be applied to kill weeds before direct drilling of seed;[3] glyphosate (**40**) is eventually broken down by soil microorganisms to CO_2, NH_3, and H_3PO_4.[16]

Wheat and other plants treated with glyphosate showed reduced levels of most amino acids and the herbicidal action probably arises from inhibition of the biosynthesis of aromatic amino acids, e.g. phenylalanine. This hypothesis is supported by experiments in which the suppression of the growth of the pond weed (*Lemna gibba*) by glyphosate (**40**) was reversed by the addition of mixtures of phenylalanine, tyrosine, and tryptophan. The reduction in the levels of these amino acids caused by treatment with glyphosate is considered to be due to the herbicide poisoning the enzyme 5-enolpyruvylshikimate-3-phosphate synthase (EPSP) which controls a vital step in their biosynthesis.[7,10,17] This enzyme is therefore concluded to be the primary target for glyphosate, although other effects have been observed:

(a) Treated plants exhibited increased levels of phenylalanine ammonia lyase (PAL) activity and
(b) Impaired chloroplast function and photoelectron transport.

The slow death of plants treated with glyphosate, however, results chiefly from the reduction in protein synthesis arising from depletion in the supply of phenylalanine.[7] Glyphosate has been a most successful herbicide; in 1984 the estimated market was $480 million.

Other organophosphorus herbicides include bensulide (**41**), introduced by the Stauffer Chemical Company (1964) for preemergence weed control in vegetables and cotton at 2.3–7.0 kg/ha.[3,10,13] It is prepared by condensation of sodium *O,O*-diisopropylphosphorodithioate with the appropriate sulphonamide (Scheme 8.33). Bensulide can be used at a higher dosage (11–22 kg/ha) to control weeds in turf; it has a moderate mammalian toxicity: LD_{50} (oral) to rats, 770 mg/kg.

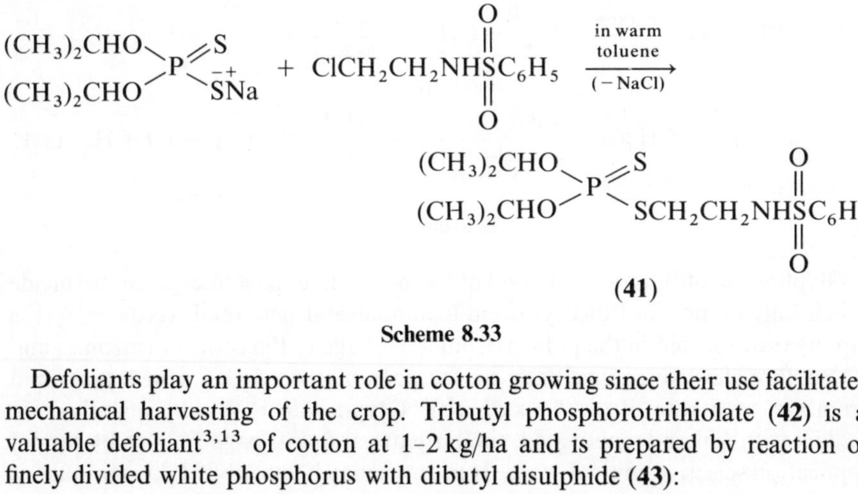

$$(CH_3)_2CHO\diagdown \atop (CH_3)_2CHO\diagup P \diagdown{S \atop SNa}^{-+} \ + \ ClCH_2CH_2NH\overset{\overset{O}{\|}}{\underset{\underset{O}{\|}}{S}}C_6H_5 \ \xrightarrow[(-\,NaCl)]{\text{in warm toluene}}$$

$$(CH_3)_2CHO\diagdown \atop (CH_3)_2CHO\diagup P\diagdown{S \atop SCH_2CH_2NH\overset{\overset{O}{\|}}{\underset{\underset{O}{\|}}{S}}C_6H_5}$$

(**41**)

Scheme 8.33

Defoliants play an important role in cotton growing since their use facilitates mechanical harvesting of the crop. Tributyl phosphorotrithiolate (**42**) is a valuable defoliant[3,13] of cotton at 1–2 kg/ha and is prepared by reaction of finely divided white phosphorus with dibutyl disulphide (**43**):

$$P + C_4H_9S{-}SC_4H_9 \ \xrightarrow[N_2]{\text{in DMSO under}} \ (n\text{-}C_4H_9S)_3P$$

(**43**)   (**42**)

SULPHONYLUREAS

The sulphonylureas are a new class of herbicides introduced by Du Pont in 1982 and within a comparatively short period they have made a major impact on weed control technology.[13,22] They are remarkably active compounds, selectively controlling many dicotyledonous weeds in cereals at dose rates of grams rather than kilograms per hectare.[21]

In the general structure of the sulphonylureas, herbicidal activity is usually restricted to those derivatives containing a methyl or methoxy group substituted in the 4- or 6-position of the heterocyclic nucleus. This is illustrated by Table 8.7 which gives the structures of some commercially useful sulphonylurea herbicides.

The sulphonylureas are prepared from the appropriate arylsulphonyl chloride as illustrated by the synthesis of chlorsulfuron (**47**) (Scheme 8.34).

The sulphonylureas are potent inhibitors of plant growth; seed germination is not generally affected but subsequent root and shoot growth is severely inhibited in sensitive seedlings. The death of susceptible plants is accompanied by chlorosis, necrosis, vein discoloration, and death of terminal buds.[13] Sulphonylureas are generally formulated either as wettable powders or water-dispersible granules, and the compounds are readily absorbed by both the roots and foliage of plants. They are translocated via the xylem and phloem. The potency of the sulphonylureas is outstanding: conventional herbicides, e.g. triazines, require dose rates of 0.5–2 kg/ha but with sulphonylureas rates of only 0.002 kg/ha are often effective.[13,21]

Table 8.7

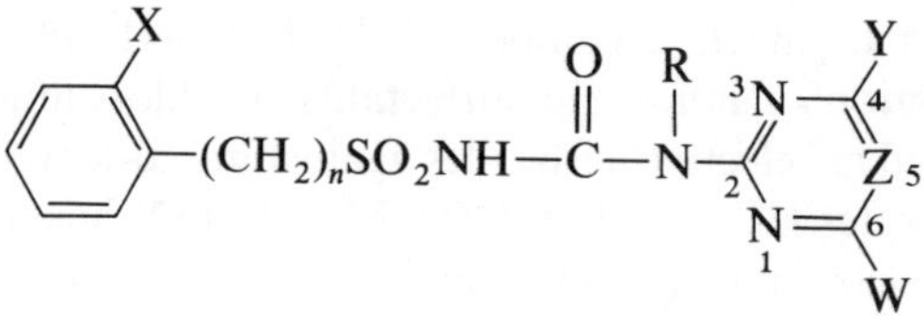

Compound	n	X	R	Y	W	Z
(**44**) Sulfometuron methyl (1983)	0	CO_2CH_3	H	CH_3	CH_3	CH
(**45**) DPX-F 6025	0	CO_2CH_3	H	Cl	OCH_3	CH
(**46**) DPX-F 5384	1	CO_2CH_3	H	OCH_3	OCH_3	CH
(**47**) Chlorsulfuron (1980)	0	Cl	H	CH_3	OCH_3	N
(**48**) Metsulfuron methyl	0	CO_2CH_3	H	CH_3	OCH_3	N
(**49**) Triasulfuron	0	OCH_2CH_2Cl	H	CH_3	OCH_3	N
(**50**) DPX-L 5300 (1985)	0	CO_2CH_3	CH_3	OCH_3	CH_3	N
(**51**) Bensulfuron methyl	0	CO_2CH_3	H	OCH_3	OCH_3	N

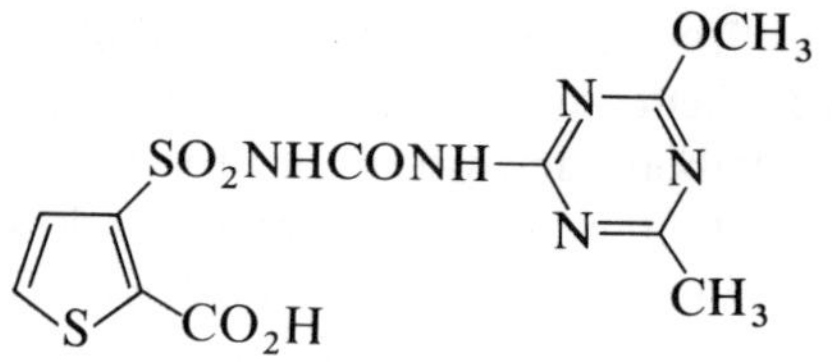

(**52**) Thifensulfuron

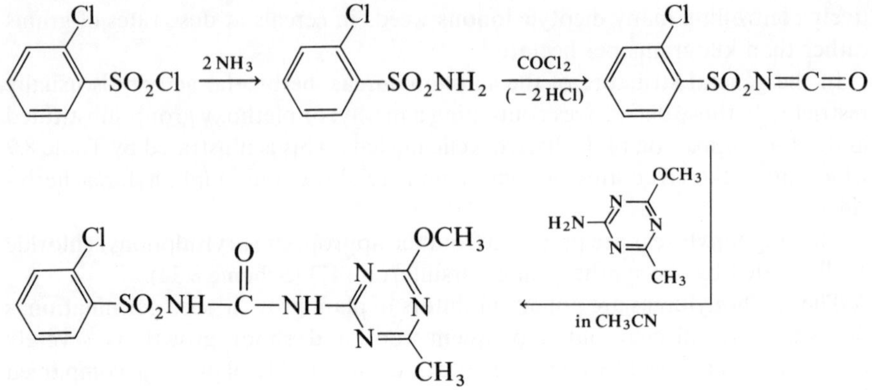

Cl
⬡—SO$_2$Cl $\xrightarrow{2\,NH_3}$ Cl ⬡—SO$_2$NH$_2$ $\xrightarrow[(-2\,HCl)]{COCl_2}$ Cl ⬡—SO$_2$N=C=O

OCH$_3$

H$_2$N—⬠(N triazine)—CH$_3$

Cl
⬡—SO$_2$NH—C(=O)—NH—⬠(triazine, OCH$_3$, CH$_3$)

$\xleftarrow{\text{in } CH_3CN}$

(47)

Scheme 8.34

Sulfometuron methyl (44) is a broad-spectrum selective herbicide against broad-leaved and some grass weeds in cereals at 4–10 g/ha, which can also be applied as a total herbicide. Compounds (45) and (46) (Table 8.7) are employed for selective weed control in soybeans and rice respectively.[15] Chlorsulfuron (47) controls the majority of broad-leaved weeds in cereals at 10–40 g/ha; it acts as a residual soil herbicide with a half-life in soil of 1–2 months and is degraded in soil by hydrolysis to inactive compounds.[13] The herbicidal activity can be increased by addition of non-ionic surfactants to chlorsulfuron. Compounds (48) to (52) are used as selective herbicides against broad-leaved weeds in cereals at very low dose rates (10–20 g/ha). DPX-M 6316 (52) has a shorter life than other sulphonylureas and is consequently valuable as a herbicide in crop rotation programmes.[21] The sulphonylureas have low mammalian toxicities (LD$_{50}$ (oral) to rats of approximately 1000 mg/kg) and are rapidly translocated from the foliage and roots of plants—in practice mainly the former, because of their short persistency in soil. The selective toxicity to certain weeds arises from their relatively rapid metabolism in cereals to inactive products—after a few days no parent compound could be detected. On the other hand, the metabolism in sensitive weeds was much slower.[16] The metabolism of sulphonylureas in tolerant plants is illustrated by the example of chlorsulfuron (47) (Scheme 8.35).

The use of safeners, e.g. 1,8-naphthalic anhydride, increased the tolerance of corn by four- to eightfold towards sulphonylureas, e.g. chlorosulfuron. A less dramatic effect was obtained by the addition of N,N-dialkyl-2,2-dichloroaceta-mide and other compounds. The safeners appeared to act by enhancement of the rate of metabolic detoxification of the sulphonylureas. With sensitive broad-leaved weeds, the safeners were much less effective in reducing phytotoxic damage.[21] Problems have been reported of damage from residues of chlorsul-furon to crops planted in the following season; sugar beet appears particularly sensitive to small residues of sulphonylurea herbicides.[22]

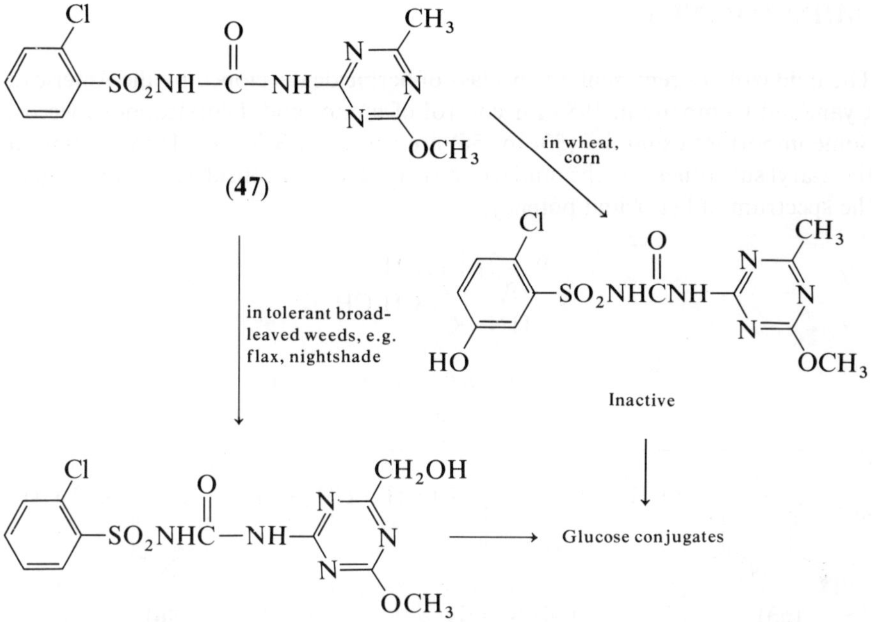

Scheme 8.35

The high activity of the sulphonylurea herbicides suggests that they act by a very specific biochemical mode of action which is probably different from that of the earlier phenylurea group of herbicides (page 239). Growth measurement studies on sensitive corn seedlings showed that chlorsulfuron (47) strongly inhibited cell division; root and shoot growth were prevented at concentrations of 0.001 and 0.01 p.p.m. respectively. At these concentrations, photosynthesis, respiration, and RNA and protein synthesis were unaffected.[13]

The herbicidal action of the sulphonylureas is therefore concluded to arise from inhibition of cell division, because this was demonstrated to be the most sensitive process. Thus treatment of the root tips of *vicia faba* with chlorsulfuron at 2.8 μM reduced cell division by 87%. Genetic and biochemical studies indicated that sulphonylureas inhibit the action of the enzyme acetolactate synthase (ALS) which catalyses the first stage in the biosynthesis of the essential amino-acids valine and isoleucine.[22]

The blocking of the enzyme ALS indirectly results in the disruption of cell division; the treated plants showed severe damage to root and shoot growth, followed by general necrosis and death in approximately two weeks.[15,21]

The primary biochemical target of the sulphonylurea herbicides is therefore concluded to be the enzyme ALS. The proposed biochemical mode of action is supported by studies that showed that addition of the amino acids valine and isoleucine at 100 μM completely reversed the herbicidal action of sulphonyl-ureas towards sensitive plants.

IMIDAZOLINES

The imidazolines represent a new class of herbicide developed by the American Cyanamid Company in 1983 for control of mono- and dicotyledonous weeds. Some important examples (53) to (56) are shown in Scheme 8.16. Variation in the 2-aryl substituent on the imidazoline ring results in considerable variation in the spectrum of herbicidal potency.

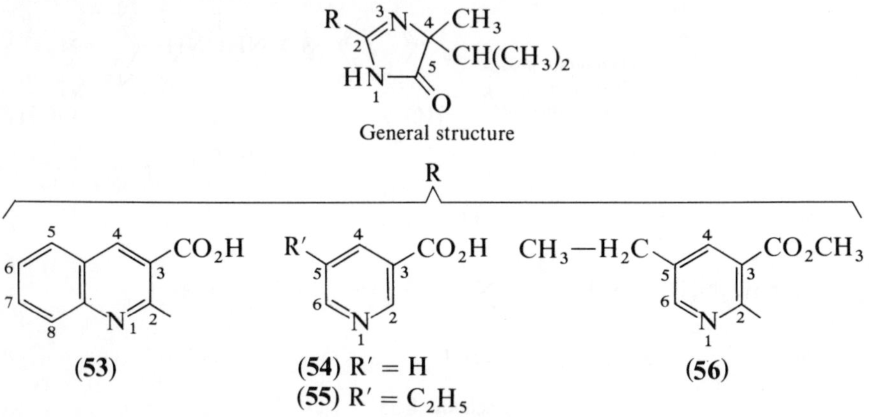

General structure

R

(53)   (54) R′ = H   (56)
             (55) R′ = C$_2$H$_5$

Scheme 8.36

Imazaquin (53) and imazethapyr (55) are effective as pre- and postemergence herbicides at doses of 70–250 g a.i./ha to control grasses and dicotyledonous weeds in soybeans and other leguminous crops. Imazapyr (54) can also be applied as an aqueous spray at 0.5–1.25 kg/ha to kill weeds in bushes and deciduous trees and at high dose rates as a total herbicide.[4] Compound (56) is used at 0.4–0.75 kg/ha as a selective postemergence herbicide against wild oats and some broad-leaved weeds in barley, wheat, and sunflowers.[15]

The most direct preparative route to the imidazolines is via an o-dicarboxylic acid anhydride and an α-aminocarboxamide, illustrated by the formation of imazapyr (54) (Scheme 8.37). The initial product is extracted with aqueous sodium hydroxide and the final cyclization is effected by heating.

The imidazolines have low mammalian toxicities (LD$_{50}$ (oral) to rats > 5000 mg/kg). The imidazolines, like the sulphonylurea herbicides, owe their herbicidal action to specific inhibition of the enzyme acetolactate synthase (ALS).[15] The enzyme catalyses the first step in the biosynthesis of the amino acids valine, leucine, and isoleucine from pyruvate. When maize seedlings were treated with imazapyr (54), DNA synthesis was inhibited after eight hours; the effect was reversed by treatment of the plants with the amino acids.[23] Imazapyr was demonstrated[23] to be a potent inhibitor of ALS enzyme which had been isolated from maize tissue, and this enzyme is clearly the primary target for this group of herbicides. Imidazolines are often formulated as aqueous solutions and

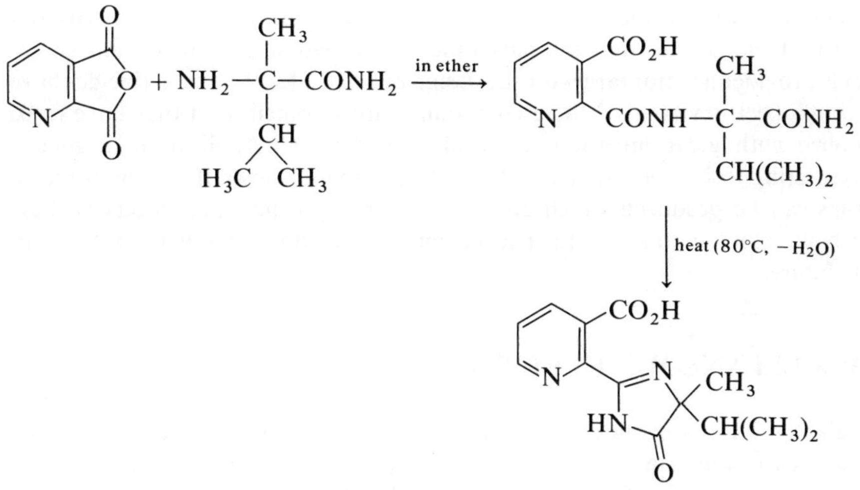

Scheme 8.37

are readily absorbed through the foliage and roots of plants and translocated to meristematic regions. Selectivity of action largely depends on differential rates of metabolism between the crop and the weed so that the levels of the herbicidally active free acids reach relatively higher concentrations in the weeds.[15]

For instance, imazethapyr (**55**) is rapidly metabolized to inactive products in non-susceptible plants, e.g. in soybeans 50% of the a.i. had gone after 1.6 days. The selective control of wild oats (*Avena* spp.) by compound (**56**) is governed by differential metabolic pathways as illustrated in Scheme 8.38.

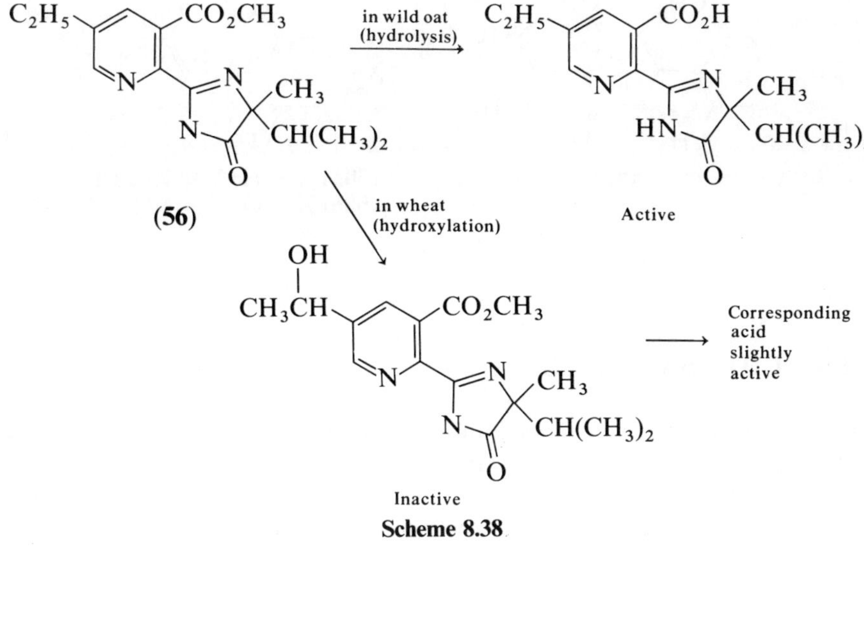

Scheme 8.38

The discovery of the highly active sulfonylurea and imidazoline herbicides clearly demonstrates that the biosynthesis of essential branched-chain amino acids provides a major target for the design of herbicides.[24] Such herbicides have recently been reviewed;[25] these compounds are so potent that they have to be applied with great precision to avoid crop damage. Application of genetic engineering techniques by scientists at Monsanto has shown that new strains of crops can be produced which are more resistant to the lethal effects of these herbicides and, in this way, their spectrum of use could probably be extended in the future.

MISCELLANEOUS HERBICIDES

Tralkoxydim, a cyclohexenone derivative (57), is a postemergence herbicide which selectively controls important grass weeds, e.g. *Avena* spp., in wheat and barley at dose rates of 150–350 g a.i./ha (Plate 25). It is used as a foliar spray (EC formulation) and is readily absorbed by the foliage and moves in the phloem to the growing points where it inhibits new growth and subsequently causes necrosis and death of the plant. Tralkoxydim has low mammalian toxicity (LD_{50} (oral) to rat $\simeq 1200$ mg/kg) and was not environmentally persistent. Several derivatives of benzoyl-1,3-diones and 1,3,5-triones (e.g. compounds (58a) and (58b)) have good activity against several weeds at 4.5 kg/ha. The herbicidal action is probably due to inhibition of the enzyme acetyl coenzyme A carboxylase; this may be an important target site for graminicides like the cyclohexanediones and phenoxypropionates, e.g. fluazifop-butyl (page 222).

(57)

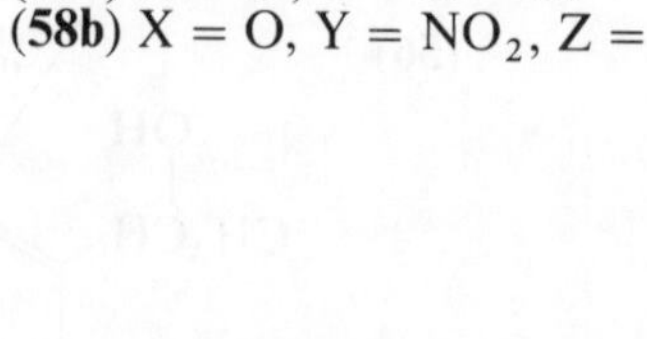

(58a) X = H; Y = Z = Cl
(58b) X = O, Y = NO$_2$, Z = H

HERBICIDE SAFENERS

These are also sometimes termed antidotes or protectants; they enhance the tolerance of the crop towards non-selective herbicides and safen the crop plants against damage by the herbicide.[26–28] The use of chemical safeners provides a method of increasing the crop selectivity of many existing herbicides; safeners can also be used to investigate the biochemical mechanism and reactive sites of herbicides. Safeners fall into four major classes:[28]

(a) *Naphthopyranone derivatives*, e.g. naphthalic anhydride (NA) (59) and phthalic anhydride (PHA) (60). For activity, the presence of the dicarboxylic anhydride group and at least one attached aryl group appear essential.

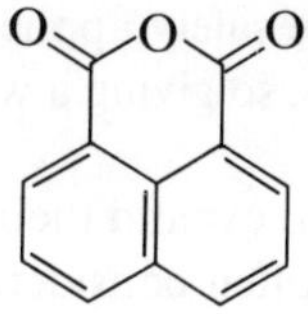

(59) 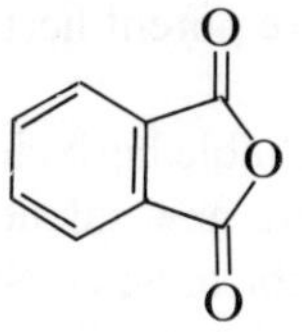   (60) 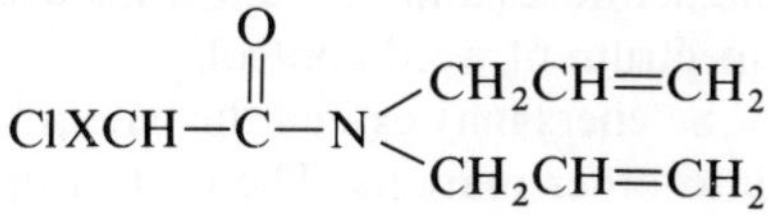   (61)

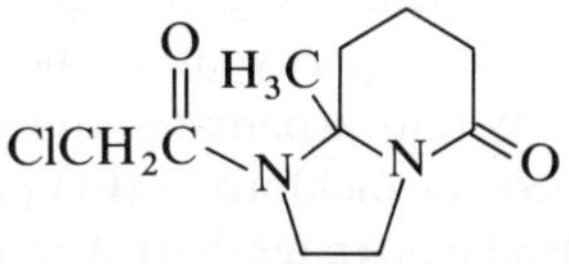

(62)

(b) *Chloroacetamides*, e.g. allidochlor CDAA (61; X = H), dichlormid DCCA (61; X = Cl) and compound (62). The presence of the dichloroacetamide moiety ($Cl_2CHCON=$) gave maximum safening activity against corn; the mono- and trichloroacetamides were less effective.

(ç) *Oxime ethers*, e.g. cyometrinil (63; $R^1 = R^2 = CN$); CGA-133205 (63);

$$\left(R^1 = CF_3, R^2 = -\overset{O}{\underset{O}{\diagdown}} \right),$$ and the pyridine aldoxime ethers (e.g. 64; $n = 1$ or 2).

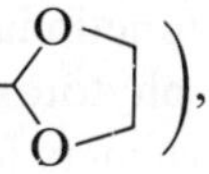

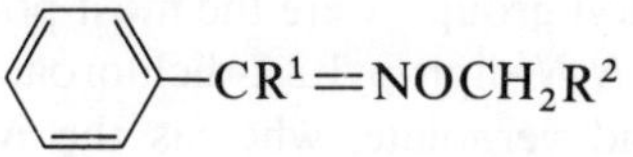

(63) 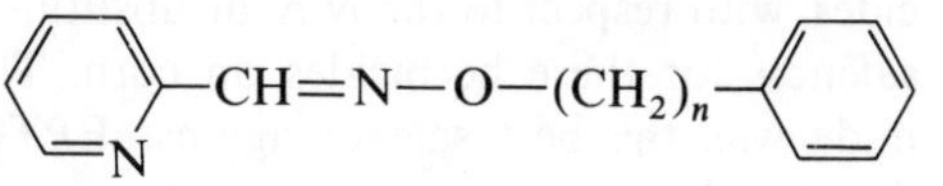   (64)

(d) *2,4-Disubstituted 5-thiazolecarboxylates*. The most effective member of this group was the 2-chloro-4-trifluoromethyl derivative, flurazole (65).

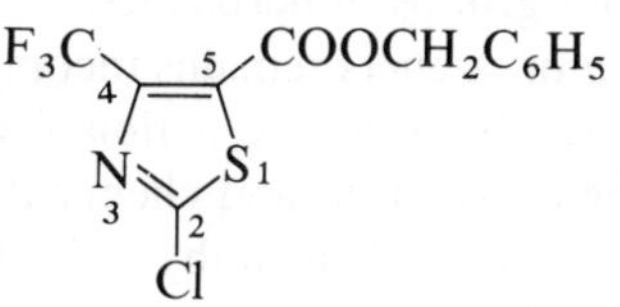

(65) 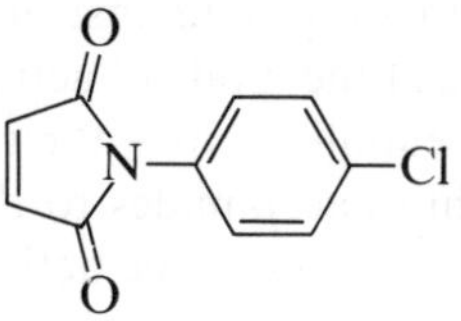   (66) 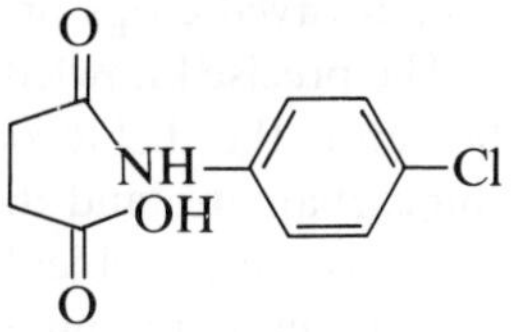   (67)

Prosafeners are chemicals that require biotransformation before they can exert their safening action, for instance N-arylmaleimides, e.g. compound (66), which is converted by *in vivo* hydrolysis into the corresponding maleamic acid (67) which is the active safener. This protected sorghum from injury by the herbicide alachlor.

The introduction of crop safeners for herbicides has opened a new era in chemical weed control. They permit existing chemicals to be safely applied in higher doses and may allow the use of more potent herbicides, so giving a wider spectrum of weed control.

Safeners may extend the patent life of reliable herbicides and expand their use into other markets. The cost of developing new safeners is often considerably cheaper than the introduction of new herbicides; consequently there has been considerable expansion of research into safeners. To date, there has been little success in discovering safeners to protect crops against damage caused by photosynthesis-inhibiting herbicides, e.g. paraquat or broad-spectrum chemicals such as glyphosate. The majority of safeners are effective against soil-applied herbicides, especially thiocarbamates and chloroacetanilides.

Naphthalic anhydride (59) or dichlormid (61) protected corn, sorghum, and rice from injury by chlorosulfuron (page 260). Application of the former resulted in an eightfold increase in the rate of the chemical tolerated by the crop, while the latter showed a two- to fourfold increase. Use of compound (59) also protected the crops from damage by imidazolines, e.g. imazaquin (page 262). Several safeners can be formulated as seed dressings; e.g. compound (59) protects rice against alachlor, allowing this herbicide to be used for the selective control of the weed red rice in rice; cyometrinil (63) protects sorghum against metolochlor.[10]

Research has demonstrated that when the functional groups associated with herbicidal activity are clearly defined, non-phytotoxic analogous molecules without these groups may often be effective safeners for the related herbicides. Dichloroacetamide derivatives most similar to selected thiocarbamate herbicides, with respect to the N,N-disubstituted alkyl groups, were the most potent safeners for these herbicides on corn. Thus, N,N-dipropyl-2,2-dichloroacetamide was the best safener against EPTC and vernolate, whereas the N,N-diisobutyl derivative was the most active against butylate.[27]

Safeners are generally most effective when mixed with the herbicide and applied to the soil. The protection is more marked with large-seeded grass crops, e.g. maize, sorghum, rice, oats. There has been only limited success in protecting broad-leaved crops or grass crops against other groups of herbicides.

The precise knowledge of the mode of action of safeners remains incomplete; however, the dichloroacetanilides owe their specific safening action towards thiocarbamates and dichloroacetamides to their structural and chemical similarity to these herbicides. Studies of the action of naphthalic anhydride (59) in counteracting the phytotoxicity of chloroacetanilide and thiocarbamate herbicides to maize demonstrated that the safener markedly increased the levels of glutathione-*S*-transferase (GST) and glutathione (GSH) in treated maize seedlings.[7]

These components are involved in the detoxification of herbicides. The GST-mediated metabolism of the herbicides in maize leads to the formation of inactive GSH conjugates:

$$R^1S-CONR^2R^3 \xrightarrow[\text{(MFO)}]{\text{oxidation}} R^1SOCONR^2R^3 \xrightarrow[\text{GST}]{\text{GSH}} GS-CONR^2R^3$$

$$ClCH_2CONR^1R^2 \xrightarrow[\text{GST}]{\text{GSH}} GS-CONR^1R^2$$

The evidence indicates that GST is the likely target enzyme of these safeners in maize.

The mode of action of safeners with sulphonylurea herbicides in wheat and maize also appears to be associated with enhanced metabolism to herbicidally inactive products (page 260). Experiments showed that several safeners (e.g. compounds (59), (60), (61) caused dramatic increases (two- to eightfold) in the rate of metabolism of chlorsulfuron in wheat and corn, but did not affect its metabolism in sensitive broad-leaved weeds. A cytochrome P-450 enzyme system is probably responsible for the metabolic inactivation of the sulphonylureas and this is the probable target for the safeners. The majority of safeners appear to act by increasing the rate of metabolism of the herbicide; however, other possible mechanisms of action include interference with the uptake and translocation of the herbicide or affecting the sensitivity of the target site.[28]

Another exciting approach to making herbicides more selective is by the application of biotechnology through genetic modification.[15]

Use of recombinant DNA technology involving gene isolation, design of a vector, and transfer of the gene into the recipient crop plant has yielded some promising results, e.g. modifications of the target site of glyphosate. Other methods of gene transfer have involved plant breeding techniques to obtain plant species more tolerant of the herbicide.

Genes which code for pesticide detoxification enzymes are possible candidates for gene transfer, e.g. glutathione-*S*-transferase (GST) which catalyses the enzymic metabolic inactivation of atrazine, alachlor, and EPTC in maize and of flurodifen in pea.[15]

Selectivity can be achieved by genetic modification of the crop species; for instance tobacco mutants that were resistant to sulphonylurea herbicides have been isolated by genetic selection in tissue culture.[24] The mutants were unaffected by treatment with chlorsulfuron at a concentration 100 times higher than produced phytotoxic damage in normal tobacco plants.

The tolerance results from the production of a different form of the target enzyme that is insensitive to the herbicide. It may be possible to isolate analogous resistant mutants in other crops; e.g. an atrazine-resistant weed that was found in a treated field has been used to transfer resistance by a genetic cross with oilseed rape.

The discovery and isolation of genes that encode for resistant variants of herbicide target proteins should enable these genes to be transferred into crop plants. In the future genetic engineering may provide the means to greatly extend the selectivity of many existing herbicides.[22]

RESISTANCE TOWARDS HERBICIDES

Resistance of weeds to herbicides is fortunately much less widespread than in the case of insects or fungi. In agriculture, the farmer is generally concerned with control of the whole spectrum of weeds competing with the crop rather just controlling one particular species, as is often the case with insect or fungal pests.[3]

The use of selective herbicides will, over a period, alter the weed species balance, e.g. continued application of the phenoxyacetic acid herbicides, like MCPA, removes competing broad-leaved weeds in cereals; consequently infestation by resistant weeds like wild oats becomes of increasing significance. Similarly, in orchards prolonged use of paraquat causes wireweed to become the dominant weed species since other competing weeds have been eliminated.

In comparison with insects or fungi, plants are slow generators. The seeds often remain dormant in the soil over a considerable time; hence resistant plant strains have been slow to develop. Resistance is most liable to emerge in short-lived annual weeds, which may go through several generations in one season and have been exposed to a single persistent chemical herbicide.

In the late 1960s annual weeds exhibiting resistance to triazine herbicides in maize were reported in the United States of America, and in 1980 the first triazine-resistant weeds appeared in English orchards. Many cases are now on record of resistance in annual weeds like grounsel, fathen, and *Erechitipes hieracifolia*; the former two weeds acquired resistance to the triazine herbicides atrazine and simazine while the latter was resistant to 2,4-D. Chickweed has become less susceptible to the action of 2,4-D after prolonged treatment.

The major weed resistance has been towards the triazine herbicides; these are very persistent chemicals and some orchards and industrial sites have been treated continuously for almost 30 years. Once resistance to simazine had appeared in one weed species, the problem can subsequently develop in other important annual weeds, like meadow grass and chickweed. In plants, resistance is transmitted in the DNA of the chloroplast and is therefore inherited along the maternal line during seed formation.

Triazines act by binding onto the chloroplast membrane where they interfere with photosynthetic electron transport. The mutant gene associated with resistance appears to modify the protein structure of the membrane preventing the herbicide binding onto it. Such a resistance mechanism once established can build up very rapidly. Optimum conditions for the spread of herbicide resistance are a wide area of monoculture and repeated use of the same chemical herbicide. Resistance can often be quickly removed by crop rotation and the use of other types of herbicides. The increasing application of herbicidal mixtures is therefore less likely to lead to weed resistance than application of a single chemical.

The phenomenon of resistance may be used to advantage, if resistance to a particular herbicide can be genetically introduced into an economically valuable crop species. In this way, new crop strains could be produced that are more tolerant to key herbicides, enabling existing herbicides to be used more selectively.

REFERENCES

1. Salisbury, E., *Weeds and Aliens, The New Naturalist*, No. 43, Collins, London, 1961.
2. Whitten, J. L., *That We May Live*, Van Nostrand, Princeton, N. J., 1966, p. 31.
3. Green, M. B., Hartley, G. S., and West, T. F., *Chemicals for Crop Improvement and Pest Management*, 3rd edn, Pergamon Press, Oxford, 1987, p. 207.
4. *The UK Pesticide Guide*, British Crop Protection Council Publications, Binfield, 1988.
5. Hassall, K. A., *The Biochemistry and Uses of Pesticides*, Macmillan Press, London, 1990.
6. Mellanby, K., *Pesticides and Pollution*, Collins, London, 1967.
7. Corbett, J. R., Wright, K., and Baillie, A. C., *The Biochemical Mode of Action of Pesticides*, 2nd edn, Academic Press, London and New York, 1984.
8. Martin, H., and Woodcock, D., *The Scientific Principles of Crop Protection*, 7th edn, Arnold, London, 1983.
9. Pillmoor, J. B., and Gaunt, J. K., 'The behaviour and mode of action of the phenoxyacetic acids in plants', in *Progress in Pesticide Biochemistry and Toxicology* (Eds. Hutson, D. H., and Roberts, T. R.), Vol. 1, Wiley, Chichester, 1981, p. 147.
10. Fletcher, W. W., and Kirkwood, R. C., *Herbicides and Plant Growth Regulators*, Granada Publishing Ltd, London, 1982.
11. Kirby, C., *The Hormone Weedkillers*, British Crop Protection Council, Croydon, 1980.
12. Fryer, J. D., and Makepeace, R. J. (Eds.), *Weed Control Handbook* 8th edn, Blackwell, Oxford, 1978.
13. Matolcsy, Gy Nádasy, and Andriska, V., *Pesticide Chemistry*, Elsevier, Amsterdam, 1988, p. 487.
14. Wain, R. L., *Chemistry and Crop Protection*, RIC Lecture Series No. 3, 1965.
15. Hutson, D. H., and Roberts, T. R. (Eds.), 'Herbicides', in *Progress in Pesticide Biochemistry and Toxicology*, Vol. 6, Wiley, Chichester, 1987.
16. Matsumura, F., and Murti, C. R. K. (Eds.), *The Biodegradation of Pesticides*, Plenum Press, London and New York, 1982.
17. Fedtke, C., *Biochemistry and Physiology of Herbicide Action*, Springer-Verlag, Berlin, 1982.
18. Hatton, L. R., *et al.*, *Pesticide Sci.*, **18**, 15 (1987).
19. Summers, L. A., *The Bipyridinium Herbicides*, Academic Press, London, 1980.
20. Grossbard, E., and Atkinson, D. (Eds.), *The Herbicide Glyphosate*, Butterworths, London, 1985.
21. Bayer, E. M., *et al.*, 'Sulfonylurea herbicides' in *Herbicides Chemistry, Degradation and Mode of Action*, Vol. 3, Dekker, 1987, p. 117.

22. Blair, A. M., and Martin, T. D., *Pesticide Science*, **22**, 195 (1988).
23. Shaner, D. L., *et al.*, *Proceedings of British Crop Protection Conference*, **1**, 147 (1985).
24. La Rossa, R. A., and Falco, S. C., *Trends in Biotechnology*, **2**(6), 158 (1984).
25. Hawkes, T. R., *et al.*, *Prospects of AA Biosynthesis Inhibiting Crop Protection by Pharmaceutical Chemicals*, British Crop Protection Monograph No. 42, 1989, pp. 23, 97, 131, 147, and 155.
26. Gubler, K., *Problems and Opportunities in Herbicide Research*, Chemical Society Special Publication, London, 1979.
27. Parker, C., *Pesticide Science*, **14**, 40 (1983).
28. Hatzios, K. K., and Hoagland, R. E., *Crop Safeners for Herbicides*, Academic Press, London, 1989.

Chapter 9

Plant Growth Regulators (PGRs)

For many centuries farmers have been selecting seeds from their best crop plant specimens and major cereals now yield substantially higher crops as compared with their wild ancestors. In the future, plant breeding techniques will be considerably extended by recent advances in genetic engineering.[1]

In higher plants, growth and development is controlled by natural phytohormones. A hormone is a substance which is produced in minute quantities and exerts its physiological effects at a site distant from the point of origin; all the hormones have specific functions.[2,3]

To produce maximum agricultural and horticultural production using modern mechanical harvesting procedures, the plant must grow in the optimum manner, e.g. wheat with short stems and fruits with long fruiting periods.

Chemists have now modified many of the natural hormones and introduced synthetic plant growth regulators (PGRs) to achieve desirable manipulation of plant growth.[4] Hormones control growth, initiate flowering, cause blossoms, fruit, and leaves to fall, induce setting of fruit, control the overall period of dormancy, and stimulate root development. Plant hormones can be divided into five major types; auxins, gibberellins, cytokinins, ethylene, and abscisic acid.

AUXINS

The principal naturally occurring auxin is indolyl-3-acetic acid (IAA) (1; $n = 1$), which probably occurs universally in plant tissue and is biosynthesized from the amino acid tryptophan. The term auxin applies to compounds that are able to stimulate the growth of excised coleoptile tissue and auxins are involved in various growth processes, e.g. internode elongation, root and leaf growth, differentiation of vascular tissues, induction of fruit growth, development and apical dominance.[5]

271

Auxin readily moves within the plant by polar transport and auxins are widely used in agriculture for promoting rooting in cuttings, fruit setting, and thinning.

The structure of auxin (1; $n = 1$) was determined in 1934 but the biochemical mode of action is still not clearly known,[3] but it involves a three-point attachment to the receptor similar to the phenoxyacetic acid and related herbicides which have auxin-type activity (Chapter 8, page 220). IAA (1; $n = 1$), due to its chemical instability, has little practical use; consequently, the more stable analogues 3-indolylbutyric acid (IBA) (1; $n = 2$), and 1-naphthylacetic acid (NAA) (2) are used instead—IBA to induce rooting of cuttings and NAA (2) to prevent preharvest drop and sucker development in fruit trees[6] and to initiate flowering in cotton.[4] Phenoxyalkanoic acids, e.g. 2,4-D, are used in sublethal doses to promote fruit set and thinning and to increase the yield of beans, corn, potatoes, and sugar beet;[5] 2-(naphthoxy)acetic acid (3) is used as a spray to enhance fruit setting in tomatoes, grapes, and strawberries.[4,6] Propyl-3-t-butylphenoxyacetate (4) is used as a foliar spray in apple and pear orchards to

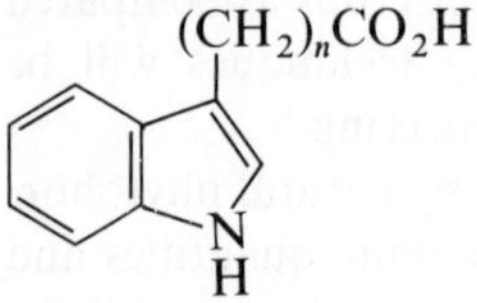

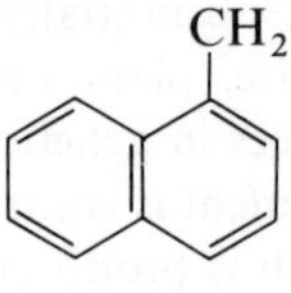

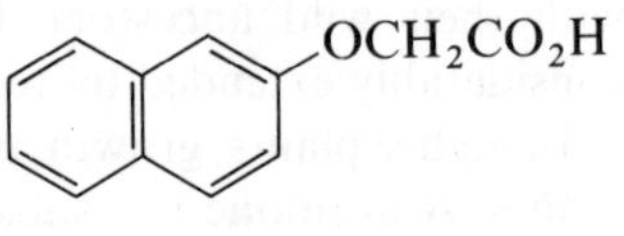

reduce the height of the trees and induce lateral growth. The action is probably due to the chemical blocking the transport of auxin to the shoot tip (Plates 26, 27 and 28, 29).

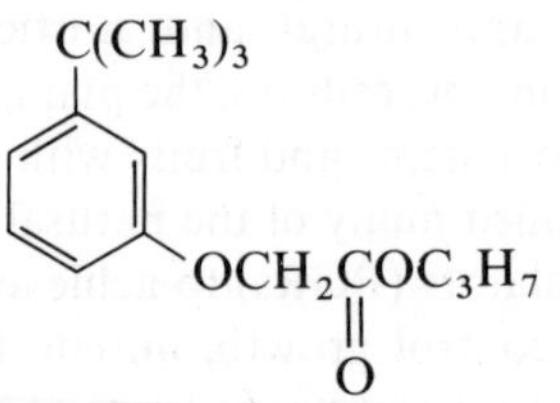

GIBBERELLINS

This second group of plant growth hormones have varied morphological effects which differ from plant to plant. They stimulate cell division or cell elongation or both. Gibberellins may possibly act by modifying auxin levels in plant tissues.[3] In 1957 gibberellic acid or GA3 (5) was introduced for a variety of purposes, the most important of which were inducing germination in barley

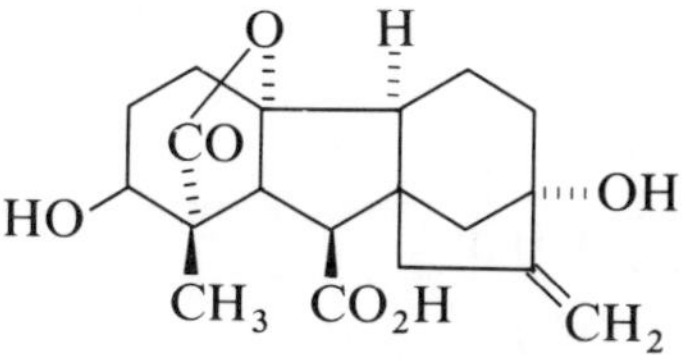

(5)

during brewing and ending dormancy of seed potatoes. Gibberellic acid (5) was originally isolated from the fungus *Gibberella fujikuroi* and caused excessive elongation of stems and leaves in many species of plants. Since then more than 50 gibberellins have been isolated known as GA1, GA2, etc., with gibberellin action. They do not affect root growth but influence the dormancy of buds and seeds. The most characteristic effect of gibberellin treatment is the stimulation of plant growth, as indicated by exceptionally long stems. Gibberellins are believed to be important in mobilizing sugars in certain plant organs and some of their effects may be due to this property, but the precise details of their biochemical mode of action are not known.[3]

The gibberellins are biosynthesized from mevalonic acid.[3] Gibberellic acid (5) has many commercial uses, for instance to break dormancy, initiate flowering, promote fruit growth and setting, delay the ripening of citrus fruits, increase the yield of sucrose from sugar cane,[4] induce sprouting in seed potatoes, and increase the enzymic content of malt to be used in brewing. The physiological action of gibberellins, like auxins, involves extension of plant growth, especially of the stem; they are transported in the xylem and phloem. Gibberellic acid is produced in the seed embryo where it initiates growth, possibly by causing changes in the genes resulting in stimulation of RNA and protein synthesis and increasing the transport of carbohydrates from the endosperm to the embryo.[2]

CYTOKININS

The third type of natural plant hormone is known to control cell division and these substances are known collectively as cytokinins. The first natural cytokinin to be isolated and identified was zeatin (6; R = OH) from immature maize grain in 1964,[2] and this is widely distributed in plants. Many other 6-N-substituted adenines have shown cytokinin-type activity; two examples of synthetic cytokinins are kinetin $\left(7; R = \underset{O}{\overset{}{\diagup\diagdown}}\underset{CH_2}{\diagdown}\right)$ or 6-benzylamino-adenine (R = C_6H_5).

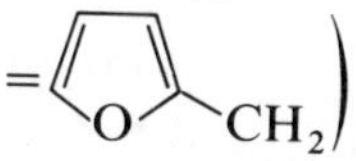

Several non-purine compounds, e.g. N,N'-diphenylurea and benzimidazole, have shown cytokinin activity. Cytokinins delay senescence and hence can be used to prolong the shelf life of fresh vegetables, cut flowers, and promote seed germination. Cytokinins appear to be involved with transfer RNA (t-RNA), and

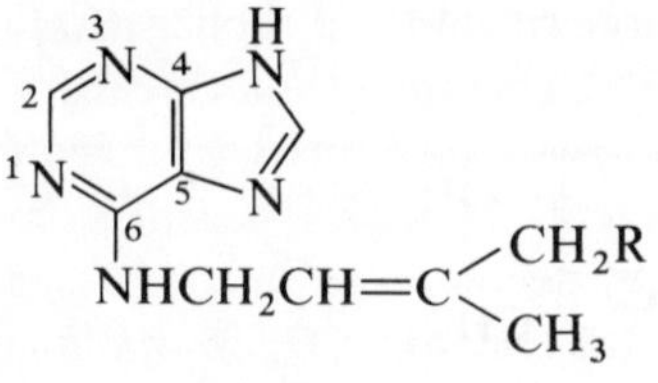

(6)

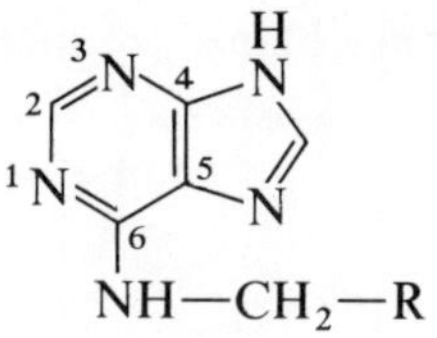

(7)

triacanthine (**6**; R = H), a very potent cytokinin, has been shown to occur in yeast RNA. Cytokinin bases may therefore possibly play a role in regulating the incorporation of specific amino acids into proteins.[2]

ETHYLENE

In contrast to the complex auxins, gibberellins, and cytokinins, ethylene is a simple gaseous organic compound.[4] It is the simplest plant hormone and there is evidence that endogenous ethylene is involved in the normal control of plant growth. The discovery of the role of ethylene in fruit ripening has been of considerable commercial importance in the banana and citrus industry;[5] it is a very active hormone and concentrations of only 0.5 p.p.m. can produce considerable effects in promoting the ripening of fruit. Ethylene has other effects on plants: it prevents the opening of the flower buds of carnations and fruit trees and stimulates latex production in rubber trees.

Ethylene is a powerful abscission agent in fruit trees and is used in thinning apples and peaches. The application of auxin has been demonstrated to promote the production of ethylene by plants; consequently some of the effects shown by auxins, e.g. stimulation of root growth and flower drop (abscission), may be due to enhanced ethylene production. Chemicals have been developed which when sprayed onto plants release ethylene and control different phases of the plant's development, depending on the time of application.[7]

Ethepon (2-chloroethylphosphonic acid) (**8**) was introduced by Amchem Products Inc. (1967) and is prepared from phosphorus trichloride and ethylene oxide as shown in Scheme 9.1. Ethepon (**8**) generates ethylene in solutions with

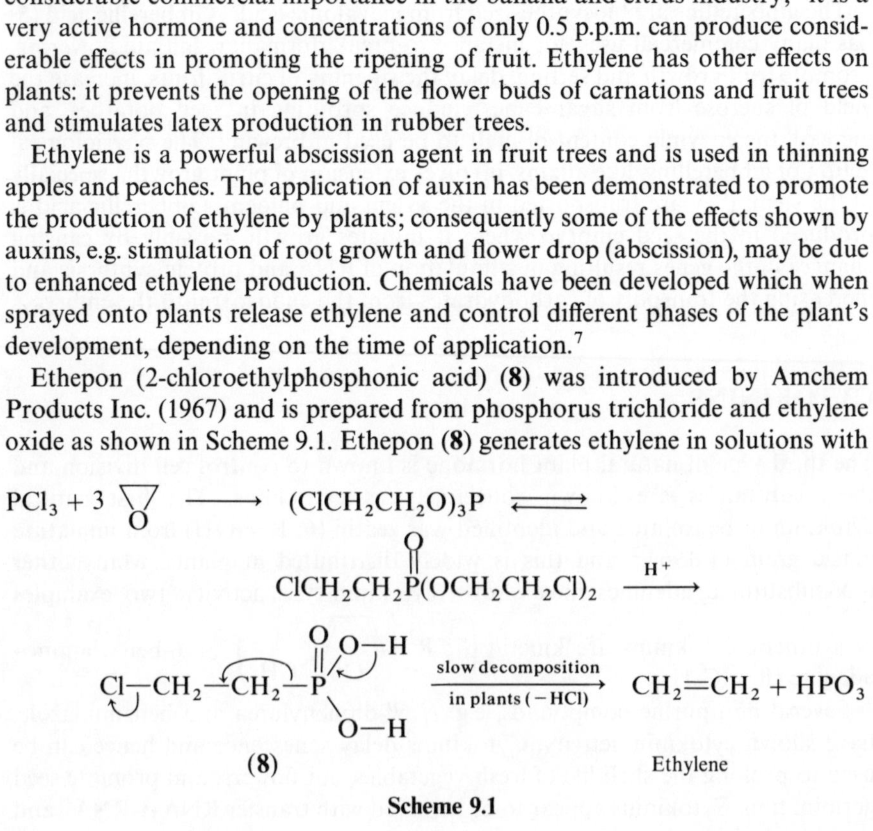

(8)

Ethylene

Scheme 9.1

pH > 4; it is sprayed onto fruit trees to promote ripening, in rubber plantations to stimulate latex production, and to improve stem strength in cereals.[1]

2-Chloroethylmethylbis(benzoxy)silane (9) was introduced by Ciba-Geigy (1972) as a chemical abscission (thinning) agent for apples, peaches, oranges, and other fruits. Like ethepon, this silane (9) decomposes in the presence of water to release ethylene; the driving force for the reaction is the formation of the very powerful silicon–oxygen double bond (Scheme 9.2).

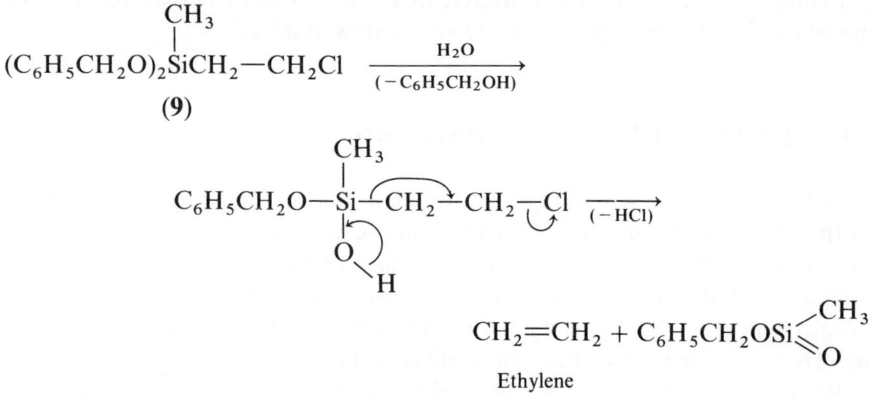

$$(C_6H_5CH_2O)_2\overset{\overset{\displaystyle CH_3}{|}}{Si}CH_2-CH_2Cl \xrightarrow[(-C_6H_5CH_2OH)]{H_2O}$$

(9)

$$C_6H_5CH_2O-\underset{\underset{\displaystyle H}{\overset{|}{O}}}{\overset{\overset{\displaystyle CH_3}{|}}{Si}}CH_2\text{—}CH_2\text{—}Cl \xrightarrow{(-HCl)}$$

$$CH_2{=}CH_2 + C_6H_5CH_2OSi{\overset{\displaystyle CH_3}{\underset{\displaystyle O}{\diagdown}}}$$

Ethylene

Scheme 9.2

The related silicon compound etacelasil (10) was introduced by Ciba-Geigy (1974) as a thinning agent in olives.[5]

$$(CH_3OCH_2CH_2O)_3SiCH_2CH_2Cl$$

(10)

Ethylene is fairly specific, since closely related compounds like propylene are much less effective; the precise molecular mode of action is obscure but probably involves bonding of the molecule to a specific receptor.[3]

NATURAL GROWTH INHIBITORS

There are many natural inhibitors found in plants, of which abscisic acid (ABA) (11) is the best known. In spite of its name ABA appears to be mainly involved in inducing bud dormancy rather than leaf shedding, but it may influence autumn

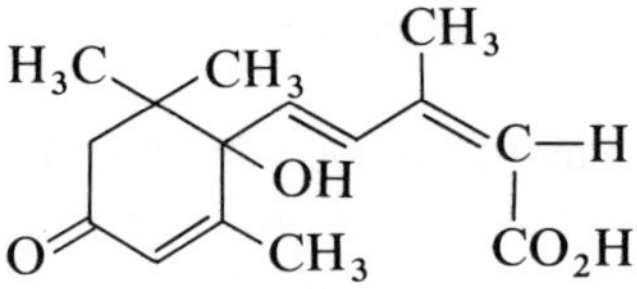

(11)

leaf fall in deciduous trees.[4] ABA has a wide range of physiological effects when applied externally to plants: it accelerates abscission in young cotton plants, leaf senescence, stomatal closure, induces flowering in species requiring short days, and has the reverse effect in those requiring long days.[5]

The natural (+) and synthetic (±) optical isomers of ABA (11) have equal biological activity on plants, but activity is restricted to the 2-*cis* geometric isomers; the ring double bond is also essential.

Endogenous ABA is biosynthesized in leaves via the isoprene route from mevalonate and is transported to the shoot apex and buds.

SYNTHETIC GROWTH INHIBITORS

The discovery of the structure of natural plant hormones and the development of the auxin-type herbicides, e.g. phenoxyacetic acids, led chemists to test many synthetic chemicals for plant growth regulatory properties. Today plant growth regulator (PGR) research is very important and no longer just an area of academic interest. The majority of the agrochemical companies are now actively involved in the search for new, more effective PGRs.

With rising research and development costs, the objectives of the chemical manipulation of plant growth have changed, and the main emphasis is now focused on PGRs formulated to increase the yields of major arable crop cereals, sugar beet, soybean, and cotton for which relatively few regulators currently exist. The small-area crops which attracted innovations in the past can no longer support the current high level of research and development costs.[8]

Synthetic growth retardants include the following compounds: maleic hydrazide (MH) (12) has been used since 1949 and is obtained by condensation of maleic anhydride and hydrazine (Scheme 9.3). It was the first synthetic PGR

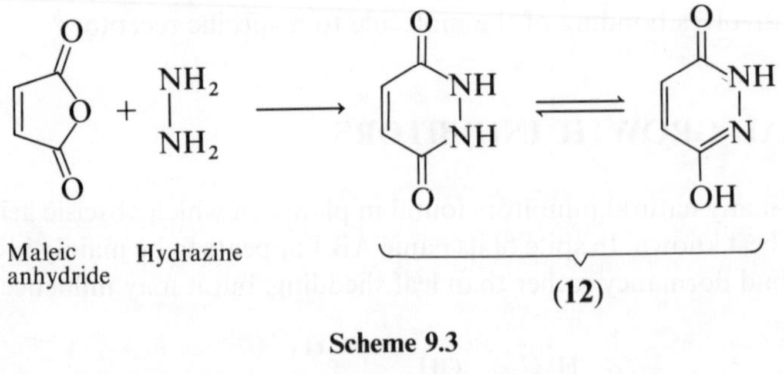

Maleic anhydride   Hydrazine   (12)

Scheme 9.3

and is used as a stunting agent to reduce the rate of growth of grass and some other weeds on verges, to suppress root and sucker growth on tomato, tobacco, and certain trees, and to prevent sprouting of potatoes and onions in storage.[1,4,6]

Maleic hydrazine (12) inhibits cell division in the actively growing tissues of treated plants, but apparently does not affect cell enlargement. It is an isomer of uracil, one of the pyrimidine bases in RNA, so it is suggested that the effect on mitosis may arise from MH becoming incorporated into the RNA molecule, and there is evidence that the compound interferes with DNA/RNA synthesis.[3]

One of the most important synthetic growth regulators is chlormequat chloride (13), introduced by the American Cyanamid Company (1959) and prepared from ethylene dichloride and trimethylamine:

$$ClCH_2CH_2Cl + N(CH_3)_3 \longrightarrow ClCH_2CH_2\overset{+}{N}(CH_3)_3Cl^-$$

(13)

Chlormequat (13) reduces stem length and lodging and produces compact plants with reduced internodes; it is used commercially in fruit trees and for reducing the height of cereals to prevent the crop bending over under its own weight so facilitating mechanical harvesting and allowing more intensive use of fertilizers.[6] It may make the crop more resistant to attack by insects and fungi. The effect of chlormequat on higher plants is competitively inhibited by gibberellin, and growth-retardant properties are probably due to the compound inhibiting gibberellin biosynthesis, as chlormequat does prevent production of gibberellin by the fungus *Gibberella fujikuroi*[3,5] (Plate 30).

Daminozide (14) was introduced by Uniroyal Inc. (1962) and is prepared by reaction of succinic anhydride and *N,N*-dimethylhydrazine (Scheme 9.4). Daminozide is used to control the growth of fruit trees and the shape and height of

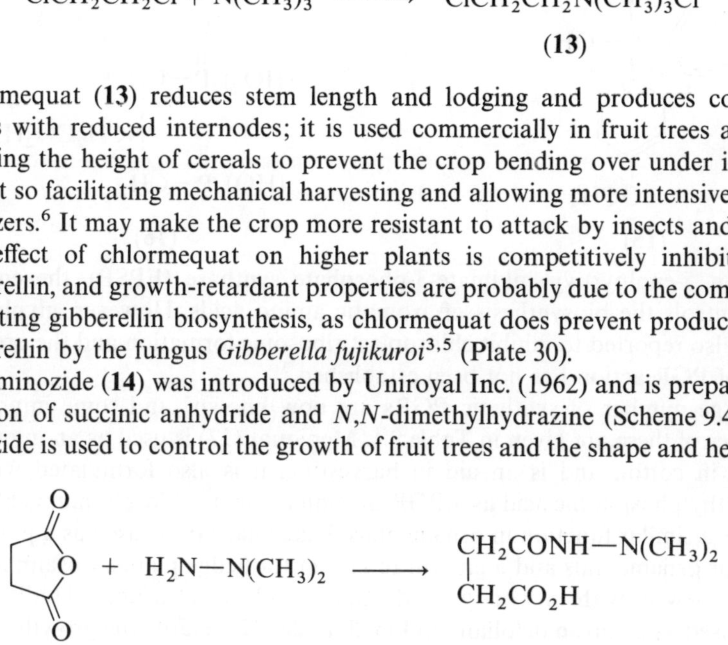

$$\text{Succinic anhydride} + H_2N-N(CH_3)_2 \longrightarrow \begin{array}{l} CH_2CONH-N(CH_3)_2 \\ | \\ CH_2CO_2H \end{array}$$

Succinic anhydride   *N,N*-Dimethylhydrazine   (14)

Scheme 9.4

ornamentals, e.g. to stunt the growth of chrysanthemums, making them more suitable for indoor cultivation.[6] Daminozide affects plant growth similarly to chlormequat and probably also owes its action to interference with gibberellin biosynthesis. The pyrimidine derivative, ancymidol (15) was introduced by the Eli Lilly Company (1971) as a plant growth retardant. It reduced internode elongation and is effective on a large range of plant species by soil or foliar application. The inhibitory effect of ancymidol (15) can be reversed by treatment with gibberellin, again suggesting that this compound acts by interference with gibberellin biosynthesis. It has delayed flower maturation in various greenhouse flowering plants (e.g. chrysanthemums, poinsettias, and Easter lilies) by up to five weeks. It is active at concentrations of only 0.025% and is structurally related to triarimol and other fungicides (Chapter 7) which also show PGR properties.

Glyphosine (16) was introduced by Monsanto Company (1972) and is used to increase the yield of sugar from sugar cane and hasten ripening.[4,5,9] Glyphosine is *N,N*-bis(phosphonomethyl)glycine and is structurally related to the herbicide glyphosate (Chapter 8, page 257). It exhibits some similar but much weaker effects, e.g. stimulation of PAL lyase; the PGR activity may arise from a weak

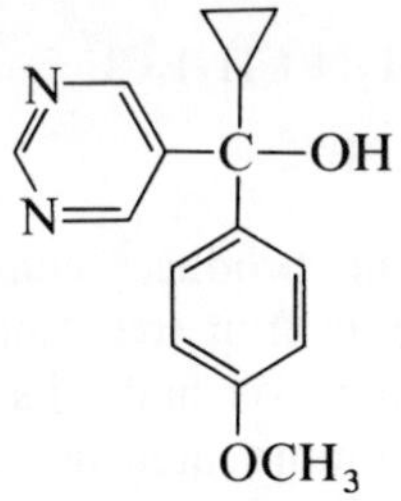

(15) 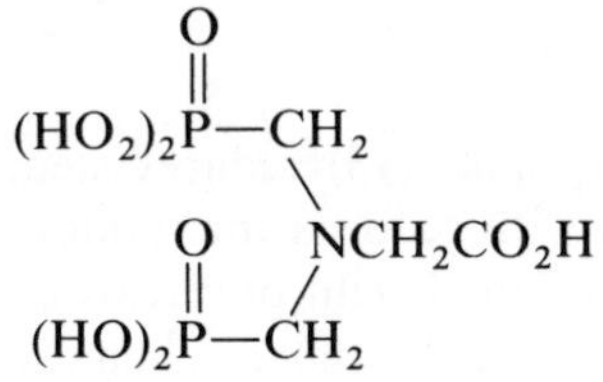   (16)

effect on 5-enolpyruvylshikimate-3-phosphate synthase (EPSP), the enzyme that controls the biosynthesis of aromatic amino acids. However, glyphosine (16) is also reported to inhibit chloroplast ribosome formation and the primary mode of PGR action has not been established.[3]

A large number of synthetic PGRs are now known[5] and some important examples of these are given in Table 9.1. Mepiquat (17) is used as an internode shorter in cotton and is an aid in harvesting; it is also formulated with 2-chloroethylphosphonic acid as a PGR in winter wheat. Chlorphonium chloride (18) has a similar function in ornamentals. Dikegulac (19) is used as a pinching agent for ornamentals and a growth retardant for hedge plants. Fosamine (20) controls new growth in woody weeds and is a chemical pruner. Thidiazuron (21) is used as a cotton defoliant and mefluidide (22) retards the growth of turf and brushwood and increases the levels of sucrose in sugar cane. A number of quaternary ammonium iodides, e.g. *N,N,N*-trimethyl-1-methyl-3-(3,3,5,-trimethylcyclohexyl)-2-propenylammonium iodide, has been developed in Japan (1978) for use as growth retardants in rice and cucumber.[4,5] The biological modes of action of most of the compounds shown in Table 9.1 are unknown mepiquat (17) and chlorphonium chloride (18) probably act by inhibition of gibberellin biosynthesis.[3]

The triazole paclobutrazol (23) was introduced by ICI (1983) as a PGR for use in fruits, ornamentals, and trees.[6] (Plates 31 to 33). It may be applied as a foliage spray or as a soil drench and is easily absorbed by the roots or foliage and translocated in the xylem. Paclobutrazol reduces vegetative growth, increases bud set and flowering, and improves fruit set and quality. In trees it can be used in chemical pruning; the structure is closely related to the triazole fungicides (Chapter 7, page 196) and shows fungicidal activity on apple scab and

Table 9.1 Some synthetic growth retardants

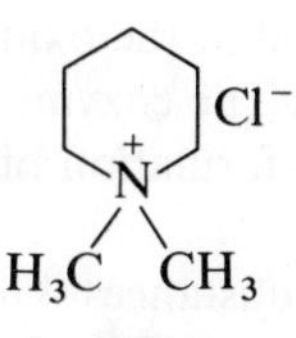

(17)

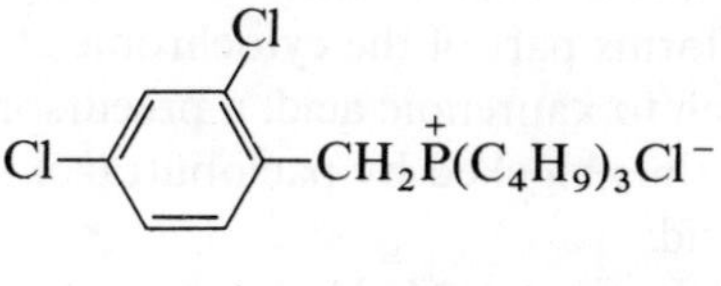

(18)

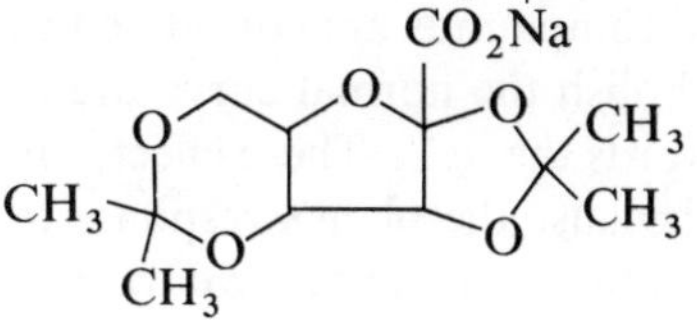

(19)

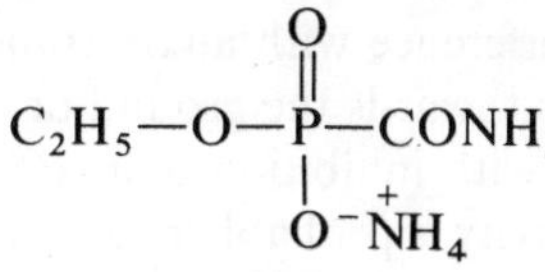

(20)

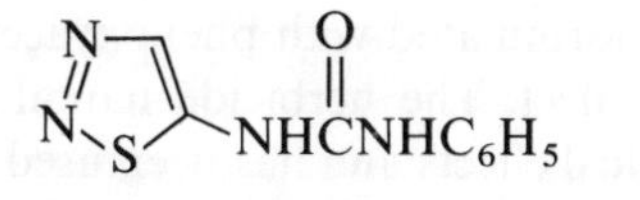

(21)

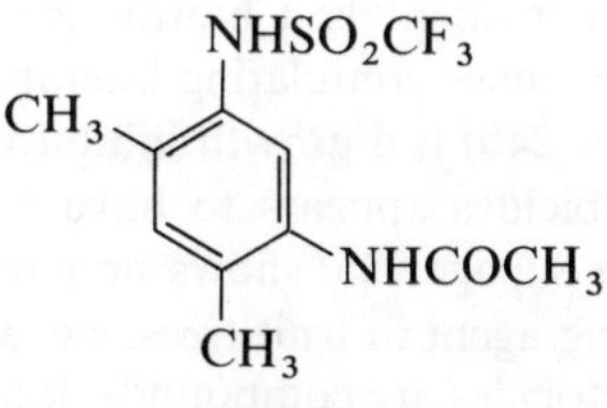

(22)

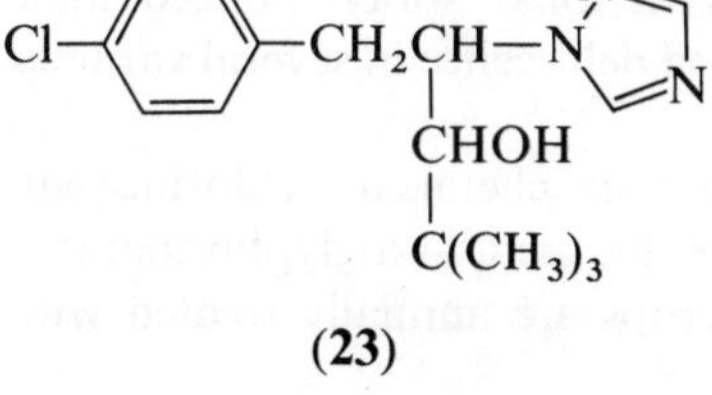

(23)

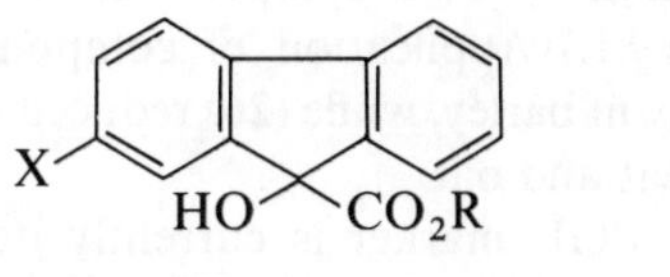

(24a) X = Cl, R = CH$_3$
(24b) X = H, R = C$_4$H$_9$

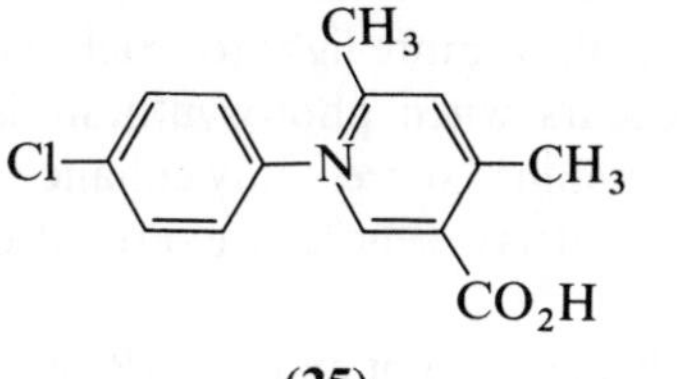

(25)

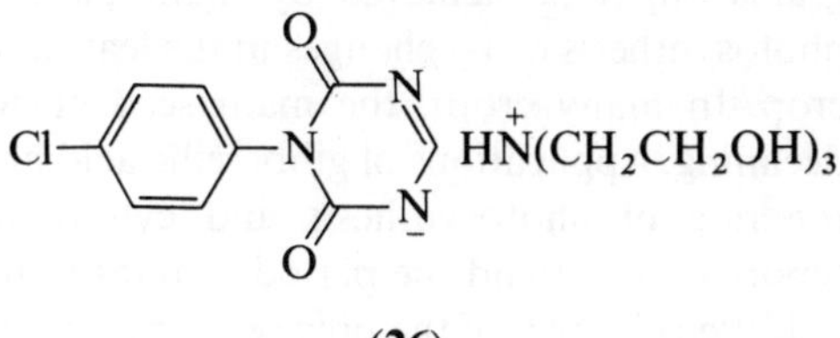

(26)

mildew. The primary mode of action is inhibition of gibberellin biosynthesis by binding a nitrogen of the triazole ring to the iron atom of a protein haem. The latter forms part of the cytochrome P 450 enzyme involved in the oxidation of kaurene to kaurenoic acid, a precursor of gibberellic acid. The enzyme binding site is thus blocked by paclobutrazol (23), preventing the formation of gibberellic acid.

The flurecols (24a,b) are morphogenetically active substances known as morphactins, which can be used either as herbicides or PGRs. They were developed by Schneider (1970) following studies of the growth-regulatory properties of fluorene-9-carboxylic acids.[5] These compounds are considered to act by interference with auxin transport; they abolish the normal curvature of plant roots towards the ground or of shoots towards the light. These effects are consistent with inhibition of auxin movement, because the plant's response to light or gravity is probably related to the redistribution of auxin to the darker or lower side respectively.

Flurecols may also suppress seed germination and the growth of dormant buds in ornamentals. Chlorflurecol-methyl (24a) is used as an apical growth inhibitor, hence stimulating lateral branching, and as a grass retardant. Flurecol-butyl (24b) is a growth inhibitor and when formulated with phenoxyacetic acid herbicides appears to have a synergistic effect. The herbicide naptalam (Chapter 8, page 235) shows similar morphological effects and has been used as a thinning agent in fruit trees, e.g. apples and peaches.

Gametocides are compounds designed to inhibit pollen production or induce pollen sterility; such compounds could play an important role in cereal breeding programmes. Two compounds with gametocidal activity are (25) and (26) (Table 9.1.) Application of compound (25) as a foliar spray induced male sterility in barley, while (26) reduced or prevented dehiscence in several varieties of wheat and oat.

The PGR market is currently dominated by six chemicals: chlormequat, daminozide, maleic hydrazide, ethephon, gibberellic acid, and glyphosine; it is estimated that some one million hectares of crops are annually treated with PGRs.[7]

Enhanced production of carbohydrates and protein to fill and expand cereal grains might be achieved by manipulation of the biochemical pathways of photosynthesis or by changes in the leaf canopy to allow more light to reach the crop. In many crops, the main seed growth occurs when photosynthesis is declining. Applications of gibberellic acid have sometimes successfully enhanced the rate of photosynthesis and cytokinins that delay senescence are also reported[8] to extend the period of photosynthesis.

Little is known of the precise biochemical mode of action of many PGRs and a better understanding of the mechanisms of action and of the structure–activity relationships should permit the introduction of more effective compounds. PGRs represent a new technology in agriculture and they appear to have great potential in improving the yields of many major agricultural crops.

REFERENCES

1. Green, M. B., Hartley, G. S., and West, T. F., *Chemicals for Crop Improvement and Pest Management*, 3rd edn, Pergamon Press, Oxford, 1987.
2. Johnson, A. W., *Chemistry in Britain*, **16**(2), 82 (1980).
3. Corbett, J. R., Wright, K., and Baillie, A. C., *The Biochemical Mode of Action of Pesticides*, 2nd edn, Academic Press, London and New York, 1984.
4. Fletcher, W. W., and Kirkwood, R. C., *Herbicides and Plant Growth Regulators*, Granada, St Albans, 1982.
5. Hutson, D. H., and Roberts, T. R. (Eds.), 'Herbicides', in *Progress in Pesticide Biochemistry and Technology*, Vol 6, Wiley, Chichester, 1987.
6. *The UK Pesticide Guide*, British Crop Protection Council, Binfield, 1991.
7. Thomas, T. H. (Ed.), *Plant Growth Regulators: Potential and Practice*, British Crop Protection Publications, Croydon, 1982.
8. Jeffcoat, B., *Chemistry in Britain*, **20**(6), 530 (1984).
9. Matolcsy, G., Nádasy, M., and Andriska, V., *Pesticide Chemistry*, Elsevier, Amsterdam, 1988.

Chapter 10
Fumigants

The process of fumigation is of great antiquity; thus the burning of aromatic resins, herbs, and incense has been widely practised for centuries as a means of disinfection. In this way, the foul smell of disease and putrefaction could at least be disguised, if not cured!

In the absence of special formulation or apparatus for application, if chemicals are to be effective fumigants they must have high volatility at room temperature. Chemical fumigants, therefore, will be gases or volatile liquids of comparatively low molecular weight, their volatility enabling the chemical to penetrate the material to be protected.

Fumigation is an important method for soil sterilization as a means of killing insects, nematodes, weed seeds, and fungi, which is widely used in glasshouses. It is also useful for protecting crops in store against attack by pests,[1-3] and for killing rodents, e.g. rats, and also moles and rabbits, by treatment of their burrows. In the treatment of stored products, it is of paramount importance that the fumigant should not leave any toxic residues.

Chemical fumigants, because of their volatility, require an enclosed space (e.g. greenhouses, food stores, warehouses) if they are to be effective; otherwise too much of the chemical will be lost to the atmosphere. For soil treatment, the surface of the soil has to be first covered with polythene or other plastic sheeting to cut down loss of the toxicant from the soil surface; citrus trees can be effectively fumigated by enclosing the tree in an airtight tent before introducing the fumigant.[4]

The earliest commercial chemical fumigant was probably hydrogen cyanide, used against scale insects on citrus trees in California in 1886 (Chapter 1, page 5). The hydrogen cyanide is usually generated by the action of acid on sodium or calcium cyanide. Although the gas can cause phytotoxic damage, this is reduced if fumigation is done at night.[3]

Hydrogen cyanide is the most effective fumigant for empty buildings or containers but has poor powers of penetration. It is therefore little used in the treatment of stored grain and is useless for soil fumigation. It has been used

283

against rabbits, moles, and in the fumigation of seeds and nursery stock.[5] It is, however, widely employed by quarantine authorities for treatment of plant materials that might carry infectious diseases. Hydrogen cyanide is a very dangerous acute poison; a concentration of 1000 p.p.m. quickly causes human death; so handling the concentrate requires proper equipment and great care. Although some people can detect the gas by its slight bitter almond smell others cannot, which increases the danger. It is now general practice to add chloropicrin or cyanogen chloride to the concentrate so that people are warned of the presence of the gas by the effects of the added chemicals on the nose, throat, and eyes.

Carbon disulphide is a very volatile and cheap organic liquid; it is flammable and hence is generally used in non-flammable mixtures with, for instance, carbon tetrachloride. These are valuable for fumigation of grain stores;[1,2] carbon disulphide is particularly good against fungi and is useful for the treatment of soils.

Methyl bromide is a widely used non-flammable general fumigant for fumigation of soil, stored grain, and buildings (see Chapter 12, page 303), but has high mammalian toxicity. Ethylene dibromide is a general fumigant;[2] it can be used for treatment of soil or harvested fruit and vegetables. The toxicity is probably due to methylation of vital sulphydryl enzymes.

Sulphur dioxide has only about one-tenth of the toxicity of hydrogen cyanide towards most pests, but is safer and cheaper. Like hydrogen cyanide, sulphur dioxide has poor penetrating properties and is only useful for fumigation of empty buildings.[1]

Ethylene oxide is a gas (b.p. 12°C) and is marketed as a liquid in pressure cylinders; it is mainly employed as a fumigant in grain stores, empty buildings, and for plant disinfection—but is rather toxic to seeds and hence should not be used in seed grain. Propylene oxide is used in fumigation of dried fruit in packets. Phosphine, generated by the action of moisture on aluminium phosphide tablets, is useful to control insects in stored grain and as a rodenticide (Chapter 11, page 289).

Trichloronitromethane or chloropicrin (1) is used as a nematicide (Chapter 12, page 300) and is also one of the best chemicals for disinfection of stored grain, but after fumigation adequate ventilation must be provided otherwise the treated grain becomes tainted. Chloropicrin (1) is a valuable general fumigant, controlling insects, fungi, nematodes, and weed seeds in soil.[4]

$$Cl_3C-N\begin{smallmatrix}O\\\\O\end{smallmatrix}$$

(1)

Formaldehyde, often as a 40% aqueous solution (formalin), is generally used as a surface dressing for seed beds to combat soil fungi, e.g. those associated with 'damping-off' diseases which cause high losses of seedling plants. Formaldehyde has poor penetration in soil which makes it rather ineffective in normal soil sterilization.[1] Methyl formate is mainly used for fumigation of animal furs and skins in store.

Soil fumigation is generally performed before introduction of the crop so complete sterilization is most satisfactory and little selectivity is required; this is also true for chemicals used in fumigation of buildings and grain stores. If fumigation is to be successful, there must be rapid diffusion of the fumigant into the soil or mixing with air in the space to be treated, which must be sealed off from the outside atmosphere. In this respect, the practice of fumigation has been substantially assisted by developments in the plastics industry—field clamps of vegetables, houses, and even food factories can be sealed off by plastic covers, enabling effective fumigation to be carried out. Agricultural land can be covered by polythene sheeting held down by being embedded in soil all round the perimeter which permits effective use of methyl bromide for fumigation. This chemical is so volatile that too much would be lost to the atmosphere without the plastic cover, whereas the less volatile ethylene dibromide (b.p. 131°C) can be used by injection into the soil. In the fumigation of grain, initial treatment with the chemical fumigant under slightly reduced pressure helps to draw the chemical into the interior of the store more effectively than when diffusion only is relied on. Grain is especially sensitive to fungal attack when stored under conditions of high moisture content and the moisture content also alters the effect of the chemical fumigant on the viability of the seed. The dry seed being much more resistant to the chemical, the moisture concentration should not exceed 16% during fumigation and storage.

Other fumigants include methyl isothiocyanate (MIT) (2) and sodium N-methyldithiocarbamate dihydrate (metham-sodium or SMDC) (3).[4] MIT (2) is applied in soil sterilization and is highly toxic to germinating weed seeds and nematodes; it is often used as the water-soluble SMDC (3) which gradually evolves the active fumigant MIT (see Chapter 12, page 302). Tetrachloroethane is valuable as a greenhouse fumigant against whitefly,[5] but its use is limited by phytotoxicity; apparently it affects the waxy coating of the larvae so that the adults do not emerge.

$$H_3C-N{=}C{=}S \qquad H_3C-NH-C\overset{\displaystyle \nearrow S}{\underset{\displaystyle \searrow SNa}{}} \cdot 2\,H_2O \qquad CH_3-\overset{\displaystyle NH_2}{\overset{|}{C}}HCH_2CH_3$$

$$\textbf{(2)} \qquad\qquad\qquad \textbf{(3)} \qquad\qquad\qquad\qquad \textbf{(4)}$$

The original fumigants were highly volatile, low molecular weight molecules, but recently the application of the method of fumigation has been considerably extended by volatilization of chemicals from thermostatically controlled heaters or by their formulation as pyrotechnic mixtures.[5] In these ways compounds of comparatively low vapour pressure at room temperature can be successfully used as fumigants. Examples are HCH, dichlorvos, and azobenzene (see Chapter 2, page 23).

2-Aminobutane (**4**) has been introduced as a fungicidal fumigant to protect harvested fruits; the activity against blue mould on orange is mainly due to the $R(-)$isomer.[5]

1,1-Dichloro-1-nitroethane is useful for the long-term sterilization of stored products and 2,2-dichloroethyl ether is used in the sterilization of soil.

REFERENCES

1. Green, M. B., Hartley, G. S., and West, T. F., *Chemicals for Crop Improvement and Pest Management*, Pergamon Press, Oxford, 1987, p. 299.
2. Woods, A., *Pest Control*, McGraw-Hill, London, 1974, p. 92.
3. Fletcher, W. W., *The Pest War*, Blackwell, Oxford, 1974, p. 95.
4. Ware, G. W., *Fundamentals of Pesticides*, Thomson Publications, Fresno, Calif., 1986, pp. 92, 162.
5. Martin, H., and Woodcock, D., *The Scientific Principles of Crop Protection*, 7th edn, Arnold, London, 1983, p. 321.

Chapter 11

Rodenticides

In certain circumstances many vertebrates, even elephants, can rank as pests, but among mammals by far the most serious are the rodents (e.g. squirrels, rats, and mice). The majority of poisons effective against vertebrate pests are also very toxic towards man and domestic animals, although there are a few examples of specific pesticides in this field. However, even with generally toxic chemicals, they can often be applied in such a way that they are not taken by man or domestic animals.[1,2]

Three rodent species which are major pests of agriculture, stored produce, and present a potential danger to public health throughout the world are the common brown or Norway rat (*Rattus norvegicus*), a major pest in temperate regions; the black or ship rat (*Rattus rattus*), a dominant pest in warmer climates; and the house mouse (*Mus musculis*). The latter is a less important pest economically, but tends to live in close association with man in his dwellings to a greater extent than rats and therefore demands effective control.

The rat is the traditional enemy of man, often living in his homes and consuming and fouling the produce of his labour. Rats are carriers of several deadly diseases (Chapter 1, page 8). In Britain black rats, the main carriers of diseases, are largely confined to ports and large towns, but the brown or common rat is much more numerous, adapting itself to both town or country habitats, and does extensive damage especially to foodstuffs such as grain. The brown rat follows man all over the world and probably came to Britain from the East on trading ships about 1740. Rats breed very rapidly and the female produces up to 50 young per year, although the majority die before becoming adults and most of the adults do not live more than a year.[3]

The high rate of reproduction and turnover makes rats a difficult pest to control. They will eat almost anything, all kinds of vegetables, and animals. In many countries, including Britain, the black rat has been driven out by the brown rat and has now become quite rare.

Rats are very destructive to the fabric of buildings and furniture where this is sufficiently soft to be gnawed in their efforts to gain access. Wooden buildings are particularly vulnerable to attack by rats; the most important defence against

287

these pests is a sound building, and perhaps the greatest contribution of modern technology has been the manufacture of Portland cement.

There is often a public outcry against the use of chemical pesticides against birds like pigeons, sparrows, and bullfinches or against grey squirrels; however, rats are so universally detested that no one objects to the use of poisons to control them.

In Britain, as in many other countries, the use of poisons against rodents and other pests is controlled by the Protection of Animals Acts 1911 to 1972. These permit the laying of poisons against 'insects, rodents and other small ground vermin', including squirrels and coypus[2] in certain cases. The fumigation of rabbit burrows and the controlled use of baits containing stupefying narcotic substances against certain bird pests is also allowed, e.g. wood-pigeons, sparrows, jays, crows, and magpies. Cinnamic acid derivatives, e.g. cinnamide and 4-hydroxy-3,5-dimethoxycinnamic acid, are bird repellents which discourage birds from feeding on treated crops. The compounds were demonstrated to significantly inhibit the activity of trypsin and chemotrypsin at 10^{-5} molar concentration by *in vitro* experiments on the quail intestine; this provides the basis for the repellent action.

In special situations (e.g. warehouses, ships' holds) rats and other animals and insects can be eliminated by fumigation with such poisons as sulphur dioxide, hydrogen cyanide, or methyl bromide (Chapter 10). More generally rodenticides are laid down as baits to control rodent pests in buildings and on farms. If the bait is to be safe and effective the following criteria must be satisfied:

(a) It must not be unpalatable to the target pests.
(b) It must not induce bait shyness, which often happens if more than one dose is needed to kill and the first feed results in painful symptoms.
(c) The poison should make the rats go out into the open to die; otherwise the rotting corpses create health hazards.
(d) It should have a much lower toxicity to domestic animals, especially cats and dogs, which often eat the poisoned rats; sometimes this problem can be reduced if the pesticide is rapidly degraded to less toxic substances in the corpse.

In many ways, the most satisfactory method of control would be by introducing an infectious disease specific to rats. At one time this appeared to have been achieved with 'Liverpool virus'—a *Salmonella* bacterium presented as an infected meat bait—but unfortunately the rats quickly developed an immunity to this not very specific disease so the procedure had to be abandoned.[2] A very successful illustration of this general method was introduction of the myxomatosis virus to control the rabbit population in 1953. During the period 1954–5 some 90% of the total rabbit population in Britain was killed by the spread of this specific disease. Rabbits have now developed significant immunity to the disease and other control methods, e.g. poisoning and shooting, will increasingly have to be used in the future.[4] The use of a mammalian sex attractant is being studied, but at present the most effective method is the introduction of a powdered mixture of sodium cyanide, magnesium carbonate, and anhydrous magnesium sulphate (Cymag) into the rabbit burrows, when the powder gradually liberates lethal hydrogen cyanide gas in contact with moist soil.[3,5] Aluminium phosphide pellets evolving poisonous phosphine gas are useful to control mole rats in paddy fields, and both sodium cyanide and aluminium phosphide are effective against moles by placement in their burrows.

Rodenticides may be classified either as acute (single-dose, rapid-acting) poisons or as chronic (multiple-dose, slow-acting) poisons. Chronic poisons are taken at present only to include the anticoagulants and have advantages in efficacy and safety, but are slower to use and more expensive as regards labour and bait material; also, resistance to anticoagulants is becoming an increasing problem. Acute poisons have the advantage of rapid kill, but may be too hazardous or not effective for use in many circumstances. Ideally, a candidate rodenticide needs the three attributes of toxicity, acceptability, and safety in use.[4]

Early rodenticides included strychnine, now only important for controlling moles in their burrows, white arsenic, yellow phosphorus, squills, and canthar-ides. By 1939, however, only red squill and arsenic remained much in use. The natural product scilliroside (page 295), extracted from the bulb of the red squill plant,[6] was a safe poison because it was rapidly detoxified in the rat's body and, if directly consumed by other animals, quickly acts as an emetic and so they do not ingest sufficient quantities of the poison to harm them. The specific toxicity of the product to rats therefore depends on the rat's inability to vomit (LD_{50} (oral) to rats, 0.45 mg/kg).

Barium carbonate, zinc phosphide (Zn_3P_2), and arsenious oxide came into prominence later. Although moist zinc phosphide emits an unpleasant smell which repels most animals it is surprisingly well accepted in baits by the generally suspicious rat.[6]

Thallium sulphate (Tl_2SO_4) has been used for a long time as a rodenticide and also to control some bird pests. It is a general cellular toxin and inactivates a number of vital enzymes, such as those containing thiol groups. Thallium sulphate has a very high mammalian toxicity and the toxicant is readily absorbed through the skin. It has cumulative and secondary, as well as acute, toxicity hazards; like barium carbonate, poisoning with thallium sulphate causes rats to seek water and consequently they die in the open. Death arises from respiratory failure and thallium sulphate must be used only with the greatest care by expert operators; it is banned in several countries including Britain.

The first synthetic organic rat poison was α-naphthylthiourea (antu) (**1**), reported in 1945, obtained by reaction of α-naphthylamine and ammonium thiocyanate[5] (Scheme 11.1).

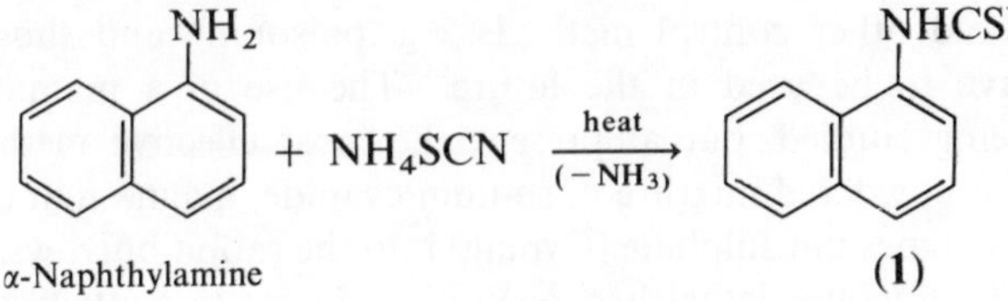

NH$_2$ NHCSNH$_2$

α-Naphthylamine + NH$_4$SCN $\xrightarrow[(-NH_3)]{\text{heat}}$ (1)

α-Naphthylamine   (1)

Scheme 11.1

Compound (1) is a somewhat specific poison for adult Norway rats; the toxic dose is 6–8 mg/kg and it is less effective against other species of rats. It suffers from the disadvantage that it induces extreme bait shyness in those rats that do not consume a fatal dose at the first feed and tolerance develops in rats by repeated administration of sublethal doses.[1,2] It has been banned in Britain and some other countries because of the carcinogenicity of occasional impurities such as β-naphthylamine.[4]

Fluoroacetic acid obtained by heating together sodium fluoride and chloro-acetic acid at high temperature and pressure was found (1945) to be an effective rodenticide either as the sodium salt (sodium fluoroacetate) or as the amide (fluoroacetamide).[2,6] The toxicity of these compounds arises from their role in the biosynthesis of fluorocitrate, which blocks the action of the enzyme aconitase in the tricarboxylic acid cycle.[6,7] Both compounds are highly toxic to rats and most other animals and there is no recognized antidote. The flesh of poisoned animals remains toxic for some time and there have been several instances of secondary poisoning. Welsh sheepdogs have been killed as a consequence of eating oatmeal baits containing fluoroacetate.[2,4] Owing to their hazardous nature, the use of fluoroacetates is restricted, but they are still employed against rats in sewers, ships, and port areas. The demand for other rat poisons was almost completely eliminated by the introduction of warfarin or 3-(α-acetonylbenzyl)-4-hydroxycoumarin (2), which seemed to make other rat poisons obsolete. It was discovered by the Wisconsin Alumini Research Foundation (1944) and was for several years the most successful of a number of hydroxycoumarin derivatives. Warfarin is prepared by condensation of 4-hydroxycoumarin with benzalacetone[2,7] (Scheme 11.2).

Spoiled sweet clover has long been known to be highly poisonous to cattle and in 1941 Dr K. P. Link and his colleagues at Wisconsin found that the active ingredient in the sweet clover was dicoumarin (3) which interferes with the action of vitamin K in the body, reducing the formation of prothrombin, the blood clotting factor, so that a minor injury can cause a fatal haemorrhage.[6] Dicoumarin was ineffective as a rat poison and its derivatives were even less active. However, derivatives of 4-hydroxycoumarin substituted in the 3-position were good anticoagulants and a range of these derivatives was examined; the forty-second on their list was selected as the most promising rat poison and became known as warfarin or WARF 42 (2).[5,7] Warfarin, and other similar compounds, function as anticoagulants so that the rats eventually die from

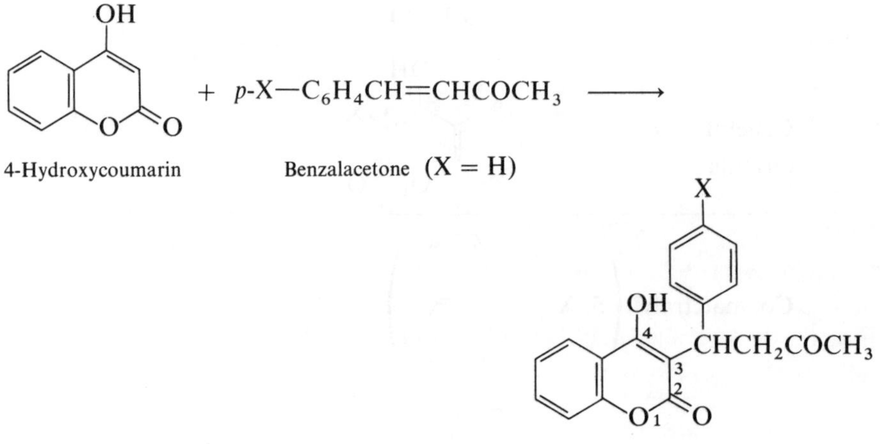

OH

$+$ $p\text{-X}-C_6H_4CH{=}CHCOCH_3$ $\longrightarrow$

4-Hydroxycoumarin   Benzalacetone $(X = H)$

(2) $X = H$
(4) $X = Cl$

Scheme 11.2

internal haemorrhage after coming into the open in search of water. Rats are very susceptible to this toxicant and do not develop bait shyness; a dose of 1 mg/kg for five days is lethal, whereas other animals are generally less affected, although the safety is mainly dependent on use of a suitable bait placement. Warfarin has also been the main weapon to control the spread of the grey squirrel, but they need a higher dosage than rats.

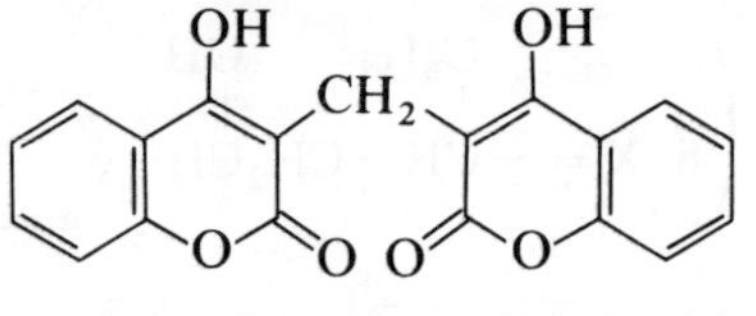

(3)

Normally these anticoagulant (chronic) rodenticides control rodent pests effectively within two to five weeks and are relatively safe, largely because other animals rarely have access to the baits over the prolonged period needed to accumulate a lethal dose; also the anticoagulatory effects can be readily reversed by vitamin K treatment.

The *p*-chloro derivative of warfarin known as coumachlor (**4**; $X = Cl$) (Ciba-Geigy, 1951) had similar anticoagulant properties and several related 3-substituted-4-hydroxycoumarins have also been introduced (Table 11.1).[2,6] Rats have developed widespread resistance to warfarin (**2**) together with some other anticoagulants, and the emergence of these 'super rats' presented a serious problem. Later studies,[7] however, showed that other coumarin derivatives, e.g. (**5**) to (**9**) (Table 11.1), were effective against warfarin-resistant rats. Difenacoum (**6**) functions as a chronic rodenticide similar to warfarin but more active (LD_{50}

Table 11.1

General formula	

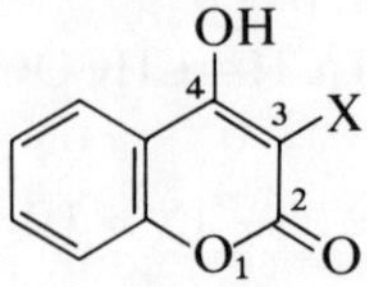

Coumatetralyl	5; X =

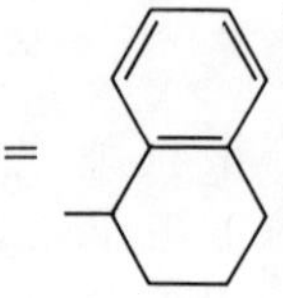

Difenacoum	6; X = , Y = H

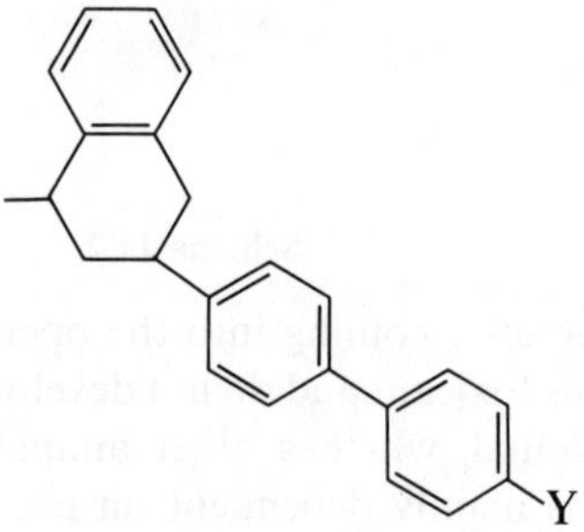

Brodifacoum (7; X as in 6; Y = Br)

Bromadiolone	8; X =

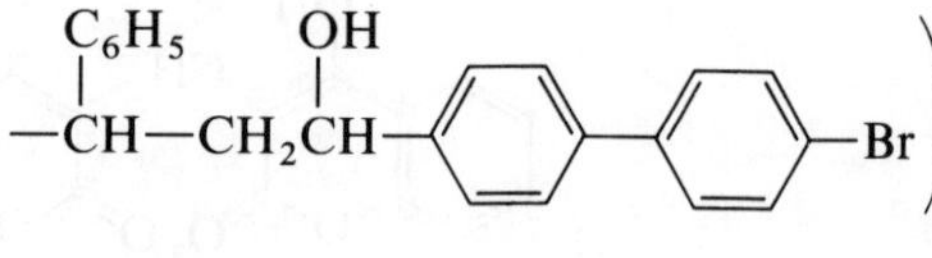

	9; X =

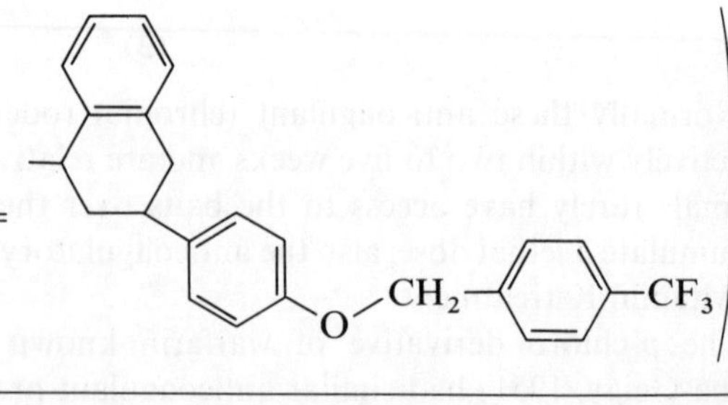

(oral) to rats is 1.8 mg/kg[7]). Brodifacoum (7) (1978) is specially useful because it combines a very high toxicity to rats and mice with low toxicity to man and domestic animals. With this compound, ingestion of a single dose of a bait containing only 50 mg a.i./kg is often fatal; this is advantageous against sporadic feeders or where the bait tends to deteriorate rapidly. Bromadiolone (8) is a very palatable and powerful anticoagulant which can achieve complete kill of rats by one day's feeding on a bait containing only 0.005% of bromadiolone (**8**). The LD_{50} (oral) to rats is 1.1 mg/kg.[7] The trifluoromethyl compound (**9**) (Shell, 1984) is more active than other members of the group—fatality is achieved by a smaller dose.[8] The normally recommended baiting technique for anticoagulants involves placing baits (ca. 250 g each) and replenishing the baits every two to four days until feeding ceases; this saturation baiting technique is very effective for the majority of anticoagulants for which repeated feeding is necessary for ingestion of a fatal dose.

However, with very active compounds like brodifacoum (**7**) the technique of 'pulsed baiting' can be employed in which only sufficient bait for a lethal dose is put out and the application is repeated after an interval of, say, one week. Trials with 'pulsed' application of brodifacoum (**7**) indicate a 75% saving in bait used as compared with the standard saturation baiting technique; good field control of rodents has been achieved with a rate of 1–3 kg/ha of brodifacoum.

The newer 'second-generation' anticoagulants, e.g. brodifacoum (**7**) and bromadiolone (**8**), are currently the most commercially important rodenticides.

The 4-hydroxycoumarin anticoagulants, e.g. warfarin (**2**) and brodifacoum (**7**), owe their activity to their ability to block the reductase-catalysed conversion of vitamin K_1 epoxide to vitamin K_1. This reduces the supply of vitamin K and hence prevents the formation of the essential blood clotting factors.

Several substituted indanediones also interfere with blood coagulation; an example is pindone (**10**), prepared by condensation of diethyl phthalate and pinacolone[2,7] (Scheme 11.3).

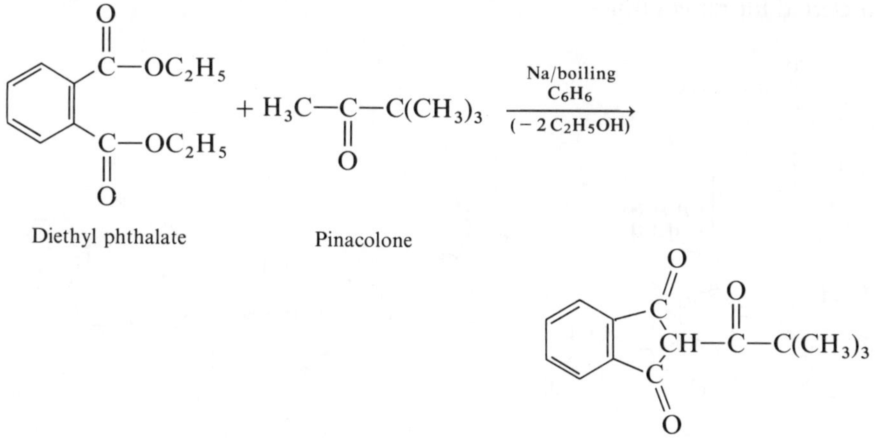

Diethyl phthalate   Pinacolone

(**10**)

Scheme 11.3

Another important example is chlorophacinone (**11**) which acts as a single-dose rodenticide; it is an anticoagulant which also uncouples oxidative phosphorylation.[5] Such indanediones can be used as a substitute for warfarin when rats become bait shy.

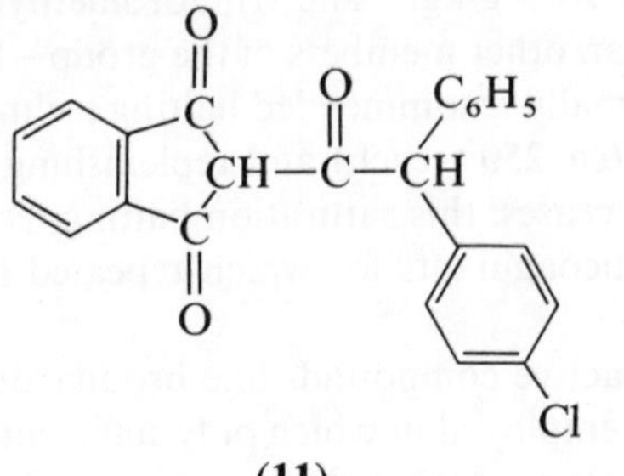

(11)

In 1964 Roszkowski and coworkers[9] reported a rodenticide called norbormide (12), obtained by reaction of the condensation product of cyclopentadiene and 2-benzoylpyridine with maleimide[7] (Scheme 11.4). Norbormide (12) is selectively toxic to rodents of the genus *Rattus* and particularly the Norway rat (*Rattus norvegicus*), to which it has an oral LD_{50} value of about 11 mg/kg;[4,9] dosages of 1000 mg/kg have produced no ill effects on cats, dogs, monkeys, or mice.[9] It is especially valuable when safety is of paramount importance during rat extermination programmes. The compound acts by constriction of the smaller peripheral blood vessels, leading to a catastrophic increase in blood pressure. Norbormide initially appeared extremely promising as a rodenticide but, unfortunately, it suffers from a low acceptability to rats and easily induces bait shyness so that non-selective, more efficacious poisons are generally to be preferred for rat control.

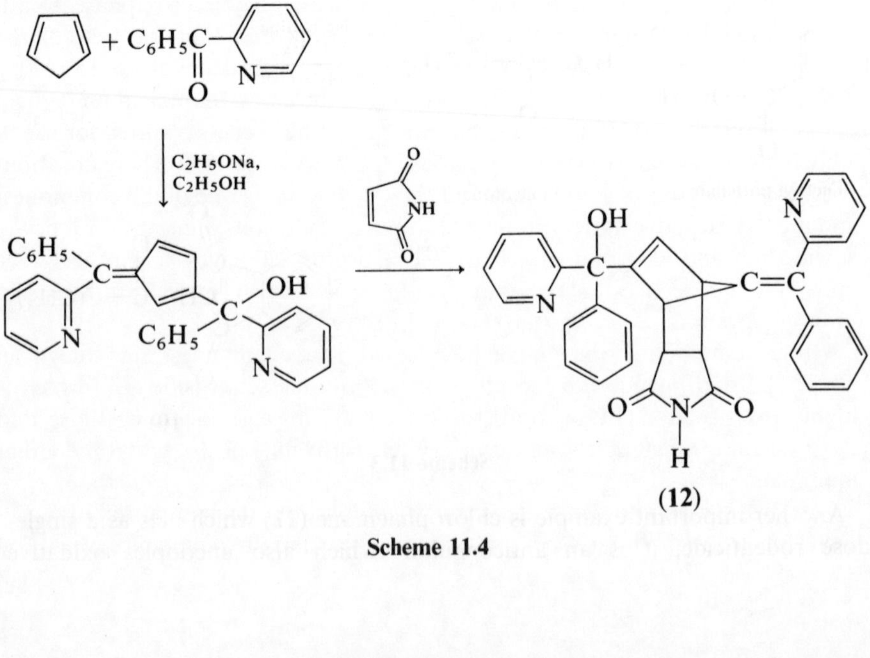

(12)

Scheme 11.4

The widespread resistance developed by rodents to the anticoagulant chemicals has encouraged the development of new chemical rodenticides; one of considerable promise is alphachloralose or chloralose (13), prepared by condensation of chloral (trichloroacetaldehyde) with glucose;[5] it is also of considerable value in bird control.

$$Cl_3C-\underset{O}{\overset{O}{\underset{\diagdown}{\diagup}}}CH\underset{\overset{|}{CH}}{\overset{\diagup O}{\underset{\diagdown}{\diagup}}}CH-CH(OH)CH_2OH$$

(13)

Chloralose probably has the chemical structure (13) and exists as two stereoisomers (α-m.p. 187°C and β-m.p. 227–230°C); both isomers have low aqueous solubility. Baits are usually formulated by incorporating about 1.5 % of the active ingredient as a dust onto moist grain[2] to make birds eating the grain sleepy. If they are left alone the birds completely recover.[4] The farmer can kill the pest birds, e.g. pigeons, sparrows, and starlings, while the other birds are released. Use of large grain such as field beans makes the operation more selective. Chloralose and other narcotics can also be applied to control birds in food stores without any danger of contamination of the food by either the poison or dead birds.[3] Bird repellents that can be used include the insecticide methiocarb and 4-aminopyridine. The latter is valuable to cause birds to keep off fruit. More effective methods of dispersing birds from such places as buildings and airports are urgently needed.[10]

Chloralose (13) is primarily a sleep-inducing narcotic drug; the hypnotic phase of its action is often preceded by a convulsive or hyperactive phase. Death of the animal generally results from hypothermia and therefore use of the compound as a rodenticide becomes unreliable at temperatures above 15°C. It is now widely employed as baits containing 4 % of the active ingredient for control of mice and other small mammals but not rats. It has been permitted for use in stupefying baits against pigeons and house sparrows since 1959. There are about 5 million wood pigeons in Britain and they now rank as one of the commonest and most destructive agricultural pests in Europe. Pigeons dig up and eat newly sown grain and in winter attack Brussels sprouts, cabbages, and other green crops—it has been estimated that pigeons cost British agriculture about £5 million per annum.

Other compounds that have been used as rodenticides are shown in Table 11.2 and include the complex natural product scilliroside (14), which is highly toxic to rats (LD_{50} (oral) to rats, 0.4–0.7 mg/kg), including those that have become resistant to warfarin, but is relatively safe to birds and other mammals.[7]

Another synthetic rodenticide is the organophosphorus compound phosacetim (15), an anticholinesterase agent that is particularly toxic to voles (the LD_{50}

Table 11.2

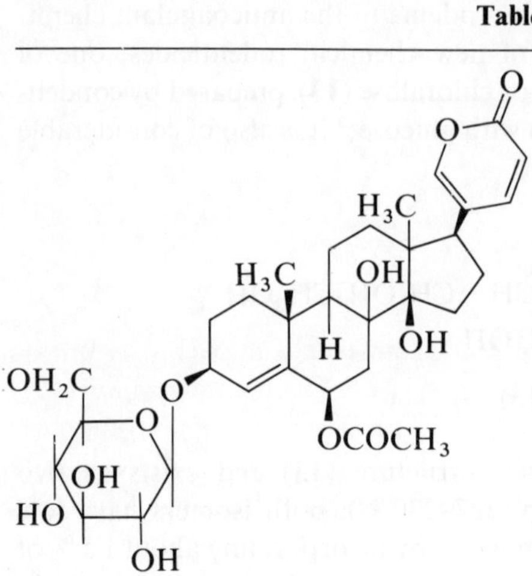

(14)

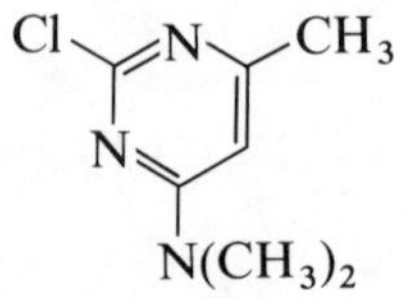

(16)

(18)

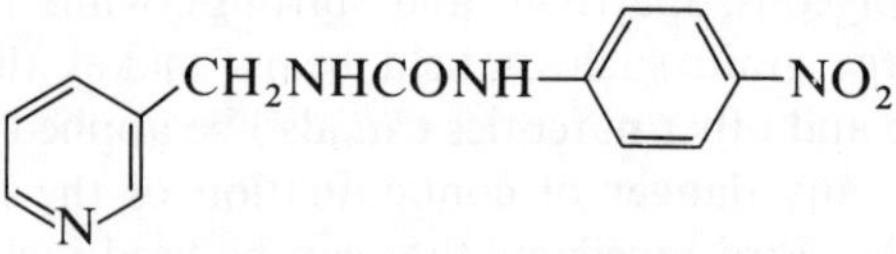

$$NH_2\overset{\overset{\displaystyle S}{\|}}{C}NHNHCH_2NHNH\overset{\overset{\displaystyle S}{\|}}{C}NH_2$$

(17)

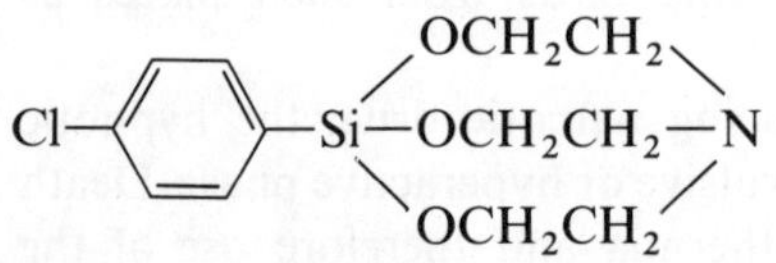

(19)

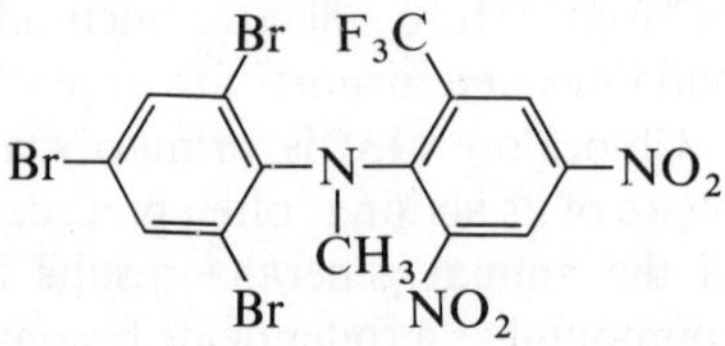

(20)

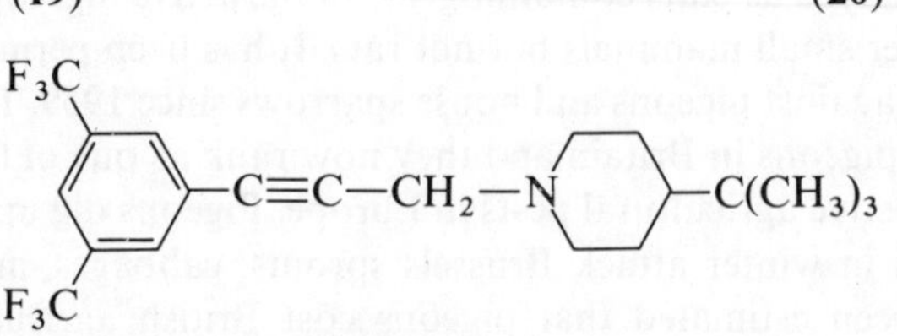

(21)

(oral) for male rats is 7.5 mg/kg). It is synthesized by condensation of *O,O'*-di-*p*-chlorophenylphosphorochloridothioate with acetamidine[2,7] (Scheme 11.5).

Many organophosphorus compounds are highly toxic to rodents, but few have proved effective rodenticides, possibly because the onset of the cholinergic effects is often so rapid that the ingestion of a lethal dose is prevented. The

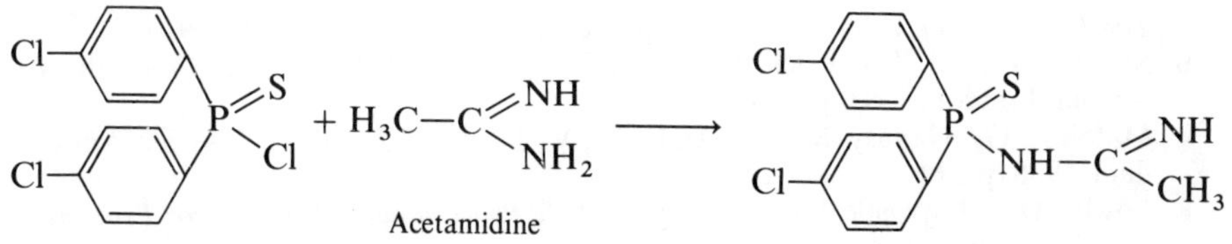

Acetamidine

(15)

Scheme 11.5

pyrimidine derivative crimidine (**16**) is useful against field mice, but is not a cumulative poison due to its rapid metabolism *in vivo*.[6] The urea pyriminyl (**17**) is a very potent broad-spectrum rodenticide (LD_{50} (oral) to rats, 4.7 mg/kg) which is effective as a single dose against rats resistant to the anticoagulant rodenticides. It has low toxicity towards domestic animals, e.g. dogs, pigeons, and chickens, although cats are more sensitive.[7]

The thiosemicarbazide (**18**) is a very fast-acting rodenticide which is effective against rats (LD_{50} (oral) to rats, 25–30 mg/kg). The silicon compound *p*-chlorosilatran (**19**) is another single-dose rodenticide (LD_{50} (oral) to rat, 1–4 mg/kg).

There is little doubt that rats will soon acquire a widespread resistance to the anticoagulants that are currently our major weapon for their control. It is unfortunate that none of the compounds listed in Table 11.2 have achieved appreciable commercial success. The discovery and development of novel, effective, non-coagulant rodenticides therefore remains a vital goal for research in this area of pest control.

The diphenylamine (**20**) developed by Eli Lilly (1979) is an acute single-dose rodenticide with LD_{50} (oral) to rats and mice of approximately 20 mg/kg.[6,7] The mode of action is unknown but the compound does not interfere with blood clotting and shows considerable promise, as does α-chlorohydrin, which either kills or induces permanent sterility in male rats.[11] α-Chlorohydrin is a chemosterilant which causes lesions in the male rat's reproductive tract. Larger doses of the compound kill male and female rats, including those that have become immune to the anticoagulant rodenticides.

Rhône-Poulenc (1987) introduced flupropadine (**21**) which is very effective against rats and mice, even when resistant to warfarin.[12]

REFERENCES

1. Woods, A., *Pest Control—A Survey*, McGraw-Hill, London, 1974, p. 105.
2. Green, M. B., Hartley, G. S., and West, T. F., *Chemicals for Crop Improvement and Pest Management*, 3rd edn, Pergamon Press, Oxford, 1987.
3. Fletcher, W. W., *The Pest War*, Blackwell, Oxford, 1974, p. 103.
4. Richards, C. G. J., and Ku, T. Y. (Eds.), *Control of Mammal Pests*, Taylor and Francis, London and New York, 1987.

5. *The UK Pesticide Guide*, British Crop Protection Council, Binfield, 1991.
6. Martin, H., and Woodcock, D., *Scientific Principles of Crop Protection*, 7th edn, Arnold, London, 1983, p. 384.
7. Matolcsy, Gy, Nádasy, M., and Andriska, V., *Pesticide Chemistry*, Elsevier, Amsterdam, 1988, p. 261.
8. Bowler, D. J., Entwistle, I. D., and Porter, A. J., *Proceedings of British Crop Protection Conference*, Brighton, **2**, 397 (1984).
9. Roszkowski, A. P., Poos, G. I., and Mohrbacher, R. J., *Science NY*, **144**, 412 (1964).
10. Wright, E. W., Inglis, I. R., and Feare, C. J., *Bird Problems in Agriculture*, British Crop Protection Council, Croydon, 1980.
11. *Chemistry in Britain*, **18**(11), 776 (1982).
12. Morgan, R. L., and Parnell, E. W., *Monograph on Stored Product Pest Control*, British Crop Protection Council, **37**, 125 (1987).

Chapter 12

Nematicides

Unsegmented eelworms belong to a group of animals termed nematodes of which certain species attack the roots of crop plants doing considerable damage. The last thirty years or so have seen a substantial increase in the number of harmful species of nematodes coupled with a greater awareness of the damage they inflict on crops. The increasing trend towards large areas of monoculture has created an environment particularly conducive to the increase of nematode populations.[1] Nematodes generally live in the soil, feeding on plant roots, although some species also invade the plants above ground, e.g. the species feeding inside chrysanthemum leaves, causing them to die progressively up the stem; another eats the stems and bulbs of narcissi.

However, the most economically damaging is the root eelworm of potato which can survive in a hard encysted form for several years in the absence of the host crop.[2] The nematode cysts can remain in a dormant state in the soil until some chemical substances secreted by the growing potato roots potentiates them to hatch, releasing their eggs; the potato cyst nematode can multiply some fiftyfold on a susceptible crop. The eelworm hatching factor has been shown to be a complex oxidized sugar, which is, unfortunately, impossible to manufacture and add to the soil.

If a suitable synthetic chemical hatching factor could be produced, it might be possible to hatch the cysts in the absence of the host and so starve them to death. Some host plants have their own built-in defences against the attack of parasitic eelworms, and recently potato varieties have been bred that are resistant to potato eelworms. When feeding on normal potato roots, the newly hatched eelworm can become either a male or female adult and the larvae develop into approximately equal numbers of both sexes. However, when eating the roots of the resistant varieties, the larvae nearly all become males; the attack on the potato in the first season is therefore not reduced but in subsequent seasons the nematode population cannot expand and the number of cysts gradually diminishes. The potato growing areas of Britain, until recently, had become heavily infested with eelworm cysts and so much of the land had to be planted with other, less economically valuable crops. It is now generally recommended that an interval of at least three years should be observed between planting potato crops on the same piece of land as a precaution against building up the eelworm population, although in some parts of the country a very early crop of potatoes can be set which may be harvested before the eelworms reach maturity. The nematode cysts which protect the eggs are resistant to chemical attack so nematode control is difficult; the principal method is still crop rotation.

Many species of nematodes only feed on one host plant or closely related plants, while others will attack almost any plant, but the latter are fortunately not very damaging. It is, however, difficult to assess the damage arising from nematodes, of which there are a large number of different varieties, because of the effect of soil fungi, some of which do not invade the plant unless it has been previously damaged by eelworm attack. Some plants, while not greatly affected by either eelworms or soil-borne fungi alone, are seriously damaged when both organisms are present. Free-living root nematodes often function as virus vectors and assist the propagation of plant viral diseases.

Plant parasitic eelworms are controlled by pesticides known as nematicides.[2–4] Chemicals are particularly effective against the more vulnerable stem and leaf eelworms and are also widely used against eelworms invading expensive greenhouse crops, e.g. tomatoes, where the value of the crop is sufficiently high to justify the cost of the massive doses of nematicide often required for effective control of the pest. Other crops often protected by chemical treatment include tobacco, especially in seed beds, and citrus fruits. Nematodes are covered with an impermeable cuticle which protects them. Effective nematicides therefore need the ability to penetrate the lipophilic cuticle.

Chemical control can be achieved by treatment of the soil with fumigants —compounds of sufficient volatility to penetrate through the upper layers of the soil where most of the insects and nematodes are to be found (Chapter 10).[1,4,5] Some nematodes have seasonal migratory habits and, in these cases, soil fumigation must be carried out when they are not residing in deep layers of the soil. Damage caused by eelworms is most serious if they attack a newly transplanted plant or seedling so elimination of eelworms from the initial rooting zone of soil may often achieve satisfactory results.

In order to achieve efficient nematode control the chemical must be well distributed in the soil[3] and fairly long lasting in its nematicidal action. Chemicals, such as dichloropropene or ethylene dibromide, that release toxic gases slowly are therefore particularly effective. Chemical fumigants are phytotoxic; hence fumigation at rates of 200–600 kg/ha must be carried out several weeks prior to planting; heavy soils require higher dose rates than light soils.

The first soil fumigant used was carbon disulphide in Germany (1871) against sugar beet eelworm.

Chloropicrin, the tear gas of World War I and obtained by nitration of chloroform, was used as a fumigant in Britain (1919) and later extensively in the United States of America. D-D, a mixture of the *cis* (Z) and *trans* (E) isomers of

1,3-dichloropropene, was introduced as a soil fumigant by Shell Limited (1943) (LD_{50} (oral) to rats, 200 mg/kg). It is an extremely valuable nematicide— injection into the soil at rates of 200–500 l/ha killed both eelworms and their cysts. D-D has been the most widely used nematicide;[6] an additional benefit arising from its application is an increase in the availability of nutrient nitrogen due to the effect of the chemical on soil microorganisms.[3] The Z isomer is more toxic to nematodes than the E isomer and D-D is not persistent since it is hydrolysed in soil to the 3-chloroallyl alcohols.[6]

Ethylene dibromide (1,2-dibromoethane) has also been used in soil fumigation and this led to the examination of other bromine compounds. Methyl bromide, which is highly volatile, was developed for soil sterilization in glasshouses, but is too volatile for general soil treatment. Methyl bromide is most effective as a nematicide when used to treat infested seed or plants in an airtight chamber. Shell Limited introduced (1958) 1,2-dibromo-3-chloropropane (DBCP) (1) as a soil fumigant and nematicide; this is prepared by liquid-phase addition of bromine to allyl chloride:

$$H_2C{=}CH{-}CH_2Cl \xrightarrow{\ Br_2\ } BrH_2C{-}CH(Br){-}CH_2Cl$$

Allyl chloride   **(1)**

DBCP is effective against many varieties of nematodes at rates of 10–125 kg/ha and it is less phytotoxic than D-D and therefore can be used for nematode control on established crops under carefully controlled conditions.[2,4,5] Thus DBCP has been applied for eelworm control in citrus orchards in California; the chemical was added to the soil on one side of the row of trees one year and the other side of the row was treated the next year. This allowed the tree roots to recover from any damage inflicted by the first treatment before the second application was made.[1] DBCP was a most valuable soil nematicide; however, it has been shown to cause male sterility and since 1985 is no longer permitted.[4]

The effectiveness of soil fumigation against nematodes depends on several factors, the most important being soil type, conditions, and temperature. A warm soil is needed so that the volatile toxicant can disperse effectively through the soil layer. Even dispersion is aided by a well-worked fine soil with a reasonable amount of moisture present. Different types of soil can vary considerably in their capacity to absorb fumigants;[7] thus soils rich in organic matter often deactivate a high proportion of the toxicant so that control of the nematodes is unsatisfactory, whereas sandy soils give much better results; e.g. on sandy soils application of D-D at 670 l/ha controls citrus nematodes but on clay soils 2800 l/ha are required for a comparable effect. Of course, one can have too rapid a dispersion of the fumigant resulting in considerable loss of the material from the top layer of the soil (down to about 7 cm), making it difficult to maintain sufficient of the toxicant to destroy the eelworms in this region. The situation can often be improved by covering the soil with a layer of polythene or using a sealant such as xylenol. The lower layers of the soil are not reached by the fumigant, so these untreated areas eventually cause recolonization of the sterilized soil by both harmful and beneficial organisms. This situation might result in a transient enhancement of some pest species, but luckily the beneficial saprophytic fungi are the most resistant species present and so these usually reinfest the treated soil first and cause some further delay in the spread of the parasitic fungi.[2]

The majority of nematicides are relatively simple molecules which are the by-products of large-scale chemical industry that often have other major uses. D-D is a special product obtained for use as a nematicide by distillation of petroleum cracking products during chlorination, but it can only be manufactured economically as part of a larger chemical programme.

The fumigation of soil is a costly operation. With volatile materials, such as those previously mentioned, the chemical is injected into the soil in a grid pattern at about 30 cm spacing and the soil needs covering. The high expense generally prohibits such treatment with the majority of field crops, but is a routine procedure with many varieties of glasshouse crops. When the toxicant is relatively non-volatile, distribution in the soil is achieved mechanically. For instance, the nematicide formulated as a dust is well stirred into the soil to the required depth by a powerful rotary hoe. Another method is soil drenching, which can only be effective on well-broken, good draining soils, so that the toxicant is washed down rapidly and reasonably uniformly. It is also important that the chemical is not appreciably adsorbed onto the soil colloids (Chapter 2, page 32).

In 1959 methyl isothiocyanate (m.p. 35°C, b.p. 119°C) was introduced as a nematicide and is a general soil fumigant for control of soil fungi, insects, nematodes, and weed seeds (Chapter 10, page 285). Methyl isothiocyanate is phytotoxic and, as with the majority of nematicides, treatment must be carried out sufficiently early to allow the toxicant to decompose before the crop is planted.[7] Methyl isothiocyanate can generally be more effectively applied as a soil drench in the form of metham-sodium (2),[6] which slowly decomposes liberating methyl isothiocyanate:

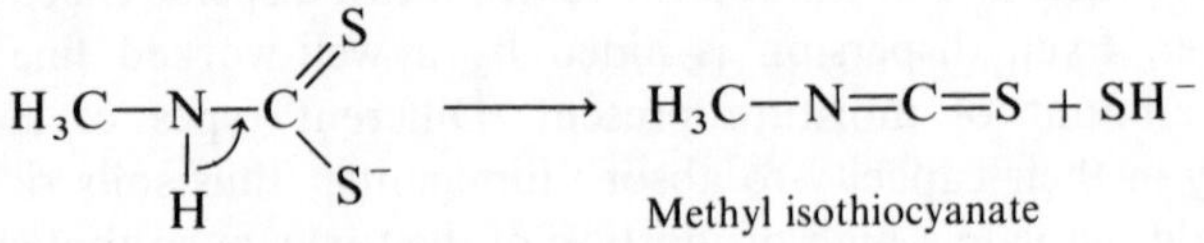

$$H_3C-N\underset{H}{\overset{}{-}}C\overset{S}{\underset{S^-}{\diagup}} \longrightarrow H_3C-N{=}C{=}S + SH^-$$

(2)   Methyl isothiocyanate

Metham-sodium (2) is prepared by reaction of methylamine with carbon disulphide in the presence of sodium hydroxide:

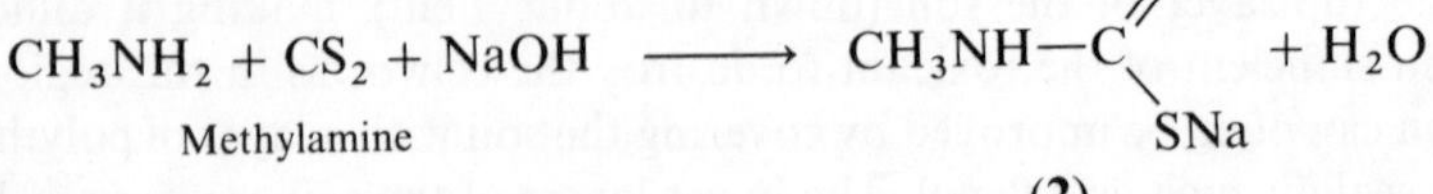

$$CH_3NH_2 + CS_2 + NaOH \longrightarrow CH_3NH-C\overset{S}{\underset{SNa}{\diagup}} + H_2O$$

Methylamine

(2)

Another useful soil sterilant is the thiadazine-2-thione or dazomet (**3**),[4,6] which is prepared by reaction of carbon disulphide, methylamine, and formaldehyde (Scheme 12.1); this is formulated as granules and is applied to the soil at rates of 400–600 kg/ha for control of soil nematodes, fungi, insects, and weeds (LD_{50} (oral) to rats, 520 mg/kg). Dazomet, like metham-sodium, owes its activity to decomposition in the soil to give methyl isothiocyanate, the rate of decomposition depending on the type of soil, the humidity, and temperature.[5,6]

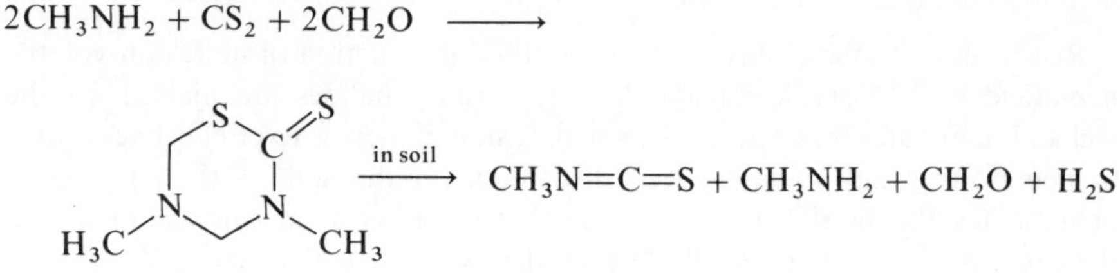

$$2CH_3NH_2 + CS_2 + 2CH_2O \longrightarrow$$

$$\xrightarrow{\text{in soil}} CH_3N=C=S + CH_3NH_2 + CH_2O + H_2S$$

(**3**)

Scheme 12.1

Several alkyl isothiocyanates, such as the vinyl thioethyl derivative (**4**), have useful nematicidal properties, as does tetrachlorothiophene (**5**).

$$H_2C=CH-S-CH_2-CH_2N=C=S$$

(**4**)

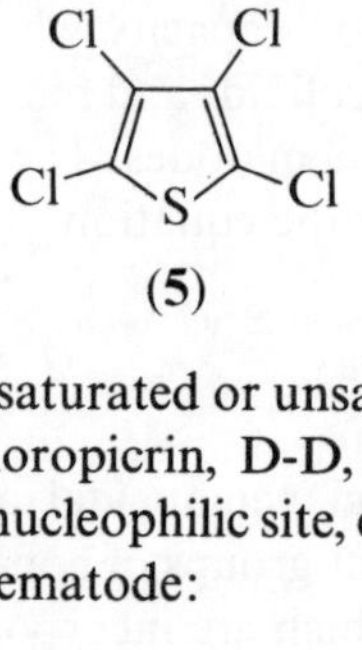

(**5**)

The majority of volatile commercial nematicides are saturated or unsaturated halides, e.g. methyl bromide, ethylene dibromide, chloropicrin, D-D, DBCP, and their activity probably depends on reaction with a nucleophilic site, e.g. OH, SH, or NH_2 groups, in a vital enzyme system in the nematode:

$$Ez-\overset{..}{S}H + R-Hal \longrightarrow Ez-SR + H-Hal \qquad (S_N2)$$

Ez = remainder of the enzyme structure

This is a bimolecular nucleophilic substitution reaction (S_N2) and there is also the possibility of a competing unimolecular hydrolysis (S_N1 reaction) by the large excess of water present in the soil which will deactivate the nematicide:

$$R-Hal + H_2O \longrightarrow R-OH + H-Hal \qquad (S_N1)$$

Therefore, compounds having high reactivity in S_N2 reactions and low reactivity in S_N1 reactions should be the most active nematicides. Nematicidal aliphatic halides induce a period of hyperactivity in nematodes, followed eventually by paralysis and death, mortality being dependent on the product of concentration and time of exposure. The alkyl isothiocyanates, including metham-sodium (**2**) and dazomet (3) owe their nematicidal properties to the ability of isothiocyanates to react with nucleophilic centres, e.g. thiol groups, in vital enzymes in nematodes by a similar type of S_N2 reaction to that operating with the aliphatic halogeno nematicides:

$$R-N{=}C{=}S + Ez-SH \longrightarrow RNH-C\underset{\displaystyle SEz}{\overset{\displaystyle S}{\diagup}}$$

Recent developments have centred on the introduction of new, non-volatile nematicides.[5,8,9] Volatile nematicides, such as alkyl halides, are injected into the soil and their vapour penetrates by soil diffusion; if there is to be effective control of nematodes, they must be well distributed in the soil. With non-volatile nematicides, the distribution has to be achieved either by mechanical mixing of the toxicant, formulated usually as granules, with the soil, or by application as aqueous drenches around the roots of the crop plants. Movement of non-volatile chemicals in soil is limited by the degree of adsorption onto soil colloids (Chapter 2, page 32). Adsorption depends on the nature of the chemical and the physical character of the soil, although it will be influenced also by soil temperature and moisture content. Soils rich in organic matter possess high adsorption capacity when chemicals tend to be strongly adsorbed on the soil organic colloids and removed from the soil solution and therefore unavailable to kill soil nematodes. The partition coefficient (Q) of a given chemical in soil is given by the equation

$$Q = \frac{\text{chemical in soil organic matter}}{\text{chemical in soil solution}}$$

For good nematicidal activity the values of Q should be fairly low.[8] Two important groups of non-volatile nematicides are organophosphates and carbamates which are incorporated into the soil at rates of 2–10 kg/ha as compared with 150–1150 kg/ha needed for control of nematodes by volatile aliphatic halides. The carbamates generally have Q values of approximately 10 (cf. 150 for the organophosphates), so carbamates tend to be more effective as soil nematicides.

Organophosphorus compounds[10] have been used extensively as insecticides, and since nematodes have an essentially similar nervous system, several organophosphorus insecticides are also nematicidal, e.g. phorate,[6] parathion, and dimethioate (Chapter 6). However, unfortunately the majority of organophosphorus insecticides are too rapidly degraded in soil so only the systemics are of much practical use. Some examples, which have been specifically developed as nematicides, include phenyl *N,N'*-dimethylphosphorodiamidate or diamidafos (6), which was introduced as a nematicide by the Dow Chemical Company (1962);[6] it is systemic in plants and hence effective against nematodes that have entered the root system.

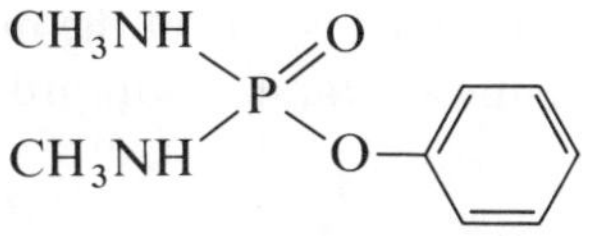

(6)

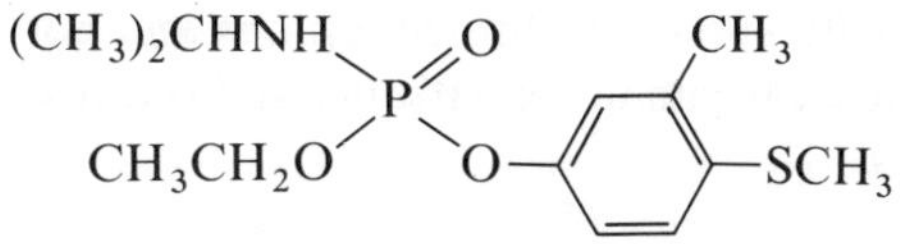

(7)

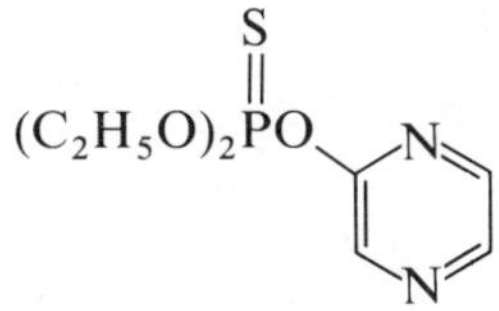

(8)

Diamidafos (**6**) is a very hydrophilic molecule with low adsorption onto soil organic matter and is very effective as a soil drench against root-knot nematode larvae. A related phosphoramidate fenamiphos (**7**) is active against nematodes by broadcast application at 5–20 kg/ha with or without soil incorporation. A disadvantage is the high mammalian toxicity: LD_{50} (oral) to rats, 17 mg/kg.[6]

Some phosphorothioates, especially those containing heterocyclic groups, have been developed as nematicides: an example is provided by thionazin (**8**), a valuable soil nematicide. Since it is systemic thionazin is especially useful against species of nematodes attacking bulbs, although it has a very high mammalian toxicity: LD_{50} (oral) to rats, 12 mg/kg.

Prophos (**9**) is a broad-spectrum nematicide and insecticide which is particularly valuable against corn rootworm (LD_{50} (oral) to rats, 61 mg/kg). It is prepared by condensation of ethyl phosphorodichloridate and propylmercaptan:[5]

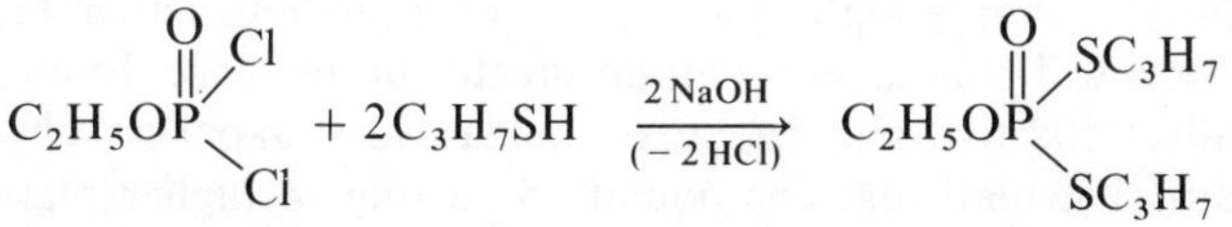

(9)

Several carbamates (Chapter 6, page 140) are valuable nematicides; important examples include carbofuran, carbosulfan (**10**) (page 147), aldicarb (page 145) and oxamyl (page 146).

Carbofuran, a systemic nematicide, is used for corn, sugar beet, peanuts, sugar cane, tobacco, and vegetables.[6] It has a comparatively short residual life.[6] Carbosulfan (**10**), formulated as granules, is used in vegetables, e.g. brassicas, carrots, and parsnips.[6]

The oxime carbamates, aldicarb and oxamyl, are both highly active systemic nematicides, but suffer from very high mammalian toxicities.[5,8] Both are formulated as granules for soil incorporation; aldicarb is used for a wide variety of crops, e.g. vegetables, strawberries, and ornamentals, whereas oxamyl is especially applied to protect onions and potatoes.[6] Oxamyl is one of the most valuable carbamate nematicides and is translocated upwards and downwards in plants.

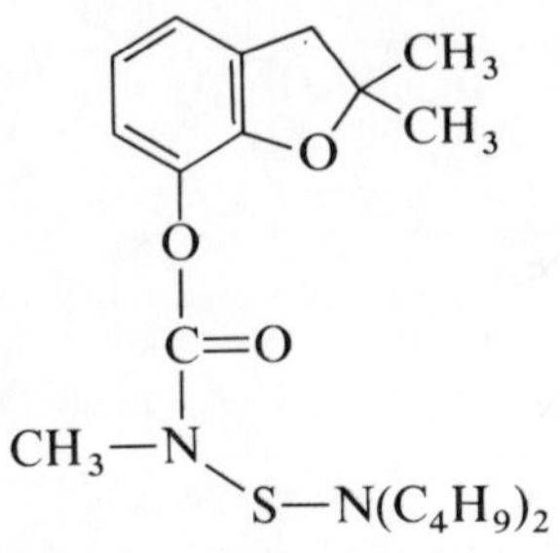

(10)

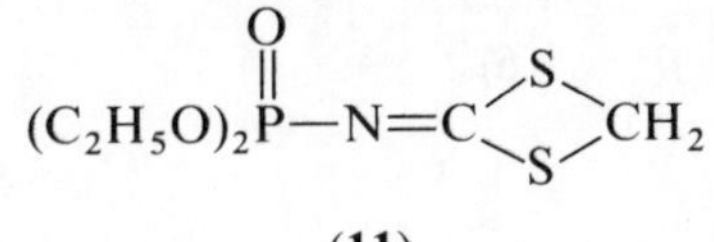

(11)

The dithioiminocarbonate **(11)** was introduced (1975)[11] by the American Cyanamide Company as a broad-spectrum nematicide effective against plant nematodes by soil incorporation; the activity is probably related to its decomposition in soil to the thiocyanate ion.

The organophosphorus and carbamate nematicides probably owe their activity to poisoning of the enzyme cholinesterase by phosphorylation or carbamoylation; the P=S derivatives being activated by conversion to the corresponding P=O compounds by the action of soil microorganisms or within the nematode (cf. Chapter 6, page 127).

The traditional halogenated volatile nematicides, after the banning of DBCP (page 301), have come under close scrutiny for deleterious long-term effects and their use will consequently decline.

The non-volatile organophosphorus and carbamate nematicides with short residual effects will consequently attain greater prominence. In an attempt to discover novel effective and less toxic nematicides, Japanese chemists have studied natural nematicidal compounds occurring in higher plants. For instance, they isolated the nematicidal polyacetylenes **(12a,b)** from the flowers of *Carthamus tinctorius*.[5] The discovery and possible structural modification of such naturally occurring compounds may provide a valuable new approach to nematode control.

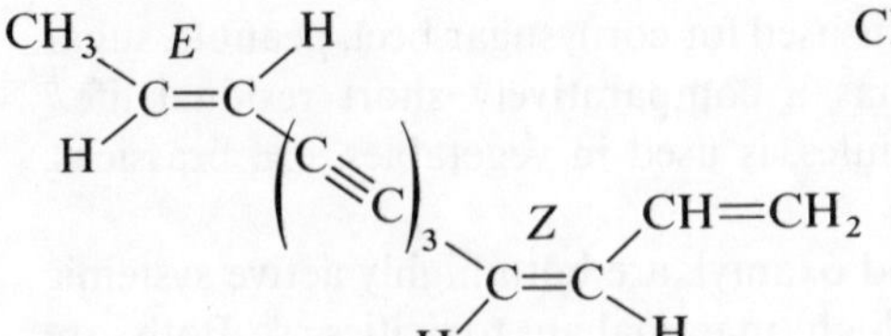

(12a) *E,Z* isomer

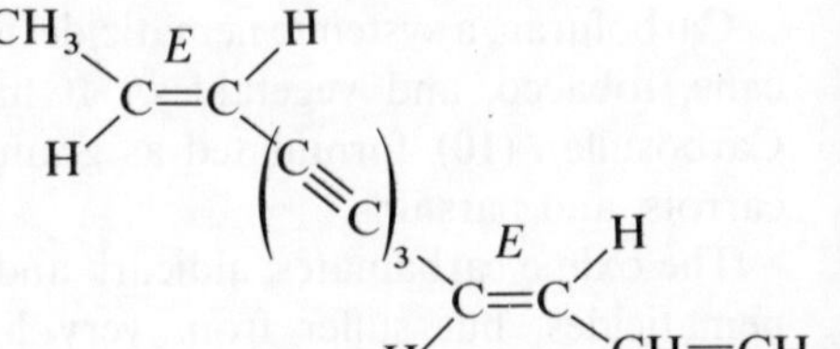

(12b) *E,E* isomer

REFERENCES

1. Hartley, G. S., and West, T. F., *Chemicals for Crop Improvement and Pest Management*, 3rd edn, Pergamon Press, Oxford, 1987, p. 138.
2. Fletcher, W. W., *The Pest War*, Blackwell, Oxford, 1974, p. 60.
3. Woods, A., *Pest Control—A Survey*, McGraw-Hill, London, 1974, p. 102.
4. Ware, G. W., *Fundamentals of Pesticides*, 2nd edn, Thomson Publications, Fresno, Calif. 1986, p. 169.
5. Matolcsy, Gy, Nádasy, M., and Andriska, V., *Pesticide Chemistry*, Elsevier, Amsterdam, 1988, p. 256.
6. Worthing, C. R., and Walker, S. (Eds.), *The Pesticide Manual*, 8th edn, British Crop Protection Council Publications, Croydon, 1987.
7. Allen, M. W., 'Nematicides', in *Plant Pathology* (Eds. Horsfall, J. G., and Dimond, A. E.), Vol. II, Academic Press, New York, 1960, p. 603.
8. Hague, N. G. M., *Proceedings of 8th British Insecticide and Fungicide Conference, Brighton*, **3**, 837 (1975).
9. Hance, R. J. (Ed.), *Soils and Crop Protection Chemicals*, Monograph No. 27, British Crop Protection Publications, Croydon, 1984.
10. Eto, M., *Organophosphorus Pesticides—Organic and Biological Chemistry*, CRC Press, Cleveland, Ohio, 1974.
11. Whitney, W. K., and Aston, J. L., *Proceedings of 8th British Insecticide and Fungicide Conference, Brighton*, **3**, 625 (1975).

Chapter 13
Molluscicides

Molluscs include land slugs and snails which do a considerable amount of damage by eating vegetable seedlings and mature green vegetables such as lettuce and cabbages; they also severely attack autumn-sown wheat and potatoes.[1] Molluscs do not directly harm mammals but are alternate hosts for some species of parasites; for instance the liver-fluke which sometimes kills its mammalian host (generally sheep) spends an essential phase of its life cycle in small *Limnaea* snails which are taken in with the grass by sheep. Lungworms, causing 'husk' in cattle, are transmitted in a similar manner, and several different species of aquatic snails act as alternate hosts for the *Schistosoma* parasites, the causal agent for the debilitating human disease of bilharzia, which is widespread in tropical countries.

Barnacles that attach themselves to the hulls of ships severely reduce the speed–power ratio, an increasingly vital factor in these times of high energy costs. Chemicals specifically designed to combat these various molluscs are termed molluscicides. No very effective method of controlling slugs and snails is available; however, one of the best is metaldehyde or meta (**1**), which is obtained by polymerization of an ethanolic solution of acetaldehyde in the presence of sulphuric acid at a temperature < 30°C:

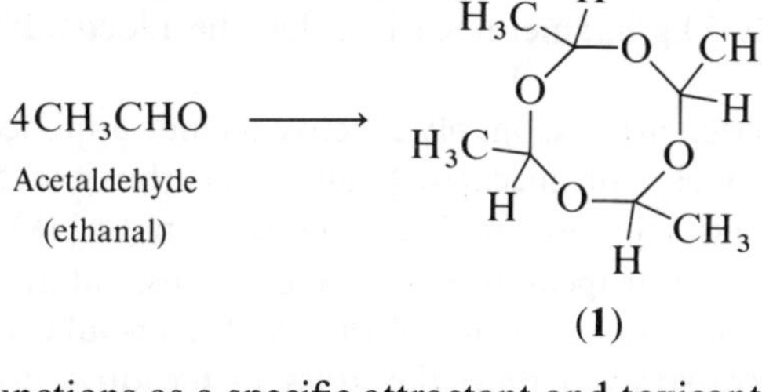

$$4CH_3CHO \longrightarrow$$

Acetaldehyde
(ethanal)

(1)

Metaldehyde functions as a specific attractant and toxicant for slugs and snails and is especially effective against slugs. It is usually formulated as baits containing 2.4–4% of the active ingredient in bran.[2] Meta is useful to control attacks by the field slug *Agriolimax reticulatus* on autumn-sown wheat at a rate

309

of 900 g of the active ingredient in 31 kg of bran per hectare. The best results are achieved when it is broadcast on the soil during a moist warm evening a few days after the end of a dry spell of weather.[3] Metaldehyde is highly flammable, burning with a non-smoky flame; this accounts for its use as a solid fuel that is very convenient for picnic stoves, and the observation that meta was lethal to slugs may have originated from its accidental spillage at a picnic party. Meta is toxic to slugs and snails both by ingestion and by absorption by the foot of the mollusc. The chemical causes an increase in the secretion of slime causing immobilization and eventual death by loss of water (desiccation). Molluscicidal activity is specific to metaldehyde (1); the trimer, paraldehyde with a six-membered ring, is not active and neither is the monomer acetaldehyde. The precise mode of action of metaldehyde (1) has not been established.[4] Metaldehyde does not give completely satisfactory control under field conditions and consequently many other compounds have been examined.

Both dinitro-*o*-cresol (DNOC) and dinitro-*o*-cyclohexylphenol (Chapter 8, page 219) have been observed to reduce the damage inflicted by slugs and snails when they were used as selective herbicides against weeds in cereals, and dead slugs and snails have been found after their use. They are also toxic to aquatic snails at 3–5 p.p.m., but have been little used as molluscicides.[5,6]

A number of carbamate insecticides (see Chapter 6, page 144) such as methiocarb[2] are very effective against snails when formulated as baits. Methiocarb (2) is obtained by reaction of 4-methylthio-3,5-dimethylphenol and methylisocyanate:[2]

4-Methylthio-3,5-dimethylphenol   (2)

Methiocarb (2) is appreciably more active against slugs and snails than metaldehyde. It is formulated as 4% pellets which control slugs and snails at a rate of approximately 5 kg/ha; methiocarb (2) has the added advantage of being a bird repellent.[2]

Other carbamates that are reasonably effective include aminocarb (3; R = H), mexacarbate (3; R = CH_3), promecarb (4), and thiocarboxime (5). The organophosphorus compound azinphos-methyl (Chapter 6, page 118) as a 4% spray was effective against the European brown snail on citrus, but not against slugs.[6]

Copper sulphate has been used to kill the snail liver-fluke vectors. Copper sulphate solutions are sprayed onto the grass in meadows where sheep are grazing. The snails tend to climb up the blades of grass and are killed by direct contact with the spray. It is, however, difficult to achieve an adequate kill of the snails without also causing substantial damage to the grass crop on account of

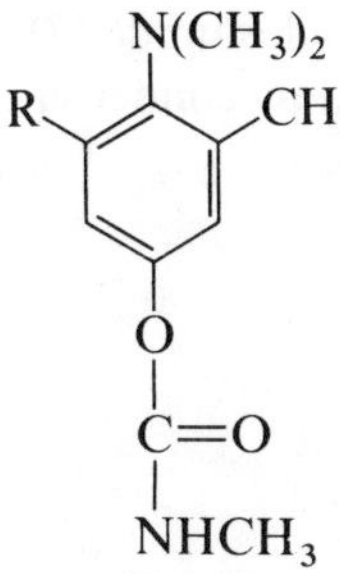

Structure (3): benzene ring with $N(CH_3)_2$, R, CH_3 substituents, and $O-C(=O)-NHCH_3$ group.

Structure (4): benzene ring with H_3C, $CH(CH_3)_2$ substituents, and $O-C(=O)-NHCH_3$ group.

Structure (5): $H_3C-C(SCH_2CH_2C\equiv N)=N-O-C(=O)-NHCH_3$

(3) 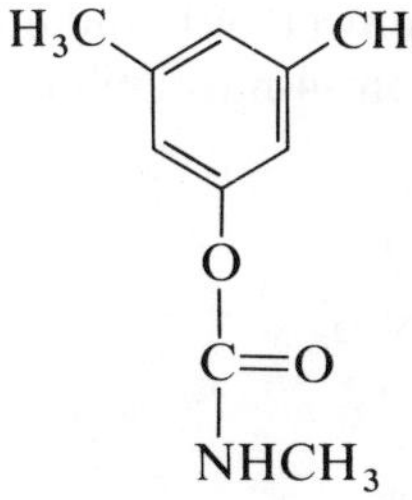   (4) 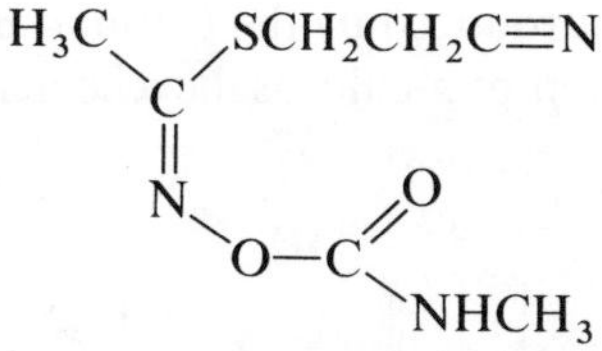   (5)

the high phytotoxicity of the cupric ion. A much more effective chemical for this purpose is trifenmorph (6), developed by Shell Research Limited (1966), and obtained by condensation of triphenylmethyl chloride with morpholine:[2]

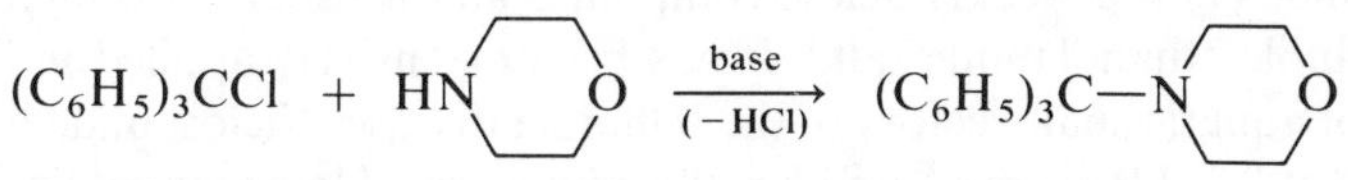

$$(C_6H_5)_3CCl + HN\overbrace{}O \xrightarrow[(-HCl)]{base} (C_6H_5)_3C-N\overbrace{}O$$

Triphenylmethyl chloride   (6)

Copper sulphate and trifenmorph are uncommon molluscicides since they are also active against aquatic snails. Trifenmorph (6) is lethal at 1–2 p.p.m., while copper sulphate at 3–4 p.p.m. kills all aquatic snails within 24 hours. Copper sulphate is the cheapest aquatic molluscicide but suffers from high toxicity to fish and algae; it is also easily precipitated by alkaline water and adsorbed by clay particles in the mud. Organic molluscicides, like trifenmorph (6), are not adsorbed by clay particles but are adsorbed by organic constituents in the mud. This is a serious problem in their use to control the aquatic snail vectors of *Schistosoma* as these occur in relatively stagnant waters containing large amounts of mud. The mode of action of trifenmorph is not definitely known; it has been associated with nucleophilic attack at the aliphatic carbon atom. This would occur via the S_N1 mechanism, involving formation of the Ph_3C^+ ion and possibly the ease of formation of this ion may be the basis of the molluscicidal activity.[4] Aquatic snails, like other aquatic species, need to process large volumes of water to obtain sufficient oxygen; hence they are susceptible to lipophilic compounds like trifenmorph (6), even when they are present in only a few p.p.m.

The organic molluscicides do, however, have the advantage of being less phytotoxic than copper sulphate, but it is often difficult to achieve effective distribution of organic compounds in the slow-moving water. Sodium pentachlorophenolate is effective at a few p.p.m. against aquatic snails, including their eggs, but suffers the disadvantage of being highly lethal to fish; acrolein has also been applied with some success.[1] Pentachlorophenol, like copper sulphate, has the disadvantage of high phytotoxicity.

Another important, and widely used, organic molluscicide is niclosamide (**7**), introduced by the German Bayer Company (1959)[2] and prepared by condensation of 5-chlorosalicyclic acid and 2-chloro-4-nitroaniline:

5-Chlorosalicylic acid   2-Chloro-4-nitroaniline

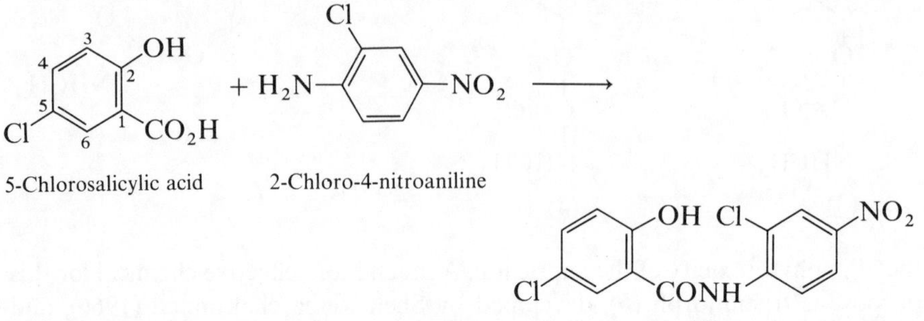

(7)

Niclosamide (**7**) is a weakly acidic compound and is usually formulated as the water-soluble ethanolamine salt.[6] It has been extensively applied in Egypt for control of aquatic snail vectors of the bilharzia disease. Niclosamide is lethal to the snails at 0.3–1.0 p.p.m.[2,6] and has the advantage of low mammalian toxicity: LD_{50} (oral) to rats > 5000 mg/kg.

Niclosamide (**7**) and pentachlorophenol both interfere with respiration by uncoupling oxidative phosphorylation; they stimulate oxygen uptake by snails in the presence of succinate substrate and this is probably the basis for their molluscicidal activity.[4]

Structure–activity studies on a series of substituted nicotinanilides indicated that the parent compound (**8**) showed potential as a selective molluscicide.[7]

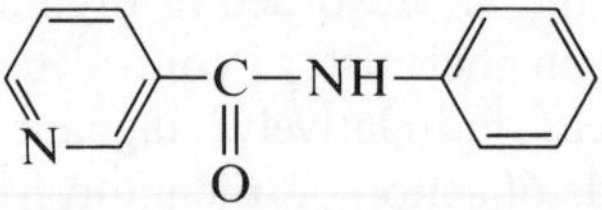

(8)

Special antifouling paints have been developed to combat the growth of barnacles on ships. These often contain such comparatively insoluble copper compounds as cuprous oxide and copper dimethyldithiocarbamate or trialkyltin compounds, e.g. tributyltin acetate. Many of the trialkyltins are extremely toxic to aquatic snails and their eggs at doses of 0.1–0.4 p.p.m., but suffer from high toxicities to fish, mammals, and other aquatic life.[1]

Various naturally occurring antifouling agents have been investigated: xanthatin (**9**) has been isolated from *Xanthium strumarium*; it has excellent repellent activity against the blue mussel. Xanthatin (**9**) was more effective than copper sulphate and of lower toxicity.[8]

Benzoylureas, e.g. compound (**10**), act as antifeedants for slugs and snails and prevented them feeding on lettuce.[9]

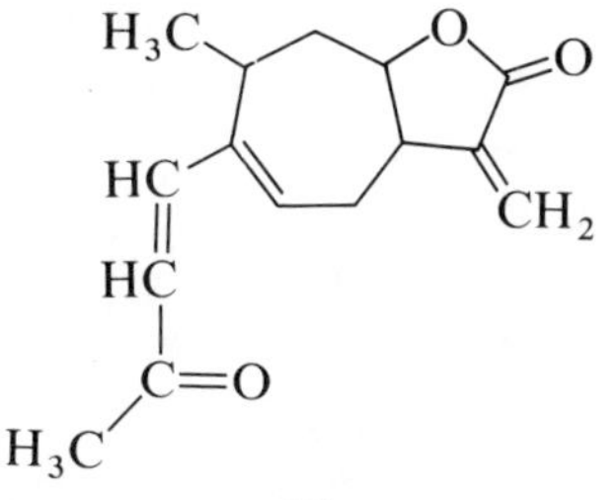

(9)

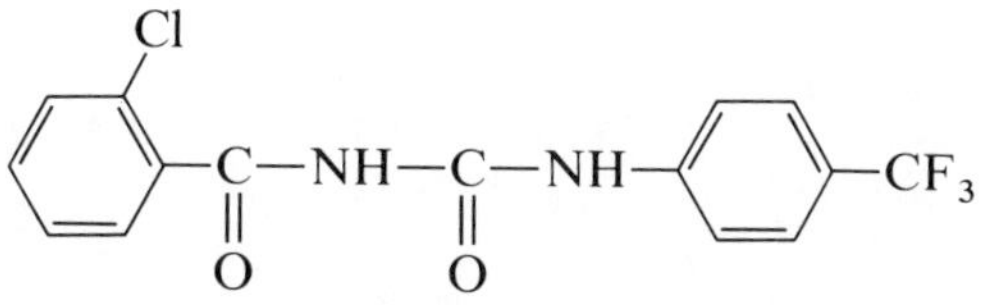

(10)

The use of chemical molluscicides is only one route to the control of bilharzia which basically arises from the insanitary habits of primitive people. For instance, if people could be persuaded not to drink, wash, and excrete in the same water, and irrigation canals were kept flowing freely with little aquatic vegetation, the incidence of the disease could be greatly reduced.

There has also been considerable progress in direct chemical attack on the disease organism when it is actually in the human bloodstream.

REFERENCES

1. Green, M. B., Hartley, G. S., and West, T. F., *Chemicals for Crop Improvement and Pest Management*, 3rd edn, Pergamon Press, 1987, p. 141.
2. *Pesticide Manual*, 9th edn, British Crop Protection Council, Croydon, 1991.
3. Woods, A., *Pest Control*, McGraw-Hill, London, 1974, p. 103.
4. Corbett, J. R., Wright, K., and Baillie, A. C., *The Biochemical Mode of Action of Pesticides*, 2nd edn, Academic Press, London and New York, 1984.
5. Fletcher, W. W., *The Pest War*, Blackwell, Oxford, 1974, p. 59.
6. Metcalf, R. L., 'Chemistry and biology of pesticides', in *Pesticides in the Environment* (Ed. White-Stevens, R.), Vol. 1, Dekker, New York, 1971, p. 124.
7. Dunlop, R. W., Duncan, J., and Ayrey, G., *Pesticide Science*, **11**, 53 (1980).
8. Harada, A., *et al.*, *Agric. Biol. Chem.*, **49**(6), 1887 (1985); *Chemical Abstracts*, **103**, 83472z (1985).
9. Schlueter, K., and Schnorbach, H. J., EP 150,031; *Chemical Abstracts*, **103**, 174067n (1985).

Chapter 14
Some Novel Methods of Insect Control

The last 25 years have witnessed spectacular advances in our knowledge of the ways in which chemicals mediate insect behaviour or physiology. Many organisms are known to excrete chemicals which induce a specific behavioural response from other members of the same or different species some distance away from their point of release. Such signalling chemicals (*semiochemicals*) are usually divided into those acting between the same species (*pheromones*) or different species (*allelochemicals*).[1,2] As potential insect control agents, the most valuable are the insect sex pheromones or attractants. These are volatile chemicals, generally released by the female, which facilitate mating, either by attracting a male insect or by inducing courtship ritual.

Considerable work has been carried out on the sex pheromones of certain moths, e.g. gypsy moth and others which are the adults of economically important caterpillar pests such as cotton leafworm and tobacco budworm.

Early in this century, it was observed that insects were capable of long-range attraction. The female Chinese silkworm moth can attract males from a distance of up to 11 km,[2] and the female gypsy moth *Porthetria dispar* releases a sex pheromone that can lure males to her from more than 3 km away.[3] This gives an indication of both the high potency of the chemical and the sensitivity of the antennae of the male gypsy moth as a receiving apparatus. The threshold amount for response is about 50 molecules or 10^{-14} μg.[2] Recently there has been increased study of sex pheromones because the development of modern analytical techniques, such as chromatography, nuclear magnetic resonance, and mass spectroscopy, has enabled some of the pheromones to be identified although only minute amounts were available. Also there have been increasing demands for more selective chemicals for use in pest control owing to a greater awareness of the dangers of environmental pollution (Chapter 15) and the spread of insect resistance.

The majority of lepidopteran (moth and butterfly) sex pheromones are even-carbon atom (C_{10}–C_{18}) straight chain alkenols; the favoured derivative is the acetate, but aldehydes or ketones may be found.[2] If two or more double bonds are present, they are usually conjugated or separated by only one methylene group and the double bonds are generally far away from the oxygen function.

Jacobson (1961) considered that the sex pheromone of the female gypsy moth was 10-acetoxy-Z-7-hexadecanol or gyptol (**1**; $n = 5$).[4,5] A synthetic homologue (**1**; $n = 7$) containing two more methylene groups, synthesized from ricinoleic acid (available from castor oil), was found to retain the attractant activity of gyptol (**1**) and was marketed as 'gyplure'.

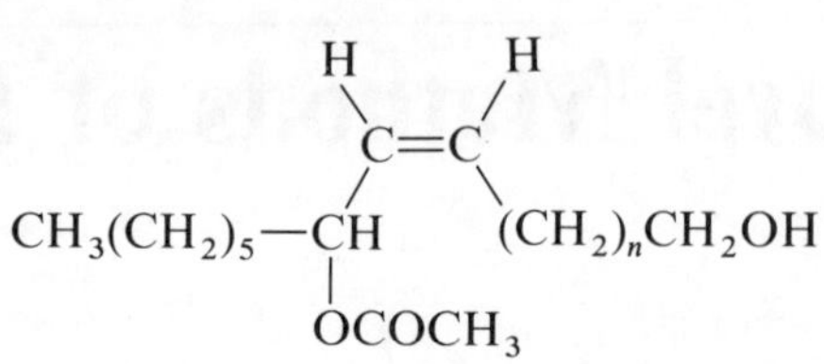

$$CH_3(CH_2)_5-\underset{\underset{OCOCH_3}{|}}{CH}\quad\overset{\overset{\displaystyle H\qquad H}{\diagdown\quad\diagup}}{\underset{\diagup\quad\diagdown}{C=C}}\quad(CH_2)_nCH_2OH$$

(1)

Gyptol (**1**; $n = 5$) is definitely released by the female gypsy moth at the same time as the sex pheromone but studies revealed that neither pure gyptol (**1**) nor 'gyplure' really have any sex attractant activity towards the male gypsy moths. The variation in the results obtained in tests with these compounds was due to their contamination with trace amounts of active substances. This showed that the true pheromone must have remarkably high attractant properties and it has now been identified as Z-7,8-epoxy-2-methyloctadecane (**2**), called disparlure.[1]

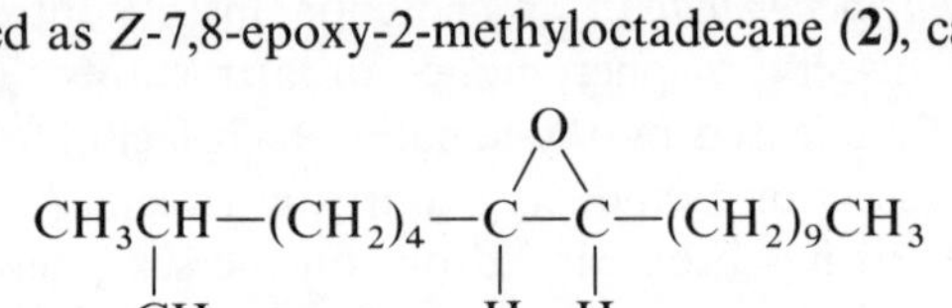

$$CH_3\underset{\underset{CH_3}{|}}{CH}-(CH_2)_4-\overset{\overset{\displaystyle O}{\diagup\diagdown}}{\underset{\underset{H}{|}}{C}}-\underset{\underset{H}{|}}{C}-(CH_2)_9CH_3$$

(2)

Other studies[2] showed that the sex pheromone of the natural silkworm moth *Bombyx mori* was 10 E-12-Z-hexadecadienol or bombykol (**3**), and that of the pink bollworm was 10-propyl-(E)-5,9-tridecadien-1-yl acetate (**4**). This has been

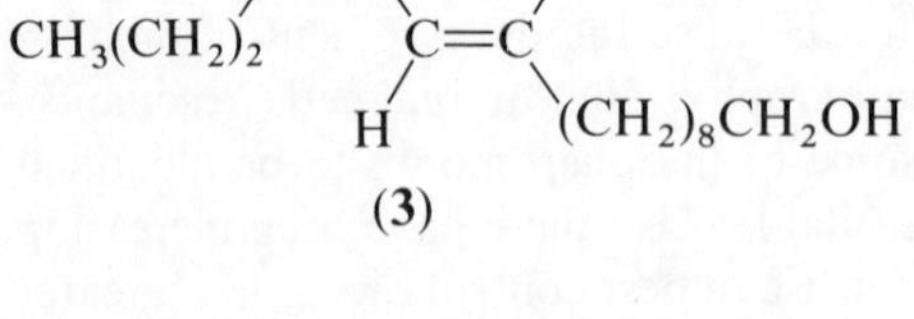

$$CH_3(CH_2)_2\quad\overset{\overset{\displaystyle H\qquad H}{\diagdown\quad\diagup}}{\underset{\diagup\quad\diagdown}{C=C}}\quad\overset{\overset{\displaystyle H}{\diagup}}{\underset{\underset{\displaystyle H}{\diagdown}}{C=C}}\quad(CH_2)_8CH_2OH$$

(3)

$$(CH_3CH_2CH_2)_2C=CH-(CH_2)_2\quad\overset{\overset{\displaystyle H\qquad (CH_2)_4OCOCH_3}{\diagdown\quad\diagup}}{\underset{\diagup\quad\diagdown}{C=C}}\quad H$$

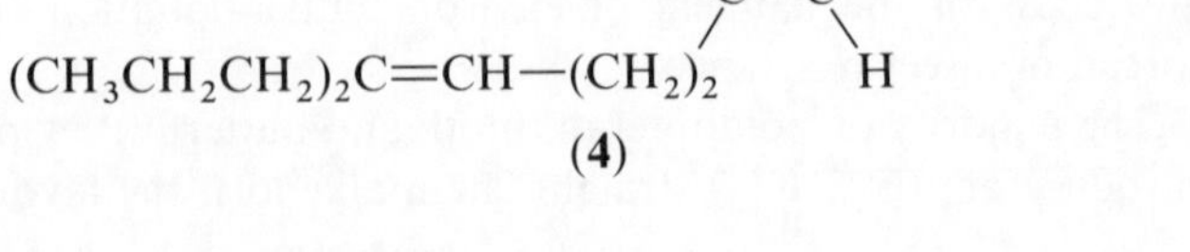

(4)

synthesized as 'propylure' (stereochemical purity is vital; 15% of the Z isomer destroys the attractant activity).

It is often found that the attractant property is critically dependent on stereochemistry; thus of the four possible geometric isomers for 10,12-hexadeca-dienol only the E–Z isomer (3) is identical to the natural sex attractant and only this specific isomer has any appreciable activity. The highly sensitive sense organs of the male silkworm moth can detect a concentration of only 10^{-16} g/cm^3 of bombykol emitted by a female.

One of the problems with synthetic pheromones is the extreme sensitivity of the insect to the precise structure and stereochemistry of the natural pheromone. Such minor structural variations as the position or stereochemistry of a double bond or changes in chain length often greatly reduces or completely removes the attractant properties.

Recent studies[6-9] indicate that the natural situation is much more complex and the majority of insects rely on multicomponent sex pheromones, sometimes a primary attractant which is synergized by the presence of other materials.

The pheromones of different classes of insects such as beetles (coleoptera), unlike those of lepidoptera, can have very diverse chemical structures (Table 14.1). For instance, compounds (5a,b), (6), and (7a) (Table 14.1) represent the sex pheromones of the American cockroach, the honeybee queen, and

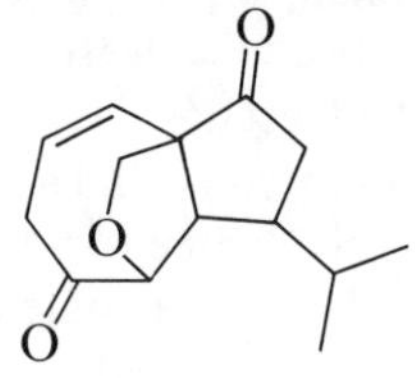

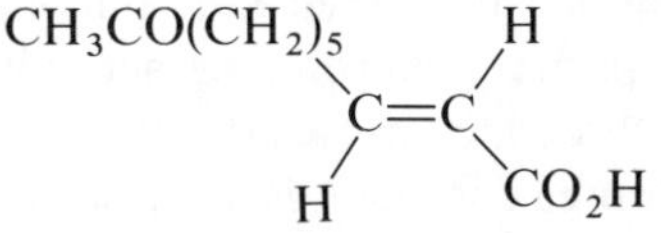

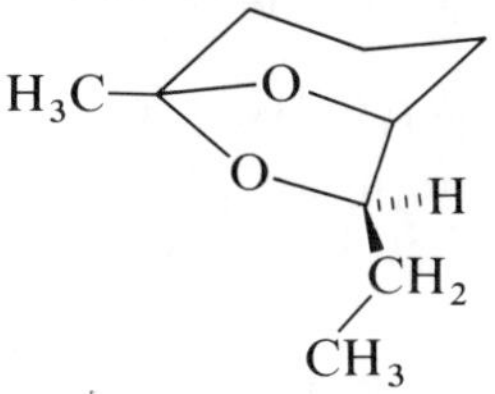

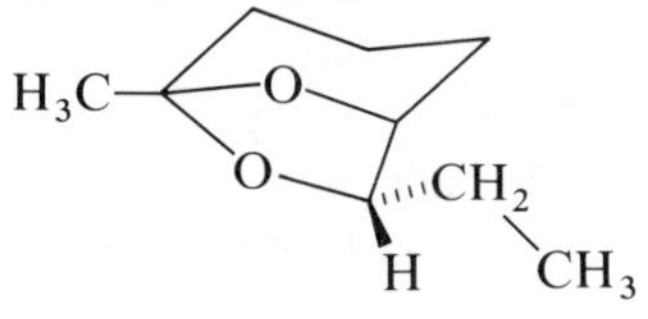

Table 14.1

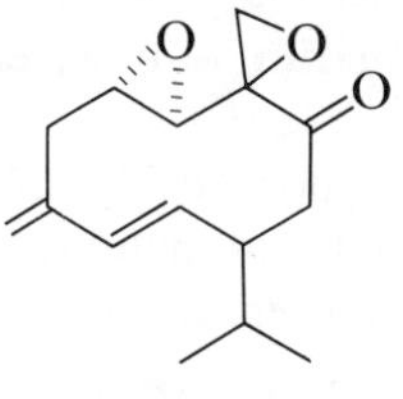

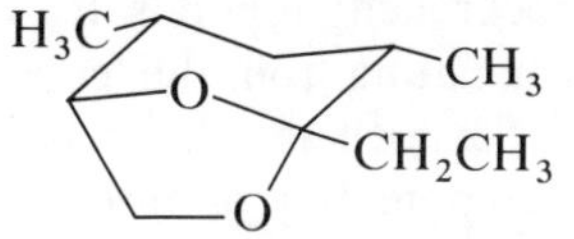

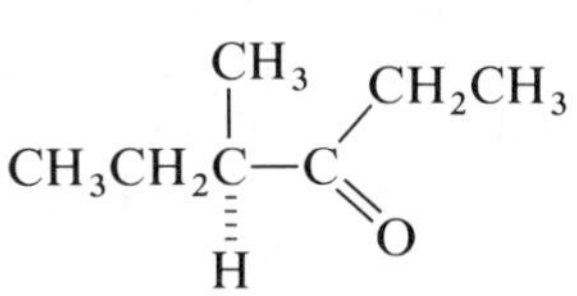

the Western pine beetle respectively. The synthetic product is a mixture of the *exo* and *endo* isomers (**7a,b**) but only the *exo* form (**7a**) has attractant properties.

Sex pheromones have received most attention, but pheromones are also involved in other insect activities, e.g. aggregation pheromones which control colonization of a suitable host for egg laying and foraging. There are also alarm pheromones to warn insects of danger and trail-marking pheromones. The colonization of a suitable host by the European elm beetle is controlled by the compounds (**8a**) and (**8b**) together with a kairomone, a semiochemical favouring the receiver, produced by a host. The process has special significance as the beetle is the vector for the Dutch elm disease. With beetles, antiaggregation chemicals have been successfully used in disruptive tests to reduce mass attack. For example, 3-methyl-2-cyclohexen-1-one (MCH) is an effective antiaggregation chemical for the Douglas fir beetle, a major forest pest in the Western USA and British Columbia. Bark beetles usually demonstrate a highly sophisticated use of pheromones for aggregation which is an essential feature of their life cycle. The Western pine beetle which feeds on the ponderosa pine in California is typical: the female beetle is first attracted by the profile of the tree and by the tree resin terpene, myrcene. After alighting on the tree, the female constructs a nuptial gallery and excretes the sex pheromone *exo*-brevicomin (**7a**) to attract the males. On arrival, the male beetles release another closely related pheromone, frontalin; the synergistic mixture of myrcene, *exo*-brevicomin (**7a**), and frontalin is highly attractive to both sexes and results in mass beetle attack (some 800 times the normal landing rate). The tree's natural defences are overwhelmed by up to 400 beetles/m^2; finally when the tree is dying and hence no longer suitable for colonization, the beetles release a mixture of deterrent pheromones from their hind gut with the faeces.[2]

The primary pheromones act only at long range and are comparatively small molecules which must possess a reasonable degree of volatility; in many cases short-range compounds are also involved once the male has been attracted to within a close distance of the female.

Insects detect pheromones with their antennae which have chiral receptor sites for the chemicals; these are acutely sensitive, e.g. the male American cockroach responds to $\simeq 10^{-14}$ µg ($\equiv 30$ molecules) of the sex pheromone emanating from the female.

Semiochemicals can be used in integrated pest management (IPM) to enhance the potential of insecticides by allowing accurate monitoring of the pest population, so that the insecticide is only applied when the pest has reached an economically serious level.[6]

The effectiveness of contact insecticides against aphids is increased by application of the aphid alarm pheromone (E)-β-farnsene (**9**) (as a stabilized derivative) to the treated plant, because the agitated aphids move out from the undersides of leaves and consequently pick up more of the toxicant.[7]

The population of foraging honeybees on a crop can be increased by 80% by spraying with the honeybee attractant, a mixture of seven terpenoids, e.g. *E*- and

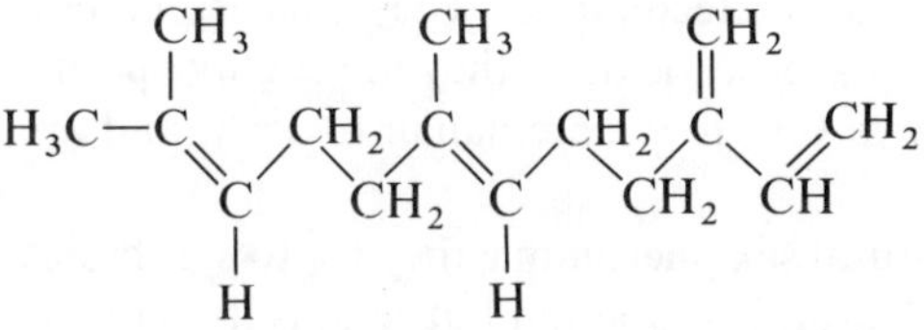

(9)

$CH_3(CH_2)_3COCH_3$

(10b)

$(CH_3)_2CHCH_2-CH_2-O-\overset{\overset{\displaystyle O}{\|}}{C}-CH_3$

(10a)

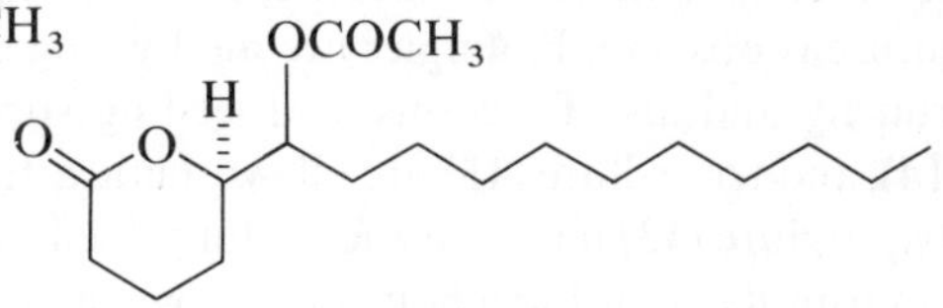

(11)

Z-citral. Conversely, spraying a crop, e.g. oilseed rape, with the honeybee alarm pheromone (**10a,b**) before spraying with an insecticide causes many bees to leave the crop and hence greatly reduces the damage inflicted on them by the chemical spray. The majority of pheromones are too volatile and unstable for direct application; however, some success has been achieved in mass trapping using controlled release formulations of pheromones which can be effective against low levels of insect pests.

The insect population can be reduced to a low level by conventional insecticides and then kept low by the use of pheromone traps containing adhesives or a fast-acting insecticide, e.g. dichlorvos. The pheromone dispenser in the trap must provide a constant slow controlled release of the pheromone; this requires specialized formulation techniques, e.g. laminated polymeric dispensers, and these are probably the key to the successful commercial use of pheromones in pest control[8] (Plate 34).

The lure and kill strategy has been successfully applied with the mosquito oviposition pheromone ($-$)(5*R*, 6*S*)-6-acetoxy-5-hexadecanolide (**11**). This has been applied in Kenya to attract mosquitoes to lay their eggs in water; the emergent larvae were then killed by a juvenile hormone (JH)-type insecticide incorporated in the pheromone formulation.[7]

Mating disruption is another procedure in which the atmosphere is permeated with the sex attractant so that the insect cannot locate a mate. For success, the attractant must be continuously emitted in the optimum concentration and to achieve this result microcapsulated formulations have been used to stabilize oxygen-sensitive compounds. Controlled release formulations, in which the pheromone is fabricated as capsules, flakes, or hollow fibre dispersers, can be applied to the crop from aircraft. Commercially such formulations have been successfully used against pink bollworm on cotton and the grape moth in European vineyards.[6,8]

The future development of pheromones in pest management would appear to largely depend on specialized formulation techniques. They possess certain important advantages over conventional insecticides—they are highly specific and only kill the target organism, have very low mammalian toxicity, and are readily biodegradable.[1]

Once chemists have identified a natural sex pheromone they try to synthesize an analogue that is cheaper than the natural compound but retains its activity; this is often unsuccessful because of the stereo and structural specificity associated with pheromone activity. The search for synthetic chemical lures has, however, produced several interesting compounds whose function, if any, in nature is obscure. Examples are methyl eugenol (12) which attracts the oriental fruit fly and also functions as a feeding stimulant, while siglure (13), medlure (14), and trimedlure (15) are all synthetic attractants for the Mediterranean fruit fly.[5] Siglure (13) has been successfully used in traps to eradicate the Mediterranean fruit fly from Florida in 1956–7 when > 50 000 baited traps were employed.

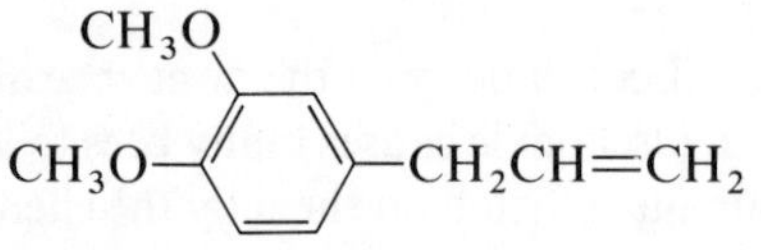

(12)

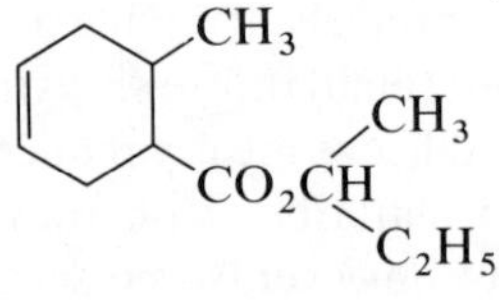

(13)

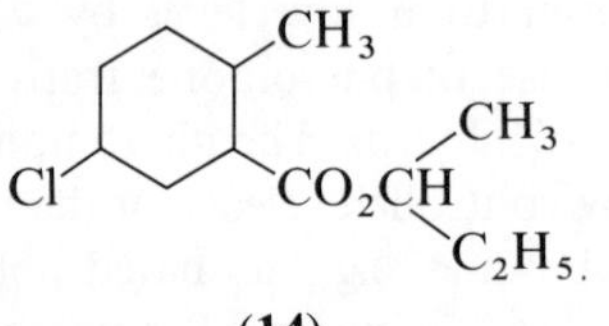

(14)

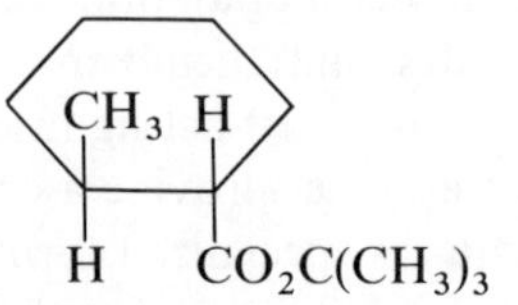

(15)

There are a few examples of synthetic attractants that are effective for insects other than fruit flies. Butyl sorbate (16) is an effective lure for the European cockchafer and methyl linolenate (17) for bark beetles. Such synthetic attractants are related to natural products which function as food or oviposition lures, but their precise mechanism of attraction is unknown. It appears that the chemical released by male bark beetles is both a sex and food attractant. When the male beetles invade a suitable tree, they burrow into it and feed on the inner tissues; then they produce faecal pellets containing the attractant which draws

$$CH_3CH=CH-CH=CHCO_2C_4H_9$$
(16)

$$CH_3CH_2CH=CH-CH_2-CH=CH-CH_2-CH=CH(CH_2)_7CO_2CH_3$$
(17)

both male and female beetles to the tree. The females on arrival are stimulated to join the males in the holes and passages they have made in the tree, while the males immediately start to construct their own galleries. The attractant thus leads to both food supply and mates and provides an extremely effective survival mechanism.[3] Most attractants, unfortunately, only affect the adults of the species and immature insects do not appear to be attracted by such food or oviposition lures and are unable to detect their food plants when they are more than a few centimetres away.[4]

The use of chemical lures in public health control does not appear promising since insects, bedbugs, fleas, and even horseflies do not seem to possess highly developed olfactory organs. On the other hand, mosquitoes are attracted by chemicals evolved by the human body—blood, sweat, and urine all stimulate yellow fever mosquitoes, and their attractiveness is enhanced by moisture and carbon dioxide. Humans secrete lactic acid in their sweat and its presence probably accounts for the appeal of sweat to mosquitoes.[3]

Oviposition lures are substances that attract gravid female insects of a species and induce them to lay eggs. For example, ammonia and ammonium carbonate exert a powerful attraction for gravid houseflies and greenbottle flies respectively. In general, food lures are not as potent as sex lures. Decomposing fruit, fermenting sugar solutions, and digested protein substances have been widely used in traps for a number of insect pests. Such chemically ill-defined materials are usually unspecific and very variable in their effectiveness. However, they are useful in preliminary survey studies, and protein hydrolysate–insecticide preparations are still used for the control of some species of fruit fly, although they have been largely replaced by specific chemical lures, e.g. compounds (**12**) to (**15**) (page 320).[5]

REPELLENTS

Chemical repellents can be either the vapour or contact type, and both must induce the insect pest to move away from them. They must be acceptable to the host, particularly for a human host, so that their application causes no discomfort. Research in chemical repellents over the last 40 years or so has indeed been largely concentrated on the production of synthetic chemicals to protect man from insect attack.[4] The use of naphthalene and *p*-dichlorobenzene in mothballs is a well-known example of repellents against the clothes moth.[5] The advent of World War II caused the United States of America to embark on a major programme of screening chemicals for repellent properties. The discovery of new repellents was vital for the success of military operations against the Japanese in the Far East; indeed at one period a soldier had a 90% chance of being attacked by a tropical disease before he encountered the real enemy. By 1945, some 7000 compounds had been screened by the United States Department of Agriculture and in 1952 the number had been increased to

11 000. The initial tests were for insecticidal and repellent properties against body lice, mosquitoes, and chiggers, and later houseflies, ticks, and fleas were included.[3]

From early times a number of herbal extracts had been recommended as insect repellents; in particular citronella oil containing various terpenes such as geraniol, citronellol, and borneol. This deters mosquitoes from alighting or coming near to objects coated with the oil, but citronella oil is too volatile and short lived to be an effective repellent.[2] An early fly repellent was dimethyl phthalate (**18**) (1929) which became an important constituent of skin preparations, but it was not effective against all species and during the War other repellents were discovered. Two of these were 2,2-dimethyl-6-butylcarboxy-2,3-dihydropyran-4-one or indalone (**19**) and 2-ethylhexane-1,3-diol or Rutgers 612 (**20**) which were widely employed together with dimethyl phthalate to give a broader spectrum of repellent activity.

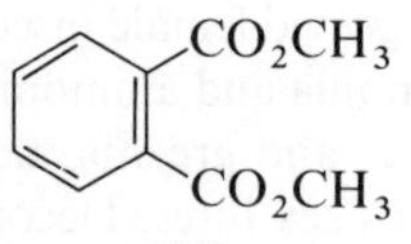
CO2CH3
CO2CH3

(**18**)

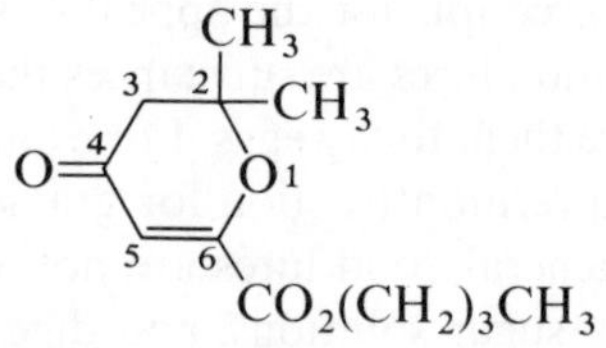
CH3
CH3
O
O1
2
3
4
5
6
CO2(CH2)3CH3

(**19**)

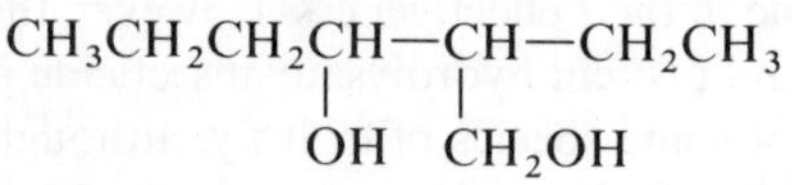
CH3CH2CH2CH—CH—CH2CH3
OH CH2OH

(**20**)

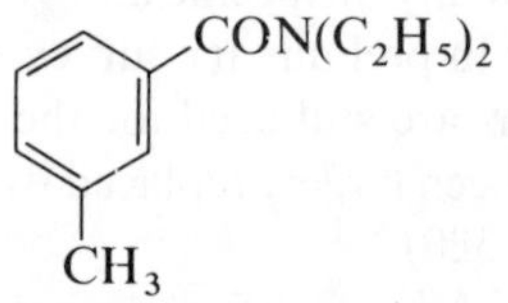
CON(C2H5)2
CH3

(**21**)

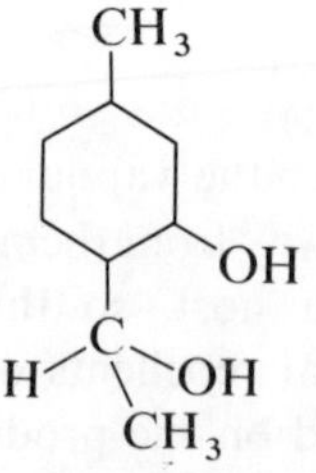
CH3
OH
C
H
OH
CH3

(**22**)

These mixtures have, however, been largely superseded by *N*,*N*-diethyl-*m*-toluamide (**21**) or deet, which was introduced in 1955. Deet (**21**) is the most widely used chemical repellent with a broad spectrum of activity against mosquitoes, flies, chiggers, and other biting insects. Recently it has been shown that *p*-menthanes, e.g. compound (**22**), are repellents for insects and mites: spraying 1% of (**22**) on rats repelled mosquitoes from attacking them. These chemicals appear to act by cancelling out the attractive stimuli emitted by the animal body at a distance; for instance they prevented mosquitoes detecting a hot wet surface. Similarly, the ancient practice of dragging elderberry twigs over germinating turnip seed to deter attack by flea beetles may be due to the smell of the elderberry masking that of the mustard oils evolved during germination of the seed rather than to a direct repellent effect on the beetles.[5]

The diseases transmitted by mites are important in the tropics and the demands of tropical warfare led to an intensive search being made for repellents against the mite vectors of scrub typhus. Impregnation of clothing, especially socks, with dimethyl or dibutyl phthalate or benzyl benzoate (23) controlled the mites, and against certain species of American ticks, Rutgers 612 (20), benzyl benzoate (23), and *N*-butylacetanilide (24) were effective repellents, while 2-hydroxy-n-octylsulphide has been claimed to repel cockroaches.[3]

(23) 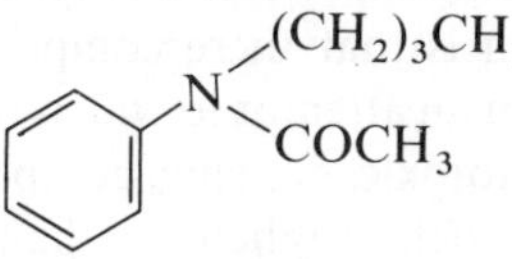   (24)

Chemical repellents are generally formulated as oils, creams, or gels for hand application or as aerosol packs. Pyrethrum preparations possess considerable repellent properties as well as being insecticidal and when a room is sprayed with a kerosene solution of pyrethrum flies tend to keep away from the treated parts of the house for a considerable time. There is a large potential market for insect repellents for the treatment of livestock against attack of ectoparasites, but the present chemicals have to be applied too frequently for convenience or economic benefit. There has recently been considerable emphasis on the introduction of new formulations to make existing chemical repellents more effective and easier to apply. One interesting area is the search for an oral repellent against such insect pests as mosquitoes and houseflies.[4]

Very little is known about the mode of action of chemical repellents or the relationship between repellent activity and chemical structure. Repellents have been little used in modern agriculture to deter pests from feeding on crops, although naphthalene has been applied on a small scale in gardens to mask the attractive smell of carrots to the female carrot fly.

ANTIFEEDANTS

Antifeedants are substances that are not necessarily food repellents but cancel out the signal to the appropriate organ in the insect to initiate feeding on the host; in contrast to chemical repellents, there has been increasing interest in chemical antifeedants over the last 20 years. In the presence of the antifeeding substance the insect may starve to death while remaining on the host plant,[2,4] possibly because it makes the host plant distasteful to the insect so feeding is inhibited. The majority of antifeedants do not directly kill the insect.

Several triazenes show antifeedant properties;[10] the most active member was 4-(dimethyltriazeno)acetanilide (25) (Table 14.2) which was the first antifeedant used in agriculture. It can be applied to the bark of trees to deter deer and rodents from eating the bark and so killing the tree.[3] Compound (25) also inhibited the feeding of caterpillars, southern army worms, and beetles, but did not affect aphids or mites. It did little harm to natural predators or honeybees and was effective when the insect population was comparatively low, but not in the presence of a large population of the pest (LD_{50} (oral) to rats, 510 mg/kg).

Other compounds active as antifeedants are shown in Table 14.2 and include the triphenyltin fungicides (e.g. fentin acetate) (Chapter 7, page 163) and some quaternary salts of heterocyclic secondary amines (e.g. compound (26)). A limitation is that these compounds will only control surface-feeding chewing insects; in an attempt to widen the spectrum of antifeedant activity some known, non-phytotoxic, systemic compounds were examined. Two of the most effective are 2,4,6-trichlorophenoxyethanol (27a) and acetic acid (27b) (Table 14.2); these showed promising activity in laboratory and field tests.[10]

Over long periods of time many plants have evolved sophisticated chemical defences against insect attack and an important approach to the discovery of new antifeedants involves the study of plants known to resist insect attack; e.g. brassicas produce allyl isothiocyanate (28)[11] (Table 14.2). However, the majority of natural antifeedants are much too weak in their antifeedant activity for successful external application in crop protection. Many antifeedants are highly complex molecules that are difficult to synthesize and are not economically available from natural sources. Some highly potent naturally occurring antifeedants are known: the Indian Neem tree produces a complex terpenoid azadirachtin and crude neem extracts are being studied against a range of insect species.[12] Azadirachtin is the most active antifeedant against lepidopterous species known; it is lethal at a concentration of only 10 p.p.m. (Plates 35 and 36). Currently efforts are being directed towards designing synthetic compounds which can mimic azadirachtin.[13,14] A promising, less complex, antifeedant is (−) polygodial (29a), isolated from the marsh pepper (*Polygonium hydropiper*). Plants treated with low rates of compound (29a) were not colonized by aphids and consequently infection by plant-virus diseases was significantly reduced. Polygodial (29a) has been synthesized by Ley's group at Imperial College, London,[12] but unfortunately the (+) isomer is phytotoxic so the synthetic (racemic) product requires a tedious resolution before it can be used in crop protection. Consequently the plant-derived material therefore appears to have more potential for use in agriculture.[7] The closely related compound warburganal (29b) has also been synthesized and shows a promising spectrum of antifeedant activity; it is especially effective against African army worms.[13]

Another group of powerful naturally occurring antifeedants contain the methylenelactone moiety, e.g. schkuhrin I (30a) and II (30b) (Table 14.2),

Table 14.2

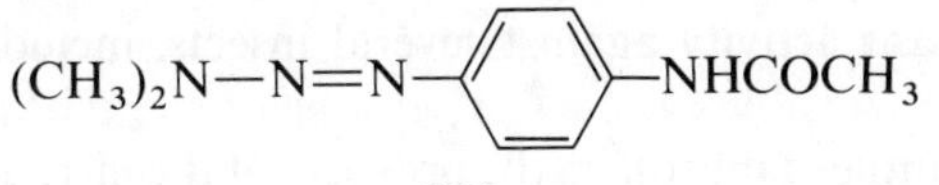

$(CH_3)_2N-N=N-\!\!\!\!\bigcirc\!\!\!\!-NHCOCH_3$

(25)

$(C_6H_5)_3SnOCOCH_3$

Fentin acetate

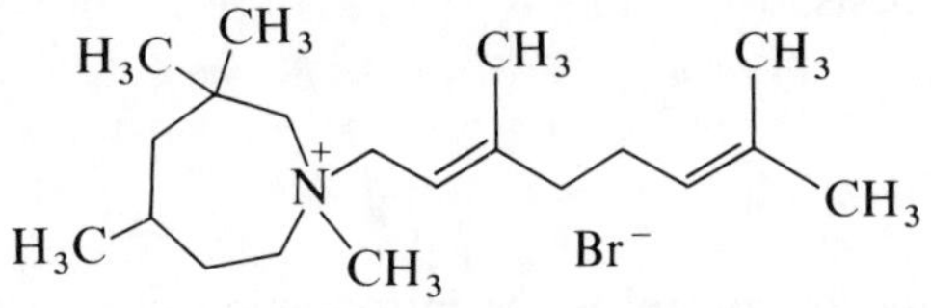

(26)

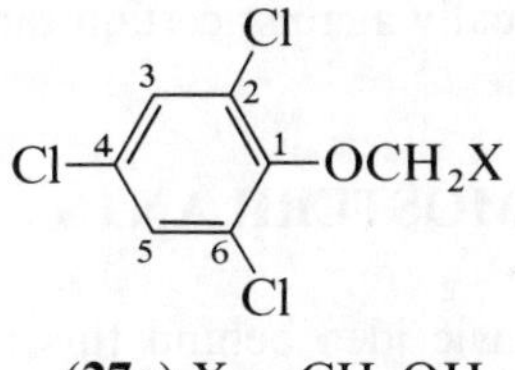

$CH_2=CH-CH_2N=C=S$

(28)

(27a) $X = CH_2OH$
(27b) $X = CO_2H$

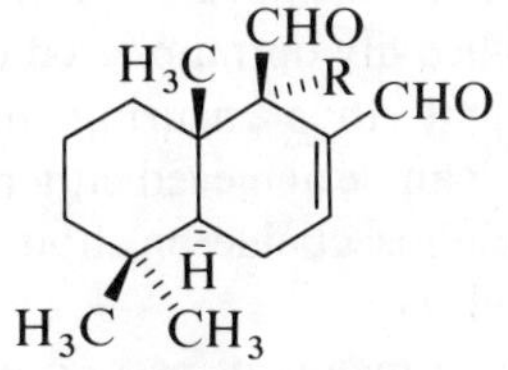

(29a) $R = H$
(29b) $R = OH$

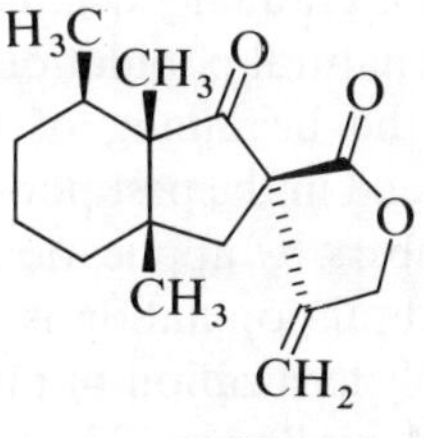

(31)

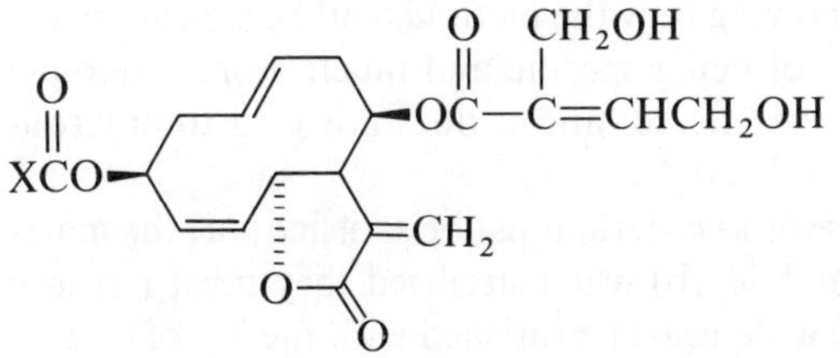

(30a) $X = CH_3$
(30b) $X = -CH(OH)CH(CH_3)_2$

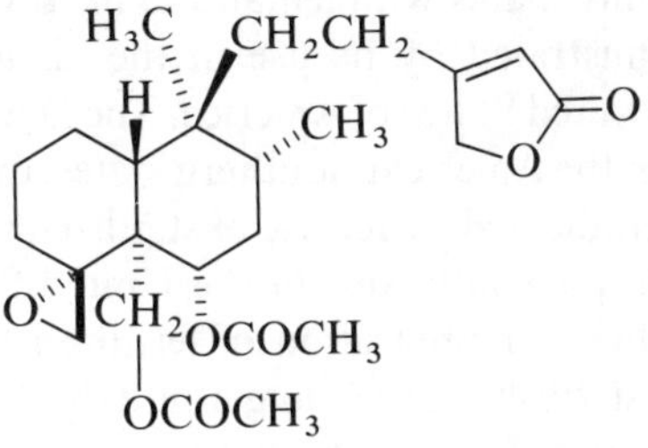

(32)

isolated from the African plant *Schkuhrina pinnata*, and the sesquiterpene bakkenolide A (**31**).[10] The lactone ajugarin I (**32**), isolated from the medicinal plant *Ajuga remota*, has antifeedant activity against several insects, including *Locusta migatoria*.[13]

The discovery of potent insect antifeedants offers the prospect of the introduction of a new group of environmentally safe insect control agents which act specifically against certain target pests.

CHEMOSTERILANTS

The basic idea behind this method of control is the introduction of a large number of sterilized males of the pest species into the area where the pest is to be controlled. The normal female population of the pest is then overwhelmed by the sterile males so that the majority of the matings are not fruitful. If the population of the sterile males can be maintained over several generations and there is no mass immigration of healthy males into the locality, the pest numbers in the area will gradually decline and eventually disappear.[4,5] For success, it is vital that the natural population of males is heavily outnumbered by the sterile males from the beginning of the experiment, thus ensuring an immediate downward trend in the pest population. This can be achieved either by reducing the pest numbers by application of a suitable insecticide or choosing a period when the natural population is at a low level.

The idea of sterilization of male insects as a means of pest control was first proposed by Knipling in 1937, but was not practical at the time owing to lack of efficient means for the mass sterilization of the pest insects. Knipling, however, developed mathematical models showing how the method could eradicate pests. The technique has the advantages of being specific and much more economic than normal chemical control by insecticides and it does not lead to environmental pollution.[15]

If the method is to be effective, several criteria must be satisfied: (a) the males of the pest species must be very mobile, (b) when sterilized they must not lose their sexual urge to mate, (c) the female must be satisfied with the act of mating rather than fertility.[4] The method is completely specific because the sterile male only mates with females of his own species, and it can be extremely effective as illustrated by its use in the virtual elimination of the screw worm from the United States of America. The larval screw worm is a parasite of cattle and sheep in the American southern states and in various South American countries. It was eradicated by release of sterilized male screw worm flies first in Florida and then in the south-west of the United States. By 1964 outbreaks in Texas were only due to immigration of fertilized females from Mexico and a barrier has been established stretching along the Mexican border patrolled by sterile male screw worm flies to cut down the danger of reinfestation.[5]

Other pests for which the sterilization technique looks promising include

Mediterranean fruit flies, melon flies, sheep blowflies, and horseflies. Thus, in 1976, more than 10^7 sterile male Mediterranean fruit flies were released over an area of $190\,km^2$ near Los Angeles to halt an invasion of fruit flies. The cockchafer was eliminated within a trial area by this technique and it looks promising for control of the codling moth, but against mosquitoes the method was unsuccessful.

Artificially reared insects can be sterilized by exposure to X-rays or gamma rays before release in the target area. Also there are several chemicals known to produce sterility in insects, which are termed chemosterilants. These are essential for the sterilization of natural pest insects in the field since the chemosterilant can be incorporated in a suitable food bait. Since 1947, the United States has embarked on an intensive search for chemosterilants and some 200 chemicals are known to sterilize certain insect species. The chief targets of the tests were houseflies and mosquitoes. Chemosterilants may be subdivided into alkylating agents, antimetabolites, and miscellaneous substances.

The alkylating chemosterilants are very reactive chemicals which replace a hydrogen atom by an alkyl group. Some of the most promising members contain aziridinyl groups in which the nitrogen atom is attached to electron-withdrawing groups, such as P=O, C=O, CN, SO, or SO_2. Important examples are apholate (**33**), tepa (**34**; X = O, R = H), thiotepa (**34**; X = S, R = H), and metepa (**34**; X = O, R = CH_3), and the diurea (**35**).[2,5,10]

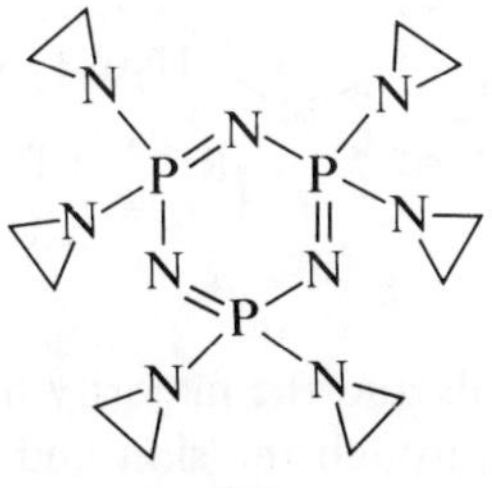

(33)

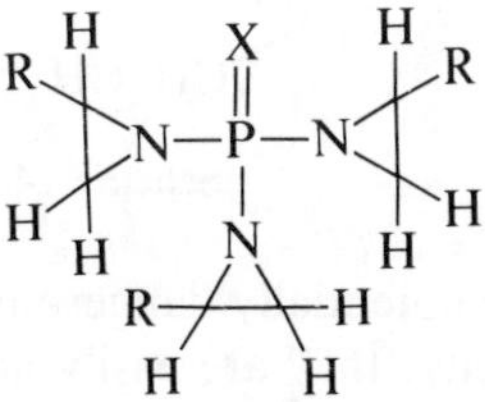

(34)

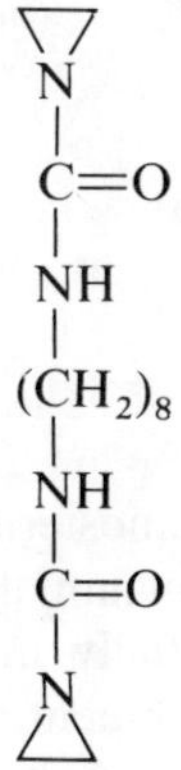

(35)

These chemicals are currently the most promising chemosterilants and effect sterilization at doses lower than those required for general toxicity. Depending on the species of insect and the dosage used, they may act by causing the insect eggs not to develop, or not to hatch, or to hatch into larvae that die before attaining maturity.[5,16] Other alkylating agents, which are promising chemosterilants, include analogues of nitrogen mustards, e.g. chlorambucil (**36**), named because of the similarity to the World War I mustard gas, and sulphonic acid esters, e.g. bisulphan (37).

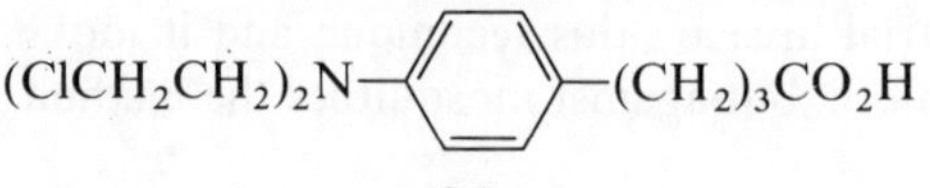

$(ClCH_2CH_2)_2N$—⟨benzene ring⟩—$(CH_2)_3CO_2H$ $CH_3SO_2O(CH_2)_4OSO_2CH_3$

(36)   (37)

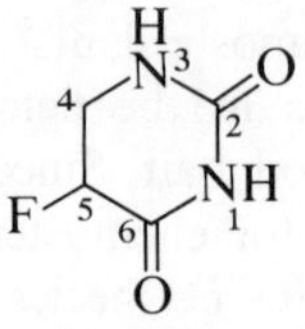

(38) 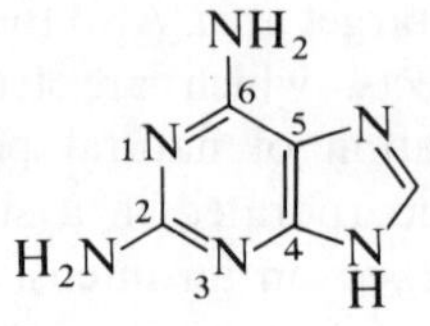   (39)

Both the aziridine derivatives and the nitrogen mustards probably function as alkylating agents by a similar mechanism, since the latter are readily converted into the cyclic aziridine ion which subsequently reacts with available nucleophiles Nu and Nu' as indicated in Scheme 14.1.

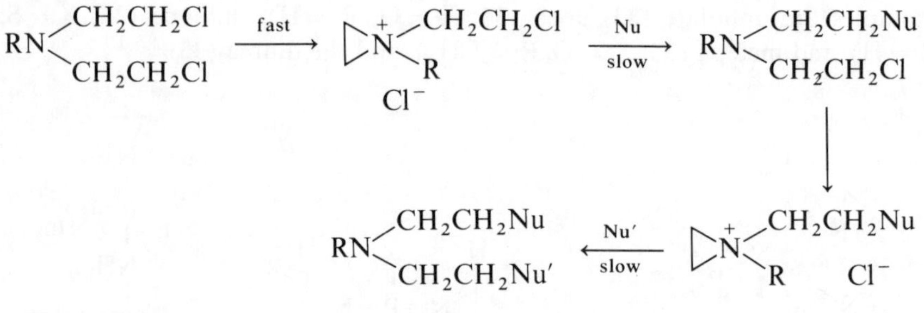

Scheme 14.1

Chemosterilants are potentially dangerous chemicals and the majority have high mammalian toxicity; they are easily absorbed through the skin and are potentially mutagenic. The mutagenicity could increase the danger of the development of a resistant strain of the treated pest insect.

Antimetabolites are chemicals that mimic natural biologically active metabolites, so they may replace them in a biochemical process with consequent alteration or inhibition. Some of the antimetabolites are chemosterilants and several owe their action to functioning as analogues of the purine and pyrimidine bases occurring in nucleic acids (Chapter 3, page 47). An example is 5-fluorouracil (38) which can replace uracil in RNA, so disrupting its normal function, or 2,6-diamino purine (39b). Antimetabolites, unlike the alkylating agents, generally function as female sterilants by disrupting the reproductive system of female insects.

The miscellaneous group of chemosterilants covers a wide variety of compounds, including antibiotics like mitomycin and cycloheximide, the alkaloid colchicine, hexamethylphosphoric triamide, triphenyltin, urea, thiourea, and derivatives of s-triazine.

ARTHROPOD HORMONES

The growth of insects and other arthropods, like that of plants (see Chapter 9, page 271), is controlled by chemical hormones. Arthropods possess tough cuticles or exoskeletons that maintain their shape and support the internal organs and muscles. The stiff outer cuticles cannot continually grow and so periodically a new soft cuticle is produced inside the old one; the latter then splits open and is discarded. The new cuticle expands and toughens. Discrete moulting operations thus separate each stage (instar) of the insect's development. The successive instars often just differ in size, but these may represent significant structural alteration which is especially true for those insects in which only the final stage is sexually active. In butterflies and moths (lepidoptera) the larval (caterpillar) form has four instars which is followed by the pupa, an immobile non-feeding stage, from which the adult winged insect emerges. This pattern of growth and metamorphosis is unique to arthropods and does not occur in vertebrates; it is dependent on the presence of two kinds of insect-specific hormones. The process of moulting or 'ecdysis', which is clearly essential for insect growth, is controlled by the *moulting hormones* (*MH*). These are steroidal compounds (ecdysones), of which the most important are α- and β-ecdysone (**40**; R = CH_3 or OH respectively), isolated from the silkworm (*Bombyx mori*) pupae.[17] It is likely that β-ecdysone is the real moulting hormone (MH).[10]

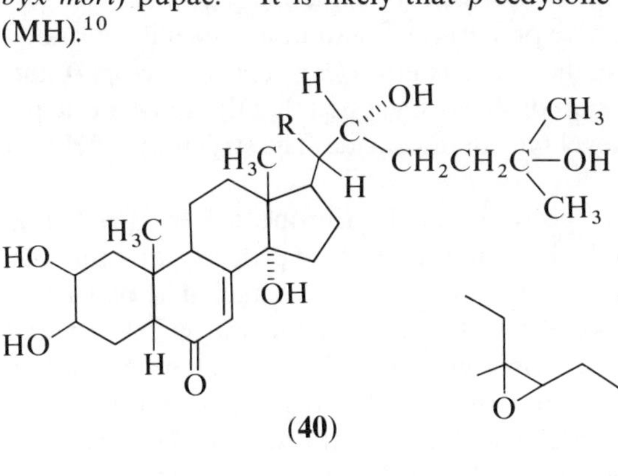

(40)

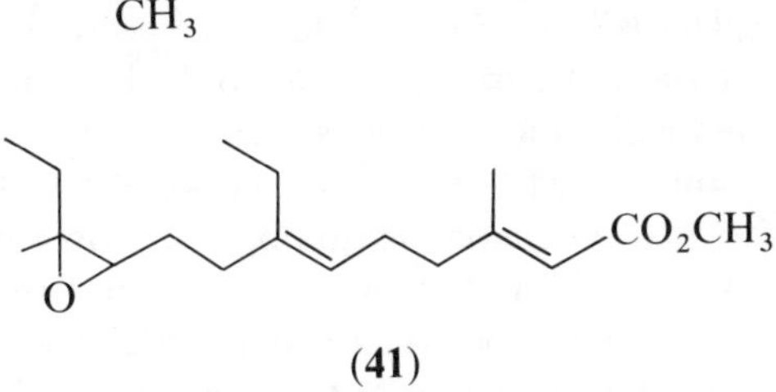

(41)

Closely related steroids have been isolated from plants, especially species of conifers and ferns, but these ecdysone analogues do not appear to have any toxic action on insects feeding on the plants, although some of them affect insect metamorphosis. Moulting hormones (MH) have not yet been commercially exploited; one problem is the high cost of their preparation. The second hormone known as the 'juvenile hormone' (JH) **(41)** controls the process of metamorphosis.

The juvenile hormone released from the *corpora allata* glands in the insect's head, together with the moulting hormones or ecdysones circulating in the insect's blood, play vital roles in the insect's growth, development, and reproduction. The amount of juvenile hormone present controls the nature of the cuticle laid down—in the larval stages when a large amount of juvenile hormone is available further juvenile cuticle is formed; if it was not present the insect would mature prematurely. On the other hand, when the larva reaches the optimum size, feeding ceases and the insect moults to the pupal stage, when almost none of the juvenile hormone must be present, so the mature or adult cuticle is deposited.[18]

With those insects that pass through a pupal stage, an intermediate quantity of juvenile hormone regulates the deposition of the pupal cuticle; therefore this hormone perpetuates immature growth and development in metamorphosis, and when it is absent allows maturation. In adult insects, the juvenile hormone controls the development of the ovaries.[2] The amount and timing of the production of juvenile hormone is vital. If it is available at the wrong period or in too large a dose gross abnormalities of growth are induced which generally kill the insect, and its presence in insect eggs will prevent hatching and normal development. Thus, if insects are treated with an excess of the juvenile hormone at an early stage in their development, their life cycle is distorted and they will remain in a juvenile (larval) stage and will not change via pupation into the adult insect.[19] The juvenile hormone may act as a coenzyme for those enzymes controlling larval development, or may alter permeability so that these enzymes are more effective, or act directly on the nuclei of epidermal cells.

It is some 30 years since an active preparation of an insect juvenile hormone (JH) was isolated from extracts of the male silk moth (*Hyalophora cecropia*), and its high ability to block insect metamorphosis suggested that the active principle could well have potential as a novel type of insecticide. The structure of JH **(41)** was elucidated in 1965.[20]

The failure of efforts to rear specimens of the European bug *Pyrrhocoris apterus* in America were due to the larva not developing properly, and this was eventually traced to the paper towelling used to line the cages. The paper was manufactured from balsam fir wood which is not used in Europe. The paper factor or juvabione inhibiting the proper growth and egg hatching of the bug was isolated and shown to have the structure **(42)** analogous to that of the juvenile hormone **(41)**. Juvabione **(42)** also prevented the development of the red cotton bug, an important pest of cotton, which belongs to the same family as the European bug, but it did not affect other families of bugs, so juvabione appears to exhibit considerable specificity. The reason for the presence of juvabione in balsam fir is obscure since the European or other related bugs are not found on this tree, though possibly they did once attack it. The isolation of the juvenile hormone mimic from the balsam fir has stimulated the search for other mimics in different plant species. The results have been rather disappointing—from examination of 60 plants only two of them gave extracts showing significant juvenile hormone activity against *Pyrrhocoris*.

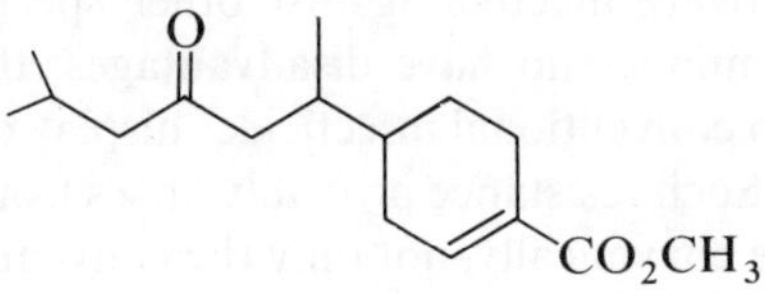

(42)

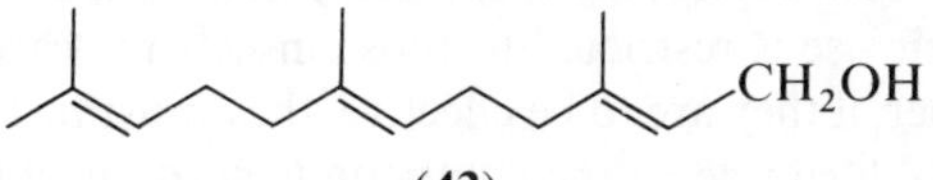

(43)

The terpene alcohol farnesol (43) and the related aldehyde were active, and further work showed that compounds containing the methylenedioxyphenyl group often show considerable activity, and are important insecticide synergists (see Chapter 4, page 71). These studies initiated a search for new synthetic compounds that can act as juvenile hormone mimics providing 'third-generation insecticides'.

A number of long-chain aliphatic compounds have JH activity, e.g. those derived from farnesol (43); however, these compounds do not apparently penetrate the insect cuticle which creates a problem in using JH mimics to control insect pests. Some analogues are active by topical application and are produced commercially (e.g. compounds (44) and (45)).

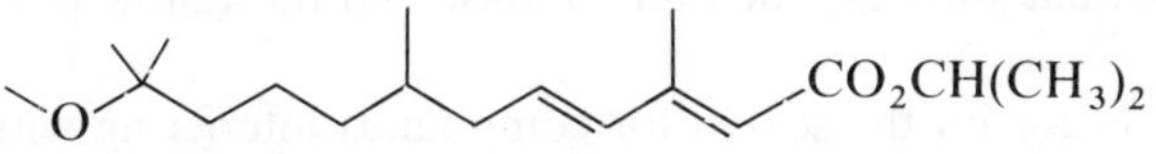

(44)

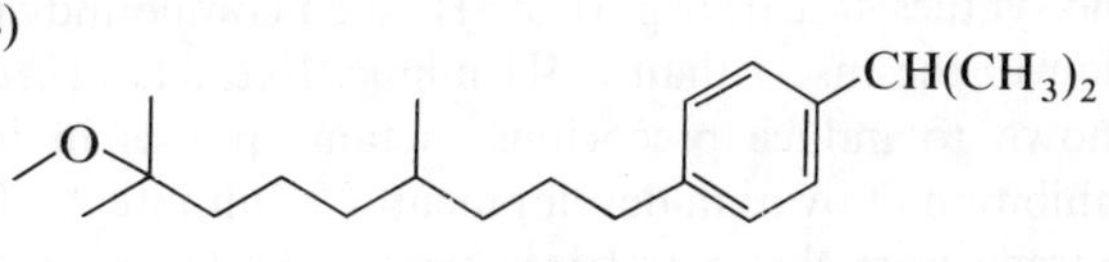

(45)

Methoprene (44) is used as a larvicide against flood water mosquitoes and, as a microencapsulated formulation, it is effective at concentrations of < 1 p.p.b.[20] Compound (45) can exterminate fire ant colonies by incorporation in a bait which is taken back to the nest where it disrupts the development of new workers.[21] These compounds act specifically on the endocrine system of insects and this different mode of action enables them to be effective against insects that have developed resistance to conventional insecticides. Methoprene has low mammalian toxicity and shows fairly specific activity against *Diptera* and has low toxicity to other insect species. Thus in its normal use against mosquito larvae, it has little effect on non-target insects such as damselflies, mayflies, and water beetles.[3]

Similarly, a recently discovered aryl terpenoid mimic (2-ethoxy-9-*p*-isopropylphenyl-2,6-dimethylnonane) is the most effective juvenile hormone mimic against livestock fly pests, but is relatively inactive against other species of insects.[20] However, juvenile hormone mimics do have disadvantages; thus a significant number of insects resistant to conventional insecticides display cross-resistance to juvenile hormone mimics. Such resistance probably arises from the ability of the tolerant insects to degrade enzymically, not only the conventional insecticides but also the juvenile hormone mimics. The mimics do not disrupt the normal development of insect larvae and do not have true larvicidal activity, so their use is restricted to those insects that are pests only in the adult phase. Further, if they are to be effective, they must be applied at the right period in the insects' life cycle so the compounds need fairly high environmental persistence for good activity against field populations of insects of varying ages, but unfortunately most of the juvenile hormone mimics break down fairly rapidly in the field. Their use is therefore confined mainly to the dipterous pests of public and animal health; juvenile hormone mimics have been used against mosquitoes. Here mimics, e.g. methoprene, have the advantages of low toxicity to mammals and non-target insects and are effective against resistant strains of mosquitoes. However, the poor persistence of the compounds is a disadvantage ; to obtain good control of Californian flood mosquitoes it is necessary to use an encapsulated slow-release formulation of methoprene (Chapter 2, page 23).

Juvenile hormone mimics may also be useful as chemosterilants (page 326); thus a dose of 1 μg or less of methyl farnesoate causes a female European bug (*Pyrrhocoris apterus*) to lay sterile eggs for the rest of her life, and other insects could be controlled in a similar way, e.g. the human louse and the yellow fever mosquito.

Another approach was based on the search for compounds interfering with the synthesis or transport of JH; such compounds might cause greater disruption to the insect than a JH mimic. Extracts of *Ageratum houstonianum* were shown to induce precocious metamorphosis in immature hemipterans and inhibition of ovarian development in adult insects. The active ingredients of the extracts were shown to be the precocenes (1 and 2) (**46, 47**). Precocenes (**46, 47**) have been shown to inhibit JH synthesis in isolated cockroach corpora allata.[9]

The scope of application of JH mimics has not been as great as was originally envisaged; however, their discovery has led to other synthetic compounds which

(**46**) R = H
(**47**) R = OCH$_3$

act by disruption of insect growth. Such compounds are not juvenile hormone mimics and consequently do not have the disadvantage of only being effective at maturation moults.[20] An example is diflubenzuron (**48**); this is not a juvenile hormone mimic and it disrupts the larval as well as the maturative stages. Diflubenzuron (**48**) is highly active against mosquitoes, livestock fly pests, cabbage white butterfly caterpillars, moths, and other larval crop pests at low doses, e.g. 30–120 g/ha.[10] The spectrum of insecticidal action is much wider than that of juvenile hormone mimics. It is reasonably persistent in its action and has very low toxicity to fish and birds.[20]

The biochemical basis for its action is considered to be disruption of the moulting process by inhibition of the normal insect cuticle formation.

The symptoms of poisoning by this compound are death of treated larvae due to their inability to shed the old cuticle at moulting time, and the distortions of any newly formed larvae suggest that the compound impairs the mechanical properties of the new cuticle. Electron micrographs of the newly formed cuticle of cabbage white caterpillars which have been treated with diflubenzuron show that the cuticle lacks the lamellar structure of the untreated insect cuticle, and the compound may interfere with the final stage of chitin synthesis, namely the polymerization of N-acetylglucosamine, but the primary target is not definitely known.[9]

Diflubenzuron (48) is prepared by reaction of 2,6-dichlorobenzamide with *p*-chlorophenylisocyanate (Scheme 14.2).

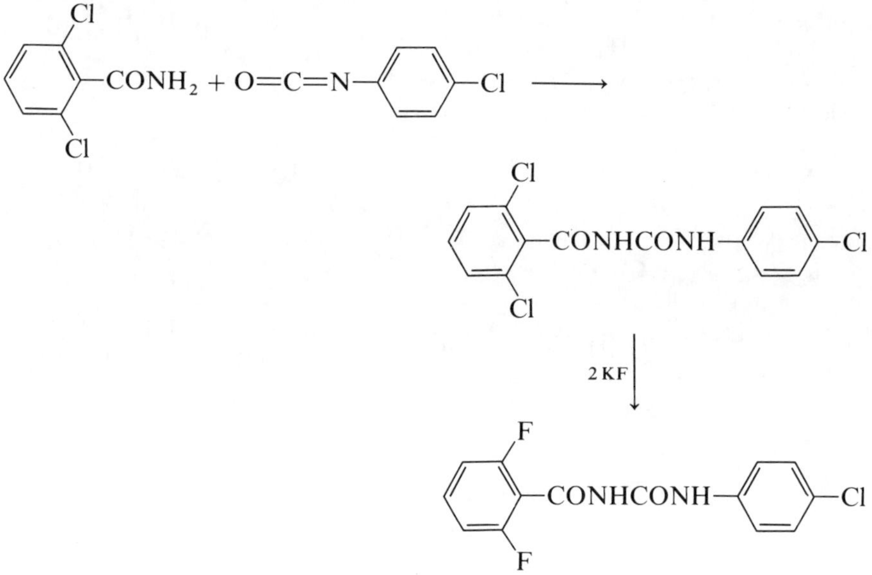

Scheme 14.2

The benzoylphenylureas (BPUs) are an important developing class of insecticides and several new members **(49)** to **(51)** (Table 14.3) with improved properties have been recently introduced.[10,21]

The trifluromethyl derivative **(49)** (Dow, 1986) has excellent insecticidal, ovicidal, and larvicidal activity.[22] Flufenoxuron **(50)** (Shell, 1986) is used to control aphids and insects in top fruit, vines, citrus, and vegetables; it shows a wider spectrum of activity than other benzoylphenylureas.[23] Flucycloxuron **(51)**

Table 14.3

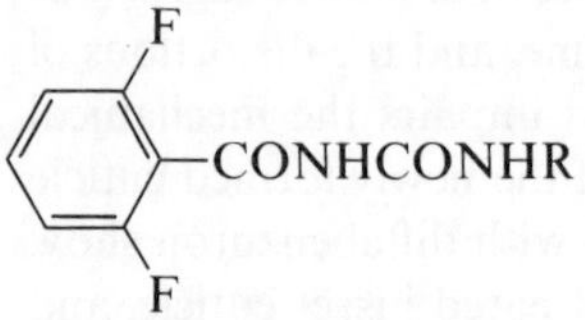

General structure

$$R \begin{cases} \textbf{(49)} & \\ \textbf{(50)} & p\text{-}C_6H_4C_6H_4OCF_2CH_2F_2 \\ \textbf{(51)} & p\text{-}C_6H_4CH_2ON{=}C \end{cases}$$

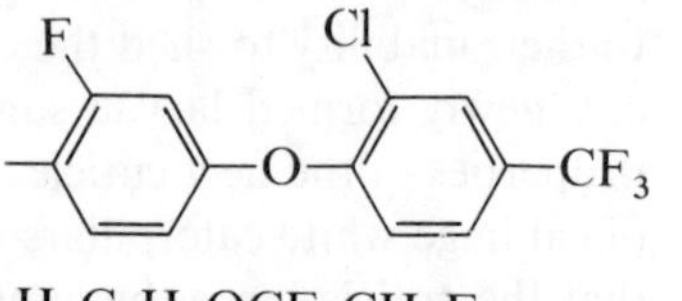

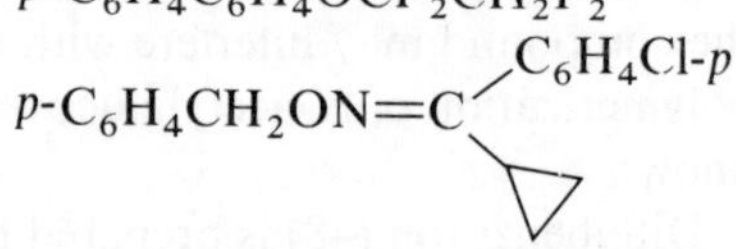

(52)

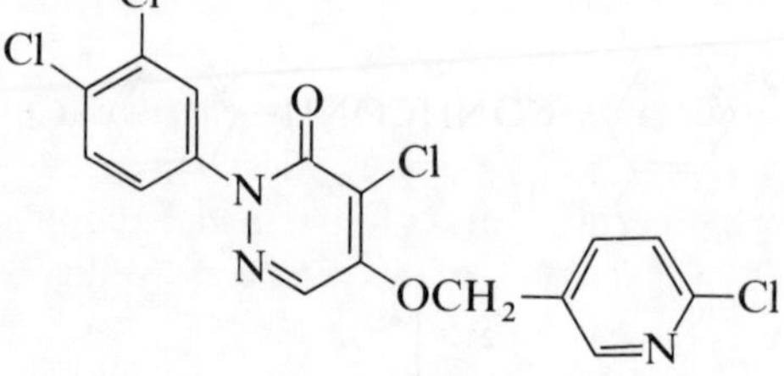

(54)

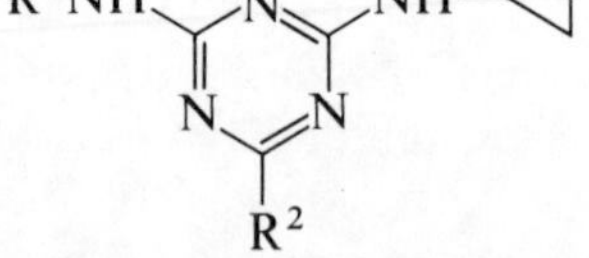

(53a) $R^1 = H$, $R^2 = NH_2$,
(53b) $R^1 = C_2H_5$, $R^2 = N_3$

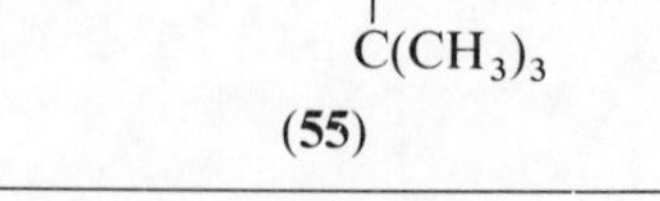

$$C_6H_5CONH{-}N{-}COC_6H_5$$
$$\underset{\displaystyle C(CH_3)_3}{|}$$

(55)

(Duphar BV, 1988) is active against mites as well as insects.[24] N-sulphenyl derivatives of benzoylphenylureas often showed better insecticidal activity, because their greater solubility in organic solvents permitted formulation as emulsifiable concentrates.[14]

The benzoylphenylureas (BPUs) all act by interference with chitin synthesis in immature insects and are highly selective insecticides with low toxicity to non-target organisms—hence their increasing importance for use in integrated pest control programmes.[21] The ovicidal activity of BPUs is also associated with disruption of cuticular formation in the developing embryo.

Benzoylphenylureas are not the only non-juvenile hormones to act as insect growth regulators (IGRs): 2,6-di-t-butyl-4-(dimethylbenzyl)phenol and 1-buten-3-yl-N-(p-chlorophenyl)carbamate specifically disrupt moults of mosquito larvae[20] while compound (52) inhibits pupal growth.

The triazine insect growth inhibitors (53a,b), e.g. cyromazine (53a) (Table 14.3), are highly active against dipterous larvae, especially leaf miners, and induce morphological aberrations during larval and pupal development. Cyromazine is systemic in plants and can be applied via the soil or the crop foliage.[25] The precise mode of action of the triazine insect growth regulators (IGRs) is not known, but apparently they do not owe their lethal effects to inhibition of chitin synthesis.[21]

The pyridazinone (54) (1988) specifically inhibits the growth of plant and leafhoppers, while the benzoylhydrazide (55) (Rohm and Haas, 1988) is a new class of non-steroidal ecdysone agonist which acts in a similar manner to the ecdysones (see page 329) and induces premature moulting in lepidoptera. This results in inhibition of feeding but, unlike the other synthetic IGRs, the benzoylhydrazide (55) does not interfere with chitin synthesis.[26] It has provided excellent control of Colorado beetle on potatoes, leaf miners on apples, rice stem borer, codling, and tortrix moths.

Natural insecticides can also be derived from insects themselves, such as spiders, snakes, beetles, ants, bees, and wasps.[27,28] Some of these toxins and venoms are highly potent and selective but the majority are only effective by injection and are not therefore of practical use as insecticides. However, arthropod venoms and toxins can serve as useful models of insecticidal action which may have novel modes of toxicity, and assist in the discovery of new synthetic chemicals with unique modes of action.[27]

MICROBIAL INSECTICIDES

Insect pests can be controlled by bacterial, fungal, and viral infections, and entomologists have long known that insect populations can be seriously affected by outbreaks of infectious diseases which was first observed with silkworm and honeybee colonies.[2,4] Some *Bacillus* species attack chafer grubs, turning their blood into milky fluid, and spore powder preparations of these bacteria have been successfully applied in America since the early 1940s for control of the chafer in grassland which has kept the beetle population at low levels.

Bacillus thuringiensis (Bt) refers to a group of bacterial isolates which have been studied for nearly a century. *B. thuringiensis* was first isolated in 1915 from diseased flour moth larvae and later several varieties were obtained from different species of infected caterpillars.[29-30] The bacteria grouped together as *B. thuringiensis* all form toxic protein crystals at the time of sporulation. The crystals are very toxic to certain insects and can be used as insecticides against many *Lepidoptera* and they do not appear to be toxic to other organisms, apart from earthworms. The toxic crystal (endotoxin) is chiefly responsible for the paralysis and ultimate death of infected caterpillars, although at least five toxic entities have been isolated from *B. thuringiensis*.[29]

The toxic protein crystal is the most studied and important of the *B. thuringiensis* toxins; it dissolves in the alkaline gut of caterpillars and other susceptible insects and the protein solution is digested by enzymes in the gut to release one or more toxins.[29] The mode of action of these is not understood, but the first symptom is paralysis of the gut followed by changes in the gut wall affecting its permeability and allowing escape of the alkaline contents with resultant general bodily paralysis and death of the caterpillar which may take up to seven days.[30] A wettable powder formulation of *B. thuringiensis* called Thuricide is used as a field insecticide and is active against a wide range of lepidopterous larvae, including many pest species. Such commercial spore/crystal mixtures are mainly used against lepidopterous larvae on leaf crops such as cabbage and tobacco. Thuricide appears harmless to mammals, birds, fishes, and most insects apart from *Lepidoptera* and crops can be harvested immediately after treatment.

With certain strains of *B. thuringiensis* a second toxin known as β-exotoxin can be isolated from their cultures. β-Exotoxin is thermally stable and possesses a much wider spectrum of insecticidal activity affecting the larval stages of many species of insects (e.g. fleas) and also showing toxicity to a range of mammals. The toxicity only becomes apparent at moulting or metamorphosis; exotoxin is a high molecular weight adenine nucleotide containing an unusual sugar allomucic acid, and there is some evidence that it acts by interference with nucleic acid metabolism and protein synthesis. In future it will be possible to develop new strains of the bacillus with more useful insecticidal potency; indeed, by 1972 strain selection and improvements in fermentation achieved a hundredfold increase in potency of Bt over early products. Crystals of some strains of Bt contain five polypeptide toxins, each with a different range of susceptible hosts, so isolation of different strains increases the spectrum of insecticide control. Application of genetic engineering techniques of Bt toxins may allow the production of superior toxin strains and the genes encoding for these may be able to be inserted into crop plants (see Chapter 16, page 369).

VIRUSES

Insects can be associated with viruses either as hosts or as vectors. Certain viruses are parasitic and pathogenic to insects and such viruses are enclosed with protein crystals, capsules, or membranes. Virus diseases in insects have long been known; the first record (1527) was of the jaundice disease of silkworm, and some 450 viruses have been identified from insect species.[5]

Insect larvae, especially those of the *Lepidoptera*, are often attacked by viral diseases described as polyhedral inclusion viruses. The viruses are embedded in a protein matrix which is generally polyhedral in shape. These viruses, known as baculoviruses, are the most promising insecticides and have been successfully sprayed onto forest trees to control sawflies.[5] However, experiments with viruses of *Lepidoptera* have not been so effective, but a polyhedral virus of the alfalfa caterpillar shows promise against this pest in America. Viruses have also been applied successfully against the cabbage looper, the cotton bollworm, tobacco budworm, and the forest tent caterpillar.[2] Some diseases are caused by non-inclusion viruses, one of which occurs in natural populations of the *Drosophila* species which makes the flies very susceptible to even small concentrations of carbon dioxide. A few viruses are known to infect mites; thus the citrus red mite rapidly succumbs to a paralysing non-inclusion virus, but this is unstable so it cannot be applied as a spray and is only effectively transmitted by release of infected males.

A granulosis virus (CpGV) has given good control of codling moth (*Cydia pomonella*), a major pest of apple and other fruit crops, over several seasons. Viruses have great potential for insect control, they are highly virulent, very specific, and can survive for years. The body of one infected insect will produce an enormous number of virus particles to infect other insects.[2]

Viruses can only be produced on living insects; hence their commercial production demands a ready supply of insects throughout the year. This has limited their use, but now these problems are being overcome and viruses appear likely to become valuable agents to control various pests.[30]

Virus diseases of higher animals have been used in certain cases; the most successful was the control of the rabbit population in Britain by the myxoma virus causing myxomatosis. The outbreak was started in France in 1952 by the release of a small number of infected rabbits inoculated with the disease; however, the effectiveness of the virus is now threatened by the emergence of rabbits resistant to the disease.[5] Attempts to control rats by a virus disease have proved largely unsuccessful (Chapter 11, page 288).

FUNGI

Several insect–fungus associations are pathogenic and hence microbial control of certain insects can be achieved by fungi. The early history of the discovery of fungi that are parasitic on insects has been discussed by Steinhaus (1956)[5] and led to the elucidation of the principles of the fungal control of insects. The muscardine disease of silkworms was shown (1835) to be caused by a fungus *Beauveria bassiana*, and this organism also attacks a number of insect pests including the codling moth and the European corn borer. The Russian, Metchnikoff, first tried (1879) to control wheat cockchafers by application of a preparation of the green muscardine fungus (isolated from diseased wheat cockchafers and grown on sterilized beer mash). This fungus is pathogenic to many insect species, including mosquitoes, that are susceptible in the larval stage, and could be a useful weapon in vector control.

The muscardine fungi have considerable potential as microbial insecticides, because they can be easily raised on artificial cultures, are pathogenic to a number of insect pests, and appear generally harmless to vertebrates. In contrast, the related *Aspergillus* species of fungi can result in severe diseases in vertebrates so they are unlikely to be of any value for insect control.

Species of the aquatic fungus *Coelomomyces* are chiefly pathogenic to mosquito larvae and have been used in field trials for control of mosquitoes; the fungus appears quite promising, but it is difficult to obtain and so far none of the species have been grown successfully on artificial media. Entomogenous fungi are sometimes of importance in the natural control of insect populations.[30] Some members of the entomophthorales are common pathogens of houseflies and in damp weather in the autumn dead flies are often seen on windows with their bodies surrounded by a halo of discharged spores. This genus of fungi attacks many kinds of insects and some members appear to be the only important pathogens of aphids. They have been applied to control, for instance, the European apple sucker in Canada, the spotted alfalfa aphid in California, and aphids on potatoes in Maine.[5]

Nematodes or eelworms are difficult to control chemically (Chapter 12) and it is therefore interesting that several species of fungi are predacious on eelworms which they trap by adhesive snares and hyphal rings. When the nematode enters the ring it contracts, trapping it fast, and then the fungal hyphae penetrate the body wall of the nematode and extract the contents. Such predacious fungi have been used in England, France, and Hawaii against nematodes when different strains of the fungi were discovered to show varying degrees of aggressiveness towards nematodes. It could therefore be advantageous to inoculate the soil with aggressive strains of the fungus, although this characteristic may not be permanent. A general problem is the difficulty of obtaining widespread growth of the pathogenic fungus artificially, before the insect pest has attained almost epidemic proportions.[5]

Several groups of insect-parasitic nematodes are known; one group, rhabditid nematodes, are lethal parasites of mushroom fly larvae and show promise for the control of these flies in mushroom houses. Their future use in the industry is largely dependent on our capacity to mass-produce them economically.

REFERENCES

1. Baker, R., and Evans, D. A., *Chemistry in Britain*, **16**, 412 (1980).
2. Kelly, D. R., *Chemistry in Britain*, **26**(2), 124 (1990).
3. Fletcher, W. W., *The Pest War*, Blackwell, Oxford, 1974, p. 137.
4. Kilgore, W. W., and Doutt, R. L. (Eds.), *Pest Control: Biological, Physical and Selected Chemical Methods*, Academic Press, New York, 1967.
5. Martin, H., and Woodcock, D., *The Scientific Principles of Crop Protection*, Arnold, London, 1983.
6. Mitchell, E. R. (Ed.), *Management of Insect Pests with Semiochemicals: Concepts and Practice*, Plenum Press, London and New York, 1981.
7. Pickett, J. A., *Phil. Trans. Roy. Soc. London*, **B310**, 235 (1985).
8. Kydonieus, A. F., Beroza, M., and Zweig, G. (Eds.), *Insect Suppression with Controlled Release Systems*, CRC Press Inc., Boca Raton, Fla., 1982.
9. Corbett, J. R., Wright, K., and Baillie, A. C., *The Biochemical Mode of Action of Pesticides*, 2nd edn, Academic Press, London and New York, 1984, p. 172.
10. Matolcsy, Gy., Nádasy, M., and Andriska, V., *Pesticide Chemistry*, Elsevier, Amsterdam, 1988.
11. Bowers, W. S., 'Phytochemical resources for plants', in *Recent Advances in the Chemistry of Insect Control* (Ed. Janes, N. F.), Royal Society of Chemistry, London, 1985, p. 272.
12. Ley, S. V., 'The synthesis of some insect antifeedants', in *Recent Advances in the Chemistry of Insect Control* (Ed. Janes, N. F.), Royal Society of Chemistry, London, 1985, p. 307.
13. Ley, S. V., and Toogood, P. L., *Chemistry in Britain*, **26**(1), 31 (1990).
14. *Recent Advances in the Chemistry of Insect Control*, Vol. II (Ed. Crombie, L.), Royal Society of Chemistry, Cambridge, 1990.
15. Knipling, E. F., *Principles of Insect Chemosterilization* (Eds. Labrecque, G. C., and Smith, C. N.), Appleton-Crofts, New York, 1968.
16. Borkovec, A. B., 'Chemosterilants for male insects', in *Insecticides* (Ed. Tahori, A. S.), Gordon and Breach, New York, 1972, p. 469.
17. Horn, D. H. S., 'The ecdysones', in *Naturally-Occurring Insecticides* (Eds. Jacobson, M., and Crosby, D. G.), Dekker, New York, 1971, p. 333.
18. Bowes, W. S., 'Juvenile hormones', in *Naturally-Occurring Insecticides* (Eds. Jacobson, M., and Crosby, D. G.), Dekker, New York, 1971, p. 307.
19. Kay, I. T., Snell, B. K., and Tomlin, C. D. S., 'Chemicals for agriculture', in *Basic Organic Chemistry*, Part 5, Wiley, London, 1975, p. 430.
20. Ruscoe, C. N. E., *Proceedings of 8th British Insecticide and Fungicide Conference, Brighton*, **3**, 927 (1975).
21. Hutson, D. H., and Roberts, T. R. (Eds.), 'Insecticides', in *Progress in Pesticide Biochemistry and Toxicology*, Vol. 5, Wiley, Chichester, 1985.
22. Komblas, K. N., and Hunter, R. C., *Proceedings of British Crop Protection Conference, Brighton*, **3**, 907 (1986).
23. Anderson, M., Fisher, J. P., and Robinson, J., *Proceedings of British Crop Protection Conference, Brighton*, **1**, 89 (1986).
24. Scheltes, P., Hofmann, T. W., and Grosscurt, A. C., *Proceedings of British Crop Protection Conference, Brighton*, **2**, 559 (1988).
25. Schlapfer, T., Cotti, T., and Moore, J. L., *Proceedings of British Crop Protection Conference, Brighton*, **1**, 123 (1986).
26. Aller, H. E., and Ramsay, J. R., *Proceedings of British Crop Protection Conference, Brighton*, **2**, 511 (1988).
27. Beard, R. L., 'Arthropod venoms as insecticides', in *Naturally-Occurring Insecticides*, (Eds. Jacobson, M., and Crosby, D. G.), Dekker, New York, 1971, p. 243.

28. Cavill, G. W. K., and Clark, D. V., 'Ant secretions and canthardin', in *Naturally-Occurring Insecticides* (Eds. Jacobson, M., and Crosby, D. G.), Dekker, New York, 1971, p. 271.
29. Angus, T. A., '*Bacillus thuringiensis* as a microbial insecticide', in *Naturally-Occurring Insecticides* (Eds. Jacobson, M., and Crosby, D. G.), Dekker, New York, 1971, p. 463.
30. Hussey, N. W., and Scopes, N. (Eds.), *Biological Pest Control*, Blandford Press, Poole, 1985.

Chapter 15
Pesticides in the Environment

Pesticides are biologically active molecules deliberately introduced into the environment to control pests, diseases, or weeds. Such chemicals may interact with the ecosystem in a harmful way and, as demonstrated in this chapter, certain pesticides have caused environmental damage, particularly to birds of prey. Both governments and agrochemical companies realize that it is vital to protect the ecosystem from potential damage and consequently the development of a new pesticide requires long-term laboratory and field studies to assess the effects of the chemical on the environment (see Chapter 1, page 14). These studies extend over a period of some seven years and involve determination of the effects of the chemical on soil microorganisms, e.g. bacteria, fungi, and protozoa, earthworms and minute creatures, like mites, springtails, and millepedes. Some of these microorganisms are pests but others are essential in maintaining the soil's fertility, and others are pollinators or beneficial predators which keep down the population of harmful pests.

Any effects on water will be examined; water, like soil, contains a vast and varied population of plant and animal species which are linked into complex food chains culminating in fish-eating birds and mammals, including man (Figure 15.1). Pesticides that are applied to the soil, plants, or water can affect creatures and their delicate interrelationship either directly or indirectly. An enormous amount of research is therefore devoted to monitoring whether the candidate chemical pesticide shows adverse effects on them and, if so, can these harmful properties be reduced to acceptable levels? The total cost of developing a crop protection chemical is now approximately £30 million and some 50% of this cost is incurred in the environmental and toxicological studies which are now demanded by the governmental registration authorities (see Chapter 1, page 15). Initially, the candidate chemical pesticide has been selected on the basis of some useful biological activity against a pest, disease, or weed. Next, analytical chemists have to develop techniques for detecting minute quantities of the pesticide so that the following questions can be answered:

(a) How much of the chemical stays on the plant or moves into the edible part of the crop?

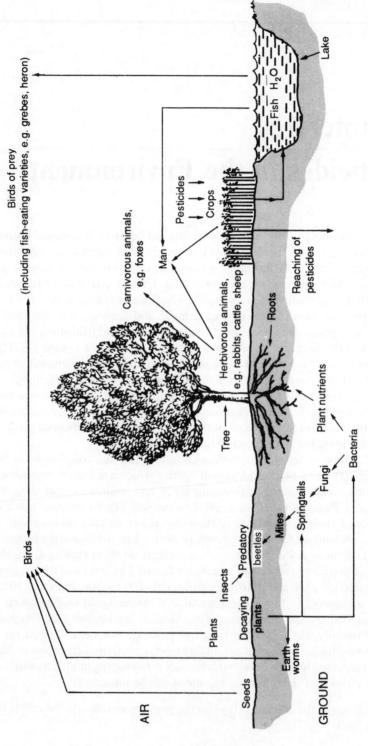

Figure 15.1 The food chain

(b) Does it vaporize or get into ground water?
(c) How does the chemical degrade and are the breakdown products toxic?
(d) Does the chemical, or its degradation products, accumulate along food chains, so that they are transferred from one part of the environment to another? (see Figure 15.1).

Additional studies must be made to determine whether the chemical shows harmful effects on some plant or animal species other than the target pest. For instance, birds, fish and other aquatic creatures, natural predators, such as spiders, ground beetles and ladybirds, bees and other pollinators. Any effects produced by the chemical on the complex balance of soil microorganisms must also be carefully monitored as this could upset a predatory chain permitting a previously controlled organism to become a pest.

The objective of all these studies is to build up a complete balance sheet of the environmental behaviour of the candidate pesticide. No matter how efficient the chemical may be against the target pest, the government will not permit it to be marketed unless it is proved to be safe to man and the environment. The toxicity must be selective to the pest and food chains must not be affected.

The increased use of various types of pesticides in the modern world has led to much greater emphasis on the possibility of serious environmental contamination arising from their use. The scientific attack on pests only dates from about the middle of the nineteenth century (see Chapter 1, page 3) and in 1963 Rachael Carson's book *Silent Spring*[1] dramatized the potential dangers of pollution arising from the mass application of chemical pesticides and greatly increased people's awareness of the problems. Pollution may be defined as fouling the environment and has many forms ranging from chemical pollution to that arising from dumping of old cars and other refuse about the countryside.

Measures to prevent damage to the environment in the United Kingdom can be traced back to the thirteenth century, but the Public Health Act of 1848 first introduced national measures to control the pollution of fresh water and the Alkali Act of 1863 attacked the worst forms of air pollution.

Since then increasing efforts have been made to protect the environment; here we are concerned specifically with pollution arising from the use of chemical pesticides. Whenever a pesticide is applied to the foliage, the seed of a crop, or to the soil, there is a possibility that some of the material is persistent and may lead to serious contamination of the ecosystem. Dangers have caused the various regulatory bodies such as the Environmental Protection Agency (EPA) in the United States to institute increasingly stringent tests which must be satisfied before candidate chemicals can be registered for use as pesticides (see Chapter 1, page 13). The trend is paralleled in other advanced countries such as Britain, e.g. The Food and Environment Protection Act 1985, with more tests devoted to examination of the relatively long-term effects of pesticides on the environment, such as accumulation along food chains.

Since the end of World War II there has been a tremendous expansion in the use of chemical herbicides as a substitute for the age-old mechanical means of weed control. This was stimulated by the shortage and high cost of agricultural labour and consequently the rate of increase in the use of herbicides is much greater than that of other types of pesticides.

Early arsenical weed killers, especially soluble sodium arsenite, were very dangerous to all forms of wild life, including man. These herbicides have high mammalian toxicity and toxic residues remain in the soil for a long time. Since 1961, the use of arsenical herbicides has been officially discouraged and they have not been employed as potato haulm destroyers.[2] Arsenic residues from the continued use of arsenic pesticides increased significantly; e.g. more than 3900 kg of lead arsenate were added to 1 hectare on a commercial orchard over a period of 25 years and the residues were mainly confined to the top 15 centimetres of soil.

The dinitrophenols DNOC and dinoseb (see Chapter 8, page 219) introduced as herbicides in the 1930s are highly poisonous to man and other mammals. Their use has damaged many forms of wild life and they have caused human fatalities; all insects and mammals caught in the insecticide spray are immediately killed. DNOC is so toxic that the quantity applied to 1 hectare (about 11 kg) would be sufficient to kill approximately 1000 people. The use of these dinitrophenols are restricted in the United Kingdom; dinoseb is still employed as a herbicide in spite of suspected mutagenic effects[3] and DNOC is used mainly as a contact insecticide on fruit during the dormant season.[4] These dinitrophenols are, however, rapidly degraded after contact with plants or soil and they do not therefore leave toxic residues. Plants killed by these compounds do not seem to harm animals which eat them unless they do so very soon after the application of the spray. Thus when spraying crops such as peas and beans to kill broad-leaved weeds and for potato haulm destruction, a minimum period of only ten days is needed between treatment and harvest. The dinitro compounds do not cause any long-term environmental pollution and do not accumulate along food chains.

The next major advance in herbicides came in 1942 with the discovery of the phenoxyacetic acid herbicides such as 2,4-D and MCPA. These hormone-type herbicides act as a mimic for the natural plant hormone indole acetic acid. This does not affect animals and these compounds have a very low mammalian toxicity; they are not very persistent and break down in the soil within a few weeks of application. There is generally little evidence that the very large-scale application of these herbicides has caused serious environmental damage.

In 1980 some 3000 tonnes of the herbicide 2,4,5-T were used in the United States and 1000 tonnes in the United Kingdom, chiefly to control weeds and bushes in plantations of young conifers; however, the brushwood killer 2,4,5-T has now been banned in the United States and several other countries, including Sweden, West Germany, Norway, and Italy.[3]

An impure form of this herbicide, known as 'Agent Orange', was used by the Americans in the Vietnam War to remove forestry cover which could conceal enemy troops. Agent Orange contained an appreciable amount ($\simeq 50$ p.p.m.) of an impurity dioxin (**1**), one of the most toxic molecules known (LD_{50} (oral) to rats $\simeq 30\ \mu g/kg$), which also has teratogenic (foetus-deforming) effects.[5] It caused several cases of deformed babies and later experiments in the United States showed that the use of 2,4,5-T on food crops resulted in teratogenic effects on rats and mice that had consumed the treated crops.

The teratogenicity of 2,4,5-T is due, at least in part, to the presence of some of the highly toxic and teratogenic 2,3,7,8-tetrachlorodibenzo-*p*-dioxin (TCDD) (**1**) produced as a by-product during the high-temperature hydrolysis of 1,2,4,5-tetrachlorobenzene to 2,4,5-trichlorophenol (**2**)[5] (Scheme 15.1).

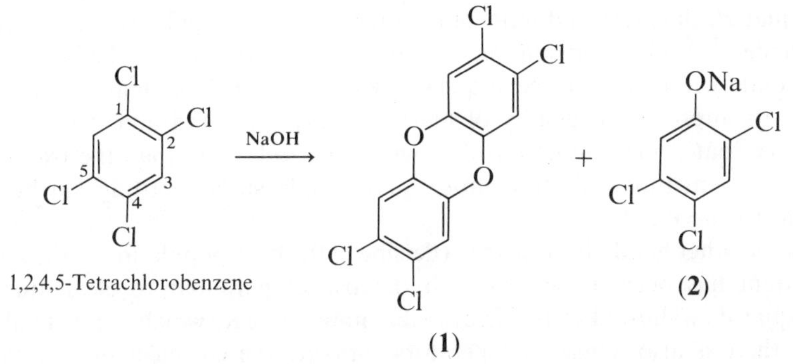

1,2,4,5-Tetrachlorobenzene (1) (2)

Scheme 15.1

The dioxin (**1**) is slowly degraded in soil but residues persist for some time in treated plants. The problem seems to have been largely eliminated by modified manufacturing processes for the phenol (**2**) that keep the dioxin content <0.1 p.p.m. However, there are disturbing indications that pure 2,4,5-T may be teratogenic at high dosages[3]—hence the removal of its use in many countries.

Urea, carbamate, and triazine herbicides (Chapter 8, page 245) act by interfering with photosynthesis in plants, causing them to be starved to death. Animals do not photosynthesize and are therefore not directly affected by these chemicals and, although persistent in their herbicidal action, they have low mammalian toxicity and do not build up along food chains. It is, however, possible that the triazines may have mutagenic effects.[6] Long-term studies of phenoxyacetic acid, urea, and triazine herbicides gave no indication of a build-up of herbicidal residues in the soil as a result of repeated annual application, nor have they shown any adverse effects on soil creatures such as earthworms, mites, and insects.[7] However, certain triazines, such as simazine and amitrole, are suspected mutagens.[3]

The application of pesticides, especially insecticides, acaricides, and systemic fungicides, has resulted in the appearance of resistant strains of insects, mites, and fungi (see Chapter 6, pages 151, 90 and Chapter 7, page 210). There have also been several instances of annual weeds becoming resistant to herbicides

(Chapter 8, page 268); thus chickweed after many years of successful treatment with 2,4-D was no longer effectively controlled by the chemical. It is probable that more examples of resistant weed strains will be continually appearing and this may cause harmful ecological effects. Consequently one cannot be absolutely certain about the potential long-term effects of even an apparently innocuous chemical like 2,4-D.

Permanent grassland, once a traditional feature of agricultural Britain, has almost disappeared—now grass survives generally as a stage in a crop rotation pattern covering two to five years, after which period it is ploughed up. The ploughing almost completely removes the surface-casting earthworms, which is unfortunate as they improve soil structure. Now with one application of paraquat all the grass and most of the other weeds can be killed, and since the herbicide is rapidly rendered inactive on contact with the soil, the land can be immediately resown. This technique known as 'chemical' ploughing can also be useful to improve the quality of marginal hilly grassland and there is some evidence that it may inflict less damage to the soil fauna than the traditional mechanical methods of cultivation; for example surface-casting earthworms appear to survive this treatment.

On the other hand, the disastrous decline in the hare population in the United Kingdom has been associated with the use of paraquat to clear stubble.[3] Paraquat if swallowed causes irreversible lung damage, which is nearly always fatal; there is also danger to operators spraying the chemical since it can be absorbed via the lungs and skin. Several countries have therefore restricted or banned the use of paraquat. It remains bound to soil colloids and therefore tends to accumulate in the upper layers of soil which could be a potential environmental hazard.

The herbicide ioxynil (Chapter 8, page 231) is currently under examination for mutagenicity in the United States and Britain.

In general, the current use of herbicides does not appear to constitute a major hazard to wild life in Britain since, on the whole, compounds of low mammalian toxicity and persistence are employed. Obviously, as with any pesticide, they must be used correctly and routine herbicidal spraying should be avoided. Herbicides should only be applied when they are definitely needed to secure an appreciable economic gain. One can never be completely sure that even a safe herbicide has no long-term deleterious effects on man and other animals that may be in contact with the chemical over many years, so it is important in the interests of the environment that they are only applied in the minimum quantity to achieve the desired result. Thus those weeds like pansies and spurreys usually appearing in a cereal crop when it is several centimetres tall need not be killed because they do not seriously compete with the crop plants for light and nutrients and do not interfere with harvesting.

Copper and sulphur fungicidal sprays have been used for a long time (Chapter 7, page 159). Bordeaux mixture (copper sulphate and slaked lime) is used against vine mildew and potato blight; lime-sulphur is used against scab diseases of fruit trees and powdery mildews. The continual treatment of crops with these fungicides leads to the formation of residues in the soil containing high concentrations of copper and sulphur, which remain in the soil for long periods and affect soil fauna and worms. They may also drain off into ditches and rivers and kill algae and fish. However, in general, fungicidal sprays and washes do not appear to have caused serious harm to wild life. An illustration of the effect of copper fungicides derives from an East Anglian orchard which had been subjected to prolonged treatment with copper fungicides. The surface soil contained 0.2% of copper which seemed to have no deleterious effect on mature apple trees but the soil contained no earthworms so the surface structure of the soil would probably be damaged over a long period. Soil fauna will be similarly affected in many orchards and potato fields. In addition soil fungi which play a vital role in breaking down leaves and other organic matter are killed by copper and sulphur residues so there may be a long-term danger to soil fertility arising from excessive use of copper and sulphur fungicides. Luckily, for many purposes, these are now being replaced by safer organic fungicides.

Organomercury compounds, such as phenylmercury acetate, are widely used as seed dressings to protect cereals from fungus diseases such as smut and bunt. These are remarkably effective in combating attack by the fungal spores. Thus normally 70 g of an organomercurial compound containing approximately 1% of the metal mixed with 28 kg of wheat seed is sufficient to prevent wheat bunt, so only about 7 g of mercury is added per hectare. Poultry may suffer when fed on the treated grain but organomercurials (mainly arylmercury compounds) as used in Britain do not appear to present a serious danger to the environment.[2,7] There is no evidence of an appreciable increase in the normal mercury concentration in the soil, although arylmercury seed dressings have been used in the UK for some fifty years.[7] Arylmercury derivatives are much less toxic than the alkylmercury compounds such as methylmercury dicyandiamide. Consequently these are now banned in several countries, notably Sweden where they have caused significant damage to wild life. Thus dead pheasants containing over 20 p.p.m. of mercury were found and the metal also appeared to have accumulated in pike, hawks, and other predators. However, mercury is also widely used in industrial processes in Sweden, so these environmental effects may not be due primarily to the use of mercury seed dressings. There is always a danger of mercury fungicides draining off into rivers and the high level of mercury in some sea fish has recently attracted much attention, but most of the metal almost certainly comes from its application in industrial processes, such as the use of mercuric chloride to inhibit the development of slime in pulp manufacture, rather than from organomercurial seed dressings.

Certain sea fish have been shown[8] to contain methylmercury and the biological methylation of mercury was first proposed by the Japanese to explain the Minamata disaster. Any discharge of metallic mercury, inorganic Hg^{2+} ions, or organic mercury compounds into the environment can be converted to methylmercury which can then be taken up by living organisms, e.g. fish;

therefore every effort must be made to reduce pollution of the environment by mercury in any form.

Mercury is a highly toxic metal and is known to bioaccumulate along food chains. Much effort has therefore been directed to the discovery of alternatives to organomercurials in the control of soil and seed-borne cereal pathogens. Recently several of the triazole fungicides (see Chapter 7, page 196) appear viable, safer chemicals for this purpose; consequently the use of organomercurial fungicides should decline in the future.

The widely used dithiocarbamate fungicides (Chapter 7, page 163) have relatively low mammalian toxicities (LD_{50} (oral) to rats, 400–8000 mg/kg). However, in chronic toxicity tests the prolonged feeding of high doses of zineb, maneb, and mancozeb to rats and dogs increased the mortality rate and caused kidney damage and teratogenic and mutagenic effects (maneb only). These toxic effects are probably due to the presence of ethylene thiourea (ETU) and the long-term consumption of foods contaminated with ETU could well damage health. There is therefore some doubt regarding the long-term safety of these fungicides and it is obviously important that the ETU content is kept as low as possible.[9]

The careful use of fungicides does not apparently cause much damage to the environment, although more long-term studies are needed to assess the effects of residues of copper and mercury on soil fungi and some of them (e.g. captan and dinocap) are toxic to fish.[2,7] Inorganic copper and sulphur fungicides are, however, being gradually phased out and replaced by organic fungicides.

A number of systemic fungicides are now available and some of these, notably the sterol biosynthesis inhibitors, like the triazoles, are highly potent so that only small amounts are needed. Such systemic fungicides are absorbed and translocated within the plant but no harmful residues are left. They are highly effective and, after several years of commercial use, no ill effects have been observed on animals, birds, or even soil microorganisms, and their introduction should reduce the danger of environmental contamination arising from fungicides.[7] However, one problem with the introduction of systemic fungicides onto the market has been the rapid development of resistant fungal strains, so to try and combat resistance, systemic fungicides are now often formulated with the older surface fungicides, like dithiocarbamates and captan.

Several organic fungicides, like captan and benomyl, have been identified as possible carcinogens and are under intensive testing in several countries; the fungicide and sprout suppressant, tecnazene, used extensively on stored potatoes,[4] appears to lead to stable residues in the tubers and hence this compound may be dangerous.[3]

Insects are more closely related to man than either plants or fungi and consequently it is perhaps not surprising that the greatest threat to the environment has come from the large-scale application of synthetic insecticides rather than from fungicides and herbicides. Many herbicides and fungicides owe their activity to action on biochemical processes specific to plants and fungi, whereas many insecticides act on the insect's nervous system, which is basically similar to that operating in mammals including man. Consequently, the greatest danger to the environment arises from insecticides rather than other types of pesticides. Early insecticides included some highly toxic materials, like arsenic and its compounds, such as Paris Green (copper arsenite) and lead arsenate. The latter was used for control of caterpillars on fruit trees and against earthworms and leatherjackets in turf, while hydrogen cyanide was used as a fumigant. The use of dinitro-*o*-cresol (DNOC) has already been discussed as a herbicide (see Chapter 8, page 219) and is also employed for controlling aphids, mites, and other insects on fruit trees in winter.[4] Sodium fluoride was formerly used against various domestic pests, such as ants and cockroaches, but it was never used extensively on crops as it could otherwise have caused serious damage as happens with fluorine from brickworks and other industries.[2]

Two natural insecticides introduced in the nineteenth century were rotenone from the derris plant and pyrethrum extracted from a species of chrysanthemum (Chapter 4). Both these substances are very safe insecticides and would be employed on a larger scale except for their relatively high cost. These were the main types of insecticides used before World War II. Obviously such highly poisonous substances as arsenic and cyanide were potentially very dangerous to the environment. However, their high toxicity was so well known that great care was exercised in their application so that paradoxically comparatively little environmental damage was caused. They have now been superseded by newer synthetic organic insecticides. These fall into three main classes—the organophosphorus and organochlorine compounds[2] and the synthetic pyrethroids. Organophosphorus insecticides (Chapter 6, page 105) were developed after World War II from work originally carried out in a search for nerve gases; not surprisingly therefore the early organophosphorus insecticides were very toxic to man and other mammals.[2,7] Parathion was the first member of the group to be widely used as a contact insecticide in agriculture and has a wide spectrum of activity, being effective against aphids, caterpillars, spider mites, and eelworms.[10] Other early examples included tetraethyl pyrophosphate (TEPP) and schradan. TEPP was indeed the most toxic material ever to be used on farms and it is estimated that 28 g of this compound could kill 500 people! All these compounds were very poisonous and their use was viewed with considerable alarm—they have caused several human fatalities and full protective clothing must be worn during application. Any birds or small mammals caught in the spray are quickly killed. However, organophosphates have the important advantage over the organochlorine insecticides that they are comparatively rapidly biologically and chemically degraded in plants, animals, and soil to nontoxic materials. The rate of breakdown depends on the nature of the organophosphate, the formulation, method of application, climate, and the growing stage of the treated plant crop.[10]

A consequence of the degradation of organophosphates is that birds or mammals entering the treated area a few hours after application generally survive and after a given period a treated crop is quite safe for consumption. Organophosphates do not accumulate in the mammalian body or along food chains and, after more than 30 years of application, there is no evidence of chronic effects being produced in the ecosystem.[2,7] Another advantage of this group of compounds is that many of them function as systemic insecticides, enabling smaller amounts of the active ingredient to be used more effectively, which reduces the harmful effects on natural predators since usually only insects actually eating the crop are killed. Research into organophosphates has resulted in the introduction of several valuable insecticides, like malathion and fenchlorphos, with very low mammalian toxicities, showing that it is possible to obtain considerable selective toxicity to insects within this group of compounds.

The very highly toxic organophosphorus compounds, like TEPP and schradan, are now not used as insecticides. Organophosphorus insecticides act by disrupting effective nervous transmission in the insect and in theory they should act in a similar manner in mammals. In compounds like malathion, the low mammalian toxicity is associated with differential detoxification processes occurring in mammals (see Chapter 6, page 129). It may therefore be concluded that provided organophosphorus insecticides are carefully applied, those currently in use are unlikely to do serious harm to the environment.[2,7,10] However, several organophosphorus compounds are currently under investigation as potential mutagens; dichlorvos, a volatile insecticide used in flypapers, is suspected of being a carcinogen.[3]

Carbamate insecticides, like carbaryl (Chapter 6, page 149), act on the insect's nervous system in a similar manner to the organophosphorus compounds, but are rather less persistent. They, like the organophosphates, readily degrade and do not accumulate along food chains. Some of the oxime carbamates, like aldicarb which is a soil-applied insecticide, acaricide, and nematicide,[4] have extremely high mammalian toxicities. Aldicarb is only available as granules for soil incorporation to reduce the dangers to operators; the use of this compound has caused the death of many species of birds—gulls, lapwings, and curlew.[3] Aldicarb (LD_{50} (oral) to rat $\simeq 1$ mg/kg) is as highly toxic as the early organophosphorus insecticides and should be withdrawn and replaced by safer chemicals as soon as possible.

The organochlorine compounds, e.g. DDT, HCH, and dieldrin (Chapter 5), represent another major group of synthetic insecticides. These also act on the insect's nervous system, but unlike organophosphorus and carbamate compounds, they alter the permeability of the axonial membrane to the passage of sodium ions. The first and most widely used member was DDT, introduced in 1943, and during the War it appeared to be the perfect insecticide, combining a wide spectrum of insecticidal activity with a comparatively low acute toxicity, about the same as that of aspirin. Indeed, the organochlorine compounds did not appear to have any harmful effects on the environment until the 1950s when some doctors became worried by the appearance of these compounds, especially when they were detected in cows' milk after the animals had fed on foodstuffs that had been treated with organochlorine insecticides. In the United Kingdom DDT has probably not caused much harm to wild life, though the related compounds aldrin and dieldrin have killed many species of birds and mammals feeding on them.[2] In Britain gross pollution of the environment by DDT has been avoided probably more by luck than judgement since the authorities considered they could not afford the massive blanket spraying programmes that were characteristic of the use of DDT and related compounds in the United States of America. The careful application of DDT at minimum dosage has successfully controlled insects in many countries without obvious harmful effects, but when used in ways now considered unwise, this compound has resulted in much environmental damage.[11-14]

In the United States attempts were made to halt the spread of Dutch elm disease by spraying the trees with high concentrations of DDT ($\simeq 28$ kg/ha). This treatment was designed to kill the bark beetles which spread the Dutch elm disease fungus. Unfortunately the disease was not halted and large numbers of birds, especially the American robin, were killed. Apparently the DDT got onto the elm leaves which fell on the ground and were subsequently eaten by worms which concentrated the DDT in their bodies; these contaminated worms were eaten by the robins so that they gradually accumulated a fatal dose of the chemical. A very serious feature of organochlorine insecticides is their ability to become concentrated along food chains, causing death to organisms at the end of the chain.[11,12] The classic example of this type of effect in the ecosystem was that of Clear Lake, 16 800 hectares of water in California. The lake was an important centre for recreation but in the summer had clouds of small *Chaoborus* gnats, which caused serious complaints to be made to the authorities; in order to control the gnats a spraying programme was initiated in 1949 using DDD (a closely related compound to DDT but rather less toxic to fish).[8,12] The first application was a success—the gnats were almost entirely eliminated and other organisms did not appear to have been seriously harmed. In 1950 the gnats were not present in sufficient numbers to cause complaints although during the next two years their numbers built up to about their original level. In 1954, therefore, the lake was resprayed with DDD and it was treated a third and last time in 1957. The second spraying again appeared to give satisfactory control, but by the time of the last spraying the gnats and almost 150 other insect species had developed varying degrees of resistance to the insecticide. In December 1954 large numbers of dead western grebes (fish-eating diving birds) were observed and this caused a public outcry. In fact the first spraying does appear to have seriously reduced the breeding success of the grebes in the lake area; indeed from 1950 to 1961 no young were hatched and reproduction remained largely unsuccessful until 1969, some twenty years after the first introduction of DDD into the area. Initially the treatment appeared highly satisfactory—one single application had lasted over several years and a safe and effective means of removing the gnats appeared to have been discovered. The death of the grebes was a striking illustration of the concentration of a toxicant along a food chain. It was found that carnivorous fish accumulated more DDD than plankton-eating fish of the same size and generally fish appeared more tolerant of the pesticide than species higher up the food chain.[15] During the period September 1949–September 1957 some 54 000 kg of DDD were applied to the lake from moving barges so as to give a concentration of DDD in water of 0.020 p.p.m. immediately after each application. This very low concentration of DDD appeared quite safe and indeed after two weeks no DDD could be detected within the water. However, later studies showed that organochlorine compounds are extremely dangerous in aqueous environments because, owing to their great chemical stability, low aqueous solubility, and high lipophilicity they become concentrated in the living components of the lake's ecosystem.[5,10,12,15]

The disappearance of the chemical from the water after two weeks therefore was not due to its degradation, but rather arose from its transport into living organisms (e.g. plankton, plants, frogs, fishes, birds, etc.). After the final treatment (1957), analysis of the lake's ecosystem showed that the microscopic plankton of the lake contained approximately 5 p.p.m. of DDD (250 times the original concentration of DDD applied to the lake); in frogs the magnification was 2000 times, in sunfish about 12 000 times, and finally some of the sick grebes contained 1600 p.p.m. of DDD in their fat, a magnification of 80 000 times the original concentration of the chemical.

The greatest danger from the use of organochlorine insecticides seems to occur when water is contaminated, because fish and other aquatic organisms have the capacity to absorb the chemical from the water and concentrate it in their fatty tissues. This is especially true of fishes because, in the course of respiration, they pass a large volume of water through their gills; thus the lipophilic organochlorine compounds are absorbed from the water into the fish. There also is some evidence that DDT and similar compounds inhibit oxygen uptake at the gills so causing fish to die from suffocation.[5,12] Obviously in such an aquatic food chain as that operating in Clear Lake the fish-eating grebes at the end of the chain obtained the maximum concentration of the organochlorine insecticide, which in many cases constituted a fatal dose.[13] Many similar examples have been documented of the concentration of organochlorine compounds along food chains, such as the study of the Long Island estuary food web.[13,14] Oysters and other shellfish obtain their nutrients by continually filtering the water, so these creatures, like fish, are very sensitive to the presence of small concentrations of organochlorine compounds and similar lipophilic molecules in the water they inhabit.

Other important food chain studies involve predatory birds, such as peregrine falcons and sparrow-hawks.[12,15] There were fewer of these species of bird in Britain in 1966 than in 1950, although the population appeared to be reasonably stable up to 1939. During the War, drastic control measures were adopted, but by the early 1950s the population of these birds was almost back to the 1939 figures. However, since then the position has become grave; in 1962 nearly all the breeding grounds of the peregrine falcon in the remote parts of the Highlands of Scotland were occupied and more than 40% of these pairs reared young birds successfully. On the other hand, in Southern England only some 20 out of the 66 breeding territories used in 1939 were occupied and only four pairs reared young successfully. Examination of peregrine corpses showed that the majority of them contained substantial quantities of organochlorine pesticide residues. In some cases the amounts could have been lethal. The corpses of other birds killed by shooting or on the roads contained rather lower pesticide residues. Eggs that failed to hatch were also found to contain substantial organochlorine residues. Experiments have shown that breeding can be inhibited by sublethal amounts of these compounds, and they may in addition cause sterility so that eggs are not laid. They may also affect bird behaviour so that the eggs are not incubated and may reduce the thickness of the egg shells—consequently more eggs are broken in the nest.[6,11,15]

The peregrine falcon is the largest falcon indigenous to Britain. The drastic fall in their population suffered in 1961 was abnormal. It was observed that in Southern England the falcon population was only 40% of the pre-War figure and this declined further in 1962. The peregrine falcon almost disappeared as a breeding species in Southern England and Wales, and the population was also substantially reduced in Northern England and Southern Scotland (Figure 15.2).

Wartime control of the falcons was carried out to prevent attacks on carrier pigeons; hence the small decrease in population from 1939 to 1945. In 1948 the

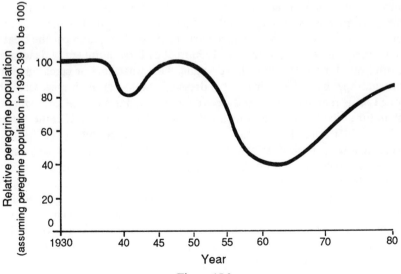

Figure 15.2

organochlorine insecticides were introduced and their use expanded until 1962 when the first restrictions were applied with further controls introduced in 1964, 1969, and 1976. Figure 15.2 shows that the falcon population increased as the use of these insecticides was curtailed.

The peregrine falcons were clearly victims of these chemicals. The birds took prey contaminated with sublethal doses of the pesticides and gradually accumulated the poison within their tissues until it killed them. Southern England had the highest proportion of arable land used for cereal production and several organochlorine insecticides, e.g. HCH, aldrin, and dieldrin, were extensively applied as seed dressings to control wheat bulb fly; dieldrin was also used in sheep dips. Hence it was in Southern England that the most drastic effects on the falcons were observed.[15]

The sparrow-hawk, until recently a common bird throughout Britain, is now quite rare and only breeds successfully in the remoter parts of the country.

There was also a marked reduction in the breeding success of golden eagles in Western Scotland during the period 1963-6. This coincided with a decrease in egg-shell thickness and consequent increase in breakage in the nest. In this area, the main food of the eagles was sheep carrion from which they derived high residues of dieldrin which was used as a sheep dip. On the other hand, in Eastern Scotland, where eagles feed mainly on wild prey, breeding success and shell thickness remained normal during this period. After the ban was imposed on the use of dieldrin as a sheep dip in 1966, there was a marked improvement in the breeding and shell thickness of eagles in Western Scotland. Both the peregrine falcon and the sparrow-hawk are also showing a slow population recovery after the restriction on the use of organochlorine insecticides.[11,15]

Since about 1948 the cyclodiene group of organochlorine insecticides, for instance aldrin, dieldrin, and endrin, were introduced as soil insecticides. In Britain dieldrin was extensively employed as a seed dressing for wheat against attacks of wheat bulb fly. Birds, particularly pigeons, dig up and eat the treated grain. Consequently in the springs of 1956 to 1961 large numbers of dead birds were discovered, particularly in East Anglia. This was a major cereal growing region and the use of dieldrin seed dressings was clearly implicated. The poisoned birds remained toxic and were eaten by predatory birds and by foxes, which in turn died by accumulating a lethal dose of the chemical. In the springs of 1959 and 1960 some 1300 dead foxes were found in areas where dieldrin seed dressings had been applied, and there is no doubt that the deaths arose from the use of dieldrin seed dressings on spring-sown wheat.[15] In 1966, dieldrin, endrin, and heptachlor were only permitted to be used as seed dressings for autumn-sown wheat in fields where high populations of wheat bulb fly eggs have been counted. This practice is much safer since birds do not dig up much wheat seed in the autumn as plenty of other food is available at that time of the year. These compounds are now not used against wheat bulb fly as they have been superseded by the much safer synthetic pyrethroid insecticides, e.g. permethrin.

The effects of DDT and other organochlorine insecticides in reducing the average thickness of birds' egg shells probably results from their interference with calcium metabolism.[7] Studies of egg shells of predatory birds in the United Kingdom showed a sharp fall in thickness during the 1945-47 period when DDT was introduced into the environment.

By 1950, the relative thickness of the eggshells of falcons and sparrow-hawks had decreased by 19.1 and 17.2% respectively; these changes continued until 1969. In the golden eagle a decrease of 9.9% in eggshell thickness was noted in 1951. A similar situation was reported in the United States of America.[15] The final removal of organochlorine seed dressings was facilitated by the introduction of two organophosphorus soil insecticides—chlorfenvinphos and carbophenothion—which, although of higher acute toxicity, did not show bioaccumulation problems. Recently these two chemicals are being replaced by much safer synthetic pyrethroid insecticides, like permethrin.

The observation that sublethal doses of DDT and other organochlorine insecticides reduced eggshell thickness was confirmed by experimental tests in which DDT and other organochlorine compounds were fed to ducks, sparrowhawks, and quail; the hatching success rates were also reduced.

Salmon in Lake Michigan accumulate quite high levels of organochlorine residues in their fatty tissues and lay eggs containing appreciable pesticide residues; this caused the death of almost 700 000 salmon fry in 1968. The successful breeding of trout has been similarly affected and rising levels of organochlorine compounds have been observed in several commercially important sea fishes like tuna, mackerel, and hake.[6,12]

DDT and other organochlorine insecticides are very stable substances which are only very slowly chemically or biologically degraded. The metabolites like DDE often contain almost as many chlorine atoms as the original compound[11,12] and consequently the whole world has become polluted with DDT and its metabolites. They are found in rainwater, antarctic snow, air, fish and animal fats—human body fat may contain some 10 p.p.m. of DDT. The universality of these compounds is a remarkable tribute to both the stability and mobility of these compounds.[6]

Organochlorine insecticides act as general nerve poisons; their greater toxicity to insects is probably largely due to the greater ease of absorption through the insect cuticle as compared with the mammalian skin. They pose a severe threat to our ecosystem because of their great stability.[13,14] Such 'hard' pesticides are relatively resistant to biodegradation. Thus 50% of the DDT sprayed on a field was still present ten years later and tended to be recycled through food chains. Chlorinated hydrocarbons are very mobile and thus they adhere to dust particles and are blown around in the dusts of the world. The organochlorine compounds illustrate the environmental consequences of the introduction of chemicals with a combination of great chemical and biological stability with lipophilicity, so that the chemicals are continually transferring from aqueous to more lipophilic media. They are concentrated in the fatty tissues of living organisms and because the aqueous solubility of DDT is only 1.2 parts per billion parts of water, it is quickly translated from the water into living aquatic organisms, often with disastrous effects (see the Clear Lake case, page 351). As a result of the hazards presented to the environment, the use of the persistent organochlorine insecticides is severely restricted in many advanced countries (e.g. Britain, Germany, France, Holland, and Japan) that can afford the higher costs of alternative compounds, such as pyrethroids, malathion, and carbaryl. The United States of America banned DDT in 1972 and Britain in 1984. In 1986, the United Kingdom stopped the use of aldrin and dieldrin as seed dressings for spring-sown wheat and in sheep dips. However, DDT remains an essential weapon for control of malaria-carrying mosquitoes in the Far East as it is the cheapest available insecticide. At the peak of the malaria control campaign during the 1960s some 66 000 tonnes of DDT, 4000 tonnes of dieldrin, and 500 tonnes of γ-HCH were applied annually. As the vectors quickly became resistant to dieldrin little of this chemical is now used, and the amount of DDT has fallen to some 46 000 tonnes each year. The World Health Organization consider that to contain malaria substantial amounts of organochlorine insecticides will still have to be used; however, by 1980, 51 out of 60 malaria-carrying *Anopheles* mosquitoes had become resistant to DDT, HCH, and dieldrin. Alternative insecticides, such as organophosphorus, carbamate, and pyrethroids, had to be applied, but these are much more expensive.[3]

The organochlorine insecticide methoxychlor does not apparently accumulate in fats because, unlike the majority of organochlorine compounds, it is biodegraded by oxidative demethylation. The use of methoxychlor therefore will probably not result in long-term environmental pollution and so the use of this compound is often permitted.

Organochlorine insecticides owe their activity to their interference with the nervous system (Chapter 5, page 84), but there is no conclusive evidence that these pesticides at the present levels of exposure cause any long-term harmful effects to man,[7,14] though they may reduce the life expectancy of those exposed to these compounds from birth (aged 40 or less). However, further studies are needed in view of experiments showing that high doses of DDT increase the incidence of liver cancers in mice;[6] there is also concern that the mutations observed in some treated animals might also occur in humans.[13] DDT stimulates production of the female sex hormone oestrogen and certainly affects the sex hormones of birds and rats; possibly the chemical might initiate hormonal changes in man.[6] It has been suggested[6,12] that there may be some correlation between high DDT levels in the brain, liver, and adipose tissues of terminal cancer patients and those who died from coronary diseases.

DDT and other organochlorine insecticides at concentrations of 10 p.p.b. inhibit the photosynthesis of four species of phytoplankton.[6] Since all life on the earth depends on photosynthesis this observation appears alarming, but lower concentrations of DDT (1 part per billion) did not affect photosynthesis and the actual concentrations of DDT and its metabolites in the sea are of the order of 1 part per trillion or less, so there is little danger of damaging photosynthesis by photoplankton, especially since there are many thousands of different species of plankton.[7]

The extensive use of the persistent organochlorine insecticides has led to the appearance of resistant strains of insects (Chapter 6, page 151); this raised the danger that larger doses of organochlorine insecticides might be applied to try and control the pest which will result in greater environmental pollution and a larger population of resistant insects. The extensive application of the persistent organochlorine insecticides has resulted in a build-up in the population of resistant insects and spider mites.

The danger of using such persistent, broad-spectrum insecticides is well illustrated by spider mites which only became a serious problem in the United States after mass forestry spraying programmes with DDT against spruce budworm. The insecticide had little action against the mites, but killed their natural predators, e.g. ladybirds; consequently the resistant spider mite population rapidly increased. The use of organochlorine insecticides against cotton pests in the United States similarly destroyed the natural predators and parasites with a consequence that cotton bollworm for the first time became a serious pest of cotton.[3] Luckily other types of insecticides such as organophosphates, carbamates, and pyrethroids are available for control of pests which have acquired tolerance to organochlorine insecticides.

In addition to the organochlorine insecticides mentioned, some 1200 tonnes of chlorophenols are employed as broad-spectrum biocides against insects, bacteria, fungi, algae, and plants. Their major use is as antirotting agents in non-woollen textiles and timber. Chlorophenols can be microbiologically and photochemically degraded but little is known about the speed of their breakdown under natural conditions, how long they persist in living organisms, or their effects on aquatic life at sublethal concentrations.[14] In Britain more than 500 000 buildings are treated with chemicals to protect timber against wood-boring insects. Over the last 30 years, these chemicals have severely restricted the bat population and safer chemicals are urgently needed for use in timber treatment.[3]

Polychorinated biphenyls (PCBs) are widely distributed in the environment due to their various industrial uses as, for instance, electrical insulators, transformer fluids, plasticizers, lubricants, and as constituents of wax polishes and sealing compounds.[16] PCBs are considered to constitute a danger to public health and the EEC regulations restrict their use to electrical insulators and certain other specified uses where there is little danger of environmental contamination. The widespread application of PCBs probably makes them a greater environmental hazard than the organochlorine insecticides; even in 1971 the world production of PCBs was still 50 000 tonnes.[16] Virtually nothing is known about the effect of PCBs on animal populations and sublethal doses may affect both the survival and reproduction of animals.[11]

In any consideration of organochlorine insecticides, as well as their pollution of the environment, the tremendous benefits obtained by the use of these compounds must be remembered. In the Far East there are few economically viable alternatives to DDT for the control of disease-carrying vectors of typhus, bubonic plague, and malaria. During malaria-control campaigns large quantities of DDT were applied to the interior walls of the homes of millions of people at concentrations of 2 g/m². To control yellow fever, DDT was introduced into domestic water supplies at 1–5 p.p.m., without causing any observed ill effects. In the war, soldiers had their clothing dusted with DDT to control body lice and employees of industries manufacturing organochlorine insecticides frequently contained as much as 200 p.p.m. of organochlorine residues in their adipose tissues. Careful monitoring of these workers over many years has not, however, revealed any illnesses that could be attributable to the presence of these residues. However, in view of their action on the nervous system of insects, which is similar to that of mammals, the use of organochlorine insecticides has rightly been drastically curtailed. There is no doubt that these compounds have caused severe damage to parts of the ecosystem, e.g. birds, fish, and some mammals.

The rodenticides, difenacoum and brodifacoum (Chapter 11, page 291), used against rats and mice resistant to warfarin, are known to accumulate in the bodies of small mammals, ultimately reaching lethal levels; these chemicals have been responsible for the death of pheasants in Hampshire. Obviously such compounds must be applied with great care and precision to avoid environmental damage.

The third major group of insecticides, the synthetic pyrethroids (Chapter 4, page 58), generally have low mammalian toxicity combined with very high levels of insecticidal potency and hence they can often be successfully applied at remarkably low dose rates. The pyrethroids, although lipophilic molecules, do not accumulate along food chains since, unlike the organochlorines, they readily degrade. The use of these compounds does not appear to cause any environmental problems and indeed has enabled some dangerous chemicals to be phased out. However, the problem of resistance remains: from 1976 to 1980, the number of insect species becoming resistant to pyrethroids increased from 6 to 17 and out of 50 species described as promising targets for pyrethroids some 15 have now become resistant.

There is much less chance of an organism developing resistance to chemicals that act indirectly; certain systemic fungicides apparently act by increasing the resistance of the host plant to the pathogen (Chapter 7, page 176). Other chemicals may control insect damage by acting as attractants, repellents, chemosterilants, or insect growth hormone mimics (Chapter 14); such compounds often combine high specific effect on the target pest with low mammalian toxicity, so their application is unlikely to cause environmental damage.

In the future, the degree of environmental pollution by pesticides should be reduced by adopting integrated pest management (IPM) programmes, involving the combined use of biological and chemical control measures, the use of pheromones to monitor insect pest levels, the introduction of improved crop strains which are more resistant to pest attack, and the discovery of new molecules which have higher activity towards the target pests.

The discovery of more biodegradable pesticides possibly based on natural products, e.g. the pyrethroids and antibiotics, would reduce environmental pollution. Other routes to such compounds include the incorporation of sugars, amino acids, or peptides into the pesticide molecule. For example, the benzimidazole derivatives of the amino acids isoleucine and valine have antifungal properties, while the phenylalanine derivative is an antiviral agent against tobacco mosaic virus. Food additives may provide another source of biodegradable pesticides; e.g. sodium saccharin and some phosphatides, like soybean lecithin, are fungicidal.[9]

The new agrochemicals will generally be highly sophisticated molecules capable of acting selectively on the target pest at very low dose rates; because of their selective toxicity they will be suitable for use in integrated pest control programmes. The adoption of these methods will reduce the quantities of chemical pesticides added to the environment. However, it is likely that chemicals will remain the major weapon against pests until well into the twenty-first century.

REFERENCES

1. Carson, R., *Silent Spring*, Hamish Hamilton, London, 1963.
2. Mellanby, K., *Pesticides and Pollution*, Collins, London, 1970.
3. Dudley, N., *This Poisoned Earth*, Piatkus, London, 1987.
4. *The UK Pesticide Guide*, British Crop Protection Council, Binfield, 1991.
5. Bovey, R. W., and Young, A. L., *The Science of 2,4,5-T and Associated Phenoxy Herbicides*, Wiley, New York, 1980.
6. Ehrlich, P. R., and Ehrlich, A. H., *Population, Resources, Environment*, W. H. Freeman, San Francisco, Calif., 1970, p. 182.
7. *Pesticides in the Modern World*, A symposium prepared by members of the Co-operative Programme of Agro-Allied Industries with FAO and other United Nations Organisations, Newgate Press, London, 1972.
8. Miller, M. W. and Berg, G. G. (Eds.), *Chemical Fall Out*, 2nd edn, Thomas, Illinois, 1972.
9. Matolcsy, Gy., Nádasy, M., and Andriska, V., *Pesticide Chemistry*, Elsevier, Amsterdam, 1988, p. 361.
10. Eto, M., *Organophosphorus Pesticides: Organic and Biological Chemistry*, CRC Press, Cleveland, Ohio, 1974, p. 192.
11. Moriarty, F., (Ed.), *Organochlorine Insecticides: Persistent Organic Pollutants*, Academic Press, New York, 1975.
12. Edwards, C. A. (Ed.), *Environmental Pollution by Pesticides*, Plenum Press, London, 1973.
13. Graham, F., *Since Silent Spring*, Hamish Hamilton, London, 1970.
14. Marco, G. J., Hollingworth, R. M., and Durham, W. (Eds.), *Silent Spring Revisited*, American Chemical Society, Washington, 1987.
15. Sheail, J., *Pesticides and Nature Conservation: The British Experience*, Clarendon Press, Oxford, 1985.
16. Green, M. B., 'Polychloroaromatics and heteroaromatics of industrial importance', in *Polychloroaromatic Compounds* (Ed. Suschitzky, H.), Plenum Press, London, 1974, p. 403.

Chapter 16
Future Developments

The original chemical pesticides were general poisons with non-specific activity; thus early herbicides like sodium chlorate and copper sulphate were total weed killers which could not be effectively applied as selective herbicides. Likewise insecticides such as hydrogen cyanide, lead arsenate, and Paris Green were highly poisonous materials with a wide spectrum of insecticidal and mammalian toxicity. Similarly, such fungicides as sulphur, Bordeaux mixture, and organomercurials tended to be comparatively non-specific in their toxicity towards fungi.

Later work led to the discovery of less toxic and more selective chemicals, e.g. the organochlorine and phosphorus insecticides, phenoxyacetic acid herbicides, and systemic fungicides like ethirimol. The organochlorine insecticides, like DDT, have now been severely curtailed due to environmental problems (see Chapter 15, page 350). However, tremendous advances have been made in producing safer insecticides: the organophosphates, e.g. malathion, carbamates, e.g. methomyl, and the synthetic pyrethroids show more acceptable environmental properties. The pyrethroids, the most recent major development, now constitute approximately one quarter of world insecticide use. Some of these pyrethroids are remarkably potent; e.g. decamethrin (Chapter 4, page 60), which is commercially obtained as a single isomer, is effective at extremely low dose rates (2-10 g/ha).[1] Pyrethroids have lower mammalian toxicities than organophosphates and carbamates and some of them exhibit selective toxicity towards different insect species.

Other modern pesticides showing remarkably high activity include the sulphonyl urea herbicides (Chapter 8, page 258), which give good weed control at 20 g/ha, and the triazole systemic fungicides (Chapter 7, page 196), which give fungal control at 100 g/ha.

There is now much greater awareness of the dangers of environmental pollution arising from the widespread application of chemical pesticides (Chapter 15), and candidate chemicals have to pass increasingly stringent tests regarding toxicity and residue formation before they can be marketed as pesticides in many countries.[2] This has caused research on new pesticides to be increasingly concerned with producing chemicals that are safer and more selective in their action (see Chapter 1, page 15).

The ideal chemical pesticide would have high specific toxicity against the target pest, should not persist longer than necessary to achieve its objective, and would not affect the rest of the ecosystem, so that natural predators and other beneficial insects are unharmed.[2] It must be remembered, however, that to achieve their objective all chemical pesticides must be toxic to some living organism.

Some well-known examples of modern pesticides approximating to these criteria are the systemic selective aphicide pirimicarb and several of the systemic fungicides, such as dimethirimol, which shows high selective activity against cucumber powdery mildew. Remarkably potent specific toxicity against flying insects has also been achieved with synthetic pyrethroids such as decamethrin (Chapter 4, page 60). However, the majority of pesticides currently in use fall short of these ideals; these are often cheap, broad-spectrum, chemicals which give adequate control for the farmers but often damage natural predators and the ecosystem generally.

The increasing emphasis on systemic rather than contact pesticides leads to more effective use of the chemicals and less danger of environmental pollution. In the case of insecticides, systemic compounds will also be more selective in their action since only those insects actually eating the treated crop will be killed. During the last twenty years a significant breakthrough in chemical pesticide research has certainly been the introduction of commercially viable systemic fungicides[3,4] (Chapter 7, page 176); the latest group are the triazoles which act by inhibition of sterol biosynthesis in the fungi.

These systemic fungicides are highly active and specific in their toxicity, but their use has rapidly led to the emergence of resistant fungal strains. The growth of resistance poses increasingly serious problems in the fields of insecticides and systemic fungicides, so there is a continual need for the development of new types of specific pesticides to combat resistant insects and fungi. At present there is little sign of the widespread emergence of weed strains showing tolerance to herbicides; however, this is likely to become an increasing problem. Certainly the development of widespread resistance in common broad-leaved weeds in cereals towards phenoxyacetic acid selective herbicides would pose a very serious threat to world cereal production.

As mentioned in Chapter 1 (page 13), the cost of introducing a new agrochemical onto the market is now very high and so the search for new chemicals will only be undertaken by international companies with sufficiently large capital reserves. The production of new agrochemicals is a high-risk enterprise and currently the chances of discovery of a commercially successful chemical are low. A chemical will only be developed provided it shows activity against a pest of a major world crop, e.g. small grain cereals, maize, cotton, soybeans, sugar cane, or sugar beet. There is little economic incentive to develop a chemical that is only active against a pest of a minor crop, however effective it may be. There is a dilemma here, because the registration authorities are looking for the introduction of more specific pesticides—the ideal already mentioned. On the other hand, for an economic return on their capital investment, the agrochemical companies require compounds with a fairly broad activity spectrum to achieve large sales.

Modern insecticides are highly effective against flies, cockroaches, forestry and timber pests, bedbugs, and fleas; this has enabled outbreaks of typhus, bubonic plague, and malaria to be checked. The phenomenon of insect resistance to insecticides is, however, an increasingly serious problem and since 1954 the number of resistant insect species, shown in brackets, has risen almost exponentially: 1954(25), 1957(76), 1960(137), 1963(159), 1968(224), 1976(364), 1986(400).

There are many individual insecticides, but all the major groups act on the insect's nervous system, e.g. the DDT group and the pyrethroids act on the sodium ion channel while the organophosphates and carbamates poison the enzyme acetylcholinesterase. New types of insecticides acting on different target sites either in the nervous system or elsewhere are therefore needed to combat the spread of resistance, because chemicals acting on the same site are cross-resistant to each other, e.g. DDT and pyrethroids.

In addition to new classes of chemical pesticides that act directly on the pest, important developments have included semiochemicals (behaviour controlling chemicals) such as pheromones, antifeedants, and insect growth hormone mimics, together with chemosterilants (Chapter 14, page 315). The best-known semiochemicals are the sex pheromones of moths produced in minute quantities by the females which attract only males of the same species; these can be produced synthetically and are valuable for monitoring pest populations. Accurate pest measurements mean that conventional insecticides are only applied when the pest population exceeds a given threshold, hence reducing the number of pesticide treatments needed. In several cases, the use of pheromones has successfully controlled the pest, e.g. the pea moth by treating the crop with a slow-release formulation of the pheromone which disrupted the moth's normal mating behaviour. Several commercial pheromone products have been introduced to similarly control the pink bollworm.[1] The aphid alarm pheromone E-β-farnsene can be applied by an electrostatic sprayer and makes aphids more susceptible to contact insecticides.

Of the novel chemical insect control agents, the sex pheromones probably achieve the ultimate in both sensitivity and specificity; indeed the remarkable sensitivity of insect sensory receptors suggests that if compounds could be found that would block these sensory organs, they might function as outstandingly valuable insect control agents.[5] Recent studies[6] into the molecular mode of action of pheromones and insect growth hormones may lead to the discovery of other molecules possessing the essential biological activity.

Another promising group of semiochemicals are the plant-derived antifeedants such as (−)polygodial (Chapter 14, page 324), obtained from the water pepper plant which prevents many insects, including aphids, from feeding on the plant. In field studies at Rothamsted, polygodial effectively controlled the transmission of barley yellow dwarf virus by the bird cherry aphid. It may be possible to synthesize simpler analogues which retain the antifeedant activity.[6]

Many chemical structures influencing insect behaviour have now been identified. Insect juvenile hormone mimics such as methoprene (Chapter 14, page 331) show high specific toxicity towards mosquitoes and related insects and act by blocking insect metamorphosis. Diflubenzuron is an insect growth regulator (IGR) which appears to inhibit chitin synthesis and so interferes with the formation of the insect cuticle. Several new benzoylphenylurea IGRs have recently been produced;[7] this is an important new group of selective chemicals for insect control—they do not damage natural predators. Another interesting approach to insect control is the use of chemosterilants. This has been applied successfully against fruit flies and there is a continual search for more effective sterilization agents (Chapter 14, page 326).

These novel insect behaviour chemicals generally have the advantages of high selective toxicity against the target insect pest and they are less likely to induce insect resistance as compared with conventional insecticides. However, within the latter field, there is still considerable scope for achieving both selectivity and high activity. For example, in the organophosphates, carbamates, and synthetic pyrethroids, remarkable selectivity can be achieved by utilizing differences in the ease of penetration of the toxicant to the site of action of different species and by exploiting variations in the nature and rates of metabolism.

The design of new insecticides should be facilitated by better understanding of the mechanisms of resistance which may be due to behavioural changes, reduced rates of penetration, enhanced detoxication, or alteration in the susceptibility of the target site. There are still many unexploited natural product leads which could provide new highly selective insecticides, hopefully with different bio-chemical modes of action, enabling them to overcome insect resistance.[8] An example is provided by the N-alkylamides based on natural insecticides found in certain plants, e.g. piperine (1) from black pepper.

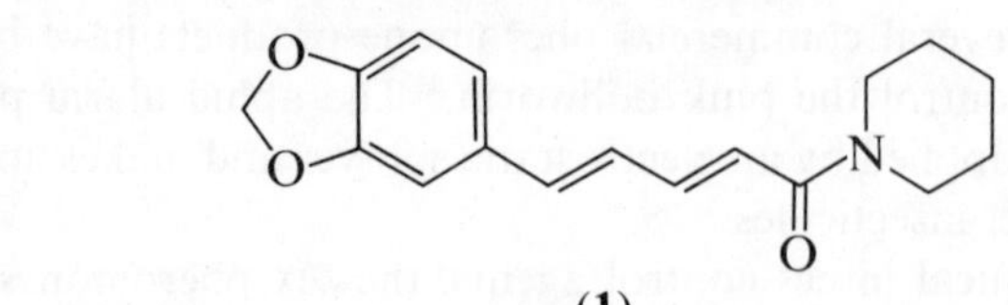

(1)

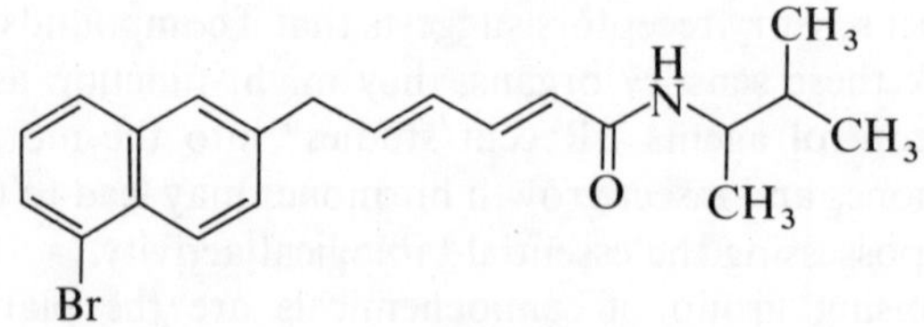

(2)

Chemical modification resulted in the discovery of more active analogues, e.g. compound (**2**); these chemicals apparently act on the sodium ion channel in the insect's nervous system but are highly effective against pyrethroid-resistant insects (see Chapter 4, page 73). In the arylalkylamides (**1**) and (**2**), the introduction of halogen atoms generally increased the insecticidal potency; the lipid amides are a potentially important novel group of insecticides.[6]

Microorganisms will continue to yield new insecticides. Current examples include the avermectins (see Chapter 4, page 74).

Corbett and coworkers[9] have suggested that as the enzyme acetylcholinesterase, the principal site of action of existing synaptic insecticides (e.g. organophosphates and carbamates), is widely distributed in the central nervous system of insects and mammals, it should be possible to select more specific nerve transmitters than acetylcholine as targets. For instance, nervous transmission in the locust may also involve relatively specific amine transmitters which would appear good targets for attack by a chemical, leading to the development of a fairly specific pesticide for control of locusts. There is substantial research at present for chemicals that affect specific insect neurotransmitters and neurohormones, e.g. the pentapeptide cockroach muscle neurotransmitter proctolin (**3**). Such compounds, together with the toxins produced by insect predators, e.g. spider venom, could provide attractive new targets for future insect control chemicals.[10]

Arg-Tyr-Leu-Pro-Thr

(**3**)

There are a wide range of commercial herbicides on the market with different modes of action, selectivity, and persistency (Chapter 8). Bipyridinium herbicides, when sprayed onto plants, kill all plant growth which has been covered by the spray, but they are immediately deactivated on contact with soil so they can be used in chemical ploughing. On the other hand, triazine herbicides, like simazine, are very persistent and retain their activity in the soil for nearly a year. Some weeds, e.g. couch grass, are difficult to control, especially without damaging other crops, so there is a continuing need for more selective herbicides and several compounds have been introduced for selective control of wild oats in cereals. The selective control of such grass weeds by new graminicides is a major research objective. The sulphonylurea herbicides show exceptionally high activity, e.g. they control broad-leaved weeds in cereals at very low doses (20 g/ha).

There is increasing emphasis on synthetic plant growth regulators (PGRs) to improve crop yields and harvesting, and further developments in this area can be expected in the future.

Hybrids of small grain cereals, like wheat, cannot be produced by crossing parents not genetically identical (F_1-hybridization) because wheat is self-fertile—the male and female parts occur in each flower. To prevent self-fertilization, the wheat flowers can be sprayed with a chemical sterilizing agent before the pollen is produced. It was discovered that fertile pollen contains much higher levels of proline than pollen from male-sterile plants. Chemists at Shell (Sittingbourne, Kent) therefore examined compounds for inhibition of proline biosynthesis. Methanoproline was active, but gave variable results under field conditions; however, the analogue (4) was an effective sterilant. This, and other chemical sterilants, may permit the introduction of commercial F_1-hybrids of wheat with improved properties, e.g. higher yields and disease resistance as compared with their parents.

$$\underset{\displaystyle \text{CO}_2\text{H}}{\overset{\displaystyle \text{N}}{\diamondsuit}}$$

(4)

Comparatively few important metabolic differences have been discovered between plants and fungi; obviously the disruption of processes unique to fungi would provide a useful basis for the design of new specific fungicides.[5] The systemic fungicides polyoxin D and kitazin (Chapter 7, pages 181, 204) are known to act by interference with chitin synthesis. Another example is provided by the latest group of azole systemic fungicides acting on sterol biosynthesis.

Screening tests have perhaps, in the past, concentrated too much in evaluating the direct effects of the chemical on the pest organism while insufficient attention has been paid to the effects of the chemical on the host plant and the environment.[11] Some chemicals owe their pesticidal activity to indirect action, by either increasing the resistance of the crop plant to the invading pathogen or by modifying the environment so as to make it less conducive to the support of the pest. Such pesticides are attractive, since their application would be unlikely to induce pest resistance as compared with chemicals exerting a direct toxic action on the pest organism.

Pesticides, too, may show harmful indirect effects; e.g. the use of 2,4-D to control broad-leaved weeds has increased the susceptibility of maize to both insects and pathogens. The reproduction of Colorado beetles and other insects was stimulated by sublethal doses of certain insecticides, e.g. parathion.

Several chemicals are already known that reduce disease levels without directly affecting the pathogen; the effectiveness of mineral oil in the control of *Sigatoka* is probably due to the oil reducing the sugar concentration in the young leaves. Many plants are known to contain natural antifungal compounds (phytoalexins), and these natural defence mechanisms can sometimes be triggered off by chemicals. The application of copper and mercury salts stimulates the production of the phytoalexin pisatin in peas, and injection of phenylalanine appears to increase the resistance of apple leaves to scab.[12] Chemicals can also modify the phenolic content or amino acid nutrition of the host plant, and this can sometimes increase resistance to the pathogen. Chemicals can also act by physically modifying the host plant by inducing the formation of thicker cuticles or starch-filled cells which form barriers to the invading pathogen. To exploit these indirect plant disease control chemicals, a greater knowledge of the differential biochemistry of host plants is needed.[5]

There will probably be further introduction of antifungal antibiotics, several of which, like blasticidin and kasugamycin, are very effective for systemic control of rice blast fungus.[11]

More precise knowledge of the various factors that govern the movement of chemicals in plants may enable the discovery of chemicals that are well translocated in shrubs and trees and might control Dutch elm and other fungus diseases of trees.[13] There is also a need for chemicals which after spraying onto the leaves of the crop plant are translocated downwards in the phloem to the roots. Such compounds could prove very effective against root pathogenic fungi and many nematodes which are not well controlled by existing chemicals—root-feeding nematodes cause serious damage to potatoes, sugar beet, and tropical crops. The majority of systemic fungicides show little downward movement from the leaves, although the synthetic plant growth regulator daminozide (5) (Chapter 9, page 277) apparently owes its activity against potato scab to downward translocation.

At present few commercial systemic fungicides control diseases caused by *Phycomycetes*, e.g. potato blight, although metalaxyl shows limited activity; however, this is rapidly being eroded by resistant blight fungi. Pyroxychlor (6), however, when sprayed on leaves is translocated downwards to the roots where it controls root-infecting *Phycomycetes*.[11] Pyroxychlor (6) formulated as granules successfully controlled foot rot in peas.

$$HO_2C—CH_2CH_2CONH—N(CH_3)_2$$

(5)

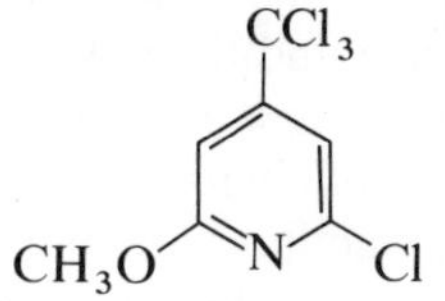

(6)

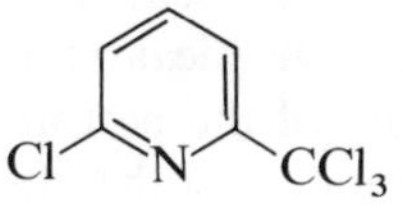

(7)

The discovery of pyroxychlor emphasizes the importance of correctly designed screening procedures. The tests must allow full interaction to occur between the host and the toxicant; this unique chemical would probably have been missed if only the seed dressing tests had been used.[12] Furthermore, as increasingly specific pesticides are needed their activity could easily not be discovered unless they are exposed to the sensitive organism in the screening tests.

The herbicide glyphosate is known to translocate from the leaves down to the rhizomes. One requirement for translocation from leaves to roots would appear to be weakly acidic molecules which are rather soluble in both lipids and water. Possibly such moieties might be attached to suitable fungicides or nematicides to confer the property of downward movement in plants.

Soil diseases can be greatly influenced by environmental effects such as those deriving from the solid, liquid, and gaseous phase components of the soil, plant exudates, microbial toxins, and crop residues. Certain diseases can be reduced by alteration of the nutrition of the host plant, e.g. by stabilizing nitrogen nutrition by addition of nitrification inhibitors like nitrapyrin (7).[12]

More attention needs to be paid to the discovery of spore germination inhibitors since various natural compounds with potent and unequivocal activity have been identified;[5] thus methyl 3,4-dimethoxycinnamate inhibits the germination of bean rust spores.

The present chemical pesticide armoury of some 1000 different active chemicals remains deficient and new compounds are urgently needed for controlling resistant pests and in other specific areas. There are no commercial antiviral agents available against plant viral diseases. Their development is difficult because it involves interfering selectively with the intimate relationship that exists between the virus and its host, but the discovery of plant viricides is undoubtedly a major target for future research effort.[5] Adequate plant bactericides are also lacking and there is considerable scope for the introduction of entirely new toxicants in this field. The problem appears to be the difficulty of finding materials that are sufficiently persistent bactericides but do not harm the host plant. Streptomycin and other antibiotics (Chapter 7, page 178) are active against certain plant bacterial diseases, e.g. cherry leaf spot, but they are expensive and there is too fine a margin between the concentration achieving adequate disease control and that causing phytotoxic damage.

The availability of more selective chemical pesticides permits them to be used in conjunction with biological control methods and it seems probable that integrated biological–chemical measures of pest control will become more common in the future. Such procedures as manipulating the ecology of host and pathogen, while reserving chemical treatment until the other measures have at least reduced the severity of the pest attack, allows the pest to be controlled effectively by smaller amounts of the often costly chemical. It would also have the advantage of reducing the danger of environmental pollution. The plant breeder has been very successful in producing new strains of crop plants with genetic disease resistance against sedentary root pathogenic fungi, but so far this approach has been less effective in combating the readily dispersible foliage fungi. The disadvantage is that the plant's resistance is generally not permanent and in time the fungus mutates to a new pathogenic strain which is capable of attacking the formerly resistant variety of the host plant. The breeding of resistant crop plants thus degenerates into a race between the plant breeder and the fungal mutation—a new aspect of the age-old war between pests and man for crop plants.

Biotechnology provides dramatic opportunities for crop improvement through tissue culture and genetic manipulation of crop plants and organisms.[14] In 1982 the annual sales of all fourteen microbial pesticides approved by the Environmental Protection Agency (EPA) in the United States was some $10 million (cf. $959 million spent on chemical insecticides). However, it has been suggested that biological pesticides could account for 50% of the world insecticide market by the year 2000 but this figure is probably exaggerated. Such 'agrobiologicals' may be specific insect infections, diseases, nematodes, protozoa, fungi, bacteria, or viruses and will become increasingly available.[15] They are selective and can often be used in conjunction with chemical insecticides (see Chapter 14, page 336). Application of modern methods of genetic manipulation should provide more virulent strains of microbial pesticides with faster action. Bioinsecticides from the bacterium *Bacillus thuringensis* (Bt) are marketed worldwide for control of specific insect pests—the larvae of certain *Lepidoptera* (butterflies, moths) and *Diptera* (flies); recently another strain of Bt has been isolated that is toxic to *Coleoptera* (beetles). The techniques are now available to transfer the gene coding for Bt-toxin into other bacteria, viruses, or plants.[16] Such gene manipulation enables the production of novel organisms, e.g. baculoviruses that produce Bt-toxin.

The techniques for the introduction of foreign genes into crop plants (genetic engineering) has been used to enhance herbicide, fungal, and viral resistance in the crop. For insect resistance, new transgenic plants containing the gene encoding for insecticidal toxins, e.g. Bt-toxin or cowpea protease inhibitor (CpTi), have been produced.[17] Biotechnology will play an important role in the battle against attacks by bacteria, fungi, and viruses on crops, because plant resistance often involves the action of single genes. The genetic engineering of resistance for the biocontrol of plant diseases is an exciting prospect. Such methods for the control of plant bacterial and viral diseases will be especially valuable because they are currently not very susceptible to chemical control. Biotechnology will have important effects on the seed industry, leading to the introduction of new crop strains with enhanced yields.[18] Stable resistant weed species arise occasionally and resistant cultured cell lines can be obtained by *in vitro* selection. The cloning and transfer into crop plants of genes conferring resistance to herbicides allows hitherto unselective herbicides to be used selectively, e.g. the introduction of atrazine-tolerant soybeans would permit the use of this herbicide for weed control in soybeans. Resistance is associated with a much reduced binding capacity for atrazine and other triazines. Plants tolerant to glyphosate and sulphonylurea herbicides have also been discovered; in some cases resistance was determined by a single dominant gene, suggesting that the herbicidal target was a single protein.[17]

The introduction of a maize plant mutant resistant to rootworm and borer would decrease the need for chemical insecticides on this crop. The impact of biotechnology can therefore either increase or decrease the application of chemical pesticides.

Applications of molecular biological techniques such as DNA cloning, *in vitro* translation, and gene manipulation have significantly extended our knowledge of the mechanisms of resistance to pests, diseases, and herbicides. These techniques will not replace plant breeding but are valuable in the study of plant gene structure and in providing genetically modified material which can be integrated into an established plant breeding programme. Plant biotechnology has provided spectacular illustrations of the design and construction of plants with induced resistance to pests, diseases, and herbicides. In the future, it may be possible to isolate the genes involved in stress responses in plants and to manipulate the synthesis and accumulation of specific secondary metabolites.[17] The biochemical and molecular biological approaches to insect resistance research have been recently reviewed.[6]

Although less spectacular than genetic engineering or the discovery of new types of chemical pesticides, the importance and need for new formulations of existing pesticides must not be overlooked (Chapter 2). Formulations (e.g. wettable powders, seed dressings, granules, or microcapsules) can substantially affect the persistency and selectivity of a given pesticide. In efforts to minimize damage to beneficial insects the timing of treatment, method of application, and formulation are adjusted to match the behaviour of the target pest. This ecological approach is capable of further exploitation and could lead to dramatic improvements in the effectiveness of many pesticide treatments;[5] thus microencapsulation can increase persistence by controlling the rate of release of the pesticide. Such slow-release formulations are essential in the practical application of pheromones. Experiments indicate that direct spraying of pesticides on pests is probably some ten times less selective than when the toxicant is taken up as vapour from surface deposits. Formulation offers many ways of altering the relative amounts of the toxicant reaching different organisms, e.g. addition of amine stearates to wettable powder formulations of DDT substantially increased the persistency of the dried deposits on the leaves without decreasing the activity. Selectivity could be attained by coating the DDT particles with hemicelluloses since only phytophagous chewing insects possessing the appropriate enzymes (hemicellases) have the ability to degrade the protective coating.

The addition of synergists to the formulation was originally only used commercially with pyrethroids (Chapter 4, page 71); however, these are now widely used, as is the incorporation of safeners with herbicides. There is also an increasing tendency to formulate mixtures of pesticides designed to control the whole spectrum of pests and diseases on a given crop; this is convenient and more economic for the farmer as it saves both time and labour costs.

Improved methods of pesticide application are continually needed so that the pesticide can be delivered to the target with greater precision. An important development has been the introduction of electrostatic sprayers; the charged droplets adhere better to the leaves of the crop and the efficiency of application can be ten times greater than is achieved by a hydraulic sprayer. Electrostatic spraying is vital for the application of pheromones or agrobiologicals.

In conclusion, there appears little prospect in the medium term that biological control measures, such as the introduction of resistant crop varieties, cultural control, genetic methods, the use of natural predators, or other agrobiologicals, will displace chemical pesticides from their dominant position. In the twenty-first century, biotechnology will have an increasing input into pest control, leading to more new crop varieties and agrobiologicals. Further research on these and other biological control measures is very necessary, to improve their efficiency and enable them to be increasingly employed in integrated pest management programmes in conjunction with the appropriate selective chemical pesticides.

At present, the most urgent task is probably to learn to use our existing armoury of pesticides more effectively so that maximum benefit is achieved with minimum risk. To achieve this goal, we need better knowledge of the ecology and population dynamics of pests so that reliable 'early warning' systems can be developed to avoid unnecessary pesticide application. Without better under-standing of the complex interrelationships between crop, pest, and environment, the optimum pest management programmes cannot be evaluated with any precision. The application of a multiplicity of pest control measures is generally preferable to reliance on any single method because this will reduce the chances of the emergence of resistant pests. The irresponsible use of some pesticides has contributed to the production of insect strains resistant to most insecticides because genetic resistance to one chemical will confer immunity to all the others acting by the same biochemical mechanism. If the pest is controlled by a combination of various methods (e.g. pesticides, natural predators, pest-resistant crop varieties, and crop rotation), it is less likely that resistant genes will survive.

The introduction of pest management programmes demand that the best available knowledge is given to farmers; they need to be aware of the advantages and limitations inherent in the proposed control measures. Such integrated pest control management (IPM) must not damage natural regulatory forces operating in the ecosystem because these help to minimize the quantity of pesticides needed. To implement these programmes, a large number of specialist advisers need to be recruited and trained to visit farmers; the present Agricultural Advisory Service would need to be greatly expanded. In addition, agrochemical companies might extend their advice to farmers so that they supply not only a product but a complete crop protection package.

Another problem with integrated pest control strategies is that effective biological control measures often require implementation over a wide area. This is particularly true for measures designed to eliminate, rather than just control, a pest. Such programmes would require state or regional funding and probably complete eradication would not be achieved, e.g. after the initially successful biological control programme to eliminate the screw-worm in the United States, it is costing some $10 million per annum to prevent reinfestation by the pest.

In conclusion, there is urgent need to discover effective methods of combating plant bacterial and viral diseases; this was emphasized in 1987 with the first outbreak of the sugar beet virus disease *Rhizomania* on a Suffolk farm. At present such virus diseases cannot be controlled by chemicals and the development of effective chemical viricides does not look promising. It is more likely that the problem of plant virus diseases will eventually be solved by biotechnology—by genetic engineering of virus-resistant crop plants.

A generally effective systemic fungicide against potato blight is still lacking and new insecticides and fungicides acting on different biochemical targets are continually needed to combat the growing threat posed by resistant insects and fungi. The pest control agents described in this book play a vital role in the unending battle between man and pests and diseases for an adequate food supply.

REFERENCES

1. Pickett, J. A., *Chemistry in Britain*, **24**, 137 (1988).
2. *Data Requirements for Approval under the Control of Pesticides Act 1986*, MAFF, London.
3. Cremlyn, R. J., *International Pest Control*, **15**(2), 8 (1973).
4. Lyr, H. (Ed.), *Modern Selective Fungicides*, Longman, Harlow, 1987.
5. Graham-Bryce, I. J., *Proceedings of British Crop Protection Conference, Brighton*, **3**, 901 (1975).
6. *Recent Advances in the Chemistry of Insect Control*, Vol. II (Ed. Crombie, L.), Royal Society of Chemistry, Cambridge, 1990.
7. Collins, M. D., *et al.*, *Proceedings of British Crop Protection Conference, Brighton*, **1**, 299 (1984).
8. Bell, E. A., *Proceedings of British Crop Protection Conference, Brighton*, **2**, 661 (1986).
9. Corbett, J. R., Wright, K., and Baillie, A. C., *The Biochemical Mode of Action of Pesticides*, 2nd edn, Academic Press, London and New York, 1984.
10. Keeley, L. L., and Hayes, T. K., *Insecticide Biochemistry*, **17**, 639 (1987).
11. Cremlyn, R. J., 'The biochemical mode of action of some well-known fungicides', in *Herbicides and Fungicides: Factors Affecting Their Activity* (Ed. McFarlane, N. R.), Chemical Society Special Publication No. 29, The Chemical Society, London, 1977, p. 22.
12. Sbragia, R. J., *Ann. Rev. Phytopath.*, **13**, 257 (1975).
13. Crowdy, S. H., 'Translocation', in *Systemic Fungicides* (Ed. Marsh, R. W.), 2nd edn, Longman, London, 1977, p. 92.
14. *Biotechnology and Crop Improvement and Production*, Monograph No. 34, British Crop Protection Council, Croydon, 1986.
15. Payne C. C., and Jarrett, B. P., *Proceedings of British Crop Protection Conference, Brighton*, **1**, 231 (1984).
16. Jetsum, A. R., *Phil. Trans. Roy. Soc. London*, **B318**, 357 (1988).
17. Lindsey, K., and Jones, M. G. K., *Plant Biotechnology in Agriculture*, Open University Press, Milton Keynes, 1989.
18. Dodds, J. H., *Plant Genetic Engineering*, Cambridge University Press, Cambridge, 1985.

Glossary

ABSCISSION   Process by which a leaf or other part is separated from the plant

ACARICIDE   A chemical designed to kill mites, including phytophagous mites and parasitic mites on animals

AEROBIC respiration   Cellular reactions in whole organisms. tissues, or cells that occur *in the presence of oxygen* resulting in the complete oxidation of organic compounds to carbon dioxide and water; this is compared with *ANAEROBIC RESPIRATION*—cellular reactions that take place in the *absence of oxygen* leading to partial breakdown of organic compounds

AGONIST   A substance that activates a drug receptor as compared with an *ANTAGONIST* which deactivates the receptor

ALGAE   Primitive green aquatic plants ranging from unicellular species to giant seaweeds

ANTHELMINTHIC   A chemical administered to an animal to kill intestinal worms

ANTIBIOTIC   A chemical produced by a microorganism (bacteria or fungi) which inhibits the growth of other microorganisms

APHICIDE   A chemical design to kill aphids, plant-sucking lice that can transmit virus diseases

APOPLAST   Semi-gelatinous material occurring outside cell walls which forms a tortuous continuum throughout the plant and is a major pathway for the diffusion of water-soluble materials

APPRESSORIUM   A swelling at the tip of a fungal hypha or spore germ tube formed prior to direct penetration of the host plant and which is normally associated with adherence to the plant surface

ARTHROPODS   The broad class of jointed-foot animals (*Arthropoda*) comprising insects, spiders, crustaceans, and millipedes

AUXIN   A substance occurring in plants that stimulates cell growth in plant tissues

BACTERIA   A group of minute unicellular prokaryotic organisms which occur in soil, water, plants, or animals

BIOASSAY   Estimation of the quantity of a toxicant in a given sample by measurement of its effect on test organisms, generally the amount needed to kill 50% of the organism

BIOTECHNOLOGY   Various industrial processes that involve the use of biological systems, e.g. the industrial applications of microbial and other biochemical processes, including the industrial aspects of genetic engineering

CARCINOGEN   A substance that induces the development of cancer tumours in animal tissue

CARNIVORE   An animal or plant that eats or feeds on animal flesh (adjective: carnivorous)

CAROTENOIDS   Polyisoprenoids containing eight isoprene units joined together; they are water-soluble plant pigments which often function as accessory pigments in photosynthesis

CHITIN   A linear array of β-linked *N*-acetylglucosamine units found in the arthropod skeleton and in some plants, especially fungi

CHLOROPLASTS   The organelles inside leaf cells containing chlorophyll where the reactions of photosynthesis occur

CHLOROSIS   Abnormal loss of the green pigments of plants due to lack of light, magnesium or iron deficiency, or to chemicals or genetic factors inhibiting chlorophyll synthesis

CHROMOSOME   Small rod-shaped, stainable structures, made partly of DNA, in the cell nucleus which are responsible for the transfer of genetic information

COLEOPTERA   An insect order comprising weevils and beetles including flea beetles

COLLOID   A substance of high molecular mass which does not readily diffuse through a semi-permeable membrane

COMMODITY PRODUCT   A substance not covered by a patent as distinct from one that is covered by a patent

COTYLEDON   The first leaf or leaves of a seed plant found in the embryo of the seed

CUTICLE   The outer protective envelope of a living organism; in mammals, or arthropods, the alternative terms skin and 'integument' are often used, but in plants cuticle is preferred

CYST A group of eggs contained in a tough protective envelope; the dormant stage of a species of nematode or eelworm

CYTOPLASM The fluid or gelatinous component of living cells which is distinct from the reproductive nucleus

DEHISCENCE The spontaneous opening of an organ or structure in a definite direction

DEXTROROTATORY The optical isomer (or enantiomer) that rotates a beam of plane polarized light to the right; often termed the (+) isomer

DICOTYLEDONS Plants containing an embryo with two cotyledons; the group includes hardwood trees, shrubs, and many herbaceous plants

DIPTERA A large order of insects comprising two-winged flies, e.g. mosquitoes, houseflies, tsetse flies, fruit flies.

DOPAMINE An inhibitory neurotransmitter in several mammals and snails

ECOLOGY The study of the interaction of organisms with one another and with the environment

ECOSYSTEM The ecological system formed by the interaction of organisms with their environment

ECTO A prefix used to imply *external* parasites, e.g. fleas and ticks, as compared with internal or *ENDO* parasites, e.g. lungworms

ELECTROPHILE A group or cation containing an electron-attracting positive centre, e.g. NO_2^+, SO_3, CHO, etc.

ENANTIOMER The optical isomer of a compound containing at least one chiral (asymmetric) atom; such isomers will rotate plane polarized light and often have different biological activity

ENDOGENOUS A chemical produced by the plant itself, as compared with one applied externally, which is termed *EXOGENOUS*

ENZYME A proteinaceous catalyst produced by a living organism which acts on one or more specific substrates

FAUNA All the animals specific to a given area

FLORA All the plants specific to a given area

GENE An individual unit of a chromosome, responsible for the transmission of some inherited characteristic

GEOMETRIC ISOMERISM   A type of stereoisomerism generally arising from non-rotation about a carbon–carbon double bond; the resultant isomers are termed *cis* (*Z*) or *trans* (*E*) forms

GRAMINACEOUS   An adjective used to describe grasses, including cereals

GRAMINICIDES   Chemicals specifically designed to kill grass weeds

GRAVID   A female carrying eggs or a pregnant uterus

HELMINTH   A large group of primitive worms, including nematodes and worms parasitic in the gut

HEMIPTERA/HOMOPTERA   An insect family comprising aphids, scales, whitefly, leafhoppers, mealybugs, suckers, capsids

HERBIVORE   An animal that eats plants (adjective; herbivorous)

HILL REACTION   The light reaction in photosynthesis which occurs in the presence of an artificial electron acceptor A: $2 H_2O + A = 2 H_2A + O_2$

HORMONES   Substances produced in living cells in one part of an organism which are transported to other parts where they exert an effect; hormones play vital roles in plants and animals, e.g. as growth regulators

HYDROPHILIC   A compound or functional group possessing a strong affinity for water and hence appreciable aqueous solubility

HYPHA (plural HYPHAE)   A tubular, threadlike, filament of a fungal mycelium

INOCULUM   Incoming fungal spores or other disease agents that initiate the development of a parasite within the host organism

INSTAR   A stage in the development of an insect between two moults of the hard exoskeleton, excluding the egg. The adult butterfly is the sixth instar; the first four instars are caterpillars of increasing size, followed by pupa

IN VITRO   An experiment on living organisms performed in the laboratory, e.g. on fungi growing on an agar plate

IN VIVO   An experiment carried out on living organisms in their natural environment, e.g. on phytopathogenic fungi growing on their host plants

ISOMER   Compounds having the same molecular formula but different chemical structures or stereochemistry

LAEVOROTATORY   The optical isomer (or enantiomer) that rotates a beam of plane polarized light to the left; this is often termed the (−) isomer

LARVA   A non-flying, but mobile, immature stage (instar) in the life cycle of most insects; structurally it is totally different from the adult form (cf. nymph)

LEPIDOPTERA   A large order of insects, e.g. butterflies and moths, stalk borers, bollworms, cutworms, loopers, leaf rollers

LIGNINS   The complex phenylpropanoid polymer that strengthens the cellulose framework of wood fibres and vascular plants

LIPID   A heterogeneous group of compounds synthesized by living cells which are soluble in non-polar organic solvents but are only sparingly soluble in water. Lipids contain long chains or rings of hydrocarbon residues and include fats, waxes, steroids, carotenoids, etc.

LIPOPHILIC   A compound or functional group promoting solubility in oil as distinct from lipophobic compounds which have very low solubility in oils and fats

MELANINS   A group of dark-coloured substances responsible for the pigmentation of the skin; they are formed by the oxidation of aromatic compounds, e.g. phenylalanine

MERISTEM   The tissues responsible for plant growth which divide mitotically; in vascular plants meristematic tissues occur principally in the shoots and root tips

MESOPHYLL   The tissue or cells in the interior of a plant leaf

METABOLISM   A sequence of biochemical reactions by which a living organism transforms a chemical into a series of other chemicals termed *METABOLITES*. In mammals, the metabolites generally have increasing aqueous solubility enabling them to be finally excreted in the urine

MITOCHONDRIA   Small subcellular organelles involved in the oxidative breakdown of molecules in living organisms

MITOSIS   The type of cell division that occurs during growth and repair of parts of the body. The cell nucleus divides into two daughter nuclei containing the same genes as the parent nucleus

MOLLUSCS   Animals with a body consisting of a head, foot, and visceral mass covered by a mantle. The majority of molluscs inhabit marine or freshwater environments but some, like slugs and snails, may be land based

MONOCOTYLEDONS   Plants possessing an embryo with one cotyledon; the group includes the palms, grasses, orchids, lilies, etc.

MORPHOLOGY   The science dealing with the structures and forms of organisms

MUTANT   An individual member of a species with some abnormality of structure, properties, or behaviour; the difference is genetic and can be transmitted to offspring

MUTATION   The process causing a change in an inherited characteristic by alteration in the amount or structure of DNA in chromosomes

MYCELIUM   The nutrient-seeking fibres (hyphae) of a fungus which are the equivalent of the roots of a green plant

MYONEURAL JUNCTION   The site of transmission of an impulse from a nerve to a muscle

NAD (nicotinamide adenine dinucleotide)   A hydrogen-carrying coenzyme. The reduced form is $NADH_2$ and is produced in many metabolic pathways (e.g. glycolysis), being reoxidized by the electron transport chain

NADP (nicotinamide adenine dinucleotide phosphate)   A hydrogen-carrying coenzyme; the reduced form is $NADPH_2$. It is involved in many biochemical reactions, e.g. photosynthesis

NEMATODES   Small, unsegmented eelworms, many of which are parasitic on plant roots

NUCLEOPHILE   An electron-donating centre or anion, e.g. $\ddot{N}H_3$, $H_2\ddot{O}$ or OH^-

NYMPH   The immature stage of insects (instar), which resembles the adult stage except for the absence of functional wings and sexual organs. It is distinct from the *larva* of other insects which are entirely different from the adult

OBLIGATE   Adjective applied to parasites to indicate those that cannot survive outside the host; the majority of parasites fall into this category

OPTICAL ISOMERISM   A type of stereoisomerism arising from the presence of one or more asymmetric or chiral atoms, e.g. carbon atoms attached to four different atoms or groups

ORGANELLES   Organized microstructures inside cells, e.g. mitochondria, chloroplasts, etc., in which certain biochemical reactions are confined

OVICIDE   A chemical that kills eggs before they hatch, especially used against the eggs of phytophagous mites, e.g. red spider

OVIPOSITION   The act of egg placement; this is achieved by an extensible tube on the abdomen of the female insect which is termed an ovipositor

OXIDATIVE PHOSPHORYLATION   The production of ATP in the electron transport chain; the process depends on oxygen to reoxidize $NADH_2$ and $FADH_2$

PARASITE   A plant or animal living on, or in, another plant or animal which is termed its host

PATHOGEN   A fungus or bacterium that invades the plant or animal host

PHEROMONE   A chemical or mixture of chemicals released by an organism that induces a response in another individual of the same species

PHLOEM   A system of interconnected, elongated tissues in plants adapted to transport the products of photosynthesis from the leaves to the growing tissues

PHOTOSYNTHESIS   The synthesis by green plants (and some bacteria) of carbohydrates from carbon dioxide and water utilizing the energy of the sunlight

PHYTOALEXIN   A group of antibiotics produced by plants which enhance the plant's resistance to bacteria or fungi

PHYTOPHAGOUS   Leaf-eating (herbivorous) organisms

PHYTOTOXIC   Toxic to plants, the term is generally restricted to chemicals that damage higher (green) plants

PLANKTON   Minute marine or freshwater plants (phytoplankton) or animals (zooplankton) which float in the surface waters and are of great economic and ecological significance

POST-EMERGENT   A chemical treatment of a crop *after* it has emerged above the soil

PRE-EMERGENT   The chemical treatment of soil *after* sowing the crop seed but *prior* to the emergence of the crop seedlings above the soil

PROLINE   A cyclic amino acid which is found in proteins and contributes to the secondary structure of proteins

PROPHYLACTIC   A treatment designed to prevent disease-producing organisms from invading the treated crop. The term *protectant* is sometimes used in the same way as *prophylactic*. It is distinct from *curative* or *eradicant* treatment of fungal crop diseases, which are designed to eliminate or control an already established disease

PROTOZOA   Microscopic, unicellular animal organisms

RACEMATE   (or a racemic mixture) An equimolar mixture of two optical isomers. The racemate ($\pm$ form) will be optically inactive, i.e. has no effect on plane polarized light and almost always results from chemical synthesis

RESPIRATION   This term generally applies to aerobic respiration in cells, but can also refer to the processes involved in extracting oxygen from the environment, e.g. breathing

RHIZOME   An organ of vegetative propagation consisting of a horizontal underground stem bearing roots, leaves, and shoots—important in plants like couch grass

RIBOSOMES   The sites of protein synthesis in the cell. Ribosomes are usually attached to the endoplasmic reticulum

RODENTS   A large group of gnawing animals, e.g. rats, mice, squirrels, beavers, which can be herbivores or omnivores

ROTATION   Applied to cropping; the practice of growing a regular recurring sequence of different crops on the same land. Rotation often hinders the growth in population of weeds, insects, and fungi and assists in the conservation of valuable soil nutrients

SAPROPHYTE   A plant (usually a fungus) living on dead tissue (adjective: saprophytic)

SEMIOCHEMICALS   Chemicals which mediate insect behaviour

SENESCENCE   The complex of ageing processes that eventually leads to the death of the organism

SPIRACLE   Small openings in the insect's abdominal segments which are connected to a system of small tubes (trachea) enabling oxygen and carbon dioxide to be exchanged with the atmosphere

SPORE   A minute reproductive unit in lower plant forms

STOMA   (plural: stomata) Small apertures in the surface of leaves which open or close depending on the light intensity, time of day, and other factors. They are adapted to control exchange of carbon dioxide, oxygen, and water vapour between internal leaf cells and the atmosphere

SYNERGISM   A mixture of two or more chemicals which is more active than the summation of their individual activities. Hence a synergist is a compound of low toxicity, the addition of which enhances the activity of the toxicant conferring economic benefit, as the synergist is generally cheaper than the active ingredient

SYSTEMIC   A chemical that enters the plant via the roots or leaves and is translocated within the plant vascular system. The degree of translocation varies considerably—some systemics have only limited movement

TAXONOMIC   Refers to the classification of the different groups of fungi

TERATOGEN   A chemical that causes birth defects

TOPOGRAPHY   The description and mapping of the configuration and structural relationships of a surface

TOXICITY   The ability of a chemical to kill a given organism by one dose (acute) or by continued dosage over a period of time (chronic)

TRANSLOCATION   The movement of chemical substances—either endogenously or applied within the plant; systemic chemicals are translocated, at least to some extent

TUBULIN   A dimeric protein composed of two closely related subunits (α- and β-tubulin) that is a major component of microtubules

VECTOR   A carrier, e.g. an animal, insect, or aphid, capable of the transmission of a pathogenic (disease-producing microorganism) from a diseased animal or plant to healthy specimens

VERTEBRATES   Animals having a backbone; this class includes fishes, reptiles, birds, and mammals

VIRUS   An infectious agent that consists of protein together with either DNA or RNA which may be surrounded by a membrane. It is smaller than a bacterium and is an obligate, intracellular parasite

XENOBIOTIC   Any chemical present in the environment that does not occur in nature, e.g. a pesticide or industrial pollutant

XYLEM   A system of almost continuous channels in plants composed of fused dead cells, which transport water and soluble minerals from the roots to the leaves. The xylem, together with the phloem, forms the vascular tissue, and is the principal water-conducting tissue in plants.

Index

Bold type indicates the more important references. Trade names have, as far as possible, been set in quotes.